2nd

EDITION

GUIDE

TO

TELECONFERENCING

& DISTANCE LEARNING

*by Patrick S. Portway
& Carla Lane, Ed.D*

**APPLIED BUSINESS
teleCOMMUNICATIONS**

P.O. Box 5106, San Ramon, CA 94583
(510) 606-5150 Fax (510) 606-9410

2nd Edition Guide to Teleconferencing & Distance Learning

Guide to Teleconferencing and Distance Learning
by Patrick S. Portway and Carla Lane, Ed.D.

2nd Edition, Copyright 1994 by Applied Business teleCommunications

Copyright 1992 by Applied Business teleCommunications (First Edition)
Technical Guide to Teleconferencing and Distance Learning

Printed in the United States of America

If you have any questions or comments concerning this book or are interested in obtaining additional copies, please call 800-829-3400.

Applied Business teleCommunications
2600 Kitty Hawk Road
Suite 110
Livermore, CA 94550

Library of Congress Card Catalog Number: 94-079338

ISBN 0-9643270-1-5

Original cover design by Peggy Williams.
2nd Edition cover design by Karen E. Myers

2nd Edition

The authors dedicate this publication to the pioneers of the teleconferencing and distance learning field as recognized in the Teleconference magazine Hall of Fame. Their efforts created the technology and applications described in the pages of this book.

Glen Rae Southworth
Lorne A. Parker
Elliot Gold
Polly Rash
J.O. Grantham
Chris E. Caplilongo
Harry S. Wohlert
Richard Jackson
Virginia Ostendorf
Lee Gunter
Ray Steele
Carl M. Marszewski
Carl Westcott
Lee Todd

Patsy Tinsley
Susan Irwin
Harold Hagopian
John Tyson
Marshall Allen
Norman Gaut
Tom Wilkins
John Brockwell, Jr.
Stanley Huffman, Jr.
Inabeth Miller
Tony Jatcko
Philip Westfall
Tim Harrington

Contents

Acknowledgements

The value of this book comes from the contribution of our industry's outstanding authorities who co-authored chapters or subsections of the book. No other publication or book in our industry has had the benefit of so many and so prestigious a group of contributors to a single project.

Much of the material included is new and reflects the latest state of the art (1994) in this rapidly changing field. A few valuable portions have appeared previously in our publication, Teleconference magazine.

A special thanks to Renee Wilmeth, vice president of ABC publications and Karen Myers, our art director, for their work on this second in a series of books from ABC on teleconferencing and distance learning.

Charles Humbert and Renee Wilmeth managed the book project and copy edited the submissions from the many co-authors of this guide to teleconferencing technology and its applications in distance learning. Renee also assisted in the research for several chapters including significant work on the desktop video chapter.

Karen Myers designed the cover and the layout of the book's contents. At our request several charts were deliberately made in a size appropriate for copying as overhead transparencies to be included in presentations by our readers.

Permission is granted for the reproduction of pages 5, 6, 7, 9, 12, and 13 with proper credit to the book and its authors as the source.

All errors and omissions are ours.

Patrick S. Portway and Dr. Carla Lane

Chapter 1

Patrick S. Portway, is president of Applied Business teleCommunications and publisher of Teleconference magazine, TeleCons newspaper and ED — Education at a Distance magazine and journal. His company runs the annual TeleCon conference, the world's largest conference and trade show on teleconferencing, and IDLCON, the world's largest conference on distance learning. He obtained his B.A. from the University of Cincinnati and his M.A. from the University of Maryland.

He was previously program manager for Satellite Business Systems, vice president of marketing for Video Systems Network, western regional manager for American Satellite, manager of strategic marketing planning for Xerox Corporation, and regional coordinator at General Services Administration. He is a leading international consultant on teleconferencing and distance learning. His numerous clients have included Oklahoma State University, the United States Social Security Administration, SIP of Italy and Tandem Computers. His company, in a joint venture with a video engineering company, supported Atlantic Richfield in the development of ArcoVision, the world's first two-way multipoint videoconferencing system.

Portway presently teaches in the graduate telecommunications program at Golden Gate University and California State University, Hayward. He is the founder and Executive Director of the United States Distance Learning Association (USDLA). He is a charter member and founder of ITCA. Portway has been listed in the Who's Who in California, Who's Who in the West, Who's Who in Industry and Finance 1993-94, and Who's Who in the World 1994-1995.

ORIENTATION

■ DEFINITION OF TERMS

Like any new and evolving technology, teleconferencing and distance learning terminology or jargon has not been precisely defined. We've included a glossary of terms as part of this text (Appendix A), but we've also included an opening discussion of common terms used differently in Europe and the United States.

"Teleconferencing" is a combination of the prefix "tele" and the word "conferencing." "Tele" meaning at a distance and "conferencing" dealing with meeting. So combined, "tele-conferencing" is to meet at a distance. To many people "tele" refers to telephone. Particularly in Europe, the word teleconferencing is restricted to telephone conferencing or conference calls. The word "videoconferencing" is used in Europe to specifically describe two-way communication including both audio and video. We've used this very specific definition of videoconferencing in this book.

Often in the U.S., the word "video" is added to "teleconferencing" or "conferencing" to specify exactly what kind of meeting at a distance you're describing. In this usage, teleconferencing is used as a generic term for all kinds of meetings conducted via communication technology — audio teleconferencing, video teleconferencing, computer conferencing, etc. We will often use "teleconferencing" in this broader definition in this book and we apologize to our European readers for any confusion.

The term "videoconferencing" has been used at times in the U.S. to describe a one-way video broadcast, usually over a satellite, where response back by audio or other non-video transmissions over a phone line provides an interactive link back to the site. As we've mentioned earlier, we've chosen to use videoconferencing to describe only two-way video communications as it's commonly used internationally. We've chosen another term, "business television," to apply to one-way video, two-way audio.

Business television describes the expanded use of television by U.S. corporations which led to the establishment of private broadcast television networks to allow corporations or other organizations to deliver their own programming to employees, customers or other private networks. This private use of television has been called "business television," a term coined by Private Satellite Networks (PSN) now merged into Convergent Media Systems of Atlanta. Unfortunately the term "business television" includes both live programs with interactive audio and simply narrowcasting or television programs directed to a narrowly defined audience that may be prerecorded and distributed over one-way video only with no interaction.

There are other terms that may cause confusion with videoconferencing. "Interactive television" has been a term used to describe interaction between a person and prerecorded instructional television program on a laser disk and does not involve transmission of the video. We now need to add to this term an array of new terms like "answer back T.V." which utilizes response terminals to respond back to live T.V. programs, and draws participants to actively interact in game shows, other broadcast television programing and video on demand. Video on demand deals with pre-stored video played out from a video server or storage device whenever requested or demanded by the user.

■ THE GROWTH OF VIDEOCONFERENCING

One of the most puzzling questions is, "Why has there been so much interest in videoconferencing over the years, yet dramatic growth in videoconferencing has only developed since 1989?"

A Video Generation

The interest in private video communications has grown with the availability of broadcast television beginning in 1940. The adults of today grew up watching six hours of television per day and have become accustomed to instant video access to world events as they happen. They have rapidly become visual communicators. Almost from the day it was first demonstrated, the general public assumed that video would eventually be added to their telephone for their own personal use.

AT&T contributed to the public's expectations of personal video communications by demonstrating the motion video telephone (Exhibit 1.1.1) at the 1964 New York World's Fair. What AT&T failed to explain to the World's Fair visitors that saw the videophone and millions who read about it, was that the communications services to support the analog video telephone would cost a thousand dollars a minute and were only available in a limited number of areas.

The Jetson's, our mythical television family of the 1960's and 70's, used the video phone routinely suggesting that we could expect one for our own use any day. The video phone was included in nearly every

projection of the future, but as we'll see, the very large and costly communications capacity required to transmit the 90 million bits per second of a standard broadcast video image was impractical.

The computer and desktop publishing industries have also contributed to increased interest in video. Computer applications in the 1970's and 80's moved from simple alpha numeric displays to sophisticated graphics and desktop publishing. In the 90's, CD-ROMs added motion video segments. No longer limited to text and numbers, computers began to easily present visual displays of graphics and charts that were more attractive and professional. The higher standard for business imagery became accepted practice in business communications.

■ IMPROVEMENTS IN COMPRESSION TECHNOLOGY

When AT&T demonstrated their video telephone in 1964, it required prohibitively expensive communication lines to transmit motion video. By 1980, a motion video image could be transmitted at 1.5 million bits per second (T-1) using digital compression. Figures 1.1.1 and 1.1.2 illustrate how digital compression of motion video has improved until today (1994). A very acceptable full screen motion image for business applications can now be transmitted at 112 kbps, two switched 56 kbps data lines, or two digital phone lines of 64 kbps each available over digital phone service (ISDN). This is a considerable achievement over the 90 million bits per second required to digitize a standard broadcast television picture. It is the equivalent of summarizing a 400-page book in a quarter page. A resulting cost savings in recurring communications costs is shown.

Availability of Communication Services to Support Video Transmission

When videoconferencing began in the U.S. in 1980, the only widely available transmission services at 1.544 mbps were satellite data communications. However, 1.544 mbps did match the data rate for terrestrial phone service of T-1 (a packet of 24 phone lines) used at that time for interexchanging communication by phone companies.

With the growth of corporate telecommunications networks, private T-span networks, made T-1 video available for alternative use or through the use of idle backup facilities. As video compression technology improved, it was possible to MUX (use an on-premises multiplexer on the T-1 line) the T-1 line into 12 phone lines and 768 kbps for video and still get acceptable video.

Exhibit 1.1.1

Exhibit 1.1.2

Beginning in 1983, systems became available that could transmit a form of near motion video at 56 kbps, another widely available data rate. The performance of video compression codecs by 1989 allowed transmission of a acceptable motion video picture over two 56 kbps lines and has continued to improve.

On the other hand, availability of fractional T-1 services at 384 kbps (6x64 kbps) and other multiples of 64 kbps from carriers has been slower than expected. Often fractional T-1 data services are available from inter-lata carriers (long distance), (Figure 1.1.3) but not available from a local phone company. As a result the use of inverse multiplexers to add 64 kbps lines together and acheive a desired data rate has grown.

ITU Standards and Interoperability

An international telecommunications standard for video compression, H.261, was accepted in 1990 by the international Consultative Committee for International Telephony and Telegraph (CCITT), part of the International Telecommunications Union (ITU) of the United Nations, and adopted by all the major manufacturers of video compression codecs.

While most of the manufacturers offer their own proprietary compression algorithms as well, CCITT H.261 provides a lowest common denominator for communicating between codecs of different manufacturers.

The same buyer confidence that resulted from a standard for FAX machines has also helped remove uncertainties from videoconferencing. While very few videoconferencing users communicate outside their own private networks today, the ability to communicate with other users in the future has created a perceived stability to the technology. The series of H.320 standards are now under ITU's Telecommunications Standards Sector (TSS).

Cost and Complexity of Facilities

The fixed facilities investment for equipment to enter videoconferencing in 1980 was $600,000 to one million dollars per site. The cost of a video compression codec was $250,000 alone and the computer controlled boardroom style facilities were custom designed for each customer.

Today video compression codecs are under $20,000 each and complete modular systems including a codec selling for under $30,000. Video desktop systems which are based on modification of a desktop computer for windowed motion videoconferencing are available for $1000-$5000.

Business Acceptance For Key Applications

The final factor in the recent burst in the growth of videoconferencing has been well publicized success stories in particular applications.

Divided Work Groups: The Department of Defense (DoD) and the Aerospace industry have managed very complex weapons systems developments involving multiple cooperations with DoD agencies via secure videoconferencing systems. Boeing Corporation estimates it saved 30 costly days in the development of the 757 through the use of videoconferencing between engineering and production groups in the same metropolitan area.

Training and Education: Distance learning, the use of videoconferencing to deliver educational and corporate training directly into the workplace, has been the most successful and fastest growing application of videoconferencing.

International Communications in a Crisis: The Gulf War in 1991 introduced several international corporations to the value of videoconferencing when travel is difficult or dangerous. Executives used public videoconferencing rooms to manage overseas operations during the war. It was a valuable first exposure to videoconferencing for many of them and they have rethought the value of conducting international meetings electronically.

As cost and availability of video communications came down, videoconferencing found a receptive video generation ready to apply the power of visual communications to business and education. In the view of Andy Grove, CEO of Intel Corp., "If the early 90's was the era of desktop publishing, the late 90's will be the era of desktop video."

■ MARKET SEGMENTS

As an overview of the $3.6 billion field of teleconferencing, we've broken it down into five major areas or markets with a brief discussion of each.
- Audio Conferencing
- Audio Graphic Conferencing (audio plus graphics)
- Business Television (one-way video)
- Videoconferencing (two-way video)
- Desktop Videoconferencing and Video Telephony including motion video windows on a computerscreen.

Audio Conferencing, or telephone-based conferencing, is the oldest and largest portion of the total market. Conference calls and the use of speaker phones to connect groups of people together is the most widely accepted form of teleconferencing, but it has not yet reached its limits of growth.

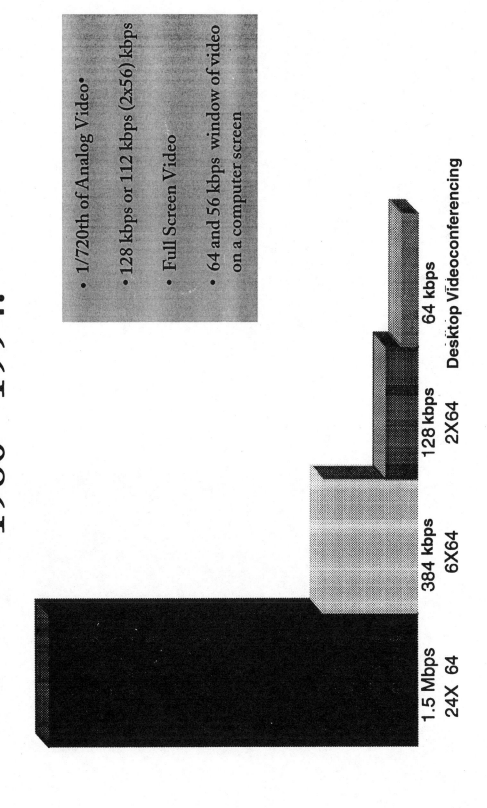

Figure 1.1.1

Video Compression Two Way Video
1889-1994

1/60th of Analog Video

1.544 kbps = 24x64 kbps
Digital Phone Lines

1.5 Million Bits Per Second

90 Million Bits Per Second

Figure 1.1.2

Commonly Available Communications Services

- POTS — 9.6 to 19.2 hbps or higher or higher with data compression

- Switched 56 Data Service — 56 kbps or 112 kbps

- ISDN (Basic Rate) — 128 (2x64) kbps

- ISDN (Primary Rate) — 384 (6x64) kbps

- T-1 (24x64kbps) — 1.544 Mbps

- Fractional T-1 — 384 kbps or 768 kbps

- DS-3 — 45 Mbps

- Satellite — Analog Video 1 or 1/2 Transponder

- Compression For Broadcast — 1.5, 3 or 6 Mbps

Figure 1.1.3

Audioconferencing has several significant advantages:

• Ubiquitous - The one means of communications available to nearly every business in the world and every home in developed countries is the telephone. Cellular phones have added cars and airplanes to teleconferencing.

• Least Costly - Of all the technologies we will discuss, audio has the lowest transmission cost and requires the least investment in additional terminal equipment beyond a common telephone hand set.

• Informational Content - Audio carries more of the actual information content than any part of a teleconference. Even in a videoconference, if you lose audio, the meeting is over; but if you lose video you can continue in an audio only mode.

• Bridging - Sophisticated audio can be mixed or bridged together so that several sites can hear each other even when they talk simultaneously. Audioconferences of several hundred sites can be fully bridged to allow all sites to hear the origination site and answer back without the need of the switching that is required in video.

• Technical Developments - The advances in digital bridging, echo suppression and echo cancelling have raised audioconferencing far above a simple conference call. Sophisticated conferencing systems allow groups to talk freely in a conference room and are able to capture audio from any corner of the room without echo or the "talking from the bottom of a barrel" effect. Sophisticated bridges allow almost any number of sites to participate in an audioconference, balance the levels of all the phones and create electronic subgroups or breakout sessions at will.

• Supplement Audio - Even without adding capability to the live transmission during an audioconference, it is possible to supplement an audioconference by sending documents associated with the audioconference by fax or overnight mail prior to the meeting. Slides can be mailed before the meeting and controlled (advanced) by remote control signaling over the same phone line used for voice.

The limits of what can be done creatively with audio are bound only by the participant's knowledge, creativity and imagination. I often cite as an example an experience we had several years ago where the editor of my magazine, Teleconference magazine, was invited to attend a press conference in New York City either in person or by audioconference. The invitation indicated lunch and cocktails would be served. Naturally, when she agreed to participate by audioconference, she assumed the refreshments referred to only those attending the press conference in person. On the day of the press conference, just prior to the audioconference, she received a box lunch by Federal Express which included a small bottle of wine.

Audiographic conferencing includes a number of different technologies that are added to the live transmission of audio including graphics. It also includes electronic blackboards or illustrators, computer-to-computer exchange by data, and graphic and still frame video. The common characteristic of this form of teleconferencing is that it still only requires a standard analog telephone line for transmission, though in some cases, more than one phone line is required. Audiographic conferencing has enjoyed new interest as bridging of audiographics and standards groups such as the Consortium on Audiographics Teleconferencing Standards (CATS) have evolved.

Business television today is satellite based teleconferencing and involves the transmission of either standard analog television or digitally compressed video at 1.5 to 6 million bits per second. Analog video normally requires a full or half a satellite transponder for transmission and can be transmitted in either C Band or Ku Band frequencies. Digitally compressed video at these megabit rates retains the appearance of broadcast television, and can be transmitted in a fraction of a satellite transponder allowing the transmission of 6 to 12 video programs over the same transponder.

Satellite conferencing can be via an ad hoc or temporarily assembled network or over a private satellite network dedicated to the use of one organization. Several industry networks for specific interest groups like the auto industry, police departments, and fire and rescue departments have developed using VSATs (Very Small Aperture Terminals or earth stations). Many companies that installed VSAT networks for two-way data transmission, like sales and ordering data, discovered they had a private television network as a bonus capability for a small additional cost.

Videoconferencing - The fastest growing segment of the teleconferencing market is, as you can see from figure 4, videoconferencing including all forms of two-way videoconferencing. We've chosen to separate this segment into two areas. 1) Videoconferencing or conference room-to-conference room conferencing at compressed video rates from 1.5 Mbps down to 384 kbps which require special data communications for transmission at T-1 (1.5 mbps) or fractional T-1 (1/2 or 1/4 T-1) rates and 2) video low data rate which is associated with Basic Rate ISDN (Integrated Service Digital Network) or digital telephone rates of 64 kbps or 128 kbps. This form of videoconferencing is appropriate for small groups or personal desktop-to-desktop videoconferencing.

While transmission costs for all levels of data communications have come down, improvements in compression technology have produced quality video at lower and lower data rates. The growth of videoconferencing has been centered in these lower transmission rates associated with ISDN (or Integrated

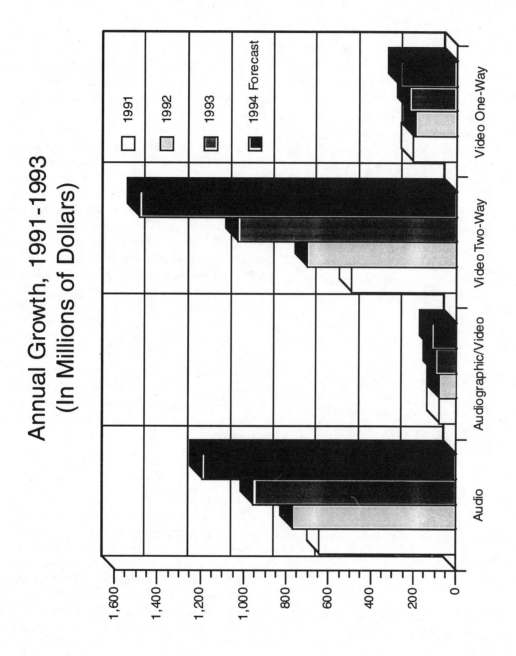

Figure 1.1.4 From the 1994 Outlook Report from Telespan & IFTF

Service Digital Network) digital phone service. While ISDN is an international standard for future phone service and has been fully implemented in Japan, only about 65 percent of all telephone company central offices in the U.S. are capable of ISDN service. However, most of the ISDN centers are in major metropolitan areas and by the end of 1994, 85 percent of the U.S. metropolitan areas will have ISDN service available. ISDN provides 128 kbps of data transmission for about the same cost as a standard phone call — $15-$30 per hour.

Fiber optics, with its vast capacity for transmission, has made possible two-way video at very high data rates like 45 mbps (DS-3) at full broadcast quality at long distance transmission costs which are now comparable to T-1.

Desktop videoconferencing, video windows and video telephony. The evolution of video communication has brought video to the desktop and finally into the home. For the business executive, his desktop terminal connected locally by a broadband local area network and world-wide by compressed video can have a live interactive video window on his computer screen. Desktop videoconferencing systems like Intel's ProShare (Exhibit 1.1.2) are available for $2500 to $1,000 and operate off a 386 or 486 PC with Windows.

The AT&T 2500 video telephone introduced in 1992 was the first commercially available motion videophone and operates over a standard analog telephone line. Priced under $1,000, this limited-motion color video image had limited resolution and applications, but it fulfilled the implied promise AT&T made 28 years ago.

More capable desktop systems, based on ISDN phone service and priced at $5,000 or more, provide better motion quality. These more costly videophones or computer-based systems conform to the H.261 standard and can interface with videoconferencing room systems.

■ THE IMPACT OF DESKTOP VIDEO AND COLLABORATIVE COMPUTING

Conferencing of all forms — video, audio and computer conferencing — was a $3 billion market in the United States in 1993 (Institute for the Future/Outlook Report, 1994) and is expected to grow to $13 billion by 1996 (F&S Market Intelligence and Communications News. (Figure1.1.5) Until 1993, audioconferencing constituted the largest segment of this growing market, primarily because of the high cost of equipment and broadband communications required for two-way video communications. In 1993, sales of equipment and services for videoconfer-

encing in the United States exceeded audio for the first time ($1.4 billion compared to $1 billion for audio). Eighty-five percent of the videoconferencing sales were for video over low data rate lines (ISDN) and utilizing modular systems priced between $20,000 and $30,000 each.

Market research conducted by The Yankee Group of New York projects the bulk of the 1994 to 1996 growth in conferencing will come in videoconferencing on the desktop computer utilizing ISDN data rates and low cost modifications to desktop computers.

Desktop videoconferencing takes advantage of an existing population of 140 million personal computers as a basic platform for videoconferencing. By adding an inexpensive camera, a video codec on a computer board and some software you can convert a 486, Pentium or Macintosh computer into a desktop videoconferencing device. The cost varies from $1,000-$5,000. Forty-three companies provide products for desktop videoconferencing as detailed in ABC's Desktop Videoconferencing Report (1994). Perhaps the most significant part of the computer-based systems is their collaborative computing capability to share screens or applications in a computer conference or virtual work group. Desktop videoconferencing has started out as a low cost extension of videoconferencing room systems to include people participating from their desk or home (telecommuting). Their promise lies in adding the personal communications of video to computer groupware applications to bridge time and distance.

■ OUR DIGITAL WORLD

The telephone, the television and the computer grew up in different worlds as illustrated here. They are now rapidly merging and digital technology is the key. Once all three technologies are digital, they are compatible and can be easily processed and translated on a computer. Computers are inherently digital where numbers that represent the information are stored in memory and processed. But the telephone, an 1850s technology, and broadcast television, an early 1940s development, are analog. They use signals that are wave forms and they are analogous or represent the sound and picture.

A good example of what happens when we convert technologies from analog to digital is the recorded music business and the compact disc (CD). Prior to digital CDs, music was recorded on records in the form of grooves. The grooves represented the music and could be read by the phonograph's needle to reproduce the music in the form of digital sound. The benefit of CD digital technology is that the CD is:

• much more precise

- less subject to damage or distortion
- able to be manipulated electronically to add, subtract or enhance the sound
- capable of more capacity for storage in a smaller area.

As a result of these factors, CDs rapidly replaced records.

Digital Phone Service

A new digital phone technology has become a world wide standard. ISDN has been installed throughout Japan and Europe. ISDN is available in most metropolitan areas in the U.S. and is rapidly becoming available in suburban and rural areas. What we gain with ISDN is a basic digital service of 128 kbps (128,000 bits per second) that compares to about 19,000 bits per second that could typically be transmitted over analog phonelines with a modem (a device that converts digital signals into signals for transmission over analog phone lines). The lines can be used for voice, data or images including highly compressed video. Cellular phone systems are converting from analog to digital service because it allows them to offer more channels of phone service.

Video is inherently a wasteful image to process or transmit. If you simply digitized analog video, it would require 90 million bits per second to transmit. That's a massive amount of transmission capacity. Video images contain a great deal of redundant information. A standard television picture is transmitted at 30 frames per second to assure that motion appears normal from frame to frame. An entirely new picture is transmitted every time even if only a small part of the picture changes from one frame to the next. However once video is digitized, it can be significantly compressed and transmitted much more effectively by transmitting only the changes rather than an entirely new picture. Digitized video can be more effectively transmitted, has less errors or snow, and can be stored on computers. The new U.S. standard for high definition television is digital and new cable services will be able to offer 500 channels over the same cable that could only offer 60 in the past. New video on demand video services that allow you to order a movie to your home whenever you want to see it will also use digital video to store and transmit video. George Gilder's book, "Life after Television," explores the broader context of how digital television and the merger of computers, video and the telephone will affect society and our lives. The chapters that follow in this book will explore in greater detail each form of teleconferencing and its application in business and education. But everything should be viewed in the context of our newly digital world.

ISDN Implementation in the U.S. Today

- 18% of the exchanges

- 50% of the major metro areas

- By the end of 1992 100% of the top 75 cities

U. S. Department of Commerce 1992

Figure 1.1.5

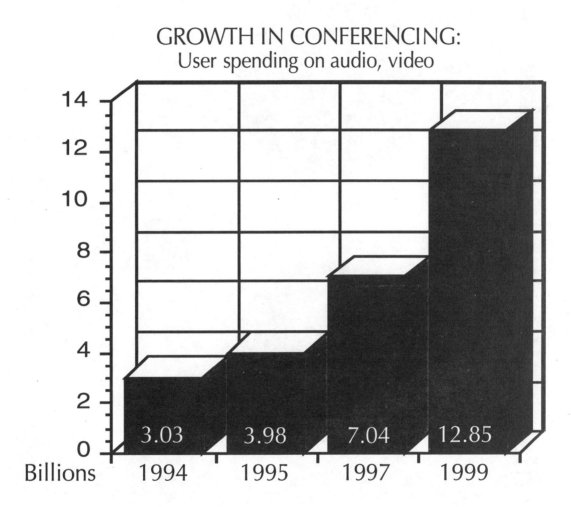

GROWTH IN CONFERENCING:
User spending on audio, video

Source: F&S/Market Intelligence and Communications News.

Evolution of the dial-up videoconferencing market

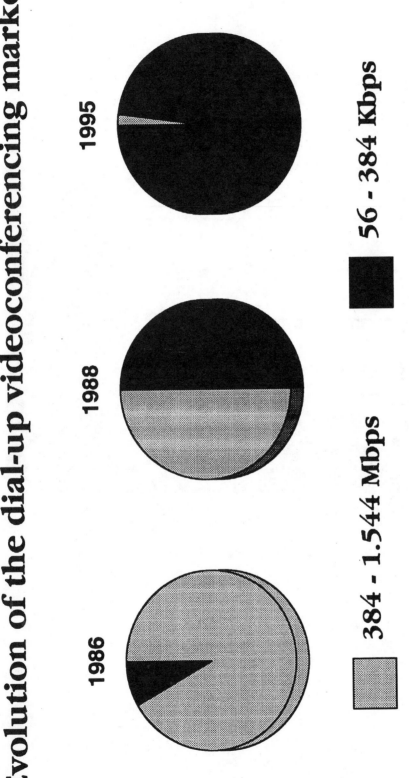

1986

1988

1995

384 - 1.544 Mbps

56 - 384 Kbps

1991 Study by Gartner Group

Figure 1.1.7

Chapter 2

Editors' Note: The use of audio technologies has increased as the cost effectiveness of the medium has been taken into consideration. This chapter is composed of four segments written by experts. The first segment is an overall technical view of audio technologies and equipment considerations. The second segment deals primarily with acoustic echo cancellation, the third segment covers audio bridging services and the fourth deals with audiographic technology.

Greg Hill, BSEET, (DeVry Institute of Technology) is a senior applications engineer for Shure Teleconferencing Systems in Evanston, Illinois. He develops and conducts product training seminars, writes applications notes, and assists integrators and large corporate users in developing appropriate applications for Shure Teleconferencing Systems. Previously, he was employed by Northern Telecom's Cook Electric Division, and worked with digital intercept announcers and mass data storage systems for call billing data. He actively follows the technological issues concerning audio teleconferencing, and has contributed articles to Sound and Video Contractor magazine, ITCA yearbook and Teleconference magazine.

AUDIO TELECONFERENCING

This segment explores some of the issues that are basic to interactive audioconferencing. Whether the goal is audio-for-videoconferencing or simple audioconferencing, the need for intelligible audio remains clear. I hope to provide the reader with an understanding of the factors that affect speech intelligibility and perceived fidelity. Topics include reverberation and noise control, microphone placement, feedback stability, echo control and transmission.

The fundamental task of audioconferencing equipment is to allow two groups of participants to converse as if they were separated by only a table and not hundreds of miles. The "transparency" of a system determines its potential for success; that is, the audio system that doesn't get noticed during a teleconference has done its job well.

To this end the audio teleconferencing system must fulfill two criteria: first, it must pick up outgoing speech from and reproduce speech back to the same point in a room without picking up excessive reverberation and noise, creating feedback or distortion, or returning echo. Second, the system must also provide appropriate user controls such as a volume control and mute function.

Advancements in technology have helped to produce superior fidelity audio in the broadcast and recording industries. Application of some of this technology to audio teleconferencing equipment has improved performance far beyond the level first set by the 4A speakerphone.

■ SPEECH PICKUP

The first step in achieving high quality audio is to obtain a clear, strong microphone pickup of all the participants. This should not be obscured by the simultaneous pickup of excessive background noise or hollow, distant-sounding reverberation. Background noise typically comes from ventilation ducts, fluorescent light ballasts, video equipment, and equipment cooling fans. The reverberant quality comes from hard wall, floor, and ceiling surfaces that reflect talkers' voices around the room many times on their way to the microphone.

These sounds can also interfere with conversation within the room itself. This interference is lessened by the "filtering" effect normal binaural hearing provides. A listener in the room can distinguish between the direct speech and the reverberation. The listener on the distant end of a conference does not have this ability. A single microphone picks up all the reverberation, noise, and direct speech and reproduces it without the directional "cues" that benefit the listener in the room. For this reason the transmitted audio must be cleaner than that in the room for the same level of intelligibility.

■ HISTORICAL OVERVIEW

In the early days of teleconferencing, speakerphones provided audio for teleconferences. These speakerphones employed omnidirectional microphones which responded equally to all sounds approaching from all directions. The omnidirectional mic allowed participants seated around it, at a uniform distance, to be heard at equal levels. This only worked well if the participants sat very close to the mic due to the amount of ambient noise and reverberation picked up in addition to the talkers' voices. This limitation reduced the number of participants to small groups.

The use of a unidirectional microphone, in place of the omnidirectional mic, improves intelligibility. A unidirectional microphone responds differently to sounds depending on their angle of approach. A sound approaching from the rear (off of the primary axis) of the microphone produces a lower output than a sound that approaches from the front (on axis). This directional character of the mic helps to reduce the amount of reverberation and noise transmitted to the distant listener. When the front of the mic is pointed toward the talker, the talker's voice will produce a stronger output than the noise and reverberation approaching from the rear and sides.

The way a mic responds to sounds arriving at different angles is described by a special graph called a polar pattern. The basic unidirectional mic has a "cardioid" (heart shaped) polar pattern. A cardioid mic (Figure 2.1.1) is about is about half as sensitive to sounds arriving from the sides as to sounds arriving from the front. Microphones are available with a variety of directional characteristics. For example, a supercardioid mic has a narrower pickup being only about 37 percent as sensitive to sounds arriving from the sides compared to sounds arriving from the front. A hypercardioid mic has an even narrower pickup, and is only 25 percent as sensitive to sounds arriving from the sides compared to sounds arriving from the front. However these narrow patterns also have rear lobes and, overall, don't have significantly less noise and reverberation pickup than the basic cardioid pattern. The cardioid microphone is the most generally suitable for teleconferencing applications. These microphones are often low-profile, surface-mounting types to minimize table reflections and visual clutter.

Replacing the speakerphone's single microphone with multiple microphones was used to try to increase the number of participants. This technique places everyone near a microphone. With the microphone closer to the talker, its pickup will have a better ratio of talker's voice to background noise and reverberation. Unfortunately, this microphone gets mixed in with all the other microphones in the room. The output of the other mics contains mostly background noise and reverberation due to their distance to the talker. The resulting signal contains more noise and reverberation than the one microphone could have picked up by itself. The use of multiple unidirectional mics produces slightly better results than multiple omnidirectional mics, but the amount of reverberation and noise picked up is still excessive if all the mics are open at the same time.

A solution to this problem is to turn on only the mic nearest to the talker(s) when they are speaking. Fitting each mic with a push-to-talk switch allows each user to select their own microphone when they want to talk. This is generally cumbersome at best and is especially difficult for new and occasional users to learn. Manually operated microphones inhibit the normal flow of conversation, limiting spontaneity and interchange.

Early systems sometimes employed simple automatic mixing devices. These automatic mixing devices used a fixed activation threshold, below which a sound would not gate a microphone on. There are several disadvantages to a fixed threshold system. First, if the system is adjusted when the ventilation is off, the system will gate on mics when the ventilation turns on. Conversely, if the system is adjusted with the ventilation on, the thresholds may be set too high for ordinary conversation. Second, a fixed threshold also permits a loud talker to turn on multiple microphones while preventing a soft talker from turning any on. Lastly, there is no provision to keep the received speech from gating any (or all) of the mics on and returning an echo to the originating site.

Modern teleconferencing systems, capable of using more than just a few microphones, employ automatic mixing with a Noise Adaptive Threshold. As its name implies, the threshold level at which a mic turns on automatically adapts to the amount of steady-state noise in the room, without any manual adjustments. The speech detection circuitry uses this to distinguish between the constant background noises and rapidly changing sounds such as speech. One system incorporates additional circuitry which automatically selects the mic nearest the talker. This mic will pick up the talker's voice quite strongly with a minimum of noise and reverberation. The result is an automatic mixer with unparalleled sensitivity and reliable operation. The Shure ST6000 Type 2 is an example of an integrated teleconference system incorporating all these features plus microphones specifically engineered for teleconferencing applications.

■ GENERAL MICROPHONE SELECTION GUIDELINES

As previously mentioned, a clear, strong pickup of the talker's voice is an essential step for good teleconferencing audio. The type of mic and its placement will determine how strong the talker's voice will be in relation to the ambient noise and reverberation. The selection process should begin with a few simple questions. Some of these could be:

What type of meetings will occur?

Where will the conferees be positioned?

Do all the conferees need the same degree of interactivity?

Is this a single or multiple use facility? Will presentation aids such as large drawings, white boards, electronic chalkboards, transparencies, or simple documents be used?

Finally, (but foremost in the client's mind), what are the aesthetic requirements?

The answers to these questions, considered with the acoustic conditions in the room will determine the type, style, and placement of the microphones.

■ MIC PLACEMENT

The Critical Distance (Dc) of the room is a good guide to mic placement when considered along with amount of ambient noise present in the room. The Critical Distance of a room is the point, relative to the source, at which sounds arriving directly from the source are equal in intensity to the sounds arriving by reflection around the room. An omnidirectional mic placed at the critical distance will have equal amounts of direct and reflected sounds in its output. This 50/50 mixture of direct and reverberant sounds makes the speech hollow sounding, fatiguing, and difficult to listen to for long periods of time. Placing the mic at half the critical distance will result in a pickup of the talker's voice with acceptable amounts of reverberation.

Virtually the same results can be achieved by using an unidirectional mic at the critical distance. With a directional mic the critical distance can be multiplied by a special number called the distance factor. The distance factor represents the improvement in critical distance a given directional pattern offers compared to the results with an omnidirectional mic. For example, a cardioid mic, with a distance factor of 1.7, can be placed 1.7 times further away than an omnidirectional mic can and still pick up the same amount of reverberation. The supercardioid has a distance factor of 1.9 and the hypercardioid has a

distance factor of 2. These mics, theoretically, give an even greater advantage over critical distance, but only if the talker is directly "on axis." The narrower pickup pattern of these mics makes staying on-axis more critical and the rear pickup lobes can often be problematic. As discussed previously, the cardioid mic is the best all-around choice.

Generally, where ambient noise is concerned, lower is better. The room noise (ideally) should not exceed 50 dBA for acceptable results. When the noise rises above this level, it causes the users to raise their voices to be heard within the room itself and also requires a higher receive level from the teleconference system.

When ambient noise levels are high, the mics must be moved closer to the participants to pick up intelligible speech. The signal to room noise ratio of the mics output will depend on the mic's distance from the talker and the level of ambient noise present. A signal to room noise ratio of at least 20 dB is desirable to prevent listener fatigue. A ratio of 10 dB is generally the absolute limit of acceptability. In a room with a 50 dBA ambient noise level, an omnidirectional mic would need to be placed 1.5 feet from the talker. A unidirectional mic in the same room could be placed at 2.5 feet for the same ratio. The minimum preferred listening level is approximately 64 dBA for an ambient noise level of 42 dBA. The preferred listening level will increase as noise increases (Figure 2.1.2) Lowering the ambient noise levels in the room allows the system to be operated at normal conversational levels, promoting natural sounding teleconferences.

■ MICROPHONE TYPE

The traditional meeting room in an organization places all the participants at a single, long, conference table. An excellent choice for this scenario is a surface-mount microphone. Unlike conventional desk stand and base station mics with their bulky appearance and high profiles, a surface-mount mic has a very unobtrusive appearance. Its pleasant styling reduces the possibility of "mic fright" and blends easily with conference room aesthetics. A surface mount mic with a cardioid pickup pattern is desirable to reduce loudspeaker to microphone coupling and pickup of ambient noise. One of the first mics to combine surface mounting, boundary effect and a unidirectional pattern in a single package specifically engineered for a teleconference system was the Shure STM 30.

A significant portion of a teleconference is often delivered from a lectern. A lectern generally limits a talker's range of movement, simplifying the mic placement. The conventional probe microphone is

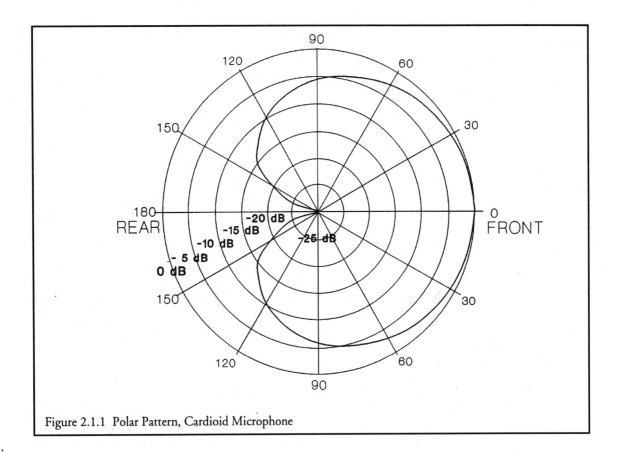

Figure 2.1.1 Polar Pattern, Cardioid Microphone

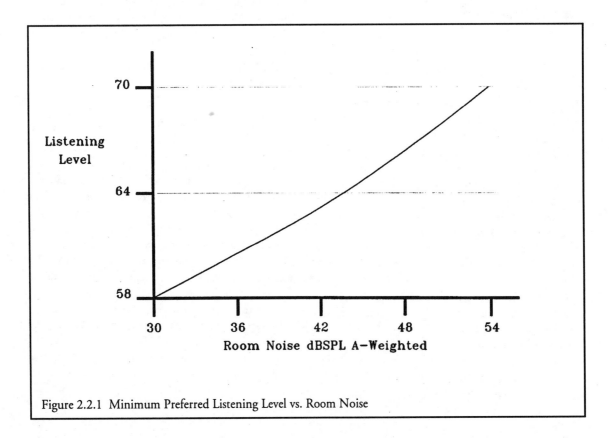

Figure 2.2.1 Minimum Preferred Listening Level vs. Room Noise

the standard for this application. A talker will expect to find a probe mic on the podium and is (generally) familiar with its use. The probe can suffer from a few problems, however. Some talkers will position the mic unnecessarily close to their mouths. This close miking can overly emphasize the low frequencies, creating a muddy sound. It also obstructs the view of the conferee's face in videoconference applications. Miniature probe mics with an integral gooseneck, such as the Shure SM99, greatly reduce these problems. The SM99 electret condenser mic has a wide frequency response with an exceptionally uniform supercardioid polar pattern. The mic employs a two-stage pop filter to control breath noses, making it especially suitable for speech pickup. A surface-mount mic, when used for podium coverage, helps to reduce these problems too. However, the mic's positioning must prevent the talker from covering the mic with lecture notes.

The conferee at the white board needs mobility. Whether facing the board or other conferees, the talker should be heard at the same level. A lavaliere mic, worn on the lapel or about the next, obtains a uniform pickup of the talker's voice since the distance from the talker to the mic never changes. If the wire becomes cumbersome and security is not an issue, a wireless mic can replace the wired one. For best operation the usage area of a lavaliere mic should be limited to well-defined areas of the conference room, away from system loudspeakers. If the conferee with the lavaliere strays too close to the loudspeaker, there may be feedback or an echo returned to the distant site. Again the surface mic comes to the rescue. Placing surface-mount mics to either side and above the board will generally give adequate coverage of the talker. This arrangement has two strong benefits. Firstly, the relative positions of mic and loudspeakers is fixed and system calibration is simplified. Secondly, fixed mics prevent accidental "walk-offs" of expensive wireless mics.

Covering the "peanut gallery" of a teleconference room presents a unique challenge since there is often no furniture other than movable chairs to affix the mics to. While mounting mics directly on ceiling surfaces often produces marginal results (especially in systems lacking automatic mic control) a mic suspended down from the ceiling often works well. The Shure SM102 is a miniature overhead condenser microphone featuring a six inch gooseneck section that facilitates setup and aiming. Its wide, smooth frequency response and cardioid unidirectional polar pattern provide outstanding "reach" with excellent feedback rejection. A grid pattern of SM102s above a gallery or audience area provides economical uniform coverage of these seating areas.

In summary, to obtain the best results, the reverberant character of the room and ambient noise levels must be controlled, since no current technology can remove the noise and reverberation a from signal once the microphone has already picked it up. The use of directional microphones reduces the amount of reverberation picked up by the microphone and thereby reduces the amount of absorption required. If the number of participants is greater than two or three, the use of multiple microphones with automatic control is required. This arrangement allows participants to be seated within the optimum (typically not more than one meter) miking distance. The direct sound from the participants is then much greater than the noise and reverberation.

■ THE TRANSMISSION LINK

The ideal audio link for interconnecting teleconference systems would provide transmission of the full voice frequency range that is uniform, stable, undistorted, and with unity gain (no amplification or loss of the signal). The practical audio links available today are bandwidth limited, of sometimes unpredictable loss, and more distorted than is desirable.

The worst of these are dialed access telephone lines. The frequency response is limited to between 300 Hz and 3.4 kHz and the loss is variable. The audio response of video codecs varies with the data rate at which the unit operates. At low data rates, the codec's audio response is limited to this telephone bandwidth. At one time 64 kbps was considered the minimum data rate for speech alone. Now codecs exist that are able to encode a video signal along with an audio channel at this data rate. These narrow band limitations reduce the amount of speech energy transmitted and affect the intelligibility and perceived loudness of the speech signal. The loss of the high frequencies is of particular concern for the perceived fidelity of the sound.

Careful frequency shaping of the narrow band signals restores much of the original intelligibility, loudness and naturalness of the received speech when compared to the full bandwidth signal. This had been well known within the communications industry for many years but most teleconference system manufacturers have not applied this knowledge. The optimum narrowband frequency shaping provides a boost in the presence frequencies (1 kHz through 3 kHz).

With the emergence of high quality standard audio coding at moderate data rates, high fidelity audioconferencing becomes possible. Wider usage of 7 kHz audio-only conferencing will become popular as transport services such as ISDN bring end-to-end digital connections with switched operation to most businesses.

■ LOUDSPEAKER SELECTION AND PLACEMENT

On the receiving side of the audio link, placement of the conference loudspeakers near the users is not as critical as is the close placement of the microphones in the transmitting room. The binaural hearing of the participant assists in differentiating the direct sound of the loudspeaker from the local background noise and reverberant sound the same as it would for someone talking in that room. Room noise should still be kept low so that the loudspeaker level doesn't have to be raised excessively. The added benefit of low room noise is that the amount of noise transmitted to the distant site is kept to a minimum.

The placement of the loudspeaker(s) should provide adequate distribution of sound to all the participants. The logical placement of the loudspeakers for videoconferencing is next to the receive video screen. This placement gives the most natural localization of the received signal and draws the attention of the participant to the screen.

The loudspeaker(s) used should have a flat, uniform frequency response throughout the available frequency range and uniform directional characteristics through at least the middle audio frequencies (to 3 kHz). These frequencies are the most critical to the intelligible reproduction of speech. Many two-way loudspeakers, particularly when mounted on a wall surface, will give less than optimum performance. The crossover frequency region (where the one transducer leaves off and the other takes over) often lies in the range critical to speech intelligibility. In many loudspeakers this crossover is not handled well. A loudspeaker that performs acceptably in a wide frequency range music application may exhibit undesirable peaks and dips in the frequency response and changes in the directional pattern over the narrowband voice frequencies.

So far, we have only discussed how to achieve the highest possible audio fidelity. We have followed the speech from where it originates in the room, to the pickup point at the microphone, its path through the transmission link, and its reproduction in the room. Since the nature of communication is bi-directional, these paths exist for the participants in both rooms. This brings us to our next topic, the feedback stability of the electro-acoustic loop formed by these two paths.

■ SYSTEM STABILITY

Figure 2.1.3 shows the separate signal paths as usually found in videoconference applications. The loudspeaker symbols may represent multiple loudspeakers and the microphone symbols represent the summation of all the microphones either conventionally or automatically mixed.

In addition to the desired signal path from the microphone to the loudspeaker, at the distant site, there exists acoustic coupling between the loudspeaker and the microphone in the same room. This predictably forms a feedback loop which behaves according to the rules similar to those that govern feedback oscillators. Given sufficient signal gain and appropriate phase shift around the loop, the system will oscillate.

When the audio system is connected by a 2-wire, dialed-up phone line, a local coupling path is created in addition to the acoustic path from the distant room. This local path is created by the action of the 2- to 4-wire hybrid that combines the two directions of signal flow onto one pair of wires suitable for the telephone network. The hybrid performance depends largely on its ability to match the impedance of the two wire line. In some cases there may be very strong coupling from the hybrid's transmit side to its receive side at certain frequencies.

To maintain stable operation, the gain around the overall end-to-end feedback loop and the local loop through the hybrid (if present) must be less than unity at all frequencies. Maintaining this stability is the most important challenge for the audio teleconference system. Many of the practices we've discussed that improve the clarity and the intelligibility of the teleconference audio also help to improve the feedback stability. Unidirectional microphones and automatic microphone control help to effectively reduce direct coupling of the loudspeaker to the microphone. Acoustic treatment helps to reduce reverberant loudspeaker-to- microphone paths. However, these practices alone are not enough to maintain feedback stability.

If a system is adjusted for feedback stability simply by reducing loudspeaker level, the system will generally be quieter than desired by 6 dB in an optimized room to well beyond that in less-than-optimum rooms. Systems without automatic microphone control would exceed these levels by even more. A system with four microphones would be 6 dB quieter than the equivalent system with just one microphone. This is a primary reason why systems lacking microphone control seldom exceed more than two or three microphones.

If the transmit level (microphone gains) are reduced to achieve system stability instead of the loudspeaker levels, the weak levels are simply transferred to the other side of the conference. Just turning down the transmit or receive gains does create the desired feedback stability but leaves the listener with either an unacceptably low listening level or the need for very close miking of the talker. Since a low listening level is unacceptable, close miking of the talker becomes the only option.

A better way to maintain feedback stability is to temporarily turn down (attenuate) the transmit or receive signal. This "feedback suppression" (not to be confused with "echo suppression") would alternately attenuate the level of the received or transmitted signal automatically in accordance with the direction of the conversation. The amount of attenuation required will vary according to the amount of acoustic coupling from the loudspeaker to the microphone, the balance of the hybrid to the line, any extra receive path make-up gain (volume control, etc.) and is equal to the amount of gain beyond unity in the "loop." If a properly integrated, adaptive, line balancing hybrid is employed in a two-way audio conference, the amount of feedback suppression may be reduced, theoretically, by an amount roughly equal to the end-to-end telephone line signal loss.

In the case of the optimized videoconference room, the amount required is often 6 dB or less. When handled in a highly interactive, conversationally-oriented manner, this level of suppression is unnoticeable. The handling of feedback suppression becomes even more critical in more challenging acoustic environments and telephone line applications. Under these conditions, the transitions of the systems from transmit to receive modes must be smooth and complimentary in each room.

■ VIDEOCONFERENCING ECHO CONTROL

There are several approaches to controlling echo in a videoconferencing system. Echo control has its roots in the telephone network where the problem was first experienced. Present day echo control techniques have drawn on the contributions of the telephone industry in solving the echo problem as it applies to teleconferencing. The type of echo control process has a strong effect on the subjective audio quality in a videoconference. Let's take a look at some of the major advances in echo control and then examine what is available today.

The Telephone Network

Long-distance telephone service became possible in the early part of this century by the invention of the vacuum tube repeater, Before the vacuum tube, telephone service was limited to small geographic areas by the loss in signal strength caused by telephone cable. A repeater is an amplifier that overcomes this loss. It boosts the level of the telephone signal to send it through the thousands of miles of cable required for a long-distance call.

The problem of echo arises from the electrical properties of the cable itself. In a typical telephone

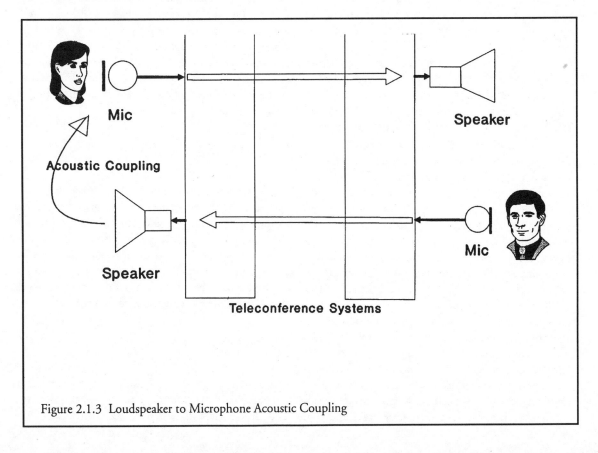

Figure 2.1.3 Loudspeaker to Microphone Acoustic Coupling

transmission line, the rate at which the signal moves down the line (referred to as the propagation velocity) is significantly less than the rate of travel of a radio signal (i.e. the speed of light). For a 22 gauge, twisted-pair, pulp-insulated cable the "velocity or propagation" is 28,800 miles/second. For a cable of 2,000 miles it would take 69.4 milliseconds (ms) for the signal to reach the other end. A radio signal would take only 10.7 ms to travel the same distance. Some of the signal is reflected back to become echo when a poorly balanced hybrid transformer couples the received signal to the sending circuit. The echo experiences twice the one-way delay (and loss) on its trip. If the connection has more than approximately 22 ms delay (one-way), the talker will hear the reflected signal as an echo.

Careful studies in the 1920s by Bell Laboratories determined the amount of echo that a talker could tolerate before it began to interfere with speech production. This amount of echo was termed the "annoyance level." It was found that the annoyance level of echo that causes disruption varied depending on the amount of delay. Generally, as the delay increased the annoyance level of the echo considered tolerable, dropped. Therefore, in lines with long delay a very small returning echo could cause interference.

Fixed Loss

One Solution to this interface was to insert enough loss in telephone connections to reduce the echo below the annoyance level. The amount of loss varies depending on the overall length and the number of separate circuits involved in the connection. The first loss plan was developed at the same time as long distance telephone service and has undergone revisions as transmission methods have changed and telephone user "opinion models" have been refined

The Via Net Loss (VNL) plan, developed in the early 1950s, is a method of calculating the optimum amount of loss that each connection must possess to reduce echo and still maintain a comfortable receive level. The introduction of digital switching systems required a restructuring of the loss plan. The result, called Switched Digital Network (SDN), is used with digital connections in the network. Although they are stated as two separate plans they are not incompatible with each other. The VNL and SDN plans are limited in the amount of echo they can control. Short connections require only small amounts of loss. In long connections (beyond distances of 1850 miles), the amount of loss required to prevent echo also reduces the receive level excessively. Circuits of this length are only encountered between the major switching offices (that carry only long distance calls) interconnecting the local switching areas and they

require a different solution.

Switched Loss

An echo suppressor is a voice-activated device designed to control echo on a circuit where the amount of fixed loss would be objectionable. Each end of a circuit has an echo suppressor that protects the other end from echo (Figure 2.1.4). The suppressor eliminates echo by inserting 35 dB or more of loss in the echo return path when the person is talking. If both persons speak at the same time, one of two things can happen: either all loss is removed from both paths or 6 dB of loss is inserted in each path until one talker yields.

Echo suppressors have been used successfully since the 1930s to control echo. However, they are only effective for controlling echo on lines with a maximum of 100 ms delay. This covered the majority of links that provided long distance service in the telephone network until the 1970s.

The late 1960s saw the launch of satellite programs to provide communications services. (Author and futurist Arthur C. Clarke had proposed such systems before the ability to place them in orbit was developed.) These satellites orbit the earth geosynchronously (stationary to a point on the surface). This type of orbit places them approximately 23,000 miles above the earth's surface where they relay television and telephone signals to and from ground stations. The time it takes for a signal to travel to and from the satellite is approximately 500 ms. Conventional echo suppressors were unable to handle this amount of delay smoothly. This long delay caused the suppressor to mutilate the speech, making it sound "choppy." Once again researchers began looking for new ways to control echo.

The Center Clipper

When speech is present in both directions, however, the center clipper tends to mutilate the speech signal, adding audible amounts of harmonic and intermodulation distortion. This distortion is often referred to as "gritch" and sounds remarkably like its name when it occurs. The transmit signal can be totally chopped out if the level of the transmit signal drops below the estimated level of returning echo.

The Echo Canceller

In the 1970s, research moved into a new area called echo cancellation. The name canceller was chosen to differentiate it from the earlier echo sup-

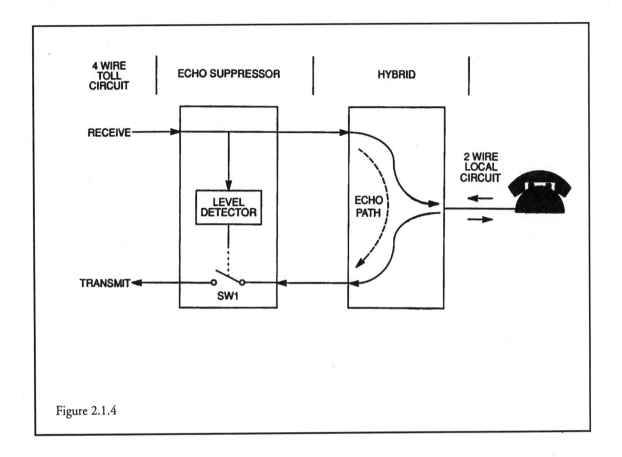

Figure 2.1.4

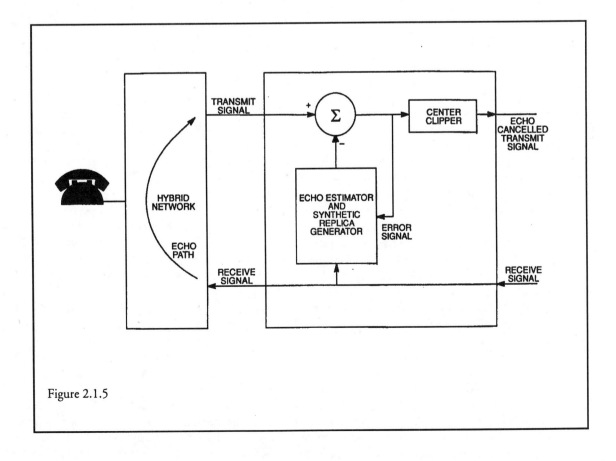

Figure 2.1.5

pression devices. A telephone line echo canceller produces a synthetic replica of the echo it expects to see returning and subtracts it from the transmitted speech (Figure 2.1.5). The replica it creates is based on the transmission characteristics of the telephone cable between the echo canceller and the telephone set. This echo path is characterized by a minimum of delay (less than 22 ms) and a minimum number of returning reflections of the original signal. Even with the relatively simple echo path of a 2-wire phone line, the echo canceller was unable to completely remove the echo under all conditions. For this reason, the center clipper was resurrected and placed after the canceller to clean up the residual echo remaining after the canceller had done its best.

Field trials by Comsat in 1973 and 1974 proved the value of echo cancellation technology for satellite-based voice circuits and set their architecture in the form we see it today. Subjective performance of satellite voice circuits was greatly improved by the canceller over the suppressor. Telephone company design practices specify the type of echo control and placement of echo control devices in the network. Echo suppressors are applied to all terrestrial lines with a length of greater than 1850 cable miles. Echo cancelers are applied to all lines routed through satellite circuits. Loss plans (VNL and SDN) provide protection on circuits less than 1850 cable miles in length.

Videoconferencing Echo Control

An audio system for videoconferencing faces the same basic echo problem (with a different cause) as found in the telephone network. The loudspeaker-to-microphone coupling in the room returns the received speech to the originating room. The time delay that arises from the coding and decoding of the audio and video signals, as well as the transmission time (satellite or digital land line), causes the returned signal to be perceived as an echo.

Initially, the only approach to solving this problem was to prevent the microphone and loudspeaker from being active at the same time. Push-to-talk microphones are an example of this approach. The push-to-talk bar would turn off the loudspeaker, preventing the return of echo. The disadvantage of this type of system is obvious; if both participants in each room activate their microphones, neither can be heard.

In the optimized videoconference room where the loudspeaker-to-microphone coupling is minimized by the use of unidirectional microphones and/or automatic microphone control, a telephone line echo canceller can be used to some advantage. However, the canceller itself does little to reduce the

echo. Most of the benefit comes from the action of the center clipper. This is due to the type of echo the canceller was originally designed to act on. A phone line returns a few, simple, reflections of the received speech. The acoustic paths present in a videoconference room return a great many more reflections over a longer period of time with different decay characteristics. The adaptive echo canceller of a telephone line echo canceller is not suited to process this type of signal.

Acoustic Echo Cancelers

Another type of echo canceller was created to attempt to solve the problem of echo due to loudspeaker-to-microphone coupling, the "Acoustic Echo Canceller." Acoustic echo cancelers can employ nearly all of the echo control processes discussed in this chapter depending on the make and model (Figure 2.1.6). All of them have some form of adaptive echo canceller that produces a synthetic replica of the potential echo to subtract from the transmit audio. Most units have a center clipping echo suppressor to remove the "residual" echo from the transmit signal. (Currently there are about a dozen different algorithms used to create the synthetic replica. New algorithms can be expected as the capacity of the digital signal processors used to implement them increases.)

The goal of the acoustic echo canceller is to reduce the amount of direct and reverberant loudspeaker coupling to the microphone to prevent echo. To achieve this task, the algorithms used in today's devices require an audio system that is feedback stable. The "Acoustic Echo Return Loss" (AERL) specification of the device describes the amount of attenuation in the loudspeaker to microphone coupling path required for proper functioning of the algorithm. A figure of 0 dB would indicate a unity gain condition, a figure of 6 dB indicates an equivalent loss in the overall coupling path (including microphone preamplifiers and power amplifier for the loudspeaker). Positive gain in this path creates significant errors in the synthetic echo replica that reduce its efficiency. We have spent some time looking at the ways to optimize an audio system's feedback stability and the compromises involving receive levels and miking practices. Current acoustic echo canceler devices require that many of these practices be used to obtain adequate receive levels.

There can be some confusion about the term "room echo" as it is used in videoconferencing. The term is frequently used to describe the hollow sound of talker reverberation. While the reverberation may sound "echo-ish" it is distinctly different from (but not unrelated to) the echo returned by loudspeaker-to-microphone coupling within the room. An acoustic echo canceler cannot remove the reverberation from local speech in the transmit signal. Once the microphone has picked up the reverberation, nothing (using present technology) can remove it. Room acoustic treatment and/or effective miking techniques (as previously discussed) are still the most effective methods of reducing talker reverberation and loudspeaker-to-microphone coupling.

Echo Reduction

A new method of echo control, developed in 1988, uses attenuation in a new way to subjectively reduce the returned echo without the mutilation (choppiness, level drops, distortion) found in suppressors or center clippers. Called Echo Reduction, this patented method rapidly (and momentarily) applies a variable amount of attenuation in between transmitted speech peaks (where the echo would be audible). The device compares the transmit and receive signals to determine the likelihood of objectionable echo in the transmit signal. It then calculates (and inserts) the appropriate amount of attenuation, for that instant, to control the echo. During the outgoing speech peaks, the echo is masked by the strong local speech and rendered inaudible to the listener, so that no attenuation is required. Thus the circuit subjectively reduces the echo by attenuating it below the level of annoyance while passing the transmit speech unattenuated. Echo Reduction produces very little of

the distortion and "gritch" caused by center clippers. As with all echo control methods, a feedback-stable environment is required since processes suited to echo control do not enhance feedback stability margins.

The choices available for echo control in videoconferencing audio systems are much wider than those of only five years ago. The performance of all types of systems has improved through application of new technology and careful attention to the lessons of the past.

■ SOME PRACTICAL EXAMPLES

The audio problems associated with teleconference rooms are varied due to the multiple disciplines that must be drawn together to create a successful videoconferencing facility. There are many disciplines associated with videoconferencing. Some of them are acoustics, audio practices, bridging, construction, corporate culture, design, digital technology, encryption, ergonomics, lighting, network design, noise control, psychoacoustics, system control, telephone, and video production. Each of these disciplines has their part in the successful completion of teleconferencing's goal: communication and personal interaction through sight, sound, data, and motion. One person rarely possesses mastery of all these areas. To do so would require a lifetime of study. Of all these disciplines, six strongly affect the audio perfor-

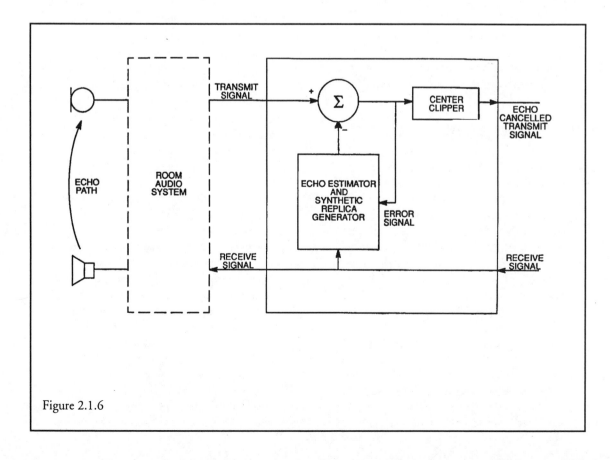

Figure 2.1.6

mance: audio practices, acoustics, noise control, psychoacoustics, bridging and telephony. Good audio performance for teleconferencing depends on a synergism between audio system and room design.

Nearly everyone involved in the field of teleconferencing has a story or two about a conference room that experienced some problems: the room with a reverberation problem caused by too much pewter panelling; the room at the end of a runway where jets taking off punctuate the flow of a conference; the HVAC system that creates more noise than airflow; the room that looks great but is not functional; the room with so much acoustic treatment that participants' ears feel "stuffy"; and others. A few examples follow:

During the construction of a videoconference room, the various systems were installed in stages. When the graphics system was installed, a ticking noise became apparent in the audio system. The technicians checked the wiring and all was connected just as it was designed. Attention focused on a diffuser for the ventilation system as the culprit. Every time someone stood on the table and held the diffuser, the ticking went away. No one seemed to notice that in order to stand on the table, the graphics pad had to be disconnected. The problem stemmed from having the mic lines and the graphics tablet on the same mic snake. The graphics pad was inducing noise into the mic line. When it was disconnected to hold the diffuser steady, the ticking went away. Moving the wiring of the graphics pad to a separate line solved the noise problem.

Too little acoustic treatment can cause even the best audio system to sound bad. A metallurgical company with a pewter panelled meeting room decided to install a teleconferencing system. The stated requirements were that the system be "full-duplex" and that there be no mics on the table. The first system that was purchased employed an acoustic echo canceler with multiple microphones. After the system was installed and adjusted, it was found that the resulting loudspeaker level was very low and the amount of reverberation picked up in the room was fatiguing to listen to for long periods of time. Since the addition of acoustic treatment was considered out of the question because of aesthetics, a system with automatic mic control was chosen and the mics were moved to the table. The closer

miking of the talkers and the mics automatic control resulted in much less reverberation pickup. Higher loudspeaker levels were possible because of the unit's feedback-controlling circuitry.

Too much acoustic treatment can be as irksome as too little. In one instance the designer of the room felt that to achieve full-duplex operation, a heavily treated room was required. Virtually every surface that could be covered with sound absorbing material was covered. The effect this had on the visual aesthetics was appalling. The room became so "dead" that the conferees had trouble hearing each other at opposite ends of the table. The audio system for the room included six mics and a simple mixer feeding the codec's input. The resulting pickup of the conferee's voices was still low in level and hollow-sounding, because all the mics were on at the same time, causing a comb-filtering effect. The solution was to change the miking system to one that permitted only a minimum number of mics to be open at the same time. This reduced the comb-filtering effect and lowered the amount of reverberation pickup. The lower pickup of reverberation permitted removal of some of the acoustic treatment leaving enough reverberation to make face-to-face conversation possible and comfortable.

A company wanted to install videoconferencing into an existing multi-media room. This room had a distributed loudspeaker system in the ceiling. To save money, the decision was made to re-use the existing speakers for the new system. The proposal was accepted and the system was installed. The problem was that during videoconferences, the conferees looked idly at the walls, at the ceiling, and at each other. This generally gave the other side the impression that they weren't getting through, or that if they were, the conferees weren't paying attention. The reason for this phenomenon is that it is common behavior to look in the direction of the sound source. In this room the sound source was overhead and very diffuse. Moving the main loudspeakers up front by the monitors while providing low level re-enforcement in the back of the room resulted in the conferees properly directing their gaze in the monitors' direction (hence to the cameras) where the sound was coming from.

This rearrangement increased the satisfaction of the presenter on the other end.

Two major corporations, each with custom-built videoconference networks, needed to conference with each other on a contract that they were jointly awarded. Each company had carefully designed their rooms, audio systems and networks for optimum performance. The rooms in each of these networks worked well with each other until the networks were interconnected. When this was done, problems with echo, low loudspeaker levels, and chopping became apparent. An analysis of the problem showed that there was more at issue than digital code compatibility between the systems. The problem stemmed from different network signal levels caused by non-standard calibration practices and audio links with non-unity gain. Transmit and receive audio signal levels did not remain consistent between different sites and different interconnections. When these problems were rectified, the audio problems were solved.

Let's take a look at these problems in more detail. As the previous example illustrates, audio and transmission adjustment practices are important to achieving interoperability between systems if videoconferencing is to become as ubiquitous as the telephone is today. Once the digital code compatibility problem is solved, the maintenance of consistent audio signal levels is the next step in achieving interoperability between systems. The level differences compound the problems associated when different

organizations, each with their own private networks and adjustment practices, try to interconnect with each other. The result is that today there are multiple private networks that are not able to talk together optimally not only because of different data rates, protocols, but also different audio system calibrations.

To ease the digital code compatibility problem, the CCITT has proposed a standard (H.261) that will allow video codecs of different manufacturers to communicate with each other. This standard, when finalized, will allow pictures to be interchanged between different brands of codecs and specifies the time slot available for audio in the bit stream. H.261 gives previously incompatible devices, running proprietary algorithms, a common language. The audio coding technique is still not established by this standard. Several audio coding algorithms are presently employed to minimize the bit rate of the coded audio signal. These include several which are proprietary to a codec manufacturer. New audio coding algorithms will continue to appear offering improved fidelity vs. bit rate tradeoffs. Of the various audio coding techniques available today the most promising is SB-ADPCM as described in CCITT recommendation G.722. This recommendation describes a method for high quality speech coding with a 7 KHz bandwidth at a data rate of 48 kbps, 56 kbps, or 64 kbps.

Since H.261 is still only a recommendation, other solutions are presently required. One approach is a gateway. A gateway is a network element (node) that performs conversions between different coding and transmission formats. The gateway does this by having many types of transmission equipment commonly used in teleconferencing and providing a means for interconnection. The video and audio signals are first brought back to their original (analog)

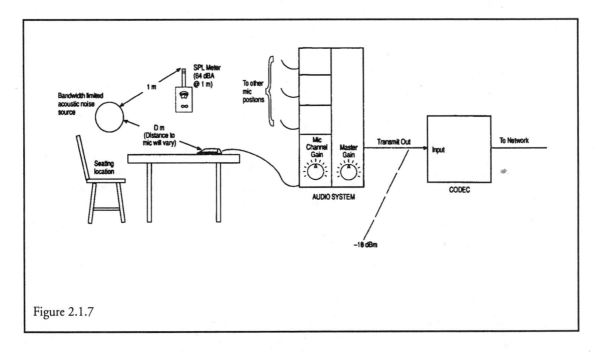

Figure 2.1.7

form and then re-coded to the different format.

The possibility for degradation of the signal exists in the gateway if equipment is not well calibrated or properly interfaced. One of the limitations is that multiple coding of the signals will decrease the signal-to-noise and signal-to-distortion ratios. The other is that extra delay is added to the connection. At longer total round-trip delays (one second or greater) interaction becomes very stilted and uncomfortable. At higher data rates, the extra amount of noise and distortion is minimum. At lower data rates, where the signal is more severely compressed, the extra added noise and distortion may seriously degrade the signal.

Another potential way of achieving interoperability between systems is the Multipoint Control Unit (MCU). Current MCU's only have the ability to support multi-point conferences on codecs of the same brand and (in most cases) model. An advanced MCU could, in theory, be constructed that would convert between different communications protocols, data rates, and television systems. The audio bridge of such an MCU would be critical to the systems' operation. It must have the ability to accept signals from a wide range of systems including open mic, gated mic, and "acoustic echo cancelers" and maintain an acceptable level of performance. The unit should not lock up on the high levels of room noise picked up by open mic systems, or be dependent on the acoustics of the teleconference rooms.

Most important for a gateway, for an MCU, and, in fact, for the entire audio transmission link is the maintenance of unity gain (no signal level increase or decrease) from the audio input of a terminal's codec to the audio output of the distant terminal's codec. (In the case of an MCU, this unity gain must be maintained at least for a single active port. It is not possible for a large number of ports to simultaneously exhibit full unity gain in both directions to all other ports without degrading overall system stability.)

With unity transmission link gain, the receive signal level at the input of one end's audio equipment will be equal to the transmit signal level produced at the other end. The audio levels of interconnected systems will then be compatible if uniform transmit levels are maintained through consistent audio system adjustment procedures at each teleconferencing site. The transmit level adjustment is aimed at producing an appropriate, consistent transmit signal level for typical speech in the conference room from at least the prime seating locations.

Perhaps surprisingly, the recommendations for optimum transmit level calibration are essentially consistent among the various audio equipment vendors, (Figure 2.1.7) and may be summarized as follows: 1. A bandwidth-limited acoustic noise source calibrated to produce 64 dBSPL at one meter is placed sequentialy at the primary seating locations (which may or may not be one meter from the conference microphones). 2. The individual microphone (and master transmit, if present) gain controls are adjusted to produce a 18 dBm transmit level feeding the codec. Some audio equipment with dedicated, calibrated microphones can provide preset gain settings for typical (one meter) miking distances.

The next step after transmit level calibration is the loudspeaker level adjustment. This can have predictable results only if consistent transmit levels at the remote sites and unity gain transmission links are now producing consistent receive input signal levels. The loudspeaker level obtainable will be limited by feedback stability considerations, though, as previously discussed.

The process used by the audio equipment to maintain the feedback stability will affect the range of environments that the equipment can operate effectively in. Acoustic echo canceller devices, that many times offer no improvement in feedback stability margins, can only be installed in very quiet, well optimized rooms. Too many open mics, long reverberation times and high room noise degrade their intellibility and signal-to-noise ratios. Devices that provide switched attenuation to address feedback stability are able to adapt to a wider range of environments. The switched attenuation (sometimes called Acoustic Feedback Suppression) makes up the differences between listening levels obtainable without suppression and desired, comfortable listening levels, while maintaining the feedback stability of the system.

To summarize, we have reviewed the basic requirements and considerations for successful videoconference audio. Speech pickup should be clear and intelligible without excessive noise and reverberation. Speech reproduction should have good fidelity and an adequate listening level. A conference site must maintain its half-loop feedback stability and control its own echo. Uniform transmit levels and unity gain transmission links must be maintained to ensure compatibility between conference sites. Only when these aspects are successfully addressed through proper room design and equipment selection, installation, and adjustment can the full potential of teleconferencing, whether audio-only, video or audiographic, be realized.

Jeremy F. Skene is vice president, business development at Coherent Communications Systems Corporation, a major worldwide supplier of audio teleconferencing systems. Skene, who has been with Coherent since December of 1985, provides the focal point for new revenue generating areas at Coherent, encompassing identification of new teleconference market and product opportunities and initiatives.

Prior to joining Coherent, Skene held a variety of technical and marketing posts at Gandalf Technologies, in Ottawa, Ontario and Chicago, Ill., where he served as senior engineer, director of corporate strategy, director of research, and marketing manager. Previous to this, Skene was a member of scientific staff at Bell Northern Research, in Ottawa, where he was involved in the integration of voice and data in various switching systems.

Skene holds a Master of Science degree in Applied Physics and Bachelor of Science degree in Physics from McMaster University in Hamilton, Ontario. Skene, 42, resides in northern Virginia with his wife and two children.

ACOUSTIC ECHO CANCELLATION

As you have read in the previous section, the audio portion of a teleconferencing system is extremely important, and can easily make the difference between a highly productive and useful teleconference call and a frustrating waste of time. In some instances, poor audio can lead to a misinterpretation of the facts and a lost business opportunity.

All too often, the audio component of a teleconferencing system is overlooked. The reason for this is that callers are accustomed to fairly good, easy to use audio when making the majority of their telephone calls, and so tend to assume that this will also be the case in teleconferencing applications. Even in a videoconference call, the majority of information transferred is actually in the audio channel. If this statement is surprising, imagine for a moment the effect of losing audio during a videoconference compared to losing video.

Not only is good, clear audio mandatory to the effective exchange of information and ideas, but it also encourages the continued use of teleconferencing, and thus leverages the benefits of teleconferencing within any organization.

While several factors contribute to the overall audio quality of a teleconference system, one of the most important is the presence of a full-duplex audio channel. Full-duplex audio is simply an audio channel which allows conversation to take place interactively and simultaneously between the various parties, without electronically cutting off one or more participants if someone else is speaking.

A normal handset telephone provides a full-duplex channel as does any face-to-face conversation. While we may not realize it, the style of our spoken communications has been shaped since birth to depend upon the presence of a full-duplex communications channel. It is vital to maintain this property in any teleconference system in order to preserve an unambiguous and uninhibited information interchange. This is not a trivial task, however, as will be explained.

■ THE NEED FOR ACOUSTIC ECHO CANCELLATION IN TELECONFERENCING

Audio in teleconferencing, whether audioconferencing, audiographics or videoconferencing, differs from normal telephony in that the participants usually use a microphone and loudspeaker rather than a handset on a telephone. This seemingly minor difference can have a major impact on the ability of the system to provide a full-duplex audio channel. The reason for this is illustrated in Figure 2.2.1.

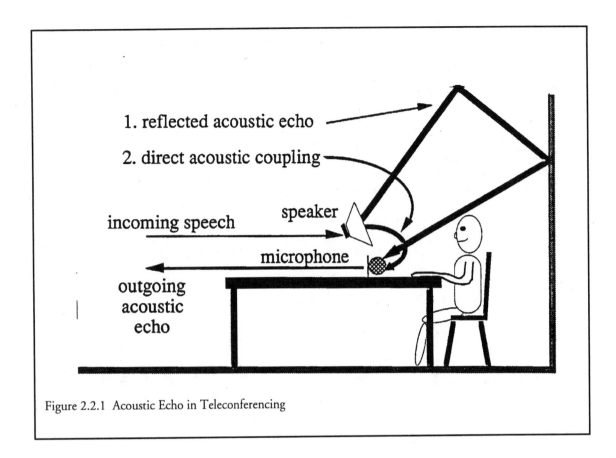

Figure 2.2.1 Acoustic Echo in Teleconferencing

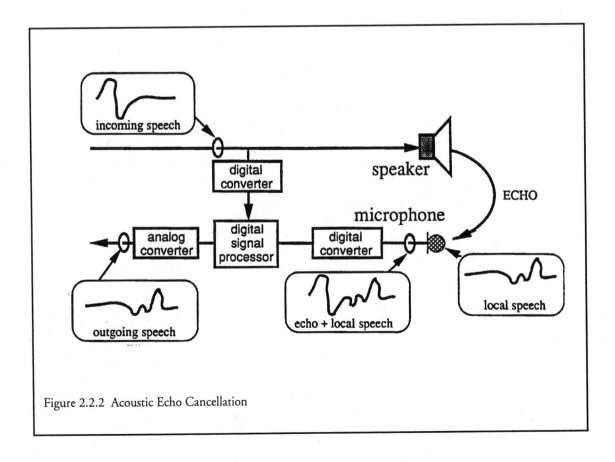

Figure 2.2.2 Acoustic Echo Cancellation

In a teleconferencing system, speech from the remote end of the conference is amplified by the conference system and exits from the local loudspeaker. Some of this sound energy goes directly from the loudspeaker into the microphone (direct acoustic coupling), and some is reflected off the walls and other objects in the conference room (reflected acoustic echo). After a time delay dependent on the size of the room and the sound reflectivity of the room's walls, this echo is also coupled into the microphone.

To the microphone, these speech signals appear to be originating in the room and would normally be sent to the remote end, where they would be heard as unwanted echo. The delayed speech signal is particularly objectionable, making it very difficult for the remote party to carry on a conversation.

In order to eliminate this unwanted echo, traditional conference systems and speakerphones simply turn off the microphone when they detect remote speech. This effectively blocks the echo from returning to the remote end. Unfortunately, it also blocks local speech from being transmitted to the remote end. More advanced systems control the loudspeaker and microphone level in a more sophisticated manner, but still block speech in one direction or the other. These older systems provide a "half-duplex" audio channel, in which only one party can speak at a time, without cutting off the other end. In such a half-duplex system, a continuous talker can monopolize the audio channel, normal interactivity is prevented, and speech syllables at the beginning and end of sentences are often cut off. The overall negative consequences of this type of system are wide ranging.

At a minimum, people dislike using such a system, because they feel frustrated at not being able to express themselves properly. In the worst case scenario, important information can be lost due to clipped syllables and lost words. There is also a feeling of lack of control during the conference, which can lead to a loss of productivity and misunderstandings of perceived intentions.

This effect is magnified in multi-way conferencing, involving three or more sites. Any speech or high level of background noise, such a cough or a door closing, at any site will cause all remote microphones to be temporarily switched off, and thus interrupt the conference.

Fortunately, a technology, known as echo cancellation, has recently been developed which eliminates the need to switch off or attenuate microphones and loudspeakers, and which provides a natural full-duplex audio channel. This technology has made a major improvement in the audio quality of both audio and videoconferencing.

■ ECHO CANCELLATION

In the echo cancellation process, the incoming audio signal from the remote site is sent to the local loudspeaker. It is also converted into a digital signal and is stored in a computer memory (Figure 2.2.2). The signal from the local microphone is also converted into a digital signal, and a digital signal processor compares the two signals.

Any similarities in these two signals will be due to components loudspeaker coupling into the microphone, either via direct acoustic coupling or from reflections of speech from the conference room walls. The signal processor has a stored image of the speech sent to the loudspeaker, and it compares this image to the received microphone signal and determines what similarities exist between them. These similarities are then electronically subtracted from the microphone input, leaving only local speech. The result, which consists only of local speech, free of echo, is then sent to the remote site.

The processor essentially builds an electronic model of the acoustic properties of the conference room. For effective echo cancellation, not only must this model be accurate, but it also must remain accurate over time, even though the acoustics of the room change.

Room acoustics are dependent upon such variable factors as the number of people in the room, the placement of microphones, and whether drapes are open or closed. Changes in these factors must rapidly be compensated for in the echo canceller if a proper model is to be achieved.

The time which the processor takes to build a correct model is called the "convergence time." A shorter convergence time will result in a more stable and robust audio channel, with less echo at the beginning of a conference or when conditions change.

The calculations which must be performed to build and maintain this model take a considerable amount of processing power, and require the use of specific, custom designed integrated circuits.

■ ADVANTAGES OF ECHO CANCELLATION

Because direct coupling and acoustic echo is removed electronically rather than acoustically, much less attention needs to be given to microphone/loudspeaker placement and room acoustics. Omnidirectional microphones can often be used, and regular conference rooms, with little or no acoustic treatment can be drafted into service as teleconference rooms. With omnidirectional microphones, there are no dead zones, and even quiet comments can be heard by all parties.

Because acoustic coupling between loudspeaker and microphone is less important, a more flexible arrangement of microphones is possible, usually resulting in far fewer microphones than with gated systems. This improves the appearance of any conference room and reduces the overall cost.

In addition, with high-quality echo cancellation a completely full-duplex, natural audio channel can be provided, without the need to resort to gated microphones and speakers or to rely heavily upon echo suppression. The full duplex nature of the audio channel can be preserved throughout multi-site networks, where the performance of gated microphone systems can deteriorate.

■ CONCLUSIONS

A high degree of audio quality in a teleconference installation will contribute greatly to the overall performance of the system. One of the important characteristics of an audio system is its ability to provide a full-duplex channel, while eliminating echo. Of the two methods available to remove echo, echo suppression and echo cancellation, only echo cancellation, provides a full duplex audio channel. Echo cancelation offers many significant advantages over older techniques of echo control, and is rapidly becoming the preferred solution.

■ A FEW ECHO CANCELLER TERMS

Double-Talk — This term describes the situation where parties at both ends of a conference are speaking simultaneously. A quality echo canceller will provide a continuous speech path in both directions during double-talk.

Acoustic Echo Return Loss (AERL) — This term describes the minimum loss experienced by a sound in traveling from the loudspeaker to the microphone in a conference room. It is expressed in dB or decibels. A 0 dB loss corresponds to a perfectly reflective room or to very close coupling between loudspeaker and microphone. In practice, AERL figures can range from 0 to -30 dB, with a poor room having the former figure.

Acoustic Echo Return Loss Enhancement (AERLE) — This term describes the maximum echo cancellation provided by the acoustic canceller. Typical figures will vary from 6 to 18 dB. The larger the number the better. It is important to note whether the figure is quoted with the center clipper

enabled or disabled. If quoted with center clipper disabled, it is a true measure of the cancellation provided by the echo canceller rather than the attenuation provided by the center clipper.

Center Clipper — The center clipper is a variable attenuator which is used to eliminate any residual echo left by the echo canceller. A key difference between one canceller and another is the manner in which this center clipper operates. In a high quality canceller, the center clipper will operate very rapidly and smoothly, resulting in no residual echo during double-talk and no clipping of syllables.

Tail Length — The tail length of an acoustic canceller, which is measured in milliseconds, corresponds to the longest echo delay which will still be cancelled. For proper operation, the tail length of a canceller should be longer than the longest echo delay in a room. If the attenuation by walls and ceiling is small, the echo delay time will include multiple reflections.

All things being equal, a canceller with a longer tail will perform better in a given room than one with a shorter tail. Conversely, a room will require less acoustic treatment if a canceller with a longer tail is used. For medium sized conference rooms, a tail length of 200 to 300 ms. is usually adequate.

Convergence Time — A high-quality echo canceller is constantly building and modifying its internal electronic model to match the acoustic characteristics of its environment. When this model is complete, the canceller will be able to cancel echo to the extent of its rated AERLE capability. The time taken to build or significantly change this model is called "convergence time," and is a key figure of merit for an acoustic echo canceller. This time should be a short as possible because some echo will be present until the electronic model matches reality.

Teleconference participants may frequently move about the room, open a door, or even move a microphone. All of these events will have a dramatic effect on the echo model which is being maintained in the canceller. Since echo will be heard until the canceller has "converged" or corrected the model, the shorter the convergence time the better. In practice, a convergence time of less than about 75 ms is required to eliminate perceived echo under changing conference conditions.

A technique which is sometimes used to assist echo cancellers with slow convergence is to inject white noise into the room to give the canceller something to converge with before the conference starts. Many people find this uncomfortable and a quality canceller should be able to converge rapidly with normal speech.

Harry R. Walls, Vice President, Sales and Marketing, ConferTech International is a native of Uniontown, Penn., a suburb of Pittsburgh. Harry received his bachelor's degree in education from California State College, in California, Penn., in 1970. He earned his master's degree in Education Administration in 1979 from the University of Akron in Ohio. Harry taught elementary school for three years and Junior High School for seven years in Ohio. He also coached Junior High and High School football, basketball and track.

He entered the telecommunications field in 1979 with a communications company that sold Northern Telecom equipment. Prior to joining ConferTech International, Inc., Harry was general manager of a telecommunications contracting company, operations/sales support manager for a US West subsidiary and national sales manager for a nationwide telecommunications firm. He has been with ConferTech for 5 and a 1/2 years and is currently vice president, sales and marketing, Systems Division.

ConferTech International, Inc., Golden, Colo., is the leading marketer and manufacturer of audio teleconferencing bridges in the United States.

TECHNOLOGY OF BRIDGING SERVICES

Audio teleconferencing bridges are telephone equipment used to connect individual telephone lines together to create a conference call. Usually two or more sites are required for a conference call. Audio bridging equipment adds what is termed "Automatic Gain Control" to the connected telephone lines to bring each line up to a normal hearing level. Audio teleconferencing bridges can be located anywhere in a telecommunications network. Most audio bridges today are what is termed, "full digital" and "full duplex." Full duplex allows two or more people to speak and be heard simultaneously. Full digital relates to a state-of-the-art application.

Audioconferencing services are service bureaus offering conference calling applications and management. These companies actually offer what may be termed "electronic business meetings." These meetings can be held from any site or location as long as a potential conferee has access to a telephone. Typical charges for this type of business meeting are based upon the type of call, long distance charges and additional services which may be offered by the conference calling company.

Most of the conferencing services today use the advanced, fully digital, full duplex audio bridge technology in their conferencing business. These businesses were actually started in the early 1980s, but did not flourish until the latter part of the 1980s. The major change in teleconferencing a actually was stim-

ulated by the "Carter Phone Decision" of 1968. This was also the start of what was known as the Interconnect Industry. More prominently known was the birth of microwave communications which today is MCI.

The breakup of AT&T, which created seven regional Bell operating companies added an additional impact on the way audio conference calling was handled. The SNAFA agreement (Shared Network Access Facilities Agreement) which the RBOC's had with AT&T utilized AT&T as their conference calling company. To date, this agreement is no longer in effect for most of the RBOC's but yet they are controlled by the MFJ (Modified Final Judgment).

The first form of audioconferencing was the AT&T, operator handled format. Along with this were the old "Willy WATTS" bridges which were push button, analog, half-duplex products. After this came the AT&T Quorum bridge, followed by PBX 3/6 way conference calling capabilities.

ConferTech International, Inc. of Golden, Colorado, actually developed the first fully digital, full duplex audio bridge. This product, completed in late 1982, was the beginning of the audio bridging "CPE" market. Around this time also was the start-up of smaller conference calling service companies who were to compete with AT&T. Today there are over 25 conference calling service companies which offer business meetings via the telephone.

The audio teleconferencing industry stayed fairly flat between 1982 and 1988. From 1988 on we have seen some very promising growth in both the audio bridging market as well as the conference calling services market. Taking a look at what sparked this interest, we need to examine the following:

Videoconferencing sparked an interest in audioconferencing.

Audioconferencing does not need to replace videoconferencing, but rather compliment it.

Audioconferencing does not necessarily need to replace face-to-face meetings, but rather can be used to make them more productive.

Operation Desert Storm forced changes in ways corporations do business.

Corporations in the 1990s are forced to cut travel budgets.

Teleconferencing became a more acceptable way of conducting business.

The economy and the on-going recession have affected conferencing.

Corporations are putting more of an emphasis on bottom line performance.

Corporations are down-sizing.

Accountability of employee performance is becoming a factor.

Board of Directors are implementing more strict controls.

There is an interest in using this technology to increase productivity.

As corporations look to implement a corporate-wide audio teleconferencing service, some of the things they need to consider are:

Centralization of usage
Emergency coverage
Confidentiality
Management reports
"Out Sourcing" or not
Departmental bill backs
Internal Promotion
Private Labeling
Convenience
Professionalism
Hours of coverage
Protocols of conducting an effective meeting
Network management and support
Tying together other needs:
　Paging
　FAX
　Voice Response
　Audio-graphics
　Video

Audio teleconferencing is an efficient, productive, and cost-effective way of doing business. It is convenient and easy to use. Customer Service/Support is a major key in whether it is done in-house or externally. A company looking to implement and use audio teleconferencing should explore alternatives that are most conducive to their business culture.

There are several different kinds of audioconferences. For example, the Tempo MBX and Allegro Audio Teleconferencing System support simultaneous use of the following conferencing modes.

Meet-Me

Operator Attended Meet-Me is where all participants dial the bridge telephone number at a designated time and they are greeted by an operator and placed in conference with other participants. Passcode Meet-Me or passcode unattended conferencing is conference where a participant dials the bridge telephone number and when prompted, enters a pre-assigned four to seven digit passcode to enter their conference without operator intervention. Another form of meet-me conferencing is the Unattended Automatic Hook. Automatic hook is a form of unattended conferencing which allows a participant to dial the telephone number to the bridge and be placed automatically into conference without operator intervention.

Dial-Out

Operator Attended Dial-Out is an attended dial-out conference where an operator dials each participant from the bridge at a designated time and placed in conference with other participants. Chairperson Manual Dial allows a chairperson to dial into the bridge, input a passcode, and access system dial tone to a conference through a series of touch-tone commands. Chairperson (Preset) Auto Dial is a feature of the Tempo MBX standard Reservation Software Program. A Tempo administrator can enter preset conference information in the Reservation Program including participant names, phone numbers, and the conference debate, time, and duration. To initiate an unattended preset conference, an attendant or users can dial the bridge at a designated time, enter an assigned Passcode, and according to reservation information, the system will begin dialing two conferees at a time until all participants are on line.

Dr. Lorne Parker is recognized as a leading authority on teleconferencing and distance education in the United States and abroad. With more than 25 years of experience, he has designed many interactive teleconferencing networks and is well-versed in system planning, management, technology and applications. He is an innovator when it comes to teleconferencing, and has been instrumental in researching industry and market trends. Prior to forming his own firms, Dr. Parker was a professor of Communications Arts at the University of Wisconsin-Madison. During his tenure there, he founded and directed the Center for Interactive Programs (CIP) which was known internationally as a leader in distance education and teleconferencing expertise. He is experienced in developing teleconferencing and distance education systems and has served as a consultant to many American and international businesses. Dr. Parker holds a Ph.D. in Continuing Education and Communications Arts from the University of Wisconsin-Madison. He is published widely on teleconferencing and distance education including his well-known book, "Teletraining Means Business."

AUDIOGRAPHICS TECHNOLOGY

This chapter examines current audiographic technologies for the primary types of audiographics teleconferencing: telewriting systems, freeze frame systems, hybrids, high resolution systems, computer conferencing, voice and electronic mail. Included are major products and services and recent developments and innovations. Other equipment that could be used as audiographic adjuncts are facsimile, voice/data terminals, optical graphic scanners and videotext systems. Because they are not usually marketed as teleconferencing products, these devices are not discussed.

The term "audiographics" refers to a range of systems used to produce and send graphical information over a narrowband transmission channel. Audiographic devices customarily use the standard telephone network or, on occasion, lower-speed (1.2 - 56 kbps) data channels.

In teleconferencing applications, audiographic systems are used as adjuncts to voice communications to provide complementary visual illustrations. This information may take several forms, including freehand sketches, paper copies of documents, stationary video images, or computer graphics. Audiographic teleconferencing is sometimes called enhanced audio or audio plus.

While the earliest uses of audio teleconferencing go back to the 1930s, audiographics devices did not appear as a teleconferencing adjunct until 30 years later. The entry of audiographics devices was supported initially by the education and government sectors that saw value in low-cost graphics systems for remote instruction or information transfer. Business firms were later users, adopting audiographics in the late 1970s for meetings among remote locations.

■ AUDIOGRAPHICS SYSTEM OPTIONS

The types of equipment included under the audiographics umbrella are:
- Electronic pens, tablets and boards for telewriting
- Freeze frame (slow scan) video systems for transmitting still images of objects, graphics or people

- Computer based and integrated systems for computer graphics, text, freeform annotation or other graphical displays
- Computer conferencing for exchange of textual information and messages

In many cases, an organization will employ more than one type of audiographics. In other cases, an organization will have more than one form of teleconferencing using audiographics as well as full-motion video or audio-only systems.

An important feature of an audiographic system is its ability to use the public telephone network which provides widely available and low-cost transmission channels. When used over regular dial-up lines, audiographic equipment provides immediate access to other compatible systems anywhere in the world.

The use of the public telephone network, however, imposes restrictions, particularly in the narrow bandwidth of voice-grade lines and, in some cases, the quality of the signal. Nevertheless, as the telephone network is upgraded with digital switching and digital transmission, this is changing.

For teleconferencing applications, telewriting systems (electronic pens, tablets and boards) are used to send and receive freehand messages and graphics. Freeze frame terminals transmit still video images captured from a video camera or other video source. Integrated graphics systems are used to produce and transmit computer graphics, alphanumeric data, still video images, and freehand graphics and annotation.

Freeze frame video units can be used to show images of people to add a "humanizing" touch to a teleconference, but their primary purpose should be to transmit pictures of graphics, objects and other materials. It is, therefore, considered to be an audiographic tool as opposed to a video teleconferencing system.

The primary options in audiographic technologies can be characterized as follows:

Telewriting systems
(electronic pens, tablets, boards)

-Freehand messages and graphics
-Transmitted over a voice-grade telephone line in real time
-Usually displayed on a monitor

- Monochrome or color
- Portable or semi-portable

Telewriting products include electronic pens, tablets, and boards. They have traditionally been used as stand-alone systems to create freehand information such as words, outlines, simple diagrams and line drawings. However, there is a current trend to interface electronic tablets to freeze frame video or computer graphic systems to annotate on images and create freeform graphics. The major limitations of telewriting systems have been display legibility, limited flexibility in generating different types of information, problems in detecting accurately the pen position, and difficulty in matching pen position to desired location on the display.

Freeze frame video system

-Still video images of illustrative materials, objects and people
-Transmitted over a voice-grade telephone line or data circuit, usually in a number of seconds
-Displayed on monitor or video projection system
    -Monochrome or color
    -Portable, semi-portable or installed
    In order to transmit a video image over a tele-

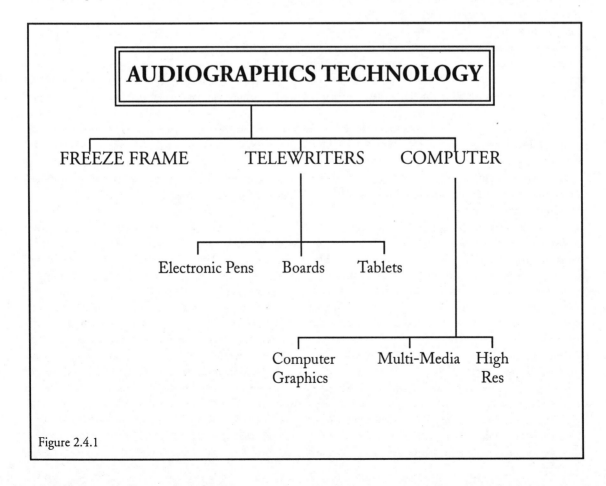

Figure 2.4.1

phone line, a freeze frame system eliminates motion, reduces picture resolution, and transmits the image during a number of seconds. The transmission time depends on the amount of picture information and channel capacity. For instance, it can take up to 75 seconds to send a monochrome picture with a resolution of 512x256 pixels over a voicegrade line. However, with the newer digital systems that are designed to be used with data circuits, a 512x256 image can be sent in about 15 seconds at 56 kbps. Other techniques include the use of multiple memories to display and receive two or more images simultaneously, or the use of a disk system to store images sent prior to a teleconference.

Because a freeze frame picture is displayed as a still image, it is most appropriately used for objects, graphics or other non-moving material.

The basic units in a freeze frame system are the transceiver (transmitter and receiver), video camera, and display monitor. Multiple cameras and monitors or large screen video projection systems can also be used, as can such peripheral devices as video pointers, video hard copiers, and disk systems for image storage.

Three of the primary problems faced by freeze frame systems are transmission time, bandwidth capacity versus display quality, and the misuse of this technology as a substitute for video teleconferencing to show images of people.

Integrated graphics systems

-Images of alphanumerics, computer graphics, freehand graphics, annotation, or still video pictures (systems differ in capability)
-Transmitted over a voice-grade telephone line or data circuit in realtime, in a number of seconds, or prior to a teleconference

Using a microcomputer or processor as a core component, these systems provide integrated, multifunction capabilities that include computer-generated graphics, freehand graphics, still video images, and annotation or images.

The information generated on the system is transmitted over a voice-grade telephone line or data circuit to remote locations and displayed on a monitor or large-screen video projection system. In the case of freehand graphics or annotation, the information is usually transmitted in realtime during the teleconference. Still video images, as in freeze frame, take a number of seconds to transmit. With more sophisticated computer graphics systems, the images are usually created and transmitted prior to the teleconference and stored at the remote sites for later display during the meeting or program.

With limited bandwidth on a voice-grade line,

there is a tradeoff between transmission time and the capacity to send images with high resolution, color and detail. Freeze frame video and integrated graphics systems, therefore, offer better performance when used with data circuits. Another alternative is to send the images ahead of time and store them at the receive site until needed.

■ TRANSMISSION FOR AUDIOGRAPHICS TELECONFERENCING

One of the primary advantages of audiographic systems is that they can use regular dial-up telephone lines to send and receive visual information. With millions of telephones in the world, voice circuits extend to virtually every part of the globe.

As discussed earlier, however, the narrow bandwidth of a dial-up, voice-grade line presents a tradeoff between image quality and transmission time. Channel impairments, such as noise, echo, and loss can also distort image information during transmission, creating drop-out or "noise" on the display.

Some users, particularly those with computer-based or freeze frame systems, may opt for private lines that have better transmission quality and, in some cases, more bandwidth. A variety of private-line services are available, including full-duplex analog circuits that are voice-grade or data-grade and digital circuits that operate at higher transmission rates.

The type of telecommunications service chosen depends on user requirements. Considerations include image content and quality, transmission time, site locations, level of teleconference usage, cost and system compatibility. Users must also determine if they should have separate transmission channels for voice and images. To be able to talk and transmit graphics simultaneously, two channels will most likely be needed-one for voice and one for graphics.

Teleconferencing should also be considered in a total communications context. Some organizations may need or already have private transmission channels for data communications that could also accommodate teleconferencing.

In general, transmission options currently available for voice and data are:

1. Dial-up, voice-grade telephone lines, usually the equivalent of up to 4800 bps
2. Private analog service with either voice-grade or data-grade channels up to 9600 bps
3. Private digital service at lower-speed data rates, usually from 2400 bps to 56 kbps
4. Private digital service for high-speed transmission, usually up to 1.544 mbps; that is, a T1 carrier

5. Packet-switched service for data communications, usually on public networks operated by value-added carriers

Many of the newest services show the merging of voice and data communications. With T1 carriers for wideband digital services, a single channel can be used for integrated voice, data and video. It can include combinations of voice, low and high-speed data, graphics, facsimile, electronic mail, and videoconferencing.

Another option is switched 56-kilobit service that allows users to send voice, graphics, data or compressed video over a dial-up telephone line.

Recent developments that affect voice and data communications and, therefore, audiographic teleconferencing include the following:

• The introduction of new bridges to interconnect multiple locations for audiographic teleconferencing. The future will also see commercial bridging services offering bridging capabilities for audiographic teleconferences.

• The expansion of long distance telephone services and the entry of new carriers into the market to provide voice and data communications.

• The availability of switched 56-kilobit service over dial-up telephone lines.

• The growth of multi-tenant shared services that provide enhanced voice, data and video services to tenants of office buildings, including audio teleconferencing, electronic mail, facsimile and so forth.

• The availability of fiber optic networks to expand telecommunications capacity for voice, data and video.

• The availability of wideband digital services for high-volume voice, data and video communications.

• An increase in satellite transponder capacity and improved access to satellites from urban areas via teleports.

■ APPLICATION CONSIDERATIONS

Like other forms of teleconferencing, use of audiographics reflects the specific needs of each organization. Systems range from a rather simple application of audio and facsimile using an internal switchboard linking a small number of locations to one that has over a dozen teleconference rooms outfitted with an array of audiographics equipment.

Audiographics systems generally are used for the same types of applications as audio teleconferencing - administrative and staff meetings, problem solving, information updates and training. There is a tendency, however, for some audiographics systems to be employed in more technical areas such as research and engineering.

It may be tempting to assume that the addition of a visual element would be advantageous to almost all teleconferences. If audio is good, would not audio plus graphics be even better: That assumption raises important questions about when and how to apply audiographic systems.

As relatively low-cost adjuncts to voice communications, audiographics can provide a visual representation of ideas to illustrate or clarify a point, direct attention to a detail, demonstrate words and numbers, or show relationships and trends among data. If used inappropriately, however, they may add nothing to the communication process and even detract from the verbal message by creating confusion or misperception. The application of audiographics, like all teleconferencing technologies, ultimately depends on communication needs and information about how to use a system effectively.

In matching audiographics technologies to user applications, several important factors need to be considered:

• Resolution. Do documents contain a great deal of detail-complicated text, charts, drawings-that would require in-depth explanation, careful modification or instant approval? In this case, higher resolution would be necessary to assure the exact image at every participating site. If documents are less complex, then resolution might be compromised in favor of other features.

• Bandwidth options. Is it important to have instantaneous exchange of information? If it is, then a dial-up 56-kilobit network would be the solution.

Take for example, NASA's use of audiographics when launching space shuttles. The "go/no" decisions that launch space shuttles with "mission critical" documents are made using audiographics teleconferencing between Cape Canaveral and Houston. The NASA network includes 23 sites that are linked via standard telephone lines that are part of a dedicated 56 kilobit network. Transmittal of these critical documents can occur in just seconds with audiographics teleconferencing. The result? NASA's Launch Systems Evaluation Advisory Team can advise the Mission Management Team whether or not to go ahead with a launch when weather or environmental conditions are questionable.

If immediacy is not so critical, then a transmission medium that results in a delayed exchange could be the answer. Additionally, many organizations will opt for using a regular, dial-up telephone circuit for audiographics and will send and store documents or graphics before a scheduled meeting or session, recalling each just as if using a view graph or slide.

• Reliability. What is the track record for the specific type of audiographics equipment you are considering? Is it proven and will it be reliable when you need it? What is its maintenance record?

• Number of locations or sites. How many sites do you need to link together? Today, special digital bridges are appearing to be used in audiographic teleconferencing. (Not all audio teleconferencing bridges will accommodate audiographic conferences.) The bridge you choose — for in-house or as a commercial bridging service — needs to support the number of sites that comprise your network.

• Facilities and staffing. Will you dedicate special rooms for audiographic conferences, provide portable units to be used in any environment or do you want to support desktop audiographics conferences? The answer to this question will dictate the type of audiographics equipment you choose. Staffing to support the use of audiographics conferencing can be as simple as delegating to one or two individuals the additional responsibility of being a "key operator" who is well-versed in operation and trouble shooting, or a full-blown staff of many individuals who can support a complex, multi-faceted network and its operation.

• Usage. Should you designate and design a room especially for audiographics teleconferencing or can any existing facility be used, including individual offices? The amount of use an audiographics system gets will dictate how a system should be established — in permanent facilities or as a flexible, portable system that might be used in any environment..

• Maintenance/Service. The customer service and support of any audiographics equipment or system is an important consideration. The type of services that an audiographics vendor offers to stand behind their equipment might include tutorials, on-line user training, user conferences, newsletters and special consultations with technical personnel. Maintenance programs are also available in which full warranty coverage is provided for a specified period of time, or other options might be on a time and materials basis. Typically, vendors will supply loaner units while equipment is being repaired.

• Cost. The range of costs for audiographics is diverse. For a simple electronic tablet with accompanying software for audiographics teleconferencing, the cost would be approximately $2,000. Mid-range systems cost between $4,000 and $12,000, while high-end, fully integrated audiographic systems cost about $34,000.

Editor's note:
Audiographics have declined in importance as the cost of low data rate two-way video has come down into the same range.

Chapter 3

BUSINESS T.V ORIENTATION

by Patrick S. Portway

Business Television (BTV) is another of those less than precise terms used to describe the private use of television technology. In teleconferencing, BTV's most widely accepted definition is to describe a broadcast of one-way video, by satellite. For BTV to be a video teleconference, some form of live interaction must be present. Usually this involves a return audio interaction by telephone. Unfortunately not all BTV programs are interactive. A large portion of today's BTV programming is prerecorded programming targeted on a specialized audience and narrowcasted by satellite directly to that select audience.

■ MEETING VS. PRESENTATION

To understand BTV and to differentiate it from two-way videoconferencing, it helps to use the analogy of a meeting vs. a presentation. BTV is directed at a large audience that primarily views and listens to the information. At a presentation you may be able to ask a question but the bulk of the information flows one way from the presenter to the audience. Your role is more passive.

In a meeting all the participants interact continuously and you are directly involved throughout. That involvement keeps your attention. In a presentation the speaker, the graphics or the message must keep your attention.

■ BTV AND PRODUCTION VALUES

This need to earn your attention is why BTV or a presentation of any kind requires more preparation, graphics or production.

In business television there is also a defacto standard applied to how people evaluate the quality of a program. It is compared in the mind of the audience to what they see on broadcast television at home.

That standard has driven BTV networks to emulate the style, graphics even the sets used on live television, news and talk shows. Often for very special programs even professional television performers are hired as moderators to give a professional look and pace to a company's own television program. As a result, the cost of production of BTV programs are often the largest single cost.

■ TRANSMISSION

The usual method of distributing a business TV program is by satellite because of the inherent advantages satellite transmission has in reaching a large number of sites in one-way communications.

■ VSAT NETWORKS

One driving factor in the early development of private business television networks was the development of very small aperture terminals (VSAT). These small earth stations were primarily installed for two-way transmissions of data, but the receive dish can be modified to receive a one-way television signal for a small addition expense.

■ SATELLITE BASICS

Without going into extreme technical detail it is useful to understand a few basic characteristics of satellite communications in order to understand how BTV programs are transmitted.

Communications satellites are usually in a geosynchronous orbit that allows them to rotate around the Earth at relatively the same speed as the Earth's rotation. This allows a communications satellite to appear to be stationary above a particular point above the earth. The orbit required to achieve this position is 22,300 miles up at the equator. A satellite dish on the earth can then be pointed at the assigned location of that satellite in order to receive its signal. The area of the earth on which the transmission energy of a particular satellite is focused is called its "footprint" on the earth (See Figure 3.1.1). Any number of earth station or satellite dishes that are located within this footprint can receive the satellite signal.

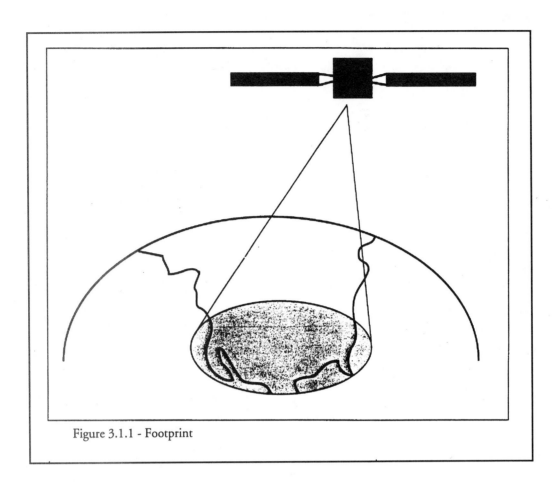

Figure 3.1.1 - Footprint

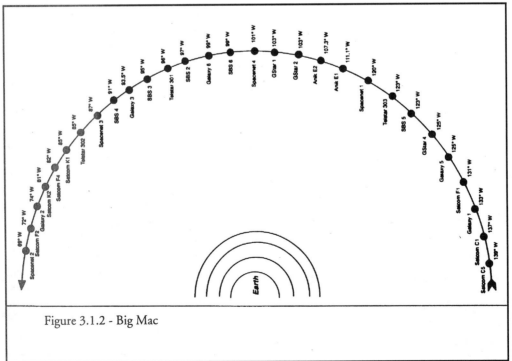

Figure 3.1.2 - Big Mac

■ TRANSMISSION FREQUENCIES

Domestic U.S., Canada and Mexican satellites usually transmit in Ku or C Band frequencies. Ku is between 11 and 14 gHz and C Band is between 4 and 6 gHz. Ku band has less terrestrial frequency interference from ground based microwave and uses small earth stations. VSAT networks are usually Ku. C Band was the first frequency used for TV and most entertainment, cable and educational networks are C Band.

Ku band (11 to 14 gHz)	C band (4 to 6 gHz)
VSATs	Entertainment
Business TV Networks	Education Networks
Corporate Programming Networks	

There are exceptions to these generalities.

■ DOMESTIC SATELLITE

The satellites viewable from North America (Mexico, America and Canada) have been called the Big MAC. (See Figure 3.1.2. for a current list and an illustration of their position around the Earth.)

■ AD HOC VS. PERMANENT NETWORKS

A temporary or ad hoc (as needed) network can be assembled for transmission of a one-time special event. Companies often try a single event like a product announcement through the use of hotels or other receive facilities as a first effort in business TV. An excellent guide to producing such an event is available. It is called "Teleguide, a Handbook on Videoteleconferencing."

Companies who successfully use BTV or have their own television studios for production of industrial television video tapes usually find it cost effective to install their own private business television networks. Data from Susan Irwin's Business Television Directory indicates approximately 80 private networks exist today.(1992)

In addition, approximately 22 programming or industry networks exist that broadcast programming of interest to a specific industry (law enforcement, automotive, etc.) and 33 private networks for Educational programming are operates (see Figure 3.1.3) combined these networks have over 60,000 receive sites nation wide. (See Figure 3.1.4.)

Widner, Doug (1986). Teleguide: A Handbook on Video-Teleconferencing. Washington D.C. Public Service Satellite Consortium.

Irwin, Susan (1992). The Business Television Directory. Washington D.C. Warren Publishing Inc. and Irwin Communications

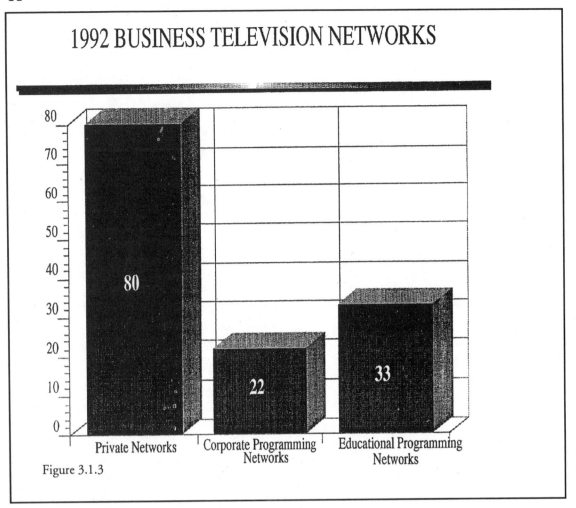

1992 BUSINESS TELEVISION NETWORKS

Figure 3.1.3

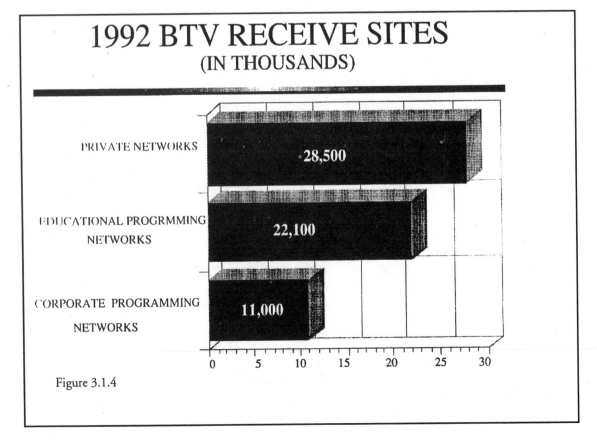

1992 BTV RECEIVE SITES
(IN THOUSANDS)

Figure 3.1.4

Blaine Colton is senior vice president, business television for Keystone Communications, a leading system integrator of business television services. Blaine is a graduate of Brigham Young University with 11 years experience in the satellite industry. He has held various senior management positions with Bonneville Satellite and Keystone Communications and presently manages Keystone's BTV division. Keystone's BTV customers include, the American Red Cross, Apple Computer, HomeBase, Nissan Motors, Travelers Insurance, Sun Microsystems, the Department of Defense, Tandy corporation and Union Pacific Railroad. Prior to joining the satellite industry, Blaine held senior management positions with Keebler Company. He is married with four children and sits on the boards of the American Red Cross and the Utah State Soccer Association.

BUSINESS TELEVISION BY SATELLITE

Business television is a rapidly developing area of corporate communications. It is a powerful tool capable of uniting widely dispersed company operations. There are literally times when a picture is worth more than a thousand words, such as a time when visual communication is required to inspire a far-spread sales force with a personal message from the president. Face-to-face videoconferences are extremely effective when an unexpected upheaval in your business makes it necessary to make immediate changes or requires you to get up-to-the minute instructions out to branch offices across the country.

Mark Porat, the former president and a founder of Private Satellite Network (PSN) coined the phrase Òelectronic campfires" referring to business television. What a descriptive term for the powerful process that allows people to communicate face-to-face regardless of geographic separation! Imagine a group of boy scouts gathered around the crackling fire, leaning in close to the camp director as he delivered his message. Regardless of where each boy sits in the circle, he feels like the speaker is sharing a secret that was meant only for him. Satellite television has the same empowering effect on audiences. In highly competitive markets these "electronic campfires" are uniting people to realize dramatic communications and productivity gains.

The ways in which business television is utilized are as varied as the companies that use it. The most popular application for BTV is skills training. On-going training in high tech companies and companies with strong service orientations is critical, particularly when the rules of business are changing daily. Televised instruction covers a spectrum as wide as a manager's imagination, including technical sessions for service representatives, new product overviews for sales staffs, management courses for executives and general enrichment classes.

There are many other compelling reasons to utilize business television for the benefit of your compa-

Figure 3.2.1 Satellite communications leverage the elements of business television.

ny: improved communications with tighter organizational objectives, providing a consistent message where everyone gets the same message at the same time (thereby eliminating filtering), more productive meetings for time-conscious participants, better attended meetings, more efficient use of employees' time and shared access to key in-house sources. A business television is like having an auditorium full of employees who can watch, listen and raise their hands to ask questions.

There are three basic components of any satellite event: 1) program origination, 2) the receive sites and 3) the transmission chain. A simple analogy which helps to clearly picture the three components is the circus strong man's barbell apparatus it consists of two large, heavy weights connected with a bar. In order to hoist the weights above his head, he had to balance himself between the weights using the bar for leverage. In our business television analogy the two heavy ends of the barbell represent the two most complicated, time-consuming and costly elements of the satellite-delivered event. The connecting bar that joins the primary elements is essential for balance and is the least complicated and costly element. This does not suggest that any one element is more important than the other. They all work in concert and are essential to success. The barbell analogy only serves to illustrate their relationship to one another.

■ PROGRAM ORIGINATION

Returning to our analogy, the heavy left side of the barbell represents program origination. The cost elements of program origination include a producer, a writer, program design and set creation, pre-production and a local production facility with staff.

Let us begin our discussion of the elements of a satellite-delivered broadcast with the issues surrounding your program. You must first ask yourself questions about the objectives of your communication as well as the look and feel you want to achieve with your show. Included at the end of this article is a list of questions that may be helpful as you plan your satellite-delivered event. A business television broadcast requires the cooperation and coordination of many individuals, departments and supplier organizations in order to enjoy maximum success. It is a productive exercise to define your primary goals in simple terms and then the strategy for accomplishing these objectives. The questionnaire can help you formalize your strategic objectives.

You should plan to develop a strong relationship with a television producer in order to organize the concepts, information and people necessary to create quality programming. The business television presen-

tation is powerful because it utilizes both sight and sound. In expertly produced programming, these elements combine to deliver a strong message that appeals to the viewer's heart and mind. Look for a producer whose approach is one of "people selling to people." To be fully effective this requires a "natural" approach to production that is basic and endlessly adaptable.

One bit of advice: don't try to provide commercial network style programming. Employees do not expect this from their company. Business television networks are not created to entertain and be flashy. They simply impart information that helps people do their jobs better. Identify the key operational issues facing your company and then allow the production team to create programming which communicates solutions and healthy attitudes about the issues. This is a strategy which will create positive results. Keep the programs simple, interesting and focused on the issues which will produce the results management desires. The bottom line is that people are best influenced by realistic portrayals of what they can expect from their company.

The script is the foundation of a business television program. Effective writing is hard work. It is also dependent on research and the energetic help of many internal company departments for a clear portrayal of the subject. Once the script is completed, the producer brings it to life through professional shooting, editing and production.

With your script developed and your communications objectives clearly defined you are ready to go into the studio. Your studio can be either a traditional broadcast quality sound stage or a remote location such as an office, hotel, store or work area. In cases where your production location is a remote venue, your "studio" consists of a video production truck hired to create the production environment. The production truck will provide the video equipment necessary to create technical results similar to what you would have at a traditional sound stage.

Plan on investing enough time in the studio to make your presenters comfortable with their material and to create the desired look and feel that will communicate your message. Program development and production can be the most costly ingredient of your satellite event. Do your homework before you get to the studio by investing in the preplanning phase and you will avoid costly delays.

■ RECEIVE SITES

The right side of our barbell analogy is the second heavy element, the receive sites. The reason that you have chosen satellite communications to deliver your message is that it is either impractical or too costly to bring your audiences viewing requirements is critical. The audience has a need to know and you have a need to share. Design your receive sites with both needs in mind.

Typically a satellite-delivered event creates an ad-hoc or temporary network of receive sites where your audience assembles to hear your message. A transportable or mobile satellite downlink is used to receive the satellite signal and deliver the programming to your viewing audience. A local satellite professional will supply all the necessary satellite equipment and the audio video display systems to transform nearly any location into a temporary viewing site.

The equipment package consists of a parabolic shaped satellite antenna or dish, low noise block converter (LNB), satellite video receiver and the associated cable, connectors and power supply. The satellite antenna concentrates the satellite signal to a focal point on the dish. At the focal point is an LNB that converts the signal to a lower range of frequencies capable of being transmitted down a coaxial video cable. The satellite receiver process the signal and outputs it to the television. If your network is to be encrypted or scrambled for confidentiality then a decoder is also required. Your local satellite vendor or network management company will supply all the equipment necessary for a transportable downlink.

Your receive location will require video display equipment to project your program. The general rule of thumb for video display is one inch of screen per person in the audience. An audience of 25 can comfortable view your broadcast with a 25 inch or 27 inch television monitor. Audiences of 50 would use two 25 inch television monitors positioned on the left and right sides of your viewing room. Audiences of 10 or more require large screen video projectors. Reinforced audio packages using speakers, amplifiers and audio mixers are recommended when the audiences are very large or the viewing rooms have inadequate house sound systems. Satellite-delivered events generally employ an interactive question and answer session that requires a return telephone line in each receive site. Temporary locations may require the installation of temporary telephone lines supplied by the hotel or our local Bell operating company. It is recommended that you have two telephone lines installed at each location.

The first line is for the program Q and A and the second is a technical coordination line for your satellite downlink technician.

Other cost factors to plan for a receive site include meeting room rental charges, set-up and rehearsal charges, catering, and duplication costs of program materials and collaterals.

■ TRANSMISSON

The third and final element of your satellite event is the transmission chain. In our barbell analogy the connecting bar uniting the two heavy ends is the transmission chain. The benefit of satellite technology as a delivery vehicle is that satellites are time and distance insensitive. Land based technologies such as the telephone are time and distance sensitive. The price of a telephone call is dependent upon the time of the day you place the call and the distance of your call. With satellite communications the costs of transmission to cover one state or 48 states, 10 people or 10,000 people are the same.

The elements that comprise the transmission chain include the first mile, satellite uplink, encryption, when necessary, and transponder space segment. The first mile is the communications link between your origination facility and the satellite uplink. The fist mile may be as simple as a cable from your remote production truck to a transportable uplink parked at your production location. Or it may require the construction of a temporary fiber optics path, microwave link, laser or telephone landline. In any case the first transmission requirement is to get from where the show is being produced to where the show is going to be transmitted to the satellite. Once the program signal has been delivered to the uplink, it is then transmitted to a geosynchronous satellite position 22,300 miles above the equator. The satellite has transponders or video channels that receive and retransmit the video signal over a geographical footprint. Authorized locations equipped with the proper receive equipment can view the event. If the broadcast is to be encrypted, the encoder or scrambler would be co-located with the uplink and each receive location would be equipped with a decoder to unscramble the broadcast.

Our analogy is now complete. To understand the relationship of the three elements of a satellite delivered event, program origination, the receive sites and the barbells ... heavy (costly) on two ends with a connecting bar that balance or leverages the elements.

Management consultant, guru and frequent user of business television Peter F. Druker noted, "Business television should be symbiotic with your other patterns of communications. This new technology is being adopted by business in a natural way just as the microphone, telephone and the computer were gradually introduced to normal business communications. Whether your company is dealing with massive sea changes, changing customer markets or the economic uncertainties that lay ahead business television is a cost effective and proven communications tool.

What is the major goal of producing programs for internal or external distribution?

■ BUSINESS TELEVISION QUESTIONNAIRE

What kind of programs do you visualize?

Who in the organization is most interested in producing and   distributing the programs?

Why do they believe it will be effective?

Who will be "on camera?"

What has been the most effective form of employee communications in the past?

Why are you considering television as a communications vehicle?

Would making the programs interactive be effective and desirable?

Who will be the primary viewing audience?

How can your program help them do their job better?

Who will have input in the program content decisions?

What image is desired?

Has a budget been developed?

What resources are already available within the company which might assist in program productions?

What is an acceptable range of time or distance for participants to travel in order to participate in a program?

Is an encrypted or "scrambled" transmission required?

Where are the remote sites where programming is to be received?

What is the average number of people in the remote sites?

Does each site have audio visual equipment for viewing the network?

Colin Boyd graduated from the Queen's University of Belfast with a First Class Honours Degree in Electronics Engineering in 1974. As a Commonwealth Scholarship holder, he attended the University of British columbia in Vancouver where he graduated with a Master of Applied Science degree in Electronics Engineering in 1976. In 1976, Colin joined Bell Northern Research in the area of software development for central office switching systems. Subsequently he moved into the strategic planning and business development of new telecommunications products for the parent company, Northern Telecom. Over the past 16 years, he has served in various senior management and strategic planning roles in product management and marketing areas. Presently, Mr. Boyd is the director for the Business Communications Product line of Scientific-Atlanta where he is responsible for worldwide marketing, sales and business development of private video business television networks.

THE DIGITAL VIDEO REVOLUTION

■ WHY A VIDEO COMPRESSION STANDARD?

The need for the adoption of a video compression standard goes way beyond the traditional markets of broadcast television over satellite, and cable and terrestrial facilities. It reaches into the consumer electronics such as VCR's, TV's, Disk players, computers and multimedia devices.

It is the strong market push to enable interoperability of these video products that has led over 152 of the world's leading companies to work together in achieving an international standard for video compression.

The objective of the standard is it can be used in a wide variety of applications ranging from video conferencing, broadcast TV, computer and multimedia, and digital recording and storage. It will also help pave the way for HDTV.

There are other important goals derived for the adoption of a standard by these industries. First, the standard will be optimized to be the best solution in terms of quality, benefits and costs. More importantly, standards will ensure high volume production by multiple manufacturers of the decompression chips that embody the standard.

Hence, production volume will be higher than if each manufacturer were producing a proprietary solution. Also, there will be solid competition from product suppliers in terms of performance, features and price for various market niches.

Finally, products based on standards result in early market acceptance as users will not fear obsolescence and incompatibility with future products or enhancements. Without the adoption of standards, many of today's products we take for granted would not have a market—i.e., the telephone, fax, PCs, LANs, WANs and VCRs.

MPEG-2 has been adopted as the international video compression standard and in the video world, could replace NTSC, PAL, and SECAM. The cable and broadcast industries have announced intentions to migrate their product lines to MPEG-2.

C-Cube Microsystems and LSI Logic are among several chip manufacturers to commit to production schedules for MPEG-2 chips. Since the last MPEG-2 committee meeting in April, chip development has moved into the fast lane — which is easy to believe since forecasters have set the demand for digital decoders to serve cable subscriber, cable handend, direct-to-home, TVRO and business learning networks at over a million units by 1994 and nearly 10 million by 1996.

These forecasts exclude digital storage and recording products and multimedia devices. However, this type of demand is only meaningful with a video compression standard.

■ WHAT IS MPEG-2?

The Moving Picture Experts Group is an international standard-setting body organization under the auspices of the International Organization for Standardization (ISO). It is composed of technical experts from around the world "to define a standard for digital compression of moving images."

Initially, the committee set the standard for low bit rate (1.2 to 1.5 mbps), non-broadcast, terrestrial applications. This standard, known as MPEG-1, is now widely used in products such as CD-1 and CD ROM storage devices.

Since 1990, the committee has focused on defining MPEG-2 for broadcast applications. The committee is divided into sub-groups to address areas such as video, audio and systems, test and verification. It meets four times a year with ad hoc meetings set to resolve specific issues.

MPEG-2 simply defines a standard set of operations—i.e., number of sample/line, line/picture sub-sampling or the quantizing rate. Essentially, it is a syntax and it should be noted that MPEG does not define decoder design.

The MPEG-2 committee defined a set of subsets taking into account the degree of picture resolution (memory) and the functionality of the decoder. The intent is to have, for example, HDTV decoders compatible with simple, lower resolution decoders. MPEG-2 goes beyond defining video compression and also defines how audio should be compressed.

The standard also defines a standard format for multiplexing many of the video, audio and data components together. This is known as the transport or systems layer and is the most complex part of the standard, as it must satisfy the requirements of cable and satellite transmissions.

Conditional access and encryption are not part of the standardization process although carried in the transport layer, but these are important components in the design of an overall network. Decoder management including security and addressability are areas where manufacturers will uniquely compare.

The MPEG-2 standard is well along the road towards market introduction with strong drive by market applications—direct-to-home, cable and entertainment program networks. The video syntax specification (a critical component) has been frozen and is now being committed to silicon. The audio for stereo is also defined and chips are currently available based on MPEG-1 (MUSICAM).

Future enhancements to the MPEG-2 will include multi-channel audio for application such as surround sound. The transport or systems layer of the standard is currently being developed and was targeted to be frozen in November 1993. In addition the standardization process has stabilized MPEG-2 enough to enable manufacturers to start ramping up for production in early 1994.

For example, Scientific-Atlanta has already began shipments to the cable industry of MPEG-based video compression systems. By mid-1994, MPEG-2 will be entrenched in many of the newly introduced video products in the various market sectors. Proprietary digital video technologies will quickly be made obsolete as a wide array of new devices and services take advantage of MPEG-2 standard.

James D. Lakin, vice president of sales and marketing for the Broadcast Products Group at Compression Labs, Incorporated (CLI), is responsible for strategic and tactical marketing, product planning, business development, as well as domestic and international sales of the Magnitude and SpectrumSaver compressed digital broadcast television systems. A professional engineer with 25 years of experience in telecommunications, he previously was president and founder of several videoconferencing and data networking companies. He was also with Contel and Rockwell International. Lakin has been nationally recognized for his contributions by the U.S. Department of Commerce for excellence in exporting. The State of Maryland recognized Lakin for his contributions to economic development, and the National Alliance of Businessmen recognized Lakin for outstanding achievements in affirmative action. Lakin graduated from the University of Houston with a degree in electrical engineering and attended Southern Methodist University and George Washington University graduate school of business. He is a member of the Institute of Electrical and Electronics Engineers, the National Society of Professional Engineers, the National Association of Broadcasters, and the International Teleconferencing Association.

COMPRESSED DIGITAL VIDEO TECHNOLOGY ACCELERATES THE DIGITAL TV REVOLUTION SUMMARY

The convergence of the telecommunications, computer and entertainment industries promises to deliver a wide array of new entertainment and information services to homes and businesses around the world. Compressed Digital Video is one of the key enabling technologies behind this digital revolution. With almost two decades of experience in advancing the state-of-the-art in Compressed Digital Video technology, CLI is playing a major role in the emerging digital TV marketplace.

• Introduction

Over the past several decades, video communication has played an increasingly important role in peopleþs lives. It has helped shape how we see ourselves and others, how we get our news, and how we spend our leisure time. And there is more to come. In this decade, video communication will emerge as a ubiquitous medium that touches almost every aspect of our daily lives. A driving force behind this emergence is Compressed Digital Video (CDV_) technology from Compression Labs, Inc. (CLI) of San Jose, California.

CDV technology overcomes the single most critical challenge in utilizing video communications to the fullest potential: bandwidth. Bandwidth þ the amount of communication channel available to transmit video signals þ is limited and expensive. Todayþs analog video signals require enormous amounts of bandwidth, limiting video communication to major players willing to make major investments.

The solution to the bandwidth challenge is digital video compression, which reduces the amount of bandwidth required by a video signal. This increases the total amount of information that can be transmitted over existing communication channels, as well as reducing the cost of transmission by allowing the cost of bandwidth to be shared. The digital conversion that is the first step in video data compression also enhances clarity and quality, eliminating ghosting, snow and other distortions that can affect analog video.

◼ How Compressed Digital Video Works

Compressing a digital video signal is more a process of elimination than compacting. Compression is achieved by eliminating all data unnecessary to achieving a transmitted image of acceptable quality.

Video is transmitted frame by frame, with each frame corresponding to a still picture. Each succeeding picture changes as motion progresses. Data that does not change between frames need not be transmitted with each frame. Similarly, data that can be anticipated not to change, such as a single white pixel surrounded by a background of white pixels, also need not be sent with every frame. The compression process is therefore a matter of selecting only that data within a picture that must be transmitted with each frame to maintain an accurate image, and then using mathematical processes to minimize the amount of data needed to represent this information.

The Compression Labs CDV process has at its center a codec (short for coder-decoder) sampling the analog signal from a video camera to produce a digital display format made up of thousands of picture elements known as pixels. Once the picture has been digitized, video frames are divided into more manageable þmacro blocks,þ consisting of 16 by 16 pixels of luminance, or brightness, and 8 by 8 pixels for each of two channels of chrominance, or color. These blocks are then analyzed by the codec to determine which picture data should be sent.

Because every pixel requires eight bits of data for transmission, a number of sophisticated coding techniques have been devised to avoid sending parts of the picture that have not changed. Interframe coding is used to transmit relatively small differences occurring from one frame to the next. Intraframe coding is employed for major scene changes, where the sending codec instructs the receiving codec to replace all previous data with the new data being sent.

When coding is complete, a sophisticated mathematical transformation called Discrete Cosine Transform (DCT) is used to reorganize the pixel information into a more compact form and generate a series of numbers that represents pixel values. These numbers are divided by a mathematical coefficient that makes them easier to send. Finally, the codec encodes and sends that data to the receiving site, where the information is decoded by a similar codec performing the process in reverse.

Formed in 1976, Compression Labs initially used its digital video compression technology for videoconferencing systems þ providing a virtual face-to-face environment for business meetings. With the introduction of the worldþs first commercially successful video codec in 1982, CLI made interactive

video communication practical and affordable. Large businesses, government agencies, and other organizations began using videoconferencing to conduct a variety of business meetings without incurring travel expenses.

◼ Compressed Digital Video and Television Broadcast

Today, industry observers agree that many of the most interesting applications of video, such as digital TV, would be inconceivable without video compression technology. Compressing digital signals into a much narrower bandwidth dramatically reduces the cost to transmit it. By making it possible to digitize and compress broadcast video from 90 Mbps þ the bandwidth of video in analog form þ to as little as 1.5 Mbps, Compressed Digital Video technology will play an indispensable role in the emerging digital TV market. CDV is integral to the cost-efficient use of video in an expanding number of applications þ from distance learning programs at major universities to business television networks, and direct-to-home satellite entertainment services.

◼ TV Gets Digitized

Television technology has not changed significantly since the 1950s when the first "picture tubes" found their way into homes across America. At that time, TV sets were built on amplitude modulated (AM) radio technologyþthe most sophisticated form of electronic mass media available. Until very recently, almost all television broadcasts relied exclusively on a form of technology called "analog" that transmits video signals as a series of wave forms. Analog technology severely limits transmission capabilities because it requires a tremendous amount of capacity or bandwidth, regardless of the transmission technology used. As an example, standard analog transmission of video requires 1000 times the bandwidth of a telephone conversation.

Conversely, the use of digital technology for television broadcast offers a number of advantages. Digital technology provides far more possibilities than analog technology because once information is "digitized" it can be compressedþresulting in efficient and economical data transmission, storage, and interaction. This leads the way to a variety of new applications for the information superhighway. The development of digital TV, in particular, and the information superhighway, in general, is entirely dependent on one common denominatorþ"Compressed Digital Video" technology.

■ The Arrival of MPEG

To support the expansion of digital television around the world and to ensure that proliferating applications work together efficiently and effectively, global standards are necessary. Beginning in the late 1980s, the Moving Picture Expert Group (MPEG), under the direction of the International Standards Organization (ISO), began work on a family of digital compression standards that would offer a strong platform for the widespread interoperability of different types of digital media. The initial MPEG effortsþreferred to as MPEG-1þcentered on regulating the storage of video images on compact disc and other media. However, it soon became clear that MPEG had far broader applicationsþranging from direct broadcast satellite services to high-definition television. This gave birth to the MPEG-2 standard, expected to be finalized in early 1994, that will further refine the standards for digital television.

MPEG-2 provides the framework for very high-quality video, CD-quality audio, and sophisticated security features. This new standard promises to open the door to mass market applications such as direct broadcast satellite (DBS) entertainment services, video-on-demand, and video dialtone (high-quality video and audio delivered over analog and digital telephone lines), and cable-headend delivery (where cable programming is delivered digitally to central sites for faster, better quality distribution). At its heart are the CDV video compression techniques pioneered by CLI. .

■ EMERGING APPLICATIONS FOR DIGITAL TV

Digital television promises to change a lot more than the programming on our televisions. It will affect satellite broadcasting, expand the capabilities of the telephone, alter cable distribution and servicesþand stimulate a whole array of new applications that haven't been possible before.

Satellites

As mentioned earlier, transmitting analog video requires a tremendous amount of capacity or bandwidth, and in the case of satellite transmission, only one or two channels of video can be transmitted per satellite transponder. In the United States, where adequate transponder space has generally been available, the expense and commitment of full-transponder analog broadcasts have limited programming options for all but the largest commercial and public broadcasting networks. In regions that are under-served by satellite coverage, the scarcity of transponder space has made it difficult to expand program offerings at any cost. Compressed Digital Video technology is making it considerably less costly to transmit video via satellite.

With CDV, many high-quality "compressed" video channels carrying audio, video, and data can be sent over one satellite transponder. As a result, digital technology promises to make more satellite transponder space available and substantially decrease satellite transmission costs on a per-channel basis.

MPEG-2-based systems also will foster new applications such as direct broadcast satellite entertainment services. As one example, it will soon be possible for consumers to select from hundreds of entertainment channels and information services "on-demand". MPEG-2-based video compression also will improve the economics and timeliness of remote news gathering. With analog transmission, a news correspondent assigned to Moscow might be required to report the most recent political developments via telephone rather than direct satellite link because adequate satellite space may not be immediately available. With CDV, the video signal would require less bandwidthþallowing it to be transmitted over a smaller, more readily available satellite space segment.

■ VIDEO DIALTONE

Telephone companies are increasingly investing in digital video and multi-media products and services. While broadcast and cable television technologies have been in existence for some time, the convergence of video, multi-media, and telephony at the central office has the greatest significance in defining the information superhighway. Almost everyone has a telephone of their own, or access to one. This fact alone makes the potential in the telephone market particularly promising. However, the telephone market also faces technology challenges associated with its existing infrastructure. The bandwidth of standard twisted-pair telephone line is extremely limitedþroughly 1.5 Mbps, an insufficient capacity for entertainment-quality video. Video compression technology overcomes this hurdle. Using CDV, an entire channel of video can be transmitted over a standard copper-wire telephone line.

Within the telephone market segment, there are two types of technologies for video transmission. First, digital signals may be sent over standard, twisted-pair telephone lines to the home using ADSL (asymmetric digital subscriber line) technologyþa modulation scheme for transmitting digital video at 1.5 Mbps for distances of up to 15,000 feet. Compression technology, working with ADSL, allows significantly more information to be carried across telephone lines. Second, digital signals may be sent over fiber and coaxial cable. Compared with twisted pair wire, fiber and coax provide wider bandwidths and increased capacity for interactivity. By using compression technology to send information over fiber and coax, service providers will be able to offer multiple, simultaneous channels and more information services to potential customers.

Regardless of transmission medium, video applications in the telephony area are generally called "video dialtone." Because the video information is in digital form, it can be readily stored on a standard computer disk, and retrieved on-demand by the computerþcalled a video serverþfor instant replay. Telephone companies can install these video servers in their telephone central offices and connect them to subscribers via either ADSL or fiber and coaxial cable. Programmed to respond to subscribersþ requests, the video server can play any digitally compressed video stored on its disk files at any time, i.e., video-on-demand. Imagine being able to watch a first-run movie or a program on how to cook a favorite dish at any timeþwhenever you want. All it would take is one phone call. Or consider the possibilities of video-on-demand for education. If your child was writing a report on mountain gorillas, she could access a TV-based video encyclopedia on the topic to enrich her learning experience. Many industry watchers believe video dialtone will be an integral part of the coming information superhighway, and will change the way we watchþand useþ our televisions. Once again, Compressed Digital Video technology will play an essential role in bringing services such as these to the market.

■ CABLE

Despite attracting more than 85 million customers in the U.S. alone, today's cable companies face tough challenges. Telephone companies and satellite service operations are encroaching on what has been the cable industry's exclusive turf. In order for the cable industry to protect its market position, it must offer more channels and new services, such as impulse pay-per-view, a form of video-on-demand. To do this, it must either lay more cable, an expensive proposition, or expand the capacity of existing cable. Digital technologies offer cable operators some clear advantages. For example, digital technology will support quick, economical and reliable delivery of programming to cable-headends. Instead of receiving programming via analog satellite feeds or tape, cable operators will receive Compressed Digital Video via satellite for re-distribution to cable subscribers.

On the consumer side, the average cable company currently offers about 36 channels. Some cable companies are investing in digital fiber transmission

systems that could provide approximately 160 channels. Adding Compressed Digital Video technology, based on MPEG-2 standards, promises to pack a number of digitally encoded channels in the space of one analog stream, bringing the potential capacity of cable systems up to as many as 500 channels, providing bandwidth for many new services.

■ COMPRESSED DIGITAL VIDEO AND DIGITAL TV: CLI

With its years of experience in digital compression technologyþthe one common denominator in the information superhighwayþ

CLI is well positioned to play a key role in the digital TV market. Unlike its competitors, CLI already has a proven track record using digital compression technology and MPEG for broadcasting applications.

For example, in a move to address the needs of the business TV and distance learning markets, CLI introduced SpectrumSaver_ in 1991. The proprietary CDV broadcast satellite system has helped revolutionize business TV and distance learning. Already, many leading business TV and distance-learning networks have migrated from analog to digital using SpectrumSaver. For example, CLI has systems installed at the Indiana Higher Education Television System (IHETS) and at nine of Wescott's business TV networks.

Drawing on this experience, CLI is developing a new generation of standards-based digital compression solutions that will be used for the delivery of entertainment and information services over telephone, cable and satellite networks. These new solutions, the Magnitude_ product family, are based on emerging MPEG-2 standards for digital video, audio, and system compression. Magnitude allows programmers, satellite operators, cable operators and telephone operating companies to economically deliver real-time television broadcast and non-real time video-on-demand services over twisted pair, coaxial cable, optical fiber and satellite. The system reduces the bandwidth required to transmit video (up to 270 megabits per second) by up to sixty times depending on the complexity of the program content and the transmission method.

The Magnitude product family, which consists of a modular encoder and a choice of decoders, was specifically designed to meet the different performance and budget requirements of television broadcasters. The flexible architecture of the Magnitude encoder economically accommodates changing broadcast requirements. Broadcasters can start with a system tailored to meet the number and type of services currently required, and then modify the system at a later date to add new services.

Magnitude features four decoders designed to meet the requirements of different broadcast applications ranging from the rigorous demands of a professional broadcast environment to a low cost consumer decoder for compressed digital video-on-demand programming delivered over twisted-pair, coaxial cable or optical fiber.

By designing the encoders and decoders as an ensemble of functional modules, CLIþs new MPEG-2-based products will serve the needs of the satellite, telephone, cable and server markets. These modules can be configured to provide Compressed Digital Video, audio and data systems with a broad spectrum of services bound for the information superhighway.

Further, with CLI's MPEG-2-based systems, broadcasters will have the flexibility to work over a variety of transmission mediaþincluding satellites, optical fiber, copper wire, and coaxial cable.

CLI is supplying Magnitude encoders to a unit of GM Hughes Electronics through an agreement with Thomson Consumer Electronics for the DIRECTV(TM) satellite entertainment service. DIRECTV is the first high power Direct Broadcast Satellite network in the United States.

In addition, CLI and Philips Consumer Electronics Company have agreed to jointly design, manufacture and market digital set-top decoders for the delivery of video-on-demand programming to the home. Philips/CLI also formed an alliance with Oracle Corporation to integrate its decoders with Oracleþs Media Server_ product line.

Bell Atlantic announced in January that it will use Philips/CLI decoders in the deployment of on-demand services beginning this year. and awarded the two companies a contract in July 1994 for the first actual commercial deployment of set-top (Digital Entertainment Terminals). The initial multi-million-dollar order to Philips and CLI will be for installations in Bell Atlanticþs Dover Township, New Jersey service area beginning later in 1994. Bell Atlantic received approval from the Federal Communications Commission on July 6 to begin construction of an interactive digital network serving 38,000 lines. Service is expected to being in early 1995.

Philips and CLI are also working with Zenith Electronics Corporation to combine their efforts to offer cable operators, telecommunications companies and other network providers essential signal delivery products for full-service digital and analog networks. They have agreed to combine their complementary technologies, product development and marketing capabilities to design and manufacture digital video-on-demand and hybrid digital/analog set-top terminals. High-data-rate digital transmissions over fiber and cable will be delivered to the new set-top decoders.

Far from slowing down, the digital revolution is gaining momentum as the 21st century approaches. The convergence of the telecommunications, computer and entertainment industries promises to deliver a wide array of new entertainment and information services to homes and businesses around the world. Compressed Digital Video is one of the key enabling technologies behind this digital revolution. With almost two decades of experience in advancing the state-of-the-art in Compressed Digital Video technology, CLI intends to play a major role in the emerging digital TV marketplace.

●●●●

CLI is a registered trademark, and CDV, Magnitude, and SpectrumSaver are trademarks of Compression Labs, Inc. (CLI).

Chapter 4

*Merrill Ray Brooksby graduated from Brigham Young University in
1982 with a degree in Television and Motion Picture Production, and
minors in Photography and Spanish. After working in broadcast television
production for a number of years, Ray joined Hewlett-Packard's Corporate
Telecommunications Group and was responsible for the early implementation
of HP's videoconference network, which included the first privately owned
and operated videoconference room outside the U.S. (HP's conference room in
Bristol, England).*

*Ray joined Compression Labs, Inc., in August 1985, as world-wide
manager of technical sales support. While with CLI, Ray has also been the
videoconference system product manager and currently manages CLI's technical training group.*

*Ray is a pilot, photographer, Scoutmaster, choir director, and sings tenor
in a barbershop quartet.*

VIDEOCONFERENCING: TWO-WAY
INTERACTIVE VIDEO

■ A BRIEF HISTORY OF TIME

The potential benefit of face-to-face meetings
without the associated inconvenience of travel (primarily lost time and related expenses), has long captured the imagination of business people, government
leaders, and educators. Captain Kirk of the starship
Enterprise routinely communicated with other starship captains and Starfleet Headquarters -- hundreds
of light-years away— by means of a face-to-face video
and audio connection. A small sensation was created
when AT&T demonstrated a "concept video telephone" at the world's fair in the early '60s.

But the technology proved far more expensive
than day-to-day business benefits could justify. The
dilemma was the amount and type of information
required to display video images. Video signals
included frequencies much higher than the telephone
network was designed to handle (particularly the telephone network of the early '60s). The only feasible
method for transmitting video signals over long distances was by satellite. The satellite industry was in
its infancy then, and the cost of earth bound equipment to use that network combined with the cost of
renting time far exceeded what could be justified for
getting comparatively small groups of people together
in a meeting.

So, through the '60s and '70s the only regular

users of face-to-face video technology were the
Captain Kirks of the universe. The potential benefit
remained— and with it a potential market.

Where Have We Come From

Through the '70s substantial progress was made
in several key fields. The various telephone network
providers began a transition to digital transmission
methods, computers came into their own with significant advances in processing power and speed, and
greatly improved methods for sampling and converting analog signals (such as those used for video and
audio), to digital bits were developed. A fundamental
transition from analog only signal processing, to
include digital signal processing occurred.

Digital signal processing offered a number of
advantages, primarily in the areas of signal quality and
analysis. Storage and transmission still posed significant obstacles. In fact, a digital representation of an
analog signal required more storage and transmission
capacity than the original. For example, digital video
methods common to the late '70s and early '80s
required transfer rates up to 90 million bits per second (or 90 Mbps)! The need for reliable digital data
compression was critical.

Beginning in the late '70s and early '80s real
progress in data compression began to occur. Video

data is a natural candidate for compression because of many redundancies inherent in the original analog signal. Redundancies resulted from the original specifications for video transmission and which were required for early television sets to receive and properly display images. A good portion of the analog video signal is dedicated to timing and synchronization of the television set. A number of video data compression methods were developed which relied entirely on the elimination of this redundant portion of the signal. Such methods typically achieved a 50 percent reduction in the amount of data required.

This compressed data rate — 45 Mbps, a 2:1 compression ratio— was a key threshold speed. Telephone network providers, in their transition to digital, developed with a number of key data rate steps. The first was 56 thousand bits per second (or 56 kbps). This was the key base rate because it was the rate required for a voice phone call (using sampling methods of the day). Groups of 56 kbps channels were gathered into a single larger channel which ran at 1.5 Mbps (normally called a 'T-1' channel). Groups of T-1 channels were gathered together into an even larger single channel running at 45 Mbps (or a 'T-3'). Thus,—using video compressed to 45 Mbps -- it was now finally possible, though still extremely expensive, to transmit live video through the public telephone network. However, compression to the T-1 rate (a compression ratio of 60:1), or better, would be required to launch a market in face-to-face communications.

Then, during the earlier '80s, several compression methods made their debut which went beyond the elimination of timing signal redundancy to an analysis of picture content for redundancy. This new generation of video codecs (codec is short for coder/decoder), not only took advantage of redundancy, but of the human vision processing system. The normal video frame rate in North America is 30 frames per second (or 30 fps). However, this exceeds the requirements of the human visual system to perceive motion. Most motion pictures run around 24 fps. The perception of smooth motion can be maintained to between 15 and 20 fps. Therefore, a frame rate reduction from 30 fps to 15 fps in itself yields another 50 percent compression gain. A 4:1 ratio has been achieved, but we are still far from our 60:1 goal.

The codecs of the early '80s pioneered a technology known as Discrete Cosign Transform (DCT) coding. A detailed treatment of DCT techniques is beyond the scope of this chapter (the October 1991, issue of the IEEE Spectrum journal provides a number of very good articles on this topic and the current state of the DCT art). Using DCT technology, video images can be analyzed for spacial and temporal redundancy. Spacial redundancy is that which can be found within a single video frame — areas of the picture which are near enough alike that they can be represented with the same sequence. Temporal redundancy is that which can be found comparing one video frame to the next — areas of the picture which have not changed in successive frames. Combining all the methods mentioned so far, the 60:1 compression ratio was achieved. The first such codec was introduced to the market by Compression Labs, Inc., and was known as the VTS 1.5. The VTS stood for Video Teleconference System, and the 1.5 stood for 1.5 Mbps — or T-1. Within a year CLI enhanced the VTS 1.5 to achieve a 117:1 compression ratio (768 kbps), and renamed the product VTS 1.5E.

Competition entered the market from Europe and Japan. The Britain based GEC Corporation, and the Japan based NEC Corporation, both made codecs that operated at T-1 (and below T-1 if the picture was relatively motionless). None of these codecs were inexpensive. The average VTS 1.5E sold for $180,000. That did not include any of the necessary audio and video equipment required to finish a face-to-face conference system, nor did it include network access costs. Many of the early face-to-face meeting pioneers spent an additional $70,000 on audio and video equipment bringing the total conference room cost to $250,000! Network access at T-1 data rates was available for approximately $1,000 per hour.

The mid-to-late '80s saw dramatic improvements in codec technology and equally dramatic network access cost reductions. CLI introduced the Rembrandt Video System which set the standard at a 235:1 compression ratio (384 kbps). Then a new company, PictureTel (originally PicTel Communications) introduced their first codec and set a new benchmark at 1600:1 compression (56 kbps). PictureTel pioneered the Hierarchical Vector Quantization (HVQ) technique to make their gains. CLI followed with the Rembrandt-56 which also operated down to 56 kbps, and took advantage of a new technique — Motion Compensation. Through the same period, network providers became aggressive in marketing access to digital circuits. Charges for T-1 access came down rapidly, and other lower data rate services were introduced.

Codec prices fell almost as fast as the compression ratios rose. By 1990, sub T-1 codecs were selling for $30,00-down over 80percent from just a few years previous. They were getting smaller as well. The original VTS 1.5E stood almost five feet tall, covered two-and-a-half square feet of floor space, and weighed several hundred pounds. The Rembrandt-56 was about 19 inches square, 25 inches deep, and weighed about 75 pounds.

Such phenomenal compression gains did not come without a picture quality penalty. The rush to 56 kbps resulted in frame rates below 15 fps, and compromises in picture sharpness. If you participated in a face-to-face meeting at 56 kbps the compression effects were not easy to overlook. Some went so far as to say the picture quality was "bad." Therefore, the face-to-face meeting business remained pretty much in the range of 384 kbps and above.

Where Are We Today?

Video codecs today range in price from $20,000 to $85,000. The bottom data rate is holding constant at 56 kbps. There is not much pressure to get below 56 kbps for group face-to-face meetings because digital network costs are close to that of a normal phone call. A full T-1 connection in today's market will cost about $50 per hour. This has allowed the codec manufacturers to focus on improved picture quality at 384 kbps and above. Some coding methods yield picture quality at 768 kbps and T-1 that is hard to distinguish from the uncompressed original.

Video and audio equipment packages tailored to face-to-face meeting applications have also become available, and range in price from $15,000 to $42,000. A complete face-to-face meeting system can be had for between $20,000 and $100,000. Many customers of older systems are upgrading to the newest technology. This has led to a secondary market for used equipment packages. Some of these can be had for as little as $15,000 complete.

The last three years have seen explosive growth in the number of systems installed. Estimates vary, but as many as 5,000 new systems could be installed during 1994. Even with all the advances in video data compression, the total package — a videoconference room — including a codec, video system, audio system and control system, must be pulled together using components from various suppliers. This integration of components requires expertise in all the related fields: audio, video, data, and control. This integration requirement has probably been as much of a barrier to market acceptance as have been costs. Not many corporate communications groups employ people knowledgeable in all these fields.

Further, proper attention must be given to the physical design of the conference room. Otherwise, even with the purest picture quality available, the room will be awkward to use and participants will be put off.

■ THREE KEY ELEMENTS

Think of a videoconference facility as being comprised of three key elements: network, codec and room. (See Figure 4.1.1)

Further subdivide the room into four important components: physical environment, video system, audio system and control system.

The Videoconference Network

A two-way, high-speed, digital connection is required between the two videoconference rooms. More often than not this connection is provided by a common carrier (such as AT&T, MCI or U.S. Sprint). These carriers have, for the most part, converted their national routes to digital equipment. Therefore, digital service is available to all but the most rural areas.

The various network options will be discussed in Chapter Seven. It is sufficient here to say that the number of possibilities is great and a particular choice will depend entirely on the requirements of the customer.

It is important to make one distinction here. Notice, in Figure 4.1.1, the codec circle does not actually touch the network circle. In fact, a barrier is shown. This is intended to represent the fact that most network providers only allow "approved" equipment to connect directly and, until recently, most codec manufacturers have not included "approved" interfaces on their equipment.

The carriers are naturally sensitive to the type of equipment attached to them. It could be that a faulty device at one customer location is causing network problems that affect many other customers. To avoid this, the carriers have very strict standards which apply to equipment intended to interface directly to them. There are manufacturers who specialize in these interfaces and the codec manufacturers have elected to connect through them. These devices are normally called Customer Service Units (CSU's), Data Service Units (DSU's), or Terminal Adaptors (TA's). Equipment installed on the customer's side of the CSU, DSU or TA, is known as Customer Premise Equipment (CPE).

A recent trend has been for carriers to include a CSU, DSU, or TA, as part of the network service. This gives them ultimate control regarding the possibility of CPE damaging their service or taking other customers "off-the-air."

By same token, a few of the codec manufacturers have included CSU, DSU or TA, options with their equipment. They are motivated by a desire to simplify the details involved in a sale, and to make

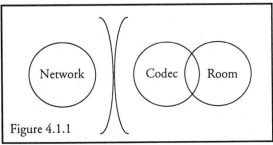

Figure 4.1.1

troubleshooting more straight forward. Codec manufacturers often get the first call when there are problems with service. Codec manufacturers can resolve the location of the problem more quickly when they are familiar with all the pieces between the two videoconference rooms.

The Videoconference Room

The videoconference room it self is the only portion of the technology that most who use the system will ever see or know anything about. Therefore, the overall level of comfort this room engenders will determine the success of the installation.

The perfect videoconference room is a room that 'feels' as much like a normal conference room -- for a given corporate setting -- as possible. Those who use the room are not intimidated by the technology required, rather, they feel completely at home. The technology is almost hidden, or transparent to their use.

1. Physical Environment

Videoconference room sizes and shapes run the full spectrum. Customers have installed systems in everything from the smallest closet to the great outdoors. The size and shape of the room must be selected to be consistent with the intended use of the room. Now, this seems a simple thing to say, however, many people have fallen into the trap of saying a videoconference room can not be smaller than 'X', or larger than 'Y'. Therefore I will try to fit my application into something between 'X' and 'Y'. It is possible to design a facility to meet any need. There are properly designed systems operating on factory floors — where airplanes are being assembled and on board ships at sea. It is also possible to generalize the typical videoconference room in a corporate meeting environment.

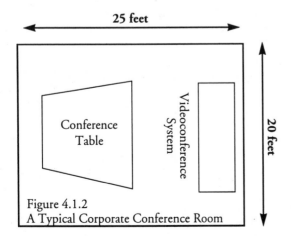

Figure 4.1.2
A Typical Corporate Conference Room

A typical videoconference room is about 25 feet deep, by 20 feet across. This will provide space for a medium sized videoconference system and conference table for approximately seven people (three on each side and one at the end of the table shown in Figure 4.1.2).

There are three elements to good lighting in a videoconference room: light levels, light angles and light color. The objective is to provide even light of the right color at levels which allow the video camera to render a natural scene.

The most common lighting mistake is made in the area of light level (either too much or too little). Most modern video cameras specify light levels between 100 and 200 footcandles (1000 - 2000 lux), but can actually function well at levels as low as 50 footcandles (500 lux). A footcandle is defined as the amount of light energy provided by a single burning candle measured at a distance of one foot.

The advantage of running at higher levels of light (approximately 125 footcandles) will be improved camera performance. Depth of field, or the camera's ability to focus, is directly related to the amount of light available to the lens. Therefore, where light levels are high it will be easier to focus. Also with plenty of light there will be little or no "noise" in the video signal from the camera (noise appears as fine-grain static on the video display). Noise is normally generated by an Automatic Gain Control (AGC) circuit in the camera which attempts to boost signal strength in low-light situations. The disadvantage of levels this high will be the additional heat generated by the light fixtures making the room more expensive (and potentially more noisy) to cool. Conference participants would probably feel uncomfortable in such a bright, hot environment.

The advantages of running at lower light levels (approximately 75 footcandles) are found in the comfort of the participants and the cost of cooling the room. However, below 75 footcandles the video camera will not be able to properly render the scene. Colors will "wash-out" and shadows will be very pronounced. The video signal from the camera will contain noise, which will likely affect the ability of the video codec to properly adapt to motion in the scene (noise is perceived as motion by the codec).

The goal is to work between 75 and 125 footcandles (shoot for 100 footcandles). In this range, camera noise levels will be acceptable, colors will be properly rendered and conference participants will be comfortable.

Light at the proper angle is also a very significant contributor to good picture quality. Unfortunately, most existing conference rooms are equipped with fixtures designed to direct their light mostly down — normally onto the surface of a conference table. This is acceptable for a normal conference room where the purpose is to provide adequate light on documents or objects on the table. Unfortunately, this type, or angle, of light casts dark shadows under the eyes, nose and chin of people at the table. It also creates "hot" areas of light on heads and shoulders. Often, normal conference room lighting is not even. A few fixtures will be scattered around the ceiling creating pockets of hot-spots and shadows.

The human eye is much more capable than even the best broadcast video cameras of compensating for this type of lighting. The range of contrast acceptable to the eye easily includes the range between these areas of bright highlights and dark shadows. A video camera is much less tolerant. Any shadows created by poor lighting angles will be much more apparent on the video monitor at the far end of a videoconference link than to the eyes of those seated in the local room itself.

To create an evenly lit scene several conditions must be met. The light source should not be a single point (such as a spotlight or single focused fixture), rather, it should be from several broad sources (such as multiple bulb 2'x2' or 2'x4' fluorescent fixtures). There is a rule of thumb for videoconference room lighting which can be applied. Generally speaking, a broad light source should be centered 45 above and ahead of the subject. Sources at angles less than 45 will be "in the eyes" of conference participants. Sources at more than 45 will leave noticeable shadows particularly below the eyes.

It is important that the video camera "see" a scene with even levels of light overall. Critical to an evenly lit scene is the amount of light reflected toward the camera from the back wall of the room. The level of light reflected from the back wall should be barely less than — and should never exceed — that reflected by the conference participants. This can be an interesting challenge because different color or texture backgrounds will reflect different levels of light. Therefore, it is not enough to install wall lighting and assume an appropriate level will be reflected.

The most accurate method of measuring reflected light levels is with a reflective "spot" lightmeter. Such meters are commonly used by photographers to selectively analyze film exposure levels in different areas of a scene. Some 35mm cameras have "spot" meters built in. Such a "spot" meter would be aimed — from the point-of-view of the camera — at someone sitting at the conference table. The meter reading would be noted then the meter would be aimed at the background. This meter reading would also be noted and compared to the first. Ideally, there will be no more than 1/2 an 'f' stop difference between the two (with the background being darker than the participant.

Standard conference room lighting is normally accomplished through a combination of two different types of light fixtures. Fluorescent fixtures which normally have a color temperature of about =5,600

Kelvin, and incandescent fixtures with a color temperature of about 3,200 Kelvin. The Kelvin color temperature scale was invented by a British physicist (by the name of Kelvin) and references the color of an iron rod as it is heated to specific temperatures. As a rod is heated it gradually changes color until it is "white" hot. Lower temperatures tend to be "red" hot. Daylight on a clear sunny day measures between 5,500 and 5,600, while a bright overcast sky measures in excess of 6,000. A tungsten light bulb measures 3,200.

There is a hidden "got-ya" in this system of measuring the color of light. Most interior decorators refer to colors in the orange-red range as "warm" colors and to colors in the blue-white range as "cool." You can see this terminology runs counter to the Kelvin scale of measurement. On the Kelvin scale the "cool" lights are the orange-reds and the "warm" or "hot" colors are the blue-whites (because the iron rod is at its hottest when it is white-hot).

As you might imagine, the color of light available in a conference room will affect how a camera perceives the color of objects (and people) in that room. Most cameras are equipped with a "white-balance" feature which electrically corrects for the color temperature of light in the room. This feature varies from camera to camera, but is generally able to correct anything between 3,200 and 5,600 light. The issue then becomes one of subjective preference. The human eye performs this adjustment automatically and very accurately, normally within a few minutes. (You can prove this to yourself by wearing a pair of yellow anti-glare ski goggles for 15 or 20 minutes. When you first put them on everything is very yellow. However, soon, white appears white, red is red, and so forth. When you remove the goggles colors are shifted to minus-yellow (green-yellow = blue). Things will look blue for a few minutes and gradually colors will appear normal again.)

Subjectively it seems that "cool" light (on the Kelvin scale) renders richer and more pleasing color than "hot" light. On the other hand, "hot" light is brighter, that is, you get higher light levels from "hot" fixtures than you will from "cool" fixtures of the same wattage. Most industrial lighting is provided via 5,600 fluorescent bulbs, though 3,200 fluorescent bulbs are available. Many conference rooms have successfully mixed the two types of bulbs about 50 percent - 50 percent with good success. This results in sufficient light levels with pleasing color.

Acoustics

There are four elements to proper acoustics in videoconference room design: ambient noise level, reverberation time, microphone and speaker placement and the method of the echo cancellation to be used.

Videoconference room audio considerations are the topic of another chapter. Here we will simply say the objective is to provide a room which is quiet with a relatively short reverberation time. Proper microphone and speaker placement will enhance the quality of sound transmitted between conference rooms. All of this combines to assist the echo canceller in it's function.

Furniture

Furniture is very much at the discretion of the owners of the conference room. Most videoconference room furniture discussions deteriorate into an argument about table shape. A variety of videoconference room table shapes have been tried, and zealous advocates of each table configuration can be found. In Figure 4.2, I presented one of the most popular shapes — a trapezoid which is wider at the end facing the videoconference monitors. This shape is popular because it allows people around the table to interact with each other as easily as with those at the other end of the conference link. It is really a matter of individual preference and should be decided with a good understanding of the various groups using the room.

The Video, Audio and Control Subsystems

Videoconference room equipment configurations are as varied as there are applications for videoconferences. All equipment packages have common subsystems: the video subsystem, the audio subsystem and the control subsystem.

We could not possibly discuss every possible equipment combination. But there are some configuration generalizations which can be made, and which will be helpful to understand.

Video System

A well designed video subsystem uses no more devices than absolutely necessary. The basic requirement is to deliver video from the cameras to the codec, and from the codec to the monitors. Beyond this are a number of other functions which range in importance and, again, depend very much of the intended use of the videoconference room.

This diagram identifies the key elements of a video subsystem. The heavy horizontal line divides the transmit (top) side from the receive (bottom) side: (See Figure 4.1.3).

The entire system can be thought of as devices that generate video, devices that receive video and devices that carry (or move) video from one point to another. The codec is unique in that it both generates and receives video.

It is appropriate to discuss the video distribution system first because it is responsible for connecting video sources to video destinations. Video sources include cameras, video slide projectors, VCR outputs for playback, the codec video outputs, etc. Video destinations include video displays (normally called video monitors), VCR inputs for recording, codec inputs for transmission, video printers, etc. The video distribution system can be as simple as a piece of cable which directly connects a camera output to a codec input, or as complicated as a complete set of video routing switchers configured to allow any video source to be connected to any combination of video destinations at any time.

Conference rooms exist at both extremes. The simplest is a room with a single camera and display directly connected to a codec. This will function without problems so long as this is all the participants require. Designs exist which include seven or eight

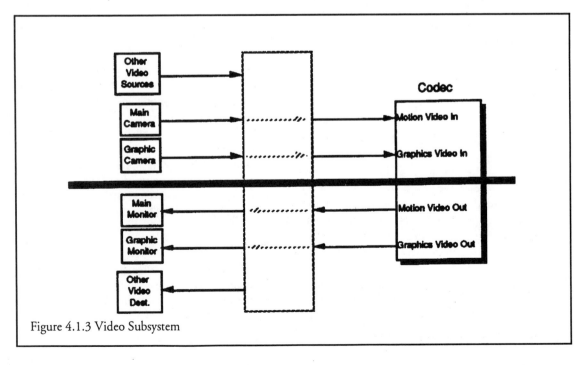

Figure 4.1.3 Video Subsystem

cameras routed through sophisticated switches to the codec and multiple video monitors.

Conference participants decide which cameras are to be seen at the far end by making selections on the conference room's control system (to be discussed shortly). Normally only one camera can be seen at the far end at any given time.

The term "Motion Video" is used to describe live — or moving — video transmitted from one videoconference room to the other. It originates with the main conference room camera and makes its way through the video distribution system to the motion video input on the codec. The codec will code and compress this video signal and pass it through the network to the codec at the far end where it will be decoded and displayed.

Virtually any video camera (or other video source) can be routed through the video distribution system to the codec for transition to the other end. Videoconference room system designs normally include a single camera located at the front of the conference room and near the main video monitor. It is located near the monitor in order to maintain an illusion of eye contact with people at the far end.

Conference participants tend to look primarily at this video monitor because there they will see people at the far end. Locating the main camera near the main monitor gives the illusion that participants are looking into the camera though they are actually looking at the display near the camera.

This illusion of eye contact is taken for granted by most conference participants. In fact, it is expected. Room designs which fail to account for this are uncomfortable for participants as eye contact is important to natural conversation.

Many videoconference rooms provide video graphic devices which facilitate the display of documents (or images stored in memory) for all participants to see — at both ends of the conference connection. The video codec has a second input — separate from the main motion video input — which is capable of transmitting a single "frozen" video image.

The most common video graphics device is a "document stand." Such a stand will have a small video camera suspended above a small tablet. Documents can be placed on the tablet within the camera's view. The camera output is routed by the video distribution system to the codec graphics input. When instructed to do so, the codec will transmit a single frozen image from the document stand to the far end. Any video device can function as a graphics source. A document stand is simply the most typical. Some conference rooms include a camera mounted in the ceiling above the conference table. Positioning a camera above the table allows participants to place documents, or larger objects, on the table in front of them to be seen at the far end. Personal computers

are sometimes used to generate charts or graphs for transmission. Devices are available for displaying 35mm slides. It is another case where the needs of participants must be considered as graphics capabilities are planned.

Motion and graphics video are the most common requirements in a video subsystem. Less obvious are other special purpose devices, and more subtle video distribution system requirements.

There are many specialized video devices which can be designed into a facility to accommodate participants needs. Some of the more common are videotape players and recorders, 35mm video slide projectors, Super 8 and 16mm video film projectors, still video cameras (such as the Sony MAVICA system), video printers, video disk players or personal computers.

If the device has an NTSC video output there is a good chance the video distribution system can be designed to accommodate it. The utility of many videoconference rooms could be enhanced through the inclusion of some peripheral equipment common to the presentation needs of the room's regular users.

A regular complaint is that a videoconference system does not let people hold the type of meetings they are accustomed to. Often this is because they are accustomed to a particular method of presenting graphics — like overhead or 35mm slides — which they can not conveniently use in the videoconference room. Such a shortcoming is the fault of the room designer. Appropriate graphics devices exist and should be included in the design.

The term "preview" is used to describe a capability of the video distribution system which allows conference participants to view images of themselves (as they are seen at the far end), or view graphic images before transmission.

Through a command to the videoconference room control system images from local video sources can be displayed on local video monitors. The preview feature is normally included to allow conference participants to get comfortable with how they are seen at the far end, and to assure themselves that the correct images are being transmitted.

Audio System

The fundamental purpose of the videoconference room audio system is to allow participants at both ends of the meeting to both hear and be heard. This is much more difficult than it sounds, and is covered in detail in another chapter. The whole issues hinges around acoustic echo suppression or cancellation. Acoustic echo is the result of sound originating with a talker in one room, for which there then exists the speaker in the other room, and enters a microphone there to make the return trip, ultimately to be heard by the talker in his room. (See Figure 4.1.4.)

It is perfectly normal to hear yourself as you

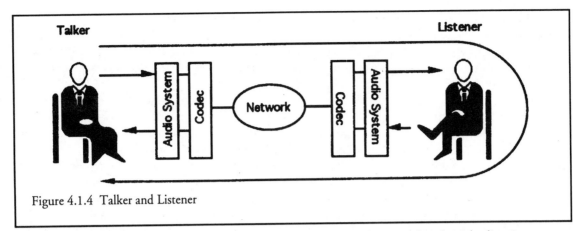

Figure 4.1.4 Talker and Listener

speak, however, the problem arises when what you hear is noticeably delayed from when you said it. Unfortunately, codecs take time to digitize and compress video signals. And there are two codecs involved, the first to code and compress for transmission, and the second to decode and decompress for reception. Round-trip transmission delay is approxi-

phones must be pointed in the right direction.

Sound waves weaken as they travel which means people further from the table will not be heard with the same clarity as people seated at the table.

The audio mixer combines all the audio sources from the conference room into a single audio signal. This would include all microphones, the audio out-

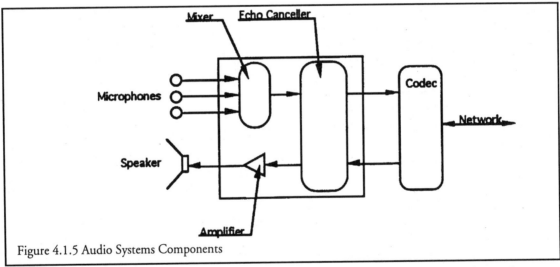

Figure 4.1.5 Audio Systems Components

mately one third of a second — not including any network transmission delay. The result is acoustic echo, that is, a talker's audio is heard in the same room noticeably delayed, after it has made the round-trip. Sophisticated echo cancellers have been developed to eliminate echo while allowing uninterrupted speech.

Major audio system components are identified in Figure 4.1.5.

In this chapter we will only briefly discuss the role of each major component.

One or two microphones are normally placed on the conference table such that the most likely locations for seated participants are covered. Directional microphones are normally chosen, meaning they "hear" better from one direction than another. This is used to reduce the amount of sound picked up from the speaker (which would return to the far end and be perceived as echo), but also means the micro-

put of a videotape player or any other audio source that needs to be heard at the far end.

The echo canceller sits in a unique position in that it monitors both outbound and inbound audio. It will try to remove signals that represent potential echo from the transmission path. The methods employed vary between manufacturers (and are discussed in another chapter). It is important to note that the echo canceller modifies audio destined for the far end (when it detects potential echo). Most echo cancellers do nothing with audio coming into a room. This means echo in my room is indicative of a problem with the system at the far end. Adjustments to my system are not likely to improve matters, though I'm tempted to make adjustments because I hear the problem.

The amplifier receives audio from the far room after it has passed through the echo canceller and boosts the signal for output to a speaker.

The speaker (or audio monitor) is the final jumping off point for audio signals into the room. It is normally located somewhere near the main video monitor to enhance the illusion of contact with the far end. It is natural to turn your head towards the direction of audio. If this is near the main video monitor — where participants from the far end are seen — the illusion of contact is strengthened.

Control System

The videoconference room control system is the heart and soul of the conference room because this is what conference participants touch and feel. There is no doubt that the quality of audio and video is directly related to the codec, and the compression mode

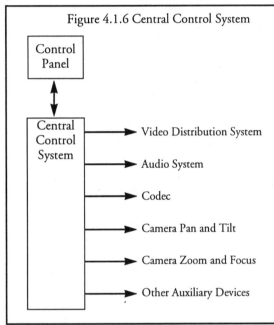

Figure 4.1.6 Central Control System

Control Panel

Central Control System

Video Distribution System

Audio System

Codec

Camera Pan and Tilt

Camera Zoom and Focus

Other Auxiliary Devices

used. However, most conference participants grow used to their level of picture quality. The control system — through the control panel on the conference table — is what they touch and use day-to-day.

A videoconference room control system has two key components: the control panel (which normally sits on the conference table), and the central control system (Figure 4.1.6).

It is through the control system that conference participants translate their wishes into actions. They select which video sources will be seen at the far end, how their cameras are to be positioned, when a videotape is to be played, etc. The central control system acts as buttons on the control panel are pressed by the conference participant. The control panel is all the participant knows about — or even cares about.

Conference participants should not be bothered with details pertaining to the control system's interface to other devices in the room. Their only concern is that the panel is easy to use and understand. Therefore, control panel design and layout become a critical function. The most technically capable room will suffer lack of use if the control panel does not simplify operation to the point that anyone can use the room with a bare minimum of training.

A typical control panel layout may look something like Figure 4.1.7.

This is the main control panel used by one system manufacturer. It is presented on a touch sensitive display. Because it is a display, other screens of buttons can be presented. The square buttons down the right side of the display are used to access screens designed to control other devices. This approach has the advantage of keeping the display simple — only presenting those features the participant has selected at the moment. However, the disadvantage is that all features are not immediately available. If I wish to control the VCR, I must first press the VCR MENU button to recall the VCR control page.

The potential for a confusing control panel grows with the complexity of the conference room. The designer of the room constantly walks that fine line between providing all the features he believes

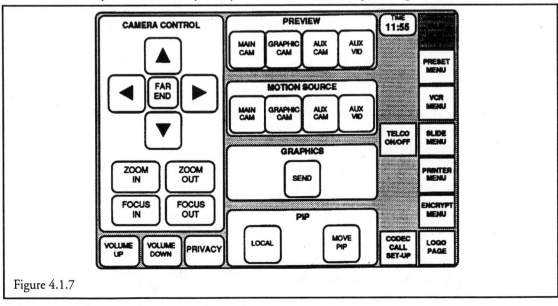

Figure 4.1.7

necessary and keeping the room simple and easy to use. This can prove to be one of the most challenging tasks when designing and installing a room.

Dedicated Rooms

A dedicated videoconference room is a room which has had physical modifications made to accommodate the videoconference equipment. Such a room would normally have a specially designed set of furniture, and a wall at the front which has locations for the cameras and monitors (video and audio) to be built in. Equipment racks for other components and the codec will be located behind this wall, or in an adjacent room.

Special lighting and carpeting may have been installed. Sometimes a waiting room is provided where conference participants may gather and wait for the room to become available.

The expense associated with a dedicated room

ed on wheels and can be moved from room-to-room with a little effort.

Systems of this type offer two very significant cost advantages. First, no physical room modifications are required. While some may still be made (such as improvements to lighting), they are not mandatory. Second, the systems are produced in quantity which translates to lower system costs. A typical dedicated room in the late 1980's could cost as much as $250,000, where a typical modular system today can be had for as little as $20,000.

Modular systems are often billed as "portable" In reality they rarely move from room-to-room because other system requirements (such as power and access to network) have only been installed at one location. The rooms chosen are typically the standard conference rooms normally found around the building. Therefore, it is common that the room will be used for meetings other than videoconferences. No

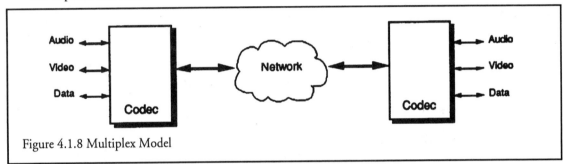

Figure 4.1.8 Multiplex Model

will vary greatly with the desires of management. Some companies take the position that the videoconference room is a technology showcase, a demonstration of their commitment to quality and effective communication. Therefore, they feel justified spending money on comfort beyond the minimum required for face-to-face meetings.

Dedicated videoconference rooms are typically not available for other uses. The table shape, while conducive to videoconference meetings, is not suitable for regular group get-togethers. Further, having spent the money to dedicate a room, it is often felt that the room should be available for videoconference meetings at all times. It should not be blocked for meetings which can be accomplished in any of the regular conference rooms around the building. Dedicated videoconference rooms were the norm up through about 1989. Beginning in 1989, a transition occurred towards prepackaged modular systems.

Modular Systems

A modular videoconference system is one that contains all the equipment necessary for a videoconference — cameras, monitors, audio, video and control systems — in a single cabinet. This cabinet is normally somewhat portable. They are often mount-

special furniture has been purchased, and people feel free moving the microphones and control panel to the side while they meet.

The Videoconference Codec

The codec's traditional role in videoconference installations is the straight forward task of digitizing and compressing video and audio signals for transmission, multiplexing those signals with whatever other data may need to be sent, and delivering that combined signal to a network for transmission.

This was the dominant role of the codec through the '80s and early '90s. Today, manufacturers are enhancing the codec to incorporate functions originally thought to be part of the external conference room equipment (that is, audio, video and control).

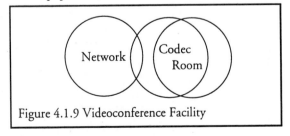

Figure 4.1.9 Videoconference Facility

Recent system introductions and announcements point to an expanding role for the codec. The original concept of the codec as one device among many required for a videoconference system is giving way to a view of the codec as the central piece of equipment incorporating many of the functions previously accomplished by external equipment.

We should revisit our original idea which separates the three primary elements of a conference facility (See Figure 4.1.9).

Videoconference system manufacturers are responding to market requirements for simpler and more cost effective products. In most cases this requires simplifying the design and incorporating previously separate functions into single devices. The result is less separation between these three rolls, illustrated in this fashion: (See Figure 4.1.1)

The most recent codec designs include many of the key components from subsystems originally managed outside the codec. The video distribution system has moved inside the codec, along with the central control system, audio mixer, amplifier and echo canceller. So far, cameras, microphones, speakers and control panels continue to sit outside the codec, but connect directly to it. Some manufacturers have moved the network data service units into the codec as well, taking upon themselves the burden of network certification and protection.

Alliances are forming between codec manufacturers and the former suppliers of these discrete components. Most codec manufacturers do not have the audio, video or network system experience necessary to do the work themselves. Therefore, audio, video and network component manufacturers are lobbying hard to have their equipment selected for inclusion in these all-in-one systems.

Put it Together

Having now been exposed to the variety of components, configurations and services involved, it should be understood that the installation of an effective videoconference system will require up-front planning. A solid understanding of the potential user base (their needs and preferences), along with knowledge of the various vendor offerings, is required. Such research is mandatory before specifying the equipment to be used in an installation.

Fortunately, as in other technical fields, there is no shortage of equipment suppliers, system designers and consultants, available to help. Many companies are using this technology with great success. Their experiences can be very beneficial. In most cases the wheel has already been invented. Much time and effort can be saved by bringing this experience to bear on an application.

Equipment vendors can detail a complete set of capabilities (though perhaps a bit jaded by prejudice).

They are more than willing to show-off their most successful customers, which can provide leads for research. It is important to be in the search by contacting an equipment vendor with a solid stake in the videoconference market. A microphone vendor would not be a good candidate because microphones represent a very small piece of the total system picture.

The codec manufacturers are a natural place to begin as they have been working for years assembling the various audio, video and control components, necessary for the codec to perform. Without these components a codec is useless. Most codec manufacturers offer modular systems tailored to their codec's capabilities. They have selected subsystems that enhance features and overcome shortcomings.

Some codec manufacturers have refined their product to the point that it only really functions well in their total modular system. Other codecs still can be integrated into a larger, more custom, system. Most of the subsystem component manufacturers have very little experience in videoconference system integration. Starting with them would not be recommended.

The term "System Integrator" has been applied to a number of companies formed for the purpose of designing and installing custom conference systems. Most system integrators are knowledgeable regarding all available codecs and subsystems. They will want to work with you to define and refine requirements, then make a total system proposal.

This proposal will include design, assembly, installation and service. System integrators can offer modular systems and dedicated installations. Some integrators are predisposed to a particular codec and set of subsystems. If you suspect this as you deal with an integrator you may wish to press the point to understand the nature of their predisposition.

Generally, integrators are interested in tailoring a design to your specific needs and will recommend equipment that best fits their understanding of your requirement. It is, therefore, very important that you be in a position to state your requirements clearly — based on earlier research into your potential user base and their needs.

Without this type of direction integrators will propose systems like they have previously installed. These may function very will having passed the test of time, but may also miss the mark when it comes to what your people need.

The one element of a videoconference system which cannot be provided by system integrators or codec manufacturers is the network connection. This remains the sole domain of the common carriers (such as AT&T, Sprint, or MCI). Unless you choose to hire a consultant (which we will discuss next), competitive quotes will need to be received from each carrier.

Again, they will need a clear understanding of requirements (bandwidth, locations and usage patterns), before they can address your requirement.

In some cases your company will already have an established relationship with a carrier. They have been selected to handle other voice and data requirements, and videoconferencing will simply be one more application for them to connect.

Most of the carriers have formed relationships with codec manufacturers and system integrators. The carriers wish to be in position to serve a client who comes to them and desires videoconferencing. They believe if they are in position to meet the need (or at least get their customer off the ground), they will not risk loosing the network business to another provider. The various network possibilities are covered in better detail in other chapters of this book.

There are a number of consulting and trade organizations well positioned to provide a complete service. Most are able to assist with the up-front research into the needs of the potential user base, help with vendor selection and network design and overall project management. Most consulting organizations do not have system integration capabilities, but are well versed regarding available integrators and their various specialties.

If you choose to hire a consultant, the money you spend will assure you do not waste time reinventing the wheel, that your research is complete, your cost justifications are close to the mark and the final product is the closest possible fit.

This is not to say a consultant is mandatory. But a thorough knowledge of audio, video and network systems, along with strong facilities skills (or available resources) will be required. If these resources are available within your organization then a team can be built. However, without previous videoconference system design experience, the risk is high that more time will be spent educating the team than moving forward to implementation.

SUMMARY

There is no doubt that videoconferencing is here to stay. There are close to 15,000 systems installed worldwide, about three quarters of which are in the United States. The growth rate is around 50 pecent per year. The digital network infrastructure is in place and continuing to reach into more rural areas. Equipment costs continue their rather steep decline, and picture quality is improving. More and more, videoconference systems are viewed as a tool that reduces the cost of doing business and increases productivity.

The world is shrinking rapidly. I have spent time in London, Hong Kong and New York City, all in the same day by using videoconference systems. Some companies are requiring their suppliers and partners to install systems. Emergency response agencies are using modular videoconference systems (designed for rapid deployment) to get to the site of an emergency and respond quickly. Videoconferencing played an important roll in the federal government's response to the October 1989 earthquake in California (the quake that interrupted the 1989 World Series).

With 15,000 systems installed it is not hard to find designs that range from excellent examples of engineering and careful planning, to those that should be completely removed and replaced. It is always valuable to find a few in your area and visit, to see how a particular company has addressed their need.

Technologies on the horizon, such as the picturephone and desktop based videoconference devices, will continue to move digital compressed video into our day-to-day activities. It is an exciting, growing field full of new opportunities!

Chapter 5

DESKTOP VIDEO

by Patrick S. Portway

■ FROM THE BOARD ROOM TO THE DESKTOP

The single most significant trend in videoconferencing over the last two years (1992-1994) has been the movement of videoconferencing from the conference room and rollabout units to the desktop. The 1980 to 1990 time frame involved an evolution from costly boardroom type conferencing facilities (Figure 5.1.1) that were custom designed to modular small group conferencing systems (see Figure 5.1.2.)

In the late 80s to 1992, small group conferencing systems utilized low data rate video (56 to 128 kbps) associated with switched data services, ISDN rates and modular cabinets that accommodated a limited number of people gathered in a less formal meeting environment.

■ DESKTOP VIDEO

In 1994, videoconferencing became a computer or workstation application. (See the discussion of the merger of computers and video under trends in Chapter 16.) Windows of motion video were displayed on a computer screen allowing two-way videoconferencing. (See Figure 5.1.1) Desktop publishing had been the computer application of the 80s, and desktop videoconferencing quickly became the application of the 90s. This merger of video and computers has been called many different things — multimedia, desktop video production, telecomputing, personal conferencing or desktop videoconferencing. Our focus in this book is on video for interactive communication rather than storage and retrieval of video on a computer or CD-ROM.

Some 43 companies including PictureTel, VTEL, CLI, Intel, IBM and AT&T have offered some kind of two-way interactive desktop videoconferencing (as of 1994). While many of these products initially focused on linking desktop units to room systems, there is much more involved here than simply video.

■ COLLABORATIVE COMPUTING

Along with the ability to link two parties visually, the computer-based products also offer a variety of collaborative computing applications or groupware capabilities.

Shared Whiteboard

A shared whiteboard feature is a simple shared space for working together where parties at both ends can add to the material as if they were both written on by the same whiteboard.

Table 5.1.1	1980-90 Board Room Facilities	1990-94 Small Group Conferencing	1994 Desktop Videoconferencing
Room and Equipment Cost	$300,000 - $1.2 Million	$30,000 - $100,000	$1,000
Data Rates	1.5 Mbps - 384 kbps	56 kbps-128 kbps	64 kbps-128 kbps ISDN
Cost Per Hour	$2000-$100/hr	$1/min-$120/hr	$1/min or less

Shared Application

Applications sharing involves sharing a document in WordPerfect, a spreadsheet or any other application that though resident in one computer can be altered from the remote computer.

■ ANALOG VS. DIGITAL TO THE DESKTOP

Desktop video is not a brand new idea.

Analog video as we've noted earlier is a very broadband signal. It has required coax or base band local area networks in the past to distribute analog video.

There are broadband local area networks that are capable of carrying analog video such as Datapoint's Minx Video Network or IBM's F Coupler.

The Datapoint Minx Video Networks have been used for years to support the Datapoint MINX 2002 terminal utilized for local video communications in several existing desktop video networks. (See Figure 5.1.2.) This local video communication can be extended to wide area communications through a shared video compression codec.

The IBM F Coupler provides analog video modulated on the coax cable like a cable television system that serves your home. Each terminal must have a video card in order to tune into 70 channels of television available over the system. (See Figure 5.1.3 for an example of a desktop videoconferencing system.)

■ DIGITAL VIDEO

Another early option was to equip each terminal with its own video codec. The low cost CLI Cameo System introduced in 1992 (see Exhibit 5.1.2) is an example of this approach. Video is transmitted as a compressed digital data message from terminal to ter-

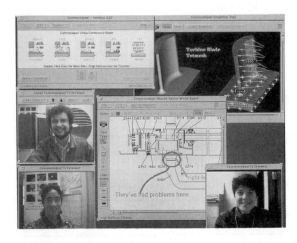

Figure 5.1.1

minal and decompressed at the receiving terminal. The 1992 CLI Cameo was a $5,000 device.

A Codec on a Board

In 1994, with the development of faster computer chips like the Intel 486 and Pentium chips, video compression could be done in a standard PC with the addition of a board and some software.

The basic idea behind the chip set takes the once bulky and expensive video codec from its box or room unit and assembles the components onto a computer board using computer processing chips. The result is not only more practical but also more affordable. CLI's John Walsh says it is possible today, just not as easy as the average user might like. The codec can be assembled with different parts depending on the application and needs of the user. A vendor might buy an entire chip set from one chip maker or several chips from different sources. The vendor assembles the chips along with proprietary chips for specialized features into a "set." The variety comes from the target markets of the various vendors.

For example, a vendor wanting to provide a low cost basic codec to a broad market might have a chip set consisting of an encoder chip and a decoder chip supporting Px64 (because to serve as broad as market as possible he would want to make the chip set as functional as possible) and any other major functions the vendor wanted to support. However, for a high end application, a vendor might want to market a higher priced, more specific set. Its encoder chips might be able to handle Px64, a proprietary algorithm and MPEG I and II, might include special processors for high resolution video and probably would include high end pre- and post-processing chips. The vendor wanting a codec offering high bandwidth or other options might fill in what was missing with their own Applied Specific Integrated Circuit (ASIC) chips. Of course, the vendor with the latter chip set would want to insure he had a large applications market to cover the cost. The options available or not available in a chip set are a trade off for both the vendor and the user.

The Chip Set and the Computer

Ultimately, says John Walsh, Vice President of at Compression Labs, Inc, there were fewer and fewer chips to do the same tasks. Their functionality has begun to be replaced by software and the processors already available in computers.

At TeleCon XIV in October 1994, a company demonstrated full compression and full decompression all by software.

Underneath it all, H.261 and other standards such as H.221 for the communications frame structure will comprehensively support video in the computer. It will be the convergence point for all other elements necessary such as compression, frame structure and connection. New developments in local area networks and backbones such as ATM and FDDI II will bring these technologies together making the computer LANs video friendly.

■ LOCAL AREA NETWORKS

In the distribution of video to the desktop terminal using a local area network, we must keep in mind that LANs were never designed to support real time information flow like video. LANs are packet oriented and designed to be used as a shared resource; anyone at anytime can access the network. Due to this flexibility, the network adjusts to accommodate its users and cannot regulate the flow of information packets. Packets may be moving at one speed and as users tap in or out of the network, the speed may change, depending on the demands of the other users. An application requiring a real time flow of information, such as video, may be subject to a broken connection for an unpredictable amount of time. For stored or tape delivered video or the retrieval of video information, buffering may be built in making it somewhat acceptable for training applications, but for a video call with the CEO to Japan, the unreliability of today's packet based LANs for desktop video is too unpredictable.

What is required is a network that can be expected to provide a certain quality of video and a reliability that the information will be delivered over the network. In order to achieve this, three criteria were recommended by PictureTel's director of research, Richard Baker: bounded latency, guaranteed throughput and isochronicity. For a company purchasing a LAN for video usage, look for these characteristics currently being focused on by vendor R&D.

Bounded Latency

Since local area networks are shared resources delivering packets of information, the first problem to be solved by tomorrow's networks would be the regulation of the information flow. The predictable, even flow of the packets is essential to getting a regular, reliable desktop signal to the desktop. Being able to predict the effect of "network hogs" on the network is the first step in the real time process.

One way video distributed today uses built in buffering patterns to give the video a built-in amount of delay. This technique helps in establishing pre-

dictability for stored video, such as tape delivered video.

Throughput

Throughput, whether or not the packet of information being sent makes it to the receiver, is not guaranteed on a local area network. Networks rely on protocols to let you know whether the information was sent or not. If usage is high or if for some other reason the information did not arrive, it is simply resent. This kind of delivery is sufficient for packet information but not sufficient for real time video. Compression only sends bits that are needed and since additional corrections are sent on the assumption that the decoder has received the original information, if any loss or information occurs, the picture falls apart.

Isochronicity

Isochronous, or nearly synchronous, information is a necessary component of video not present in local area networks. While phone networks, for example, are synchronized to one clock for real time voice transmission, LANs have no synchronous system due to their packet orientation. For video, LANs must guarantee some form of reasonably constant delivery.

Under Development

None of these needs for video are currently available on any one network. New protocols and technologies are under development and will emerge in many forms soon, but any user looking to invest in desktop video using a local area network should ask their vendor what they can provide regarding buffering, throughput and regular flow of video signals.

Under development today, two main paths seem to be the most trod in the race to give users the video quality they expect with the reliability.

Some vendors are concentrating on trying to build a more accommodating network. The application is much more tolerant of poor service from the network if the quality is good, studies have shown, however, the threshold for productivity requires a very steady and reliable picture. In order to keep the video interactive, researchers are working on the latency. If small buffers are developed, improving the picture quality, there is a chance the packets could get overrun. Occasionally, the channel will get lost causing the picture to freeze or worse yet, the audio to go. At this point, pictures have not been too acceptable. Use of the network itself is a factor in this problem as well. A lightly loaded network should be more reliable. Products are expected in the LAN area in the next one to two years.

The second and possibly more popular path is being taken by researchers looking for what works right, not wrong, on the network. Concentrating on making the network more reliable, engineers are developing transmission protocols and backbones such as Asynchronous Transfer Mode and FDDI II to include all service criteria. For network use, the technology is looking at backbones and the network itself for video and other reasons as well. Once that happens, we can have a possibility of reliable, predictable, high quality live video directly over the network. Products are expected in a two to three year minimum.

■ COMMUNICATING TO THE DESKTOP

Video communications is moving from the boardroom to the desktop, and the vehicle accelerating this move is the desktop computer. One of the reasons the TeleCon conference had Professor Nicholas Negroponte as a major speaker in 1991 and 1992 was to expose the professionals in our field to Negroponte's vision of the future. Professor Negroponte as director of the world-famous MIT Media Lab has been the principal prophet of the merger of computers and video. He confronted the audience in 1992 with a prediction that other computer-based video standards such as Intel's Indeo video would obsolete H.261.

■ STANDARDS AND INTEROPERABILITY

Initially the computer manufacturers also believed they could use their own video standards for two-way interactive video and ignore H.261. They based their market on about 140 million installed PCs. However, early adopters of desktop videoconferencing in large organizations were concerned about being able to link their desktop units to the existing 30,000 videoconferencing room systems.

Intel offered a forum of interoperability based on VTEL's compression technology and H.261 in late 1994. However, users could only implement videoconferencing and they lost the collaborative computing functions in this low cost approach. VTEL and PictureTel offered full H.261 capability and PictureTel offered Lotus Notes for collaborative computing functions.

A variety of solutions will evolve from a number of sources.

The Personal Conferencing Specification effort, focused on ISDN and was lead by Intel, while the Packet Switched Forum of companies focused on LAN-based video. Among the most promising approaches is transcoding where dissimilar systems exchange each others decoding software and are able to decode each others' transmissions.

A much more exhaustive discussion of desktop videoconferencing is included later in this chapter and a complete comparative study of all desktop products is available in ABC's Desktop Videoconferencing Report.

■ VIDEOPHONES

A discussion of videoconferencing would not be complete without a discussion of the evolution of the video telephone, a device that Andy Rooney categorized as one of those future products that just never happened.

In 1964, at the New York World's Fair, AT&T demonstrated the picturephone shown in Chapter 1. Ever since 1939 when broadcast television came along, the American consumer assumed that some day the telephone and television would be combined into such a videophone.

The public automatically included the video telephone in their view of the future and it seemed perfectly natural for the Jetson's, our mythical family of the future, to have one in their home. So, Andy Rooney asks, why hasn't it happened if the prototype was shown in 1964. Why haven't we got one in every home by now? Skeptics in our industry call the videophone more vapor than reality, a product idea that sounds good but fills no known need. What the critics overlook are examples of other products that apparently served no real need in the view of many people of their time.

When Alexander Graham Bell offered his invention to the Western Union Company, their experts could not envision a need for people to talk to each other when they could get a printed message so easily by telegraph.

When Chester Carlsen brought Xerography to IBM the experts could not understand why offices would want poor quality copies when they had excellent offset printing available.

Empowering the Individual

The factor that drove these two devices to success was a development that made it technically possible to place in the hands of an individual capabilities that previously were only available from a centralized source as a service.

Yes, you could walk down to the company printing plant and get excellent quality offset printing but it was a lot of trouble for a few copies. Xerography didn't replace printing; it replaced carbon paper and stencil machines.

The personal computer delivered the power of computers down to the level of individual use. Central computers can still do more than a PC, but a PC is ours to use as we want. A telegram still has tremendous impact as a permanent record of communications, but we're free to say anything on the phone, most of which would never warrant the trouble of sending a telegram.

The videophone is to videoconferencing what the telephone was to the telegraph, what Xerography was to printing and what the PC was to the main frame computer. The videophone is a lower quality device that puts into the hands of an individual a tremendous new technical capability that appears to the experts to have no real critical application other than to empower the individual to personally communicate with video.

Where Does the Videophone Stand Technically?

When we discuss videophone technology, you need to distinguish between digital videophones operating at low data rates over ISDN or switched 56 kbps circuits like desktop videoconferencing systems on computers from a video telephone that operates over the analog public switched telephone network. These video telephones are often referred to as analog videophones. They are really a digital phone operating at data rates that can be transmitted over a modem attached to a standard analog phone line.

Digital Videophones

Digital videophones operate at 64 to 384 kbps data rates and utilize ISDN (Integrated Switched Digital Network) or similar dial-up switched data services at 56 kbps or multiples of 56 kbps. These low data rate devices are priced at about $5,000 depending on the capabilities and data rates at which they operate. We'll call these "low data rate" or executive videophones as opposed to a "video telephone" built to operate over an analog standard phone line by a consumer. The latter is the low cost video telephone for the consumer market.

Consumer Video Telephones

A motion video telephone with a built in modem that can simply plug into a standard phone jack in the wall is what AT&T implied when it showed the picturephone in 1964. Or at least Andy Rooney and I thought so. But if we are going to be limited by the capability of a phone line to carry data, what can we expect as an available data rate over the public switched telephone network?

Transmission

A standard copper phone line modified with the loading coils taken out of the line can transmit 250 kbps over short distances. A normal phone line can carry data at between 9.6 kbps and 19.2 kbps without data compression, but if the signal goes through several switches it can degrade significantly, particularly in overseas calls. So, what can you count on as available data rates in a video telephone call?

Dennis Bonnie, operations manager for GPT Video Systems in the United Kingdom, felt 9.6 was

Figure 5.1.2 Datapoint Minx 2002

Figure 5.1.3 Desktop Videoconferencing System

the rate you could reasonably target internationally. Dennis also made some very good points about the importance of graceful degradation if you tried to operate at 19.2. Modems designed for this higher rate data transmission automatically default to lower rates when 19.2 is not available. Dennis pointed out that the result could be devastating to an already highly compressed audio and video transmission if the transmission line dropped from 19.2 to 9.6 suddenly.

If we accept these limitations, we have the AT&T videophone, a low cost video telephone ($1,000) product that failed because of poor picture quality.

What Would a Successful Video Telephone Look Like?

A description of a typical product might include a three to six inch color monitor screen. The detail or resolution of the image would not allow for much in the way of graphics or alpha numerics. It would primarily deliver a moving picture of the person you're talking to on the line. Given a limit of 9.6 kbps in the non-ISDN world, about 6 kbps would be assigned to the video and 2.4 to a highly compressed audio.

If we move to a 19.2 kbps model for the transmission networks, we could envision a 16 kbps video image as Professor Negroponte's group at MIT is focusing on in their research. Regardless of what you may think of the usefulness of such a product, all the necessary research and development is completed and the necessary chips already exist to build a low cost video telephone. However, it is much more likely that the PC will become the platform that will make personal video communications practical.

Special thanks to Renee Wilmeth of ABC, John Walsh of CLI, Richard Baker at PictureTel and Dennis Bonnie from GPT Video Systems.

John Walsh, Senior Vice President of Planning and Strategy for Compression Labs, Incorporated, is a widely-recognized leader in the telecommunications industry. He is responsible for the corporate business planning process and for new business development in the personal video segment of CLI's business. In his previous position as executive vice president of sales and marketing for videoconferencing, Walsh was responsible for all strategic and tactical marketing, product planning and business development, as well as domestic and international sales. Prior to joining CLI, Walsh served as vice president of product management in the Computer Systems Division of AT&T Information Systems, responsible for marketing computers and communications equipment and enhanced network services in the Pacific region. Before joining AT&T, John spent fifteen years in management positions at IBM. Walsh holds a master's degree in Electrical Engineering from Rutgers University and a bachelor's degree in Electrical Engineering from the New Jersey Institute of Technology.

DESKTOP

■ VIEWS ON THE FUTURE OF VIDEO TELEPHONES

Our discussion will focus on the factors that make personal video a reality. Before advancing too far in the argument, it is critical that we establish the importance of visual communication.

Human beings are visually oriented. From the cave walls in Lascaux, France which served as canvas for early man some 40,000 years ago, to today's demand for graphics-based interfaces, images not only remain the most effective communication medium, but pack an information wallop when compared with written words or conceptual ideas.

In fact, according to authors David Lewis and James Green, who have written about improving memory, the mind retains vivid or unusual images better than words, numbers or abstract concepts.

Given the sophistication of the human visual system, our predilection for images is hardly surprising. According to Richard Mark Friedhoff in "Visualization, The Second Computer Revolution," not only is a large proportion of the brain devoted to vision and visual analysis, but the information-carrying capacity (the bandwidth) of our visual system is greater than any of our other senses.

■ FACE TO FACE COMMUNICATION

Of all the pictures and images we are exposed to on a daily basis, the human face reigns supreme as an information source. When we talk face-to-face with another person, we get as much — and arguably more — of our information from that person's facial expressions than from his or her words or voice qualities combined, according to Mele Koneya and Alton Harbour in "Louder Than Words ... Nonverbal Communication."

In fact, psychologists have determined that when we talk to someone face-to-face, only about seven percent of what is communicated is conveyed by the meaning of the words themselves. Another 38 percent comes from how the words are said — the intonation we use. That leaves fully 55 percent of communication to take the form of visual cues.

The problem is that in today's global business environment — not to mention our penchant for moving hundreds or thousands of miles from our families — face-to-face communication has become a costly, time-consuming and oft-omitted practice. We turn, instead, to group videoconferencing, the telephone, fax machine or modem to meet our corporate communication needs, and send the occasional snapshot to friends and family when words simply won't do.

■ BRINGING THE "PERSONAL" BACK TO COMMUNICATING

The obvious answer to this dilemma — and a solution proffered by industry innovators since at least the early 1960's — is the picture phone. In addition to linking people located around the globe via voice, the picture phone also has long promised to deliver interactive images of associates, customers, experts and friends.

Despite nearly three decades of hype and bally-hoo, videophones have remained about as accessible to individuals at home or business as Star Trek's "Beam me up, Scotty" transporters.

In 1992, however, all this changed. In the first quarter, AT&T introduced a consumer videophone for the home market. Likewise, Compression Labs, Inc. (CLI), an industry leader in group videoconferencing, introduced the industry's first affordably priced, computer-based business videophone.

Both of these first-to-market products are based on Compressed Digital Video (CDV) technology. In comparison with voice or data, video images require a huge amount of data to be transmitted very quickly. In fact, video consists of 90,000,000 bits of information, and in order to send video over normal analog telephone lines, that data must be compressed to 10,000 bits — representing a 99.98 percent reduction of the original information. Innovations in video compression technology have dramatically reduced the amount of data required to transmit video images and systems-level innovations have reduced the size of a video codec by several orders of magnitude — from a refrigerator-sized device just ten years ago to a PC board-sized codec today. It is because of these innovations that inexpensive, personal video systems are now available.

The arrival of personal videophones for office and home is expected to fuel growth in the interactive video communications market, and to create a new product category that potentially will offer individuals in the 1990's as much utility as the telephone/fax combination brought to market in the 1980's. While the focus in this chapter is on the personal video market, it is valuable to examine the group videoconferencing market because it serves as an analog for predicting what might happen in the personal video industry.

■ COST VERSUS UTILITY THRESHOLD

It is important to note that the market success of any application is really based upon the achievement of the cross-over point of the perceived utility of that application versus the cost to acquire it. For example, in the early 1970's and 1980's, interactive video communication remained a high art of sorts. Using expensive, high-bandwidth transmission media and equipment, a select group of corporations, government agencies and universities began setting up videoconferencing rooms so that groups of people in geographically dispersed locations could share ideas and communicate strategies without having to travel to one anothers' sites. These high-end users could justify the expense of such systems by leveraging savings

in travel, lodging and increased worker productivity.

In the past decade, technology advances have driven down costs, making group videoconferencing more economically practical. As a result, many businesses and institutions now employ group videoconferencing as an integral part of their corporate communications strategy. Videoconferencing has reached the cost versus utility threshold for a large number of companies. Given that premise, it is valuable to examine the cost versus utility equation for the personal video market.

■ PERSONAL VIDEOCONFERENCING: WHY NOW?

Clearly, group videoconferencing technology has been able to answer users' utility and cost requirements. The same, however, has not been true of personal videoconferencing technology. In fact, until CLI and AT&T's recent offerings, no product has been able to deliver high picture quality and wide-scale connectivity at a non-prohibitive cost to individuals desiring to communicate one-on-one. The cost versus utility threshold has simply been too high.

We can already speak on the telephone person-to-person; we can send faxes back and forth readily and we can share data on our workstations. The single element that is added with personal video is the ability to see the other person. This may be very important for some people, but debatable for others.

As a result, the few products that have hit the market — from AT&T's 1964 PicturePhone to shared desktop systems costing up to $10,000 per seat — have failed to catch the imagination of users or fuel a personal video market segment. It is clear that personal video must reach a very low cost point in order to generate a reasonably sized market.

Even if you accept the importance of visual communications, there are the other elements that must be achieved before personal video has utility. First, you must have connectivity — whether in the form of switched digital services, local area networks or the public switched network. The other critical element is picture quality. Is the picture quality good enough to accomplish the goals of the interaction?

■ EARLY ATTEMPTS AT THE VIDEOPHONE

Now let's examine some of the earlier attempts to accomplish personal video and see if they achieved the appropriate elements in the cost versus utility equation.

The world-famous 1964 World's Fair Videophone is an excellent example. The connectivity was relatively low, as there were not very many places in which you could connect the Videophone. The picture quality was medium and the cost was high.

In the mid 1980's, shared resource video telephony was attempted. In this scenario, an expensive video coding device was shared across a large number of desktop workstations. A good example is the Datapoint-Minx system. Within the facility, users sent video over analog coax, and therefore, the video did not require compression and connectivity was good. Once a user tried to videoconference outside of the facility however, he would have to compete for the use of the codec. This made effective connectivity poor. The average price per workstation was about $7,000, making the cost point medium. The picture quality was high within the facility and low outside the facility.

Another attempt at video telephony came in the form of still-image systems. These sent a still image frame every 10 seconds. The connectivity was high because the system used the public switched network. The picture quality was low, and the cost was low.

These three examples illustrate the premise that a certain balance must be achieved in the cost versus utility equation. Since the 1964 World's Fair, the picturephone still is not widespread 30 years after its introduction. We can assume that the appropriate balance was not achieved for the shared resource and still image systems.

■ COMPONENTS OF A PERSONAL VIDEO SYSTEM

From more than two years of experience in the desktop videoconferencing market with its Macintosh based Cameo personal video system, CLI has identified four components of a personal video system that must be met to achieve widespread usage. These requirements are:multipurpose applications, network options, interoperability and low cost.

Multipurpose applications

To justify a desktop videoconferencing solution, a user must have a compelling need for both real-time, face-to-face communications with distant group or team members and collaboration on documents, presentations, graphics or other computer applications without the added cost of real-time video communications. On the other hand, if videoconferencing is needed primarily to improve staff communications, streamline project reviews for widely dispersed groups, provide training or reduce travel costs, then a low-cost (less than $20,000) small-group videoconferencing system is likely to be a more appropriate choice.

Network options

Desktop videoconferencing solutions must ultimately support a variety of network options including LANs, ISDN, Switched 56 digital services, analog transmission (POTS), and even new alternatives such as networks based on Asynchronous Transfer Mode (ATM) technology. Because of the expense and effort required to install digital services such as ISDN and Switched 56 lines, many corporations favor desktop systems that operate over LANs or analog telephone lines. However, using existing LAN-based technology, bandwidth-intensive applications like videoconferencing create numerous problems for heavily-used LANs — in areas such as packet prioritization, synchronization and latency — that must be addressed for desktop video solutions to work.

Interoperability:

The third requirement for desktop videoconferencing systems is interoperability with group videoconferencing systems. Desktop video users need easy and transparent access to critical group videoconferencing discussions. Interoperability requires standards that not only provide video, audio and other protocols for real-time videoconferencing, but also define how collaborative computing software will operate when handling tasks such as file transfers, document annotation, whiteboard usage and application sharing. The current set of standards, as defined by the ITU-TSS, focus on the issues and requirements of group videoconferencing and do not address desktop videoconferencing requirements.

Low cost:

Not surprisingly, affordability is perhaps the most crucial requirement for potential users. Studies by leading industry analysts such as the Yankee Group, Gartner Group and Personal Technology Research show that the majority of corporate MIS and telecommunications managers will spend no more than $1,000 on a single-user video communications solution. A small percentage of corporate buyers are willing to purchase a desktop videoconferencing system priced between $1,000 and $5,000. But, virtually no one will expect wide deployment at approximately $5,000 for single-use desktop solution.

■ A MULTIFACETED MARKETPLACE

With no single system able to meet all the

requirements for wide user acceptance, the emerging market for desktop videoconferencing more closely resembles the multifaceted personal computer marketplace than the group videoconferencing market with its handful of dominant vendors.

Today, there is a proliferation of desktop videoconferencing products. There are products that operate over a variety of computer platforms and operating systems (IBM-compatible with MS-DOS or Windows, IBM PS2, Apple Macintosh, Sun, Silicon Graphics, and more), and use different audio and video compression algorithms , ITU-TSS's desktop video products can be divided into two categories: those that conform to existing audio and video standards for group videoconferencing (H.320-based) and those that offer alternative approaches designed to better meet the unique needs of desktop users.

H.320

Developed by the ITU-TSS, an international standards body, H.320 is a family of worldwide standards that primarily pertain to group videoconferencing systems. H.320 has been adopted and integrated by many vendors into CLI ensuring interoperability among other H.320-based group systems. In the past year, a relatively small number of expensively-priced desktop systems have been introduced that comply with the H.320 standard.

These H.320Compliantdesktop videoconferencing systems offer two advantages over non-H.320 based systems. First, they offer interoperability with an installed base of more than 30,000 H.320-based group videoconferencing systems around the world. (Source: Personal Technology Research.) Secondly, when used with a digital bridging device, H.320-based desktop systems can participate in multipoint videoconferences, in which multiple group and desktop videoconferencing systems are simultaneously connected, allowing the video to automatically switch from location to location as people speak.

The biggest barrier to widespread usage is the high cost of H.320-compliant desktop systems. The average price for an H.320-based system ranges from $4,500 to $15,000 (excluding the computer,) greatly exceeding the $1,000 most users say they are willing to pay. An additional consideration for the desktop market is that H.320 systems are designed for use over switched digital networks, not LANs or analog lines, over which most desktop systems today are connected. Last, there is a lack of data conferencing standards, meaning that H.320-based desktop systems from different vendors may not be fully compatible for sharing files or working collaboratively on documents.

Alternatives to H.320

The majority of desktop videoconferencing systems in the market today do not comply with the H.320 standards described above. Alternative products exist that cost far less and offer much greater network flexibility than H.320-compliant desktop videoconferencing systems.

These non-H.320-based systems can be purchased for as low as $1,000. They operate over digital networks, analog networks, local area networks or a combination of the three.

The only real disadvantage of these nonH.320 desktop systems and group systems.

In addition, the PCS workgroup, a consortium of computer hardware, computer software, communications and videoconferencing companies, including CLI, is working toward standards for desktop videoconferencing.

■ THE BEGINNING OF A NEW ERA: LEVERAGING THE VISUAL MEDIA

It is easy to imagine a telephone user who could benefit by having access to an interactive, personal videophone's concerns, an instructor answering questions of a student in the field, a human resource representative discussing company benefits with a new employee, to an author negotiating publishing rights with an editor. In effect, any scenario involving one-on-one communication that cannot be conducted in person is a target for personal videoconferencing.

Along with the recent introduction of powerful Pentiumbased Macintosh systems, industry analysts estimate that there will be more than 60 million high-end computers and workstations capable of supporting desktop videoconferencing by the end of 1995. Once the obstacles created by user requirements have been overcome, desktop video may well take its place along with group videoconferencing as one of the great enabling technologies of the 1990s — resulting in ubiquitous video communications that will restore the personal side of business communications in an ever more impersonal world.

David Boomstein joined Applied Business teleCommunications (ABC) in
January 1994 as the senior vice president for desktop programs. He is a fre-
quent speaker at trade shows and conferences and is an editor of the Desktop
Videoconferencing Report published by ABC. Prior to joining ABC, David
spent five years with Boeing Computer Support Services, Inc on a mission ser-
vices contract for the National Aeronautics and Space Administration
(NASA) at the George C. Marshall Space Flight Center in Huntsville, AL.
As the product development manager of the Information Services Engineering
group, he specialized in investigating and developing new technologies and
applications in multimedia communications, electronic imaging and telecon-
ferencing. From 1984-1988, David was the product marketing manager at
Compression Labs, Inc. In that position he was responsible for project man-
agement on CLI's initial Rembrandt Video System (RVS).David has a B.A.
in Communications Arts from the New Institute of Technology, a Masters of
Professional Studies from New York University and was founding president of
the International Teleconferencing Association (ITCA).

DESKTOP CONFERENCING

■ INTRODUCTION

Desktop conferencing is the fastest growing seg-ment of the teleconferencing market place. Desktop conferencing includes videoconferencing from per-sonal computing platforms, standalone videophones and various forms of data conferencing. Cooperation, the process of individuals working sepa-rately to reach a common goal, and collaboration, the process of individuals working together to achieve a common goal, are both germane to the growth of desktop conferencing. The current growth of desktop conferencing is due primarily to the convergence through cooperation and collaboration of both tech-nical and social factors.

The cooperation between developers and manu-facturers in the teleconferencing, communications and computing industries has led to the development of guidelines and standards that are beginning to allow desktop conferencing systems to communicate as freely as today's telephones. These technological advances are helping the convergence of the desktop's three cultures; telephony, computing and the way people work to conduct their business. Now that computers are powerful enough to provide acceptable video and audio quality, we are finding that more and more computers are equipped with some type of net-work connection. In 1994, for the first time there were more fax modems operational than standalone facsimile units. In 1992, worldwide shipments of modems was 3.5 million units. (Fortune, April 4, 1994.) According to a1994 Communications News

survey, 61 percent of business were looking at imple-menting some kind of desktop conferencing in the next year.

Business practices have advanced to a level where there are very few single contributors in companies today. Workers are both consumers of and providers of services with others in their companies. The "para-digm of interdependence — (where) we can combine our talents and abilities and create something greater together" (Stephen R. Covey. the Seven Habits of Highly Effective People. pg 49) within the workforce along with shorter project time frames and reduced financial resources has led to a stronger dependence on the "telecommunications tool kit". Every worker performs a sort of triage on each task he needs to per-form. Based upon a task's level of urgency, complexi-ty and cost, we dip into our tool kit and either place a phone call or audio conference, fax or email a mes-sage, go on-line to access information, or schedule a video conference. This chapter addresses the tool kit's newest tool — desktop conferencing.

■ WHY DESKTOP CONFERENCING ?

From the end user's perspective it appears unavoidable; to many the question is not will we be doing desktop conferencing, but when and in what form? Countless studies are showing that corporate workers are being taxed. Whether its TQM (Total Quality Management), downsizing or re-inventing the organization, everyone seems to have more work, tighter deadlines and fewer resources. The PC has become almost everyone's personal assistant. The PC is allowing companies to eliminate layers of support staff.

A large segment of office communications is PC generated; reports, memos, electronic mail messages or facsimiles. By some accounts about 90 percent of a senior manager's communications are PC generated. According to Intel Corporation's Jim Johnson, assistant general manager, vice president, Intel Products Group, " ... during the mid to late 1990's, we can expect to see an evolution of the PC from our primary personal assistant to our primary communications tool. ... the goal is to allow us to work together apart ..."

■ APPLICATIONS OF DESKTOP CONFERENCING

There are various applications to desktop conferencing. Initial use by many is an upgraded phone conversation or substitute for the group or room conferencing systems. In a number of the busier conferencing networks, desktop conferencing is seen as a way to displace many conferences where only single users use the rooms, thus leaving the conference rooms for large groups. One of the promising applications for desktop systems is telecommuting. According to a Yankee Group prediction, by the mid 1990's, 80 percent of all employees will adopt some sort of remote work in order to compete in world markets. A Department of Transportation study showed that in 1992 there about 2 million telecommuters. that number is expected to grow to 3.1 to 6.2 million by 1997, about 4.5 percent of the US labor force. Another promising application for desktop conferencing systems is telemedicine. There are a number of hospitals adding desktop conferencing for patient-doctor consultations, teaching and remote diagnostics.

The benefits seen by using desktop conferencing are the ability to "move the work, not the people" and overall save time and resources — people, office space and costs.

■ DESKTOP CONFERENCING COMPARED TO GROUP CONFERENCING

At initial glance there appear to be many similarities between the two types of systems. In fact many of the initial offerings in the desktop arena are group systems repackaged for the desktop environment. There are a few instances where the desktop system, i.e. computer based conferencing system, is supplied in a cart with a large monitor and sold for small group usage.

While there are many parallels between group and desktop systems, to effectively target the office worker desktop systems need to support the different work habits and support requirements.

This article provides a comparison between group versus individual interactions and group and desktop conferencing systems.

The Group

By its nature most groups are not formed spontaneously. When a meeting is required the various individuals needed to participate are contacted, a meeting room's availability is checked and schedules are coordinated. Depending on locations, organizational positions of the required participants and other factors, convening a meeting can be a laborious exercise.

In the group, interaction tends to include numerous formalities. In many cases formal agendas, a good meeting practice, are distributed — sometimes as an enticement to attract participants. Handouts and overheads are created. All of this necessary preparation reduces the spontaneity of and time to complete the required task.

In size, about half of the meetings scheduled using group videoconferencing systems are multipoint and include small to medium sized groups at each participating location. In many instances group system users have upwards of 8 rooms per multipoint conference. Group conferences tend to be fairly long — in many cases over two hours in duration.

The most common metaphor for group conferencing systems is the group face-to-meeting. Group videoconferencing tends to be both an adjunct to face-to-face meetings as well as a travel substitution. Many feel that the extra amount of preparation done before videoconferences enhances the outcome when compared to face-to-face meetings.

The Individual

The individual's requirements are quite different from the group's.

The individual tends to be more spontaneous in his interaction with others. In many instances the need to interact is not preplanned but conceived based upon the current task at hand.

Interaction with others from the desktop tends to include one to three other parties in most instances. The PBX's three-way calling capability is one of the heaviest used features by the office worker.

The most common metaphor for desktop conferencing is the phone call. Many current users of desktop conferencing feel that as availability of systems increases, the desktop conferencing system will begin to replace normal phone calls.

Group systems by design are usually provided with larger monitors to support viewing by more participants and tend to look similar in packaging to the corner video tape player. Many group systems tend to operate at higher bandwidths to provide a better perceived image quality. Due to higher costs group systems are usually shared corporate resources. Following the face-to-face meeting metaphor many group systems mirror the tools available in a basic conference room — audio, video and freeze frame graphics.

Desktop systems tend to packaged within or as adjuncts to the desktop computer. Supporting only one or two simultaneous users, the desktop system tends to have smaller displays sometimes with lower resolutions than the group system. Since many office workers will operate from open cubicles, many desktop systems do not have open audio systems and include a headphone or use a handset. Many do use speakerphones. Most desktop systems incorporate some type of computer-to-computer interface to allow collaboration by users. Most users of desktop conferencing systems expect to share computer data during their meetings.

While there are many differences between group and desktop conferencing systems, the differences allow the two types of systems to better suit the needs of users. It can be expected that once users become accustomed to collaboration at the desktop, one of desktop conferencing's strengths, users will want these capabilities included ion their group systems.

■ WHAT IS A DESKTOP CONFERENCING SYSTEM COMPRISED OF?

There are four broad categories of desktop conferencing systems; hardware-based motion video systems, software-based motion video systems, collaborative computing systems and hybrid systems. The majority of the systems to be discussed in this chapter are based upon personal computing platforms. There are a number of systems being sold as standalone picturephones that will be also addressed.

The current wave of desktop conferencing systems began showing up on desktops in 1990. Feature wise, most were designed to parallel the functionality of the group conferencing systems. While many were limited to point-to-point audio and video only, some included a freeze frame graphics capability. In an attempt to utilize some of the computer's features, many early systems were later enhanced to included a snap shot capability that utilized the computer's video display card and clipboard.

Early systems operated over switched circuits; either analog phone lines using modems or digital circuits including ISDN andSwitched 56. Upgrades to these systems allowed sharing of snap shots with the distant party during a conference and the transfer of files during a conference. Due to bandwidth limitations and simplistic multiplexing schemes the video and audio each used a single 56 kbps portion of a digital transmission circuit. During file transfers also know as ftp (file transfer protocol), motion video was frozen.

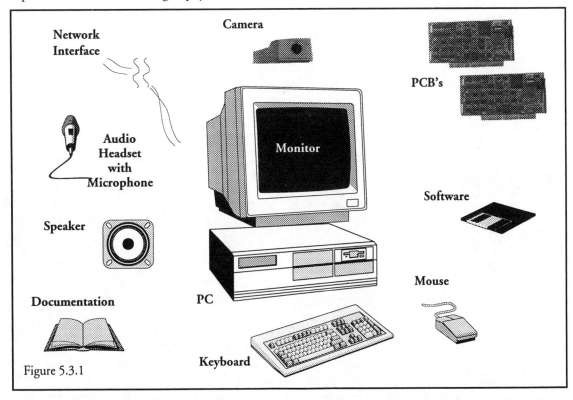

Figure 5.3.1

Desktop conferencing systems were initially comprised of the computer platform; PC's running DOS or Windows, Apple Macintosh or Unix workstations like Sun Microsystems SPARC stations, add-on video capture and display cards that grabbed incoming analog video from a monitor top camera, an external video coder, and a communications interface. (Figure 5.3.1)

In the case of some ISDN based systems, an analog telephone plugged into the computer's ISDN terminal adaptor card and was used to either supply the conference audio or act as a standalone ISDN telephone.

Early adopters to desktop conferencing experienced high costs. One manufacturer's system requiring a computer with a specialized video display card and monitor cost $5,000 in addition to the camera and outboard coder $1,500 and ISDN interface card $1,500, a total of $8,000. Today, systems of similar capability cost about half ($2,000 for the conferencing systems, $2,000 for the computer) including the cost of the computer.

While these early systems functioned, they did not provide the best video quality (112 lines vertically with 128 pixels per line) and did not integrate well into the office desktop: most implementations had the user's "normal" phone connection for everyday phone conversations in addition to the conferencing system's phone for use while on-line. At times, users became confused by ringing from the two phones and the computer.

A second approach taken by a number of manufacturers and system integrators to supply desktop technology was to use a low level of computer integration. In this approach the video codec, audio subsystem and transmission interface were all housed in a chassis that sat beneath or next to the desk. The computer was used simply as a display device. Since these were simply a repackaged group system the quality and cost were high, while the level of computer integration was low.

While the early conferencing systems were being promoted by the teleconferencing system's leading video codec manufacturers, development was under way by many data communications, computer and software companies to develop hardware/software hybrids, software-only coding schemes and data sharing approaches. The current trend in desktop systems is to migrate away from the specialized hardware coder/decoders to software based approaches.

In the first half of 1994, the initial step in this migration has been to integrate a hybrid hardware compression engine with software decoding. Many developers of software-only approaches have found that the amount of processing power necessary to provide acceptable video — resolution and motion quality, was not obtainable on a standard 386/486 PC. They found that the combination of add-on boards in the PC with chips dedicated to coding and using software to decode provided the image quality while running on existing computers. According to some estimates the hybrid approach leaves approximately 150 million PC platforms suitable for desktop videoconferencing.

The later half of 1994 shows extensive interest in software-only systems. Software-only systems are desirable because the elimination of specialized hardware significantly reduces costs. The draw back is that many of these systems require top of the line processors like the Pentium and the PowerPC. The good news is that the cost for many of these computers begins at below $4,000.

■ WHAT VIDEO QUALITY IS ACCEPTABLE?

Acceptable video quality is purely subjective. Depending upon the task at hand — general business communications, distance learning, telemedicine, or crisis management, users may rate a particular system's video resolution and motion handling capability differently. It appears that the more discretionary

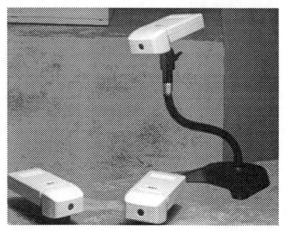

Figure 5.3.1 and Figure 5.2.1 Desktop Videoconferencing Cameras

the use of the system, the more critical some users tend to be.

Depending upon the type of system — hardware, software, hybrid, the quality of the coding and compression scheme employed and the bandwidth available, video frame rates range from a low of 1-3 frames per second (fps) in some of the analog/modem based systems to a high of 15-30 fps in some of the baseband analog and high end switched circuit and packet based systems. Many system designers assume the on-screen image will be a single head and shoulders shot of the participant seated at a desk, the assumption is that motion will be limited thus allowing the system to operate at its maximum frame rate most of the time.

A May 1994 survey of teleconferencing professionals (ABC Advanced Seminar) asked "What minimum level of video quality would be acceptable?" 66 percent of the respondents answered that the ITU/TSS' (International Telecommunications Union/Telecommunications Standards Sector) current H.261 standard of QCIF to FCIF (144 lines x 176 pixels per line to 288 lines x 352 pixels per line) was a minimum. Other factors that effect a user's quality assessment include the size of the video window on the computer screen and the overall color, brightness and contrast of the video image. Video window size in many systems is a factor of the computer's CPU power, video display card and monitor resolution. Picture "fidelity" in most cases is effected by the location where the system is being used. Standard office fluorescent ceiling lighting systems are not color balanced for video cameras and rarely are the lighting fixtures positioned to provide an even dispersion on the conference participants.

■ USAGE OF DESKTOP CONFERENCING

Practitioners in the business and the conferencing fields continue to debate the value desktop video as a tool in the office. The question of motion video's necessity continually arises. When a group of teleconferencing professionals were asked, "Is there a value to the motion video if I can do collaborative computing?" 79 percent responded that there was value to the video.

Many early adopters of desktop video find that they use desktop video as a replacement to their phone. When the group of teleconferencing professionals were polled on expected usage, 60 percent expected 1-2 hours of conferencing per day, assuming cost of use was not an issue.

During a Communications Week user call-in poll (July 18, 1994), 61 percent of the respondents felt there was a need to widely deploy desktop video-conferencing products in their enterprise networks.

■ CAMERAS

Most desktop conferencing systems sold today are provided as packages that include the camera and audio unit. Most cameras are single chip CCD (charged coupled device) and are designed to mount on top of the pc's monitor. Many include a simple mount that allows some degree of side-to-side panning and up-down tilting. Due to the close proximity of the viewer to the screen, usually 18 - 24 inches, most monitor top cameras are a fixed focal length. Some camera systems provide the ability to switch between an electronic close-up and wide angle view through controls in the desktop system.

Most desktop conferencing systems allow external cameras to be connected to the system. It is quite common to see desktop installations where a higher quality video camera or camcorder is used in place of the system's main camera.

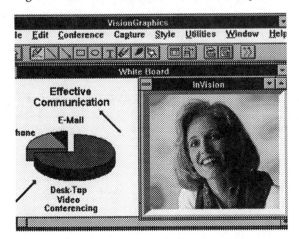

Figure 5.3.3

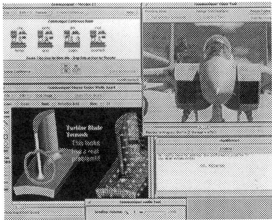

Figure 5.3.4

◼ AUDIO

As in group conferencing, audio intelligibility is necessary to hold a conversation. Many systems have taken a closed mic approach to audio that eliminates the need to account for poor room acoustics and direct feedback caused by the interaction of the microphone and speaker systems. Closed mic implementations range from the use of a telephone handset to standard telephone headsets to custom ear pieces with integrated low profile microphones (Note: show diagram of Intel ear piece). Common to all of these implementations is that the conferencing system does not require any type of echo cancellation or suppression circuitry thus reducing the overall system cost.

Many closed mic systems allow you to convert to an open mic system by using a set of computer speakers and a standalone microphone. Since the speakers and microphone are external to the system, an echo cancellation unit can be added if required. There are a couple of systems currently offered like the PictureTel PCS 100 Live that include a speakerphone configuration and echo cancellation. These systems work well in small to medium sized offices. Many of these systems provide excellent audio quality. Some provide wider band audio, 7 khz, versus a normal telephone line's 3.5 khz.

Where most desktop conferencing system's audio subsystems fall short is the lack of integration at the desktop. Aside from the systems that do not have integrated audio and require the use of the standard telephone for the audio, today none of the systems offered allow the user to have one interface for both normal telephony and conferencing audio. Many users find this lack of integration bothersome and are not satisfied with the requirement to switch between multiple audio systems. It is felt that this is a sign that the systems are not yet integrated into the total desktop technology and work environment.

◼ COLLABORATIVE COMPUTING AND GROUPWARE

Users of desktop conferencing have definite ideas of the capabilities they want from their systems. A key requirement is the ability to share information currently residing on their computers. There were two main forces driving the use of collaborative tools; the trend toward globalization and the shift towards team-based management.

Many companies like Vis-a-Vis offer standalone data conferencing software. Called collaborative computing or groupware, these computer applications allowed multiple users to view and in many cases interact with the same native application files residing on one or both of the conferees computers.

Today the majority of the collaboration is performed through separate groupware applications. Many desktop systems either include their own collaborative tools or have a third party's integrated into the overall desktop conferencing package. Future lab's highly popular TalkShow can be found integrate with four different desktop systems.

Intel's Jim Johnson's assessment on computing today is that most applications are "focussed vertically", towards the single user. Johnson feels that while the long term goal is to provide full workgroup interactivity, an "intermediate step" is required. Products like ProShare address this step by creating an "intersection" with its one-on-one focus. Groupware has become so popular in business that some feel it may become its own industry segment. It is felt that while there will be a place for groupware packages designed for specific tasks, key PC applications — word processing, spread sheets, databases, project scheduling and management and presentation programs, will ultimately be enhanced by the manufactures to operate in a workgroup shared data environment.

The most prevalent collaborative metaphor displayed in desktop systems is the shared whiteboard. Shared whiteboards allow users to see a common screen and annotate or type on them. Many shared whiteboards operate in a "object-oriented" architecture and allow the user to "cut and paste" objects from other applications like spread sheets, drawing programs and word processors into the whiteboard. Once the object has been pasted from the pc's clipboard all conferees can see the image and annotate on it. Intel's ProShare provides a graphical "spiral notebook" with numerous pages and tabs to allow multiple images to be shared. Most shared whiteboards include a tool palette similar to a simple graphic package's that allows writing with a mouse, or graphics tablet if the pc is so equipped, highlight similar to a yellow marker, type and circle or box an object in the whiteboard. The more advanced whiteboards can store objects and annotations a separate "layers" of data that can either be locked to the image for archival purposes — as when an electronic signature is provided on a document, or separated so the original object is left intact.

There are three generations of groupware. The first are basic messaging systems. As standalone applications they may be part of electronic mail or on-line server based scheduling systems. As part of desktop conferencing systems they provide a simple background "chat box" between on-line users. While a number of conferencing systems include this chat capability along with other collaborative tools, most users find them as the sole collaborative tool it is too limited. One use of the chat capability is the ability to provide background communications from one conferee to another during a presentation to a third conference participant.

The second generation of collaborative tools provides "over-time" collaboration. These systems operate in a client-server or server-only architecture. Operating similar to an audio meet-me bridge, users of these applications log into a server in their data networks and interact with each other's applications and data. Some of these systems also require an application package to be running on the users computer, i.e. the client software. Systems in this category usually allow many users to interact with the same application in either real time or on an as needed non-interactive basis. Servers tend to have more storage that a user's pc and a long term "over time" record of all interaction can be maintained if the software allows. Some of these systems like Lotus Notes, the largest groupware system, provide a data replication feature that duplicate files between the various user's servers. This allows users in different locations to interact with servers close to their location, thus minimizing wide area data traffic. When updates to files are made the servers update each other.

The third generation of collaborative applications include systems like the ForeFront Group's Virtual Notebook System (VNS). These systems provide both real-time and over-time collaboration similar to the second generation systems but also operate as object-based information repositories. They are designed to integrated a multimedia open architecture and many will allow pc's with Windows, Macintosh and Unix operating systems to interact with each other's data.

■ NETWORKING DESKTOP SYSTEMS

Many corporate networks are not one but numerous networks. Most companies are operating separate networks for their packet data. In many cases they are operating one network for their X.25 packet data and another multiprotocol router based network for TCP/IP, IPX and AppleTalk computer protocols. In the office LAN environment many companies have their LAN's configured where upwards of 200 desktops are sharing the same 10 MB on the Ethernet.

Collaborative computing, desktop video conferencing and video broadcasting to the desktop will strain many LAN's capacity. In the testing of one PC/LAN-based desktop video system a single point-to-point conference utilized approximately 8 percent of the LAN's overall capacity. If there are only a limited number of conferences taking place overall data activity on the LAN will not be effected, but its easy to see how just a moderate number of conferences, point-to-point or multicasted, during a peak computing period can severely effect the computing of an organization.

In the near term we'll see numerous POTS (Plain Old Telephone System) like ShareVision Technology's ShareView 3000 providing immediate solutions. Everyone has POTS. In most cases it is the least expensive communications alternative and it is ubiquitous. POTS based systems are quickly moving from data rates of 9.6 - 14.4 to include the new V.FAST modems operating at 28.8kbps. Many collaborative applications operate over modems. Systems like TalkShow provide point-to-point and multipoint sharing of files and slides over systems using dial-up and networked based communications solutions. While POTS based systems can communicate to almost anyone, their down side is that they take longer to transmit their images. Many users become frustrated by low video frame rates and long file transmission times. Some manufacturers have configured their systems to use three phone lines to maximize throughput.

Many of the products operate over switched circuits: Switched 56, fractional T1, ISDN. When compared to POTS approaches the additional bandwidth switched circuit based systems provide truly makes a difference in video quality and collaborative response times. Most switched circuit based systems operate at 56kbps to 112kbps. Some using multiple circuits, ISDN or fractional T1 operate to 128kbps, 384kbps or higher data rates.

Basic rate ISDN (2B+D) promises to be the most prevalent approach used. ISDN is currently available in about 70 percent of the nation and it is highly prevalent in many other countries. Companies like PictureTel report that a majority of early sales of their desktop systems are selling overseas into countries where ISDN is prevalent.

It is felt that desktop conferencing may be the application that drives ISDN. 68 percent of respondents to ABC's Advanced Seminar poll responded that their would run their desktop video on either basic or primary rate ISDN. 11 percent of the respondents switched 56 would be their backbone for desktop conferencing and 20 percent of the respondents answered digital packet-based systems like asynchronous transfer mode (ATM) and fiber distributed data interface (FDDI).

Virtually all desktop systems provide on-screen dialing via mouse controlled keypads along with auto-dialing through an address book or speed dial utility. Some of the address books also allow the user to store snapshots of the person in the address book entry. While it is felt that this will aid in quick identification of people in the address book, many users look at it as a non-essential capability that uses additional hard disk storage space.

A number of the packet-based LAN/WAN systems like Bolt Beraneck and Newman's (BBN) PictureWindow and ViewPoint's Personal ViewPoint have the ability to operate over the Internet. While the Internet is limited in its ability to guarantee end-to-end bandwidth, it is predicted that over the long term the packet based systems will go farther to meet both the user and network administrator's desire to have one network supporting the desktop for both computing, conferencing and basic telephony needs.

A benefit seen in the packet approach is that in most corporations and government agencies packet data networks are already identified, budgeted and paid for as an overall corporate expense. this accounting approach lowers the cost each user needs to budget to begin desktop conferencing. Additionally, since the majority of information you will want to share will be created, or is already on your pc or server, the networked approach, versus the telephony approach, provides an instant integration and access to your files for collaboration.

Currently most packet networks do not have the capacity for heavy audio, video and graphics traffic. Whereas the heavy amount of client/server, CAD/CAM, e-mail and other database traffic has identified the need to upgrade corporate networks, the pending impact of desktop video is tending to solidify the requirement and add a level of urgency on that need. LAN/WAN capacity and connectivity are continually improving. Isochronous ethernet (isoEthernet), packet network switching and hubbing arrangements should overcome packet network's "quality of service" (QoS) problems: throughput, error and delay.

IsoEthernet has been developed to operate on user's existing LAN cabling; in many cases twisted wire pairs. In addition to the existing 10 Mbps their LAN provides, isoEthernet will provide an additional 96 B-channels. Instead of an average of 200 users sharing 10Mb on the LAN they will share a total of 16.144 Mbps. IsoEthernet also has the benefit of being modular. Not all desktops need to be upgraded at the same time and additional capability can be added at the switching hub, ie the "phone closet" as required. In a campus environment multiple hubs in multiple buildings can be interconnected. As ATM (Asynchronous Transfer Mode) and other broadband networking techniques become available on the backbone, the switching hubs are designed to integrate in the WAN architectures.

Asynchronous Transfer Mode (ATM) is considered on of the most promising technologies that will aid users of desktop conferencing. ATM is a technology that was developed with the assumption that a single network will carry four different types of traffic; voice data video and image. In many of today's networking approaches separate circuits and switching systems are dedicated to the different types of data;

voice, computer data and video. While this approach is functional, dedicating network resources, ie bandwidth, to non-full period activities is wasteful. In most networks, packets are transmitted regardless if there is actual data to transmit. ATM takes all of the data traffic on the network and breaks it up into small cells of data. If there is traffic to send the cell carries the data. If there is an idle period no cells are transmitted. ATM offers many advantages not present in today's networking environment. Foremost is the ability to utilize one network architecture for both local and wide area networks. ATM has a hierarchical approach that is supported by the equipment providers, telephone companies and computer manufacturers. The approach provides for data from various services including Frame Relay, SMDS (Switched Multimegabit Data Service) and T1 to be concentrated at premises level switches with a 155 MB capacity. The premises switches supply data to the ATM "cloud" at rates of 45 MB to 2 gigabites. Wide area interconnection of ATM switches are based upon DS-3 and SONET (Synchronous Optical Network). SONET operates over fiber optic-based circuits and operates in multiples of 51 Mbps up to 48 Gbps.

One of the biggest limitations to the spread of desktop conferencing is the inability of the different types of networks to easily communicate with each other. Carriers like AT&T, Sprint and MCI have been working to provide connectivity between their network's dial-up services: switched 56 and ISDN. Today it is quite common to receive domestic or international desktop conferencing calls.

■ MULTIPOINT

Since users expect their desktop conferencing systems to operate like their telephones, multipoint is a must. When asked "how important is 3-4 way multipoint video to your anticipated first application?" 57 percent responded it was critical while only 5 percent felt it was not important. In the same poll, 50 percent felt 3-5 other conferencing units needed to be bridged together.

Most desktop conferencing systems operate around a "telephony" model for networking. This is not surprising since many of the desktop systems available are built by the leading codec manufacturers. In this approach multipoint conferencing is conducted through the use of a bridge. the requirement to reserve bridges work well for large conferences, but fall short meeting users requirement for spontaneity. Ultimately, many users want their desktop conferencing systems to have a three-way calling feature similar to a PBX's. Users want the ability to have a "video dial tone" and convert a point-to-point call to a multipoint conference with the same ease as in voice-only calls.

STANDARDS

In the standards arena we are seeing developers and manufacturers in the teleconferencing, communications and computing industries working cooperatively to develop guidelines and standards that work and will be scalable. Companies like Intel are leading the way with the Personal conferencing Specification (PCS). Industry groups are concentrating on audiographics - Consortium for Audiographics Teleconferencing Standards (CATS) and technologists are developing solutions; ITU/TSS - H.320, Packet Video Forum (PVF) and the Internet Engineering Task Force's (IETF) Real Time Protocol (RTP).

The outcome of all this work will result in multiple standards; industry governed or de facto, that are targeted towards specific classes of systems and applications that will have the capability to allow interoperability between desktop systems, low bandwidth portable systems and dedicated room systems. In the near term users will have to rely on bridges, gateways and hardware and software transcoders. The long term vision has software agents operating much like today's fax machines and print drivers that "read" the received packets and automatically interpolate and decode them. How far off is this? There are a number of companies vigorously working towards an interoperability goal. Currently the ITU/TSS H.320 approach is considered the approach to providing interoperability.

While most manufacturers of desktop video will include H.261 compatibility, some are also including proprietary algorithms to have product differentiation. Since not all desktop conferencing systems are H.261 based, today operation is similar to the way facsimile usage was in the early 1980's; almost everyone had access to one, but prior to a transmission you had to ask what type it was and set your unit to be compatible. Users have been vocal enough where a number of manufacturers whose systems were not standards based have come back and committed to upgrading to standards.

HOW MUCH DESKTOP CONFERENCING IS GOING ON?

Most manufacturers of desktop systems are still selling relatively small numbers of systems. They report that most buyers are purchasing small quantities, 3-10 on average. These small quantities allow the early adopters to begin a corporate trial and test desktop conferencing's effectiveness. Most manufacturers report sales of under 2,500 units total with many customers coming back to purchase additional systems to add more work groups. Sixty-four percent of respondents polled at ABC's 1994 Advanced Seminar answered that their company is currently doing desktop video? Of those not doing desktop video, 77 percent anticipated initial deployment of desktop conferencing in the 12 to 18 months time frame.

Chapter 6

Gayle D. Gordon has vast experience with the technically complex, multi-faceted field of communications transmission. Gordon has expansive expertise in the development and marketing of two-way video communication systems and was the Southwestern Bell Telephone Subject Matter Expert for video before his retirement in December 1991. He was the 1991 recipient of a recognition award from the Mid-America Association for Technology in education for outstanding leadership in pioneering two-way video for schools and business in Kansas. Gordon has served in various positions with AT&T and Southwestern Bell for 31 years including many engineering and marketing positions with Southwestern Bell. In 1989, he was assigned to the St. Louis Headquarters "Skunkworks" team as the Video Champion to explore two-way video as a telephone company service offering. Gordon also designed and installed the first four two-way videoconferencing rooms in Southwestern Bell. These rooms served as a model for the 14 rooms now operated by SWBT. In addition, he developed, introduced and sold a nine-point education two-way full motion video network on fiber for SWBT. He has attended Wichita State University and also has 106 weeks of Technical Electronics and Communications training. He has written several articles for major industry publications on two-way video rooms and public switched video networks and is a successful consultant in Topeka, Kansas.

TRANSMISSION

Telephone transmission theory and the associated electrical characteristics will not be discussed here, except as it applies to the understanding of the video user. In order to understand **transmission,** we will start with a simple definition:

A transmission system consists of three essential parts: a source of energy, a medium over which it is desired to transmit energy to a receiving device, and the receiving device itself, which usually converts electrical energy into some form more useful.

In a power transmission line, an electric generator may be the source of energy; high voltage lines with transformers at either end may be the transmission medium; a motor, a lamp or heater may be the receiving device for converting the electrical energy into some other useful form. In a long distance telephone connection, a transmitter may be considered as the source of energy; the line from the speaking party, with all its conductors, coils, and connections may be thought of as the transmission medium, and the telephone receiver at the distant end may be considered as the receiving device that converts the electrical energy to audible sound waves.

The key to any transmission system is the amount of power that is delivered to the receiving end to do the intended job. Therefore, transmission systems of any type must be designed to deliver a specified amount of power to the receiving end, which may be a great distance from the transmitting end. In any kind of electrical transmission device, "resistance" within the transmission system, especially the transport medium, causes the transmitted power to drop steadily until the power may no longer be usable for its intended purpose. Telephone local loops have to be designed so that the power used to deliver a telephone signal will be great enough to operate the telephone set. If the distance is too great, amplifiers installed somewhere within the loop to boost the signal strength so that the telephone set will operate.

In this chapter we will concentrate on the transmission medium, the technology that connects the transmitter and receiver together. We will discuss some commonly used terms, and give some definitions and explanations of the terminology used in discussing video transmission. Also discussed here will be the methods of stacking voice channels to create larger transmission carrier systems, and ways that these larger "transmission pipes" can be used to carry data and video signals that require greater bandwidth.

■ TRANSMISSION MEDIUM

Voice Channels

The most common transmission medium is metallic wire. In telephone practice, two wires are used to carry the telephone signal, one for each direction. These two wires are called a cable pair, and is made by twisting two wires together, each being insulated from the other by an insulated covering on each wire. Pairs of wires are bundled together to form cables. Neighboring pairs are twisted with different pitch (twist length) in order to limit the electrical interference between pairs called crosstalk. The pairs are stranded into units, and the units are then cabled into cores. The cores are then covered with various types of sheaths depending on the intended use. Polyethylene-insulated cables (PIC) are made in sizes from 6 to 900 pairs while pulp-insulated cables come in sizes from 300 to 2700 pairs. The common wire sizes are 19, 22, 24 and 26 gauge, (the smaller the gauge number, the larger the wire).

There are some basic electrical characteristics of cable pairs that should be discussed here, capacitance, inductance and resistance. Capacitance (C) is an electrical effect that is constant in a given cable. It is caused by the electrical interaction of the signals and voltage in each wire, and is affected by the length of the cable. Inductance (L) is affected by the frequency of the signal, the higher the frequency, the larger the inductive effect. Resistance (R) of the wire in the cable is a function of the wire size, length and signal frequency. The overall effect is called to as attenuation.

Since the electrical characteristics interact with each other, substantial reduction of attenuation can be accomplished by increasing the inductance (L) in the cable pair. A load coil (a coil of wire that increases the inductance) can be placed strategically somewhere in the cable to reduce the electrical effects that limit the distance a voice channel can be sent over a cable pair (Figure 6.1.1).

Plain Old Telephone Service (POTS)

Copper twisted pair cable is usually found in the local loop, that is, the transmission line between the telephone and the nearest telephone office. Today, copper cable is used to connect almost every telephone, computer and data set. This terminal equipment has been designed to fit within the transmission bandwidth limits of the transmission line.

A voice telephone circuit was designed to have a bandwidth of 3500 cycles per second, or 3.5 kilohertz, to carry the most important range of frequencies used by the human voice. All transmission equipment that is connected to a POTS line, is designed to operate in this voice bandwidth range.

When a POTS line is used to connect anything but human voices, other problems, of a technical nature, have to be overcome. A POTS line works very well for carrying voice signals because these voice (analog) signals are not as sensitive as the very specific "language" required by machines. Local loop cables may have noise, which is a condition similar to weeds in a yard, that is, any plant that is not supposed to be there is a weed. Noise is somewhat similar in that electrical signals that are not a part of the intended signal is out of place and can cause interference. Machines such as computers and data equipment cannot tolerate much noise.

Another irregularity frequently found in cables is the bridged tap. This consists of another pair of wires which are connected in shunt to the main cable pair at any point along its length. This pair may or may not be used at some future time, depending on the way in which service demands develop. In any case only one of the pairs going away from the bridging point is likely to be used at any given time. The other hangs open-circuited across the working pair and introduces bridging loss. In order to limit this transmission impairment, there are rigid rules concerning the number, length, and location of bridged taps allowable on pairs assigned to various kinds of service.

Two Wire-Four Wire Circuits

Telephone instruments and the voice frequency facilities to which they connect may be either two-wire or four-wire circuits. The economic advantage of using only half as much copper dictates the use of the two-wire circuits in many voice-frequency lines and in most loops from a Telephone Central Office to a user location. Some special applications require four-wire loops. The use of four-wire loops is becoming greater with the use of more sophisticated data equipment. The four-wire loop provides a more electrically stable circuit, and can more easily be used to extend the circuit a greater distance.

The measurement of the relative strength of a signal that is widely accepted is called the decibel. It is defined only for power ratios (signal input strength compared to output strength). These ratios can be measured in voltage or current. A reference point, say 1 milliwatt measured across a resistance of 75 ohms, can be used as the 0 decibel (0dB) reference point. A measurement at the other end of the circuit at a similar 75 ohm resistance point, would read lower or higher depending on whether there has been signal loss or gain from one end to the other.

In a transmission system, there are a string of losses and gains from cable and equipment from one end to the other. It is necessary to account for the signal magnitude in the system from one point to the

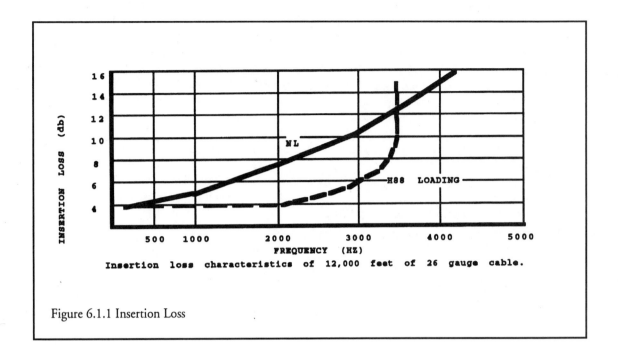

Insertion loss characteristics of 12,000 feet of 26 gauge cable.

Figure 6.1.1 Insertion Loss

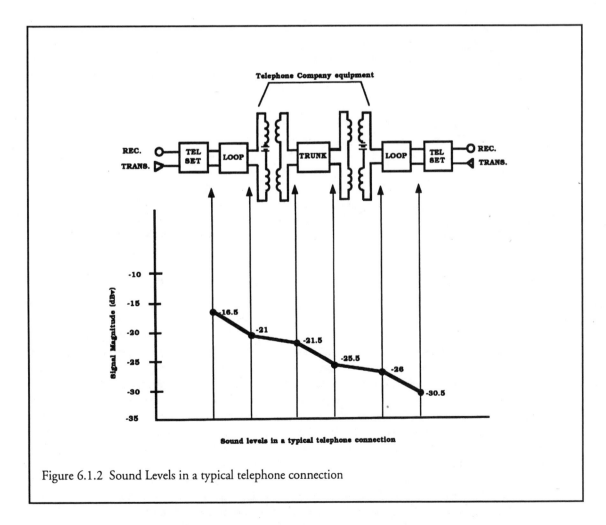

Sound levels in a typical telephone connection

Figure 6.1.2 Sound Levels in a typical telephone connection

other. This measurement is called Transmission Level Point, and is measured in dB. The resulting summation of the losses and gains are called the Transmission Level. The reference point is usually considered as "0-dB Transmission Level Point" (0 TLP).

These measurements are important when considering connecting non-telephone industry standard equipment to a standard telephone network line. Since most terminal equipment manufacturers are well versed in the industry standards, this is not usually a problem.

The possibility of electrical interference (crosstalk) between cable pairs has already been mentioned. When very high signal frequencies are being transmitted on a cable pair, say one million hertz (1 Megahertz) or higher, crosstalk coupling between adjacent cable pairs may become a dominant design consideration. In such cases, the transmission medium may be isolated by shielding.

Shielded pairs are frequently included in twisted pair cable in order to provide a satisfactory medium for the transmission of high frequency signals such as baseband television (video), . In video service the signal spectrum extends from near 0 to 4.5 MHz. The shielded pair (called 16 PEVL) consists of two 16 gauge conductors insulated with expanded polyethylene and surrounded by a longitudinal-seam copper shield to reduce the crosstalk.

At higher frequencies, the isolation between transmission paths can be achieved very efficiently by the use of coaxial conductors. The coax cable consists of a center conductor surrounded by a concentric outer conductor. At high frequencies, the coaxial outer conductor provides excellent shielding against extraneous signals. However, at low frequencies, the shielding is ineffective. Therefore, at low frequencies coax is not usually used.

Modulation

Communications signals must be transmitted over some medium separating the transmitter and the receiver. The information to be sent is rarely in the best form for direct transmission over this medium. Efficiency of transmission requires that this information be processed in some manner before being transmitted. Modulation may be defined as that process whereby a signal is transformed from its original form into a signal that is more suitable for transmission over the medium between the transmitter and the receiver. It may shift the signal frequencies for ease of transmission or to change the bandwidth occupancy, or it may alter the form of the signal to optimize noise or distortion performance. At the receiver this process is reversed by demodulation methods.

In order to transform the analog signal intelligence to a modulated signal, a carrier frequency is used to combine with the signal. The resulting of this combining is a modulated signal.

A carrier is a continuous frequency that can be modulated or impressed with an information-carrying signal. An example of this is a radio station signal, say 97.5 FM. This 97.5 means the carrier signal is 97.5 mb per second, (97,500,000 Hertz). The intelligence, voice, music, etc., is impressed on the carrier to modify it to be transmitted.

The most used of the many types of analog modulation is (1) Amplitude Modulation, (2) Frequency Modulation and (3) Pulse Modulation.

Amplitude Modulation

In amplitude modulation the amplitude of the carrier wave is varied in accordance with the variations of the signal wave. This modulation process produces upper and lower sidebands that are each capable of carrying the signal intelligence by itself. Usually all the sideband frequencies except one are removed from the transmitted signal.

Here are four types of amplitude modulated carrier systems that can be employed.

1. Double Sideband with Transmitted Carrier
2. Double Sideband Suppressed Carrier
3. Single Sideband
4. Vestigial Sideband

Of these four, the single sideband is used by the telephone industry for a majority of analog transmission beyond the local loop, that is, when voice channels are combined to form carrier systems. For television, the most efficient modulation method is Vestigial sideband transmission because it provides the most efficient transmission of the low and high frequencies required by television signals.

Frequency Modulation

The use of frequency modulation is confined to radio systems operating at very high frequencies, where it has certain advantages over amplitude modulation, that is, it's characteristics minimize "static" interference from extraneous signals. It depends upon varying the frequency of a carrier wave of fixed amplitude above and below a central frequency in accordance with the amplitude of an applied voltage. The amount of frequency change that is produced by the signal is called the frequency deviation, and, ideally, this should be as high as possible in order to obtain the maximum **signal to noise ratio**. (Signal-to-noise ratio is a measurement of the electronic noise that is generated by equipment within a system such as amplifiers. This noise buildup is a limiting factor in analog systems.)

Pulse Modulation

In pulse modulation systems the unmodulated carrier is usually a series of regularly recurrent pulses. Modulation results from varying some parameter of the transmitted pulses, such as the amplitude, duration, shape or timing. If the baseband signal is a continuous waveform, it will be broken up by the discrete nature of the pulses. In considering the feasibility of pulse modulation, it is important to recognize that the continuous transmission of the information describing the modulating function is unnecessary, provided that the modulating function is band limited and the pulses occur often enough.

It is usually convenient to specify the signaling speed or pulse rate in bauds. By definition, the speed in bauds is equal to the number of signaling elements or symbols per second. Thus, the baud denotes pulses per second in a manner parallel to hertz denoting cycles per second. Note that all possible pulses are counted whether or not a pulse is sent since no pulse is also usually a valid symbol. Since there is no restriction on the allowed amplitudes of the pulses, a baud can contain any arbitrary information rate in bits per second. Unfortunately, the term bits per second is often used incorrectly to specify a digital transmission rate in bauds. For binary symbols the information rate in bits per second is equal to the signaling speed in bauds. In general, the relation between information rate and signaling rate depends upon the coding scheme employed.

With sampling, if a message is sampled at a rate at least twice the highest message frequency, and at regular intervals, then the samples contain all of the information of the original message. The amplitude-modulated pulse that results from sampling the input message may be transmitted to the receiver in any form which is convenient or desirable from a transmission standpoint. To reconstruct the message at the receiver, it is necessary to generate from each sample a proportional impulse, and to pass this regularly spaced series of impulses through a filter to remove the carrier frequency. The result is identical to the original message.

In **Pulse Amplitude Modulation (PAM)**, the amplitude of a pulse carrier is varied in accordance with the value of the modulating wave.

PAM is a modulation in which the value of each instantaneous sample of the modulating wave is caused to modulate the amplitude of a pulse. Signal processing in time division multiples terminals often begins with PAM although further processing usually takes place before the signal is launched onto a transmission system.

In **Pulse Code Modulation (PCM)**, instead of attempting the impossible task of transmitting the exact amplitude of a sampled signal, suppose only certain discrete amplitudes of sample size are allowed. Then, when the message is sampled in a PAM system, the amplitude nearest the true amplitude is sent. When this is received and amplified, it will have an amplitude slightly different from what was sent anyway, because of the disturbances encountered in transmission. If the system noise and distortion are not too great, it will be possible to tell accurately which amplitude of the signal was transmitted. The signal can then be reformed, or a new signal created which has the amplitude originally sent.

Representing the message by a limited number of signal amplitudes is called quantizing. It inherently introduces an initial error in the amplitude of the samples, giving rise to quantization noise. But once the message information is in a quantized state, it can be relayed for any distance without further loss in quality, provided only that the added noise in the signal received at each repeater is not too great to prevent correct recognition of the particular amplitude each signal is intended to represent. Note that in

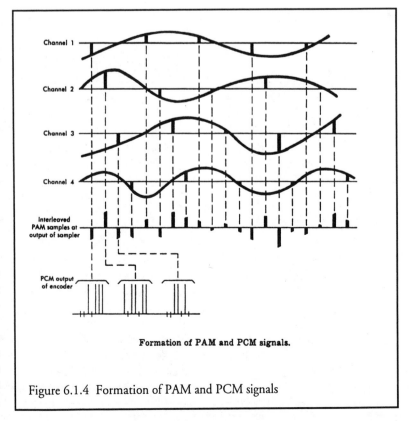

Formation of PAM and PCM signals.

Figure 6.1.4 Formation of PAM and PCM signals

quantized signal transmission the maximum noise is selected by the number of bits in the code, while in analog signal transmission, it is determined by the repeater spacing and characteristics of the medium.

Pulse Duration Modulation (PDM), sometimes referred to as pulse length modulation, is a particular form of pulse time modulation. It is a modulation of a pulse carrier in which the value of each sample of a continuously varying modulating wave is caused to produce a pulse of proportional duration. The pulses are the same amplitude but vary in duration. In PDM, long pulses expend considerable power during the pulse while bearing no additional information. If this unused power is subtracted from PDM so that only transitions are preserved, another type of pulse modulation, called pulse position modulation, results. The power saved represents the fundamental advantage of pulse position modulation.

Pulse Position Modulation (PPM) is a form of pulse time modulation in which the value of each sample of a modulating wave varies the position of a pulse. Practical applications of PPM are few. Both PDM and PPM fall short of the ideal when used for multiplexing ordinary telephone channels.

A quantized sample can be sent as a single pulse having certain possible discrete amplitudes or certain discrete positions with respect to a reference position. If many discrete samples are required (one hundred for example), it is difficult to design circuits that can distinguish between this many amplitudes. It is much easier to design circuits that determine whether or not a pulse is present. If several pulses are used as a code group to describe the amplitude of a single sample, each pulse can be present (1) or absent (0).

These codes in fact represent the binary notation scheme that is commonly used. It is possible, of course, to code the amplitude in terms of a number of pulses which have discrete amplitudes of 0,1, and 2 (ternary or base 3), or 0,1,2, and 3 (quarternary or base 4), etc., instead of the pulses with amplitudes 0 and 1 (binary or base 2). Therefore, systems using codes to represent discrete signal amplitudes are called pulse code modulation or PCM systems.

Signal Multiplexing

Multiplexing in transmission systems is a means of utilizing the same transmission medium for many different users. Before being placed on the transmission medium, each user's signal may have to be modified in some unique way so that it can be separated from all of the other signals at the distant end of the transmission path. This separation involves basically the inverse of the original modification. There are several ways signals can be multiplexed, the most important being space divi-

sion multiplex, frequency division multiplex and time division multiplex.

Space Division Multiplex

Space division multiplex is simply the bundling of many physically separate transmission paths into a common cable. A telephone cable consisting of hundreds (or thousands) of twisted pairs constitutes a space division multiplex system since many conversations can be carried in the single cable although each is assigned a different pair in the cable. Such a scheme is economical when it is remembered that the transmission right-of-way absorbs a substantial part of the cost of any transmission system. The advantages of space division multiplex do not come free, however. First of all, the traffic must be combined into specific routes to achieve the large channel cross sections. Secondly, achieving true isolation between transmission media separated by long distances is difficult. As a consequence, such systems are subject to interference resulting from coupling between channels. Space division multiplex is not confined to voice frequency circuits. Many high capacity transmission systems (either frequency or time division multiplexed) can in turn be space division multiplexed on parallel facilities sharing the same right-of-way.

Frequency Division Multiplex

In frequency division multiplex (FDM) each channel of the system is assigned a portion of the transmitted frequency spectrum. Thus, many narrow bandwidth channels can be accommodated by a single wide bandwidth transmission systems. The multiplexing operation involves a frequency shift of each channel before launching it on the broadband facility. At the receiving end, each channel has to be shifted back to the original frequency spectrum.

There are many applications using frequency division multiplex. For example, frequencies are allocated among broadcast stations using this form of multiplex plan in which many stations can broadcast simultaneously over the same medium and yet be separated by a receiver for individual reception. Likewise, the frequency allocation of microwave stations is a form of FDM. The practical implementation of frequency division multiplex may involve many steps of modulation and demodulation. The basic building block of the hierarchical plan is the message channel. The message channel can be used for the transmission of voice or data. Several narrow band data signals have components below 200 Hz and can be multiplexed into a single message channel which occupies the bandwidth spectrum between 200-3400 Hz. Data signals requiring more bandwidth than that provided by the message channel are called wideband data, and more than one message channel may be required to support it.

Analog transmission systems are being phased out in favor of digital transmission systems. Therefore, the multiplexing hierarchy for analog transmission will not be discussed here.

Time Division Multiplex

Time division multiplex (TDM) is the third common type of multiplexing. As the name implies, it is simply the sharing of a common facility in time — an extension of the principle of "taking turns." For signals that are not full time, TDM is often used in telephone communications. For example, most telephones are in use only a small portion of the time; thus several telephones can time share a common line to the nearest telephone office. Speeding up the time scale, the same principle can be used to TDM several speech channels by taking advantage of pauses between words and statements. Utilizing this principle, a Time Assignment Speech Interpolation (TASI) system is used on many overseas channels to effectively increase the capacity of the channels. The switching scheme must be very rapid, and the resulting complex equipment is not cost effective except for use on expensive channels such as overseas channels..

The fact that a bandwidth limited signal can be sampled at discrete times, while preserving all of the information, leads to the most popular method of processing for TDM.

Each signal is sampled; the resulting PAM signals are interleaved; and finally, each PAM sample is encoded into binary PCM. As the number of message channels is increased, the time interval that can be allotted to each must be reduced since all of them must be fitted into the sampling interval. The allowed duration of a coded pulse train representing an individual sample must be shortened, and the individual pulses moved closer together as the number of time division channels is increased.

The sampling rate is typically 7 to 9 times higher than the highest frequency bandwidth required for direct transmission. The increased bandwidth required is offset by the noise advantage resulting from the regeneration of a new, essentially noise-free signal at each repeater point.

A plan has developed for time division multiplexing.

Digital Transmission

So far we have discussed transmission systems in which the signals applied to the transmission media are continuous functions of the message waveform. It has been shown how a carrier can have either its amplitude or phase continuously varied in accordance with the message. In digital transmission systems the applied signals are discrete in time and amplitude. Signals that are discrete only in time, such as pulse amplitude mod-ulated (PAM) signals, or discrete only in amplitude, such as facsimile data, will not be considered here as digital signals since they cannot be carried by digital transmission lines containing regenerative repeaters. In the simplest case either a pulse or a space (no pulse) is transmitted in each unit of time. The stream of pulses and spaces can be thought of as binary numbers that represent analog signals to which sampling and appropriate coding rules have been applied. Although binary signals are easiest to regenerate, the actual signals usually found on cable media have more than two symbols (i.e. positive pulse, negative pulse, and no pulse) and quarternary using two different amplitude pulses for each polarity.

The advantage of converting message signals into digital form is the ruggedness of the digital signal. Quantizing noise associated with analog-to-digital conversion occurs at the terminal where it can be controlled by assigning a sufficient number of digits for each sample. Once these signals are in digital form, additional impairments are negligible in a properly engineered system since each repeater transmits a new waveform in response to the received symbol and will almost always do so correctly in spite of noise and interference in the media. This process is called regeneration and provides the primary advantages to digital transmission. The price paid for this ruggedness is increased bandwidth relative to that required for the original signal.

Analog signals, including message signals, are converted into digital form in steps depicted in Figure 6.1.5. First the signal, which has been bandlimited by a filter process, is sampled. These samples are called PAM pulses.

Although discrete in time, PAM pulses are not suitable for digital transmission because they are not discrete in amplitude. The next step is analog-to-digital conversion and is accomplished by a coder. The coder converts each PAM sample into a binary number called a code word. Binary digits as generated by the coder are then translated into ternary or other multilevel digital signals acceptable to the transmission facility.

At the receiving terminal the binary digits are recovered, and the code words are decoded into a PAM form by a digital-to-analog converter. The pulses are then passed through a filter to recover the original message signal. Due to the finite number of code words available at the coder, the decoded PAM samples can have only a finite number of discrete amplitudes; therefore, some error between the original and the recovered signal may be expected. This error is called quantizing distortion, which is the controlling signal impairment encountered in digital transmission.

Digital Hierarchy

In addition to being characterized by the discrete nature of the signals and the use of regenerative

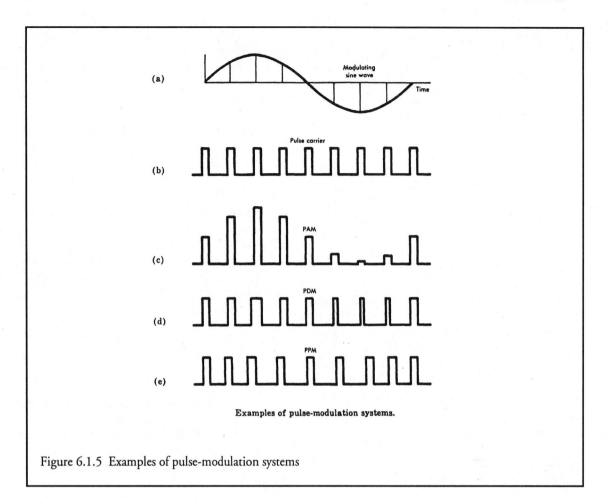

Figure 6.1.5 Examples of pulse-modulation systems

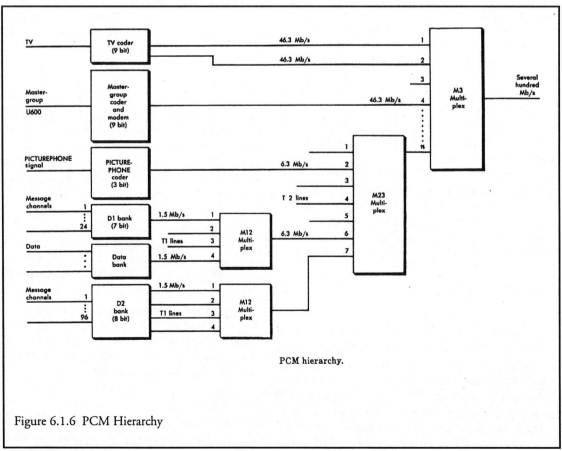

Figure 6.1.6 PCM Hierarchy

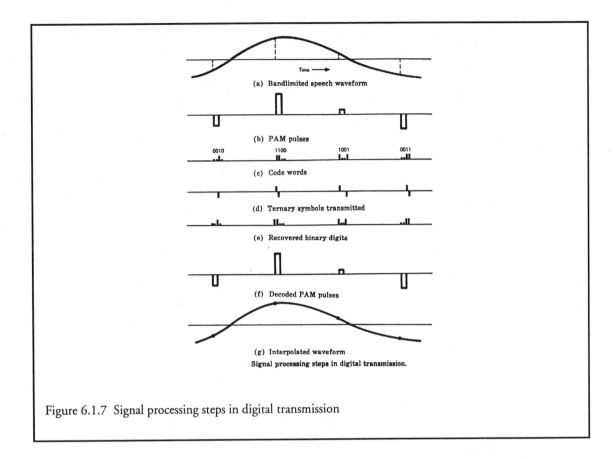

(a) Bandlimited speech waveform

(b) PAM pulses

0010   1100   1001   0011

(c) Code words

(d) Ternary symbols transmitted

(e) Recovered binary digits

(f) Decoded PAM pulses

(g) Interpolated waveform

Signal processing steps in digital transmission.

Figure 6.1.7 Signal processing steps in digital transmission

repeaters, digital transmission is also characterized by the use of time division multiplexing. Multiplexing of signals in digital form allows interconnection of digital transmission facilities with different signaling speeds. The digital hierarchy shown here forms the basis of an interconnected digital network.

The interface between the digital system and the analog system is made by digital terminals which convert the incoming analog signals to a digital form suitable for application to a digital transmission facility. Terminals that multiplex many message channels for application to a single digital line are called digital channel banks. Digital multiplexers form the interface between digital transmission facilities of different rates. They combine digital signals from several lines in the same level of the hierarchy into a single pulse stream suitable for application to a facility of the next higher level in the hierarchy.

Channel Banks

Digital channel banks multiplex many voice frequency signals and code them into digital form. The incoming message signal is applied to the transmitting portion of the channel bank. The signal is passed through a filtering process that limits the bandwidth to less than one-half of the sampling frequency. The signal is then sampled at an 8-kHz rate. Samples from many message channels are applied to a common bus. Since the gates operate sequentially, the

voltage on the common bus is seen to be a time division multiplexed version of all the PAM samples. The voltage on the common bus is processed by the coder, which is shared by all the channels to produce PCM code words.

Before this digital signal can be applied to a transmission line, additional digital processing is usually necessary. First, the signaling information for each incoming message channel is multiplexed with the coder output. Next, in order to permit identification of the PCM code words at the receiver, framing information is inserted. Finally, the binary signals are converted to a form acceptable to the digital transmission line.

The D1 channel bank converts 24 message channels to digital form. Each channel is coded into a 7-digit binary word. A signaling digit is then multiplexed to each word associated with the channel. Finally, a framing digit is multiplexed, resulting in the format shown in Figure 6.1.9.

Thus, the total number of digits per sampling cycle is (7+1)x24+1=193. The block of 193 digits is called a frame. Since there are 8000 frames per second, a digital capacity of 1.544 mbps. The receiving portion of the channel bank performs the inverse operations. The incoming signal from the digital line is first converted to binary form. A framing circuit searches for and synchronizes to the framing bit pattern, which insures that the locally generated timing

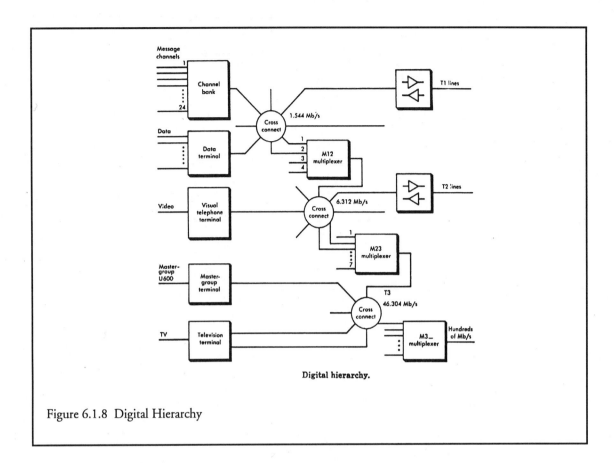

Digital hierarchy.

Figure 6.1.8 Digital Hierarchy

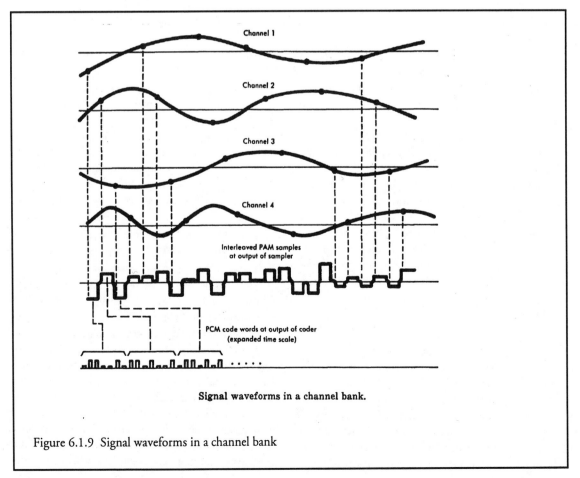

Signal waveforms in a channel bank.

Figure 6.1.9 Signal waveforms in a channel bank

pulses are in synchronism with the incoming pulse train. The signaling digits are sorted out and directed to the individual channels, and the PCM code words are delivered to the decoder. The output of the decoder is a series of quantized PAM pulses which are demultiplexed and applied to the individual interpolation filters. Interpolation produces an analog signal which is applied to the receiving voice-frequency line.

Single Channel Terminals

When the bandwidth of the signals to be transmitted is such that after digital conversion it occupies the entire capacity of a digital transmission line, a single channel terminal is provided. One example of this is a full motion video signal. The signal processor provides frequency shifting and d-c restoration. Sampling rate for a full band video signal is 10.2 MHz. To meet the transmission requirements, nine binary digits per sample. The digital rate is therefore 92 mbps for the video signal. In addition, the 92 mbps signal must be split into two 46 mbps signals since only the 46 mbps speed belongs to the digital hierarchy.

Compressed Video Terminals

This may be a good place to discuss video compression in terms of digital sampling characteristics. To permit both adequate detail and contrast resolution in a compressed video signal, some specific characteristics of the signal and some subjective effects on motion pictures can be exploited. In regions of low detail in a picture, such as simple backgrounds, the eye is sensitive to abrupt brightness changes and insensitive to brightness colors. The eye is also rather sensitive to position error or jitter of the edges of the detail. These signal characteristics and subjective effects suggest the use of differential pulse code modulation as means of bandwidth conservation in the digitizing of visual signals. In this form of modulation the difference between the present sample and the previous sample is coded. Since these differences are expected to be smaller than the samples themselves, a smaller number of digits is needed for coding. The differences between the present sample and the corresponding sample in the last frame is expected to be even smaller. To exploit frame-to-frame correlation, however, requires storage of a full picture.

Data Terminals

An increasing portion of communications traffic involves signals other than voice. Most are data signals which are binary in form with transitions between levels occurring randomly. The function of a data terminal is to accept a serial periodic or nonperiodic data signal and convert it to a synchronous stream acceptable to the digital transmission facility. Data signals could be sampled directly; however, to preserve transition time accuracy the sample rate would have to be quite high, resulting in an excessive bit rate, especially for a signal with infrequent transitions.

A more efficient format would be one which codes transition time. When there is no transition, a signal of all ones is transmitted. A first zero, called the address bit, indicates the presence of a transition. Following the address bit is a code bit which indicates in which half of the address timing interval the actual transition took place. This reduces the time quantizing error by a factor of two compared to the direct sampling method. The last bit is a sign bit that indicates whether the direction of transition is zero to one or vice-versa.

The efficiency is about three transmitted bits for each transition or data bit. A 50 kbps data signal thus displaces three 64 kbps digital voice signals; a 250 kbps data signal displaces 12 voice channels. Even at one-third efficiency, these data banks are more efficient in the substitution of data for voice signals than are analog systems, which require 60 message channels to transmit 250 kbps data. Ultimately, data terminals will be designed which operate with better than 90 percent efficiency, there may be some available even today.

■ DIGITAL MULTIPLEXES

In the same way that various analog systems are used to transmit analog signals of different bandwidths, there are various digital transmission facilities designed to transmit digital signals of different rates. These facilities must be interconnected into a network that allows a digital signal to reach its destination using one or more of these facilities. Interconnections must be flexible enough to provide for alternate transmission paths in case of equipment failures, changing traffic patterns, and routine maintenance. Interconnection of facilities of the same digital rate involves either manual patching or automatic switching. Interconnection of different digital rates require multiplexers that combine several signals to share a higher speed digital facility.

Time division multiplexing of several digital signals to produce a higher speed stream can be accomplished by a selector switch that takes a pulse from each incoming line in turn and applies it to the higher speed line. The receiving end will do the inverse of separating the higher speed pulse stream into its component parts and thus recover the several lower speed digital signals. The main problems involved are the synchronization of the several pulse streams so that they can be properly interleaved and the framing of the high-speed signal so that the component parts can be identified at the receiver end. Both of these operations require elastic stores, which constitutes important parts of a multiplexer.

Elastic stores are also called data buffers. Information pulses arriving at the multiplexer must wait their turn to be applied to a higher speed transmission system. Due to the delay variations of the incoming lines and to the framing and synchronization operation of the multiplexer terminal, this wait is variable in time.

The major problem of multiplexer system design is synchronization. Digital signals cannot be directly interleaved in a way that allows for their eventual identification unless their pulse rates are locked to a common clock. Because the sources of these digital signals are often separated by large distances, their synchronization is difficult.

There are four methods used for digital synchronization. The Master Clock concept times all digital systems to one timing source. Mutual Synchronization is a method whereby each Telephone Central Office has its own timing source whose frequency is the average of all the incoming frequencies on the digital system. Third is the Stable Clock method which relies on the atomic clock and must be reset periodically. A combination of these three is likely to be used in the future. The final method is pulse stuffing. This is being used in the design of all multiplexers. The concept is to have the outgoing digital rate of a multiplexer higher than the sum of the incoming rates by stuffing in additional pulses. All incoming digital signals are stuffed with sufficient number of pulses to raise each of their rates to that of the locally generated clock signal.

Microwave Radio Systems

Analog microwave radio systems were a very important part of the Telephone Network from the 1950s through the 1970s. These radio routes served about half of the long distance telephone network during this period. These systems had a capacity of about 22,000 message circuits. They were primarily FM type systems. The typical microwave route had microwave tower spacing between 20 and 30 miles apart. These systems were traditionally between 250 and 4000 miles long. In the late 1970s, digital radio systems were developed that began to replace the analog radio network. As we discussed earlier, there are certain advantages to digital transmission. By the mid 1980s, most of the microwave network had been converted to digital.

Digital microwave uses the same multiplex scheme and is compatible with digital cable systems. There are 24 message channels slotted to a T1 (1.5 mbps) system. There are four T1 systems slotted to a T2 (6.2 mbps) system. There are seven T2 systems

slotted to a T3 (45 mbps) system, and there can be three T3 systems slotted to a radio channel. Therefore there can be 2016 voice channels to one digital microwave system. There can be 8 to 12 microwave channels assigned to a radio route, the limitation being the frequency slots available assigned by the FCC.

Microwave transmission is subject to atmospheric interference and therefore is a less desirable form of transmission than fiber. Microwave systems require a line of sight path between towers. This path is subject to change due to building erection etc. Fading, which is a condition caused by such things as ground fog or extremely cold air over a warm earth, causes the line of sight path to become obstructed, and a partial or total loss of the signal. The duration of this signal interruption can be from less than a second to several hours.

Microwave protection switching systems that monitor the receiving signal strength, are used to parallel the working channel. If the signal loss is great enough, the protection switching system tests the alternate antenna or channel to see if the signal strength is better. If it is better, the traffic is switched automatically to the better path. These systems have become very sophisticated in recent years, and the reliability is high. The frequency spectrum typically used for microwave transmission ranges from 2 GHz to 18 GHz. There will continue to be a need for microwave systems, especially in areas where it is not economical to bury fiber.

■ HIGH BIT RATE DIGITAL SUBSCRIBER LINE

HDSL technology will provide an alternative to repeatered T1 lines to support repeaterless DS1 rate access over copper loops that conform to Carrier Serving Area (CSA) design rules. There is an increasing demand for DS1 rate access in a cost effective and timely manner, and to improve the utilization of the embedded base of copper plant. HDSL will deliver services such as High Capacity Digital Service (HCDS), ISDN Primary Rate Access, Switched Multi-megabit Data Service, and Switched-DS1/Switched Fractional DS1 (SWF-DS1).

HDSL should have a synergistic impact on fiber deployment. HDSL can complement fiber growth in the distribution plant, allowing fiber deployment to take place in a prudent manner while taking advantage of the existing copper base. An example may be seen in the case where DS1 service is carried over fiber in the feeder portion of the loop to a fiber hub, and copper T1 pairs are distributed from the hub to the customer. In such a case, HDSL could be used as an alternative to the copper repeatered line. Repeaterless

HDSL DS1 transport can save a great deal of engineering design and construction needed to place T1 line repeaters in the outside plant and can aid telephone companies in providing service in a more timely manner after receiving a customer order.

HDSL is an outgrowth of the Digital Subscriber Line (DSL) technology that supports ISDN Basic Rate Access at 160 kbps over nonloaded loops up to 18 Kft. The current telephone environment, including bridged taps and mixed wire gauge loops, intended for voice frequency transmission, presents a complex environment for wideband transmission. Equipment design and CSA guidelines for loops should be followed. Transceivers at each end of the line use digital signal processors or equalize the signal loss at all frequencies and cancel signal reflections (echoes) resulting from line imperfections. HDSL also uses adaptive filtering and the 2B1Q line coding scheme to counteract noise and crosstalk effects on the line.

The massive investment in copper facilities throughout the exchange network make it essential to expand the use of copper already deployed. Supporting 1.544 mbps transmission in a repeaterless fashion with an HDSL could save time and cost compared to current service provisioning using engineered T1 repeatered lines, and could position the embedded copper base as a viable access capability for future new services. In addition, until such time that fiber deployment is ubiquitous throughout the distribution network, supporting early broadband services at the DS1 rate with an HDSL will allow a telephone to establish an early presence.

Timely Service Provisioning

Fiber and copper facilities are presently used to meet demand for DS1 based services. Fiber facilities are used when (1) they have a lower first installed cost than metallic lines, (2) they are the only choice available to deliver the needed bandwidth, and (3) they meet strategic future needs. Where customers are scattered, copper is often used to provision T1 lines. Although the technology is mature and provides reliable service, building a T1 line requires a labor-intensive design and installation process: pairs must be selected and probably conditioned by removing bridged taps. As an alternate to conditioning existing loop cable pairs, new screened cable may be installed. The pairs for each direction of transmission are usually placed in separate binder groups in the cable to reduce crosstalk. Repeaters must be engineered and placed. The process might be further slowed when manholes or poles need to placed or reinforced to accommodate line repeaters. Such construction can add weeks to the service provisioning interval.

With HDSL, the provisioning process for DS1 rate access will not require a complicated design and the service interval will be relatedly short. As much as 95 percent of the implementation time of T1 service can be reduced by using HDSL, because it will have a POTS-like provisioning process; for example, there will be no need to condition the selected loops. Loops would be prequalified in bulk as conforming to CSA design rules and then assigned as needed to DSL. Bridged taps do not need to be removed, and binder group separation will not apply. The beneficial impact of HDSL on provisioning is due mainly to the elimination of the time and labor consuming activity associated with the engineering and installation of T1 line repeaters. HDSL is expected to result in a much reduced service installation time.

Description of HDSL Technology

The High-bit-rate Digital Subscriber Line (HDSL) provides a full duplex (bidirectional) 1.544 mbps transmission capability over 2 non-loaded metallic cable pairs which are governed by CSA design rules. Specifically, bidirectional data at a rate of 784 kbps is transmitted and received on each pair. The data from both pairs are combined to form a DS1 compatible format in an architecture referred to as "dual duplex". The 784 kbps rate includes performance monitoring, framing and timing functions.

Each of the High-bit-rate Terminal Units (HTU), at either the Central Office (HTU-C) or the Remote Distribution (HTU-R) location, consists of two pairs of digital transmitter/receivers (transceivers), each connected by a two wire CSA compatible loop.

The HTU-C unit has a standard DSX-1 interface.

The HTU-R unit has a customer Network Interface meeting ANSI T1.403-1989 standards.

Each HTU will accept a DS1 signal, add any needed HDSL overhead for synchronization and maintenance purposes, perform digital signal processing and generate a line signal that will be placed on the loop. The opposite HTU will perform digital signal processing to minimize the effects of impairments on the loop and recover the original DS1 signal.

HDSL Technology as a T1 Line Alternative

During the past few years, the annual growth of DS1 rate services has been as high as 40 percent. A strong growth pattern is forecasted for the future. It is reasonable to assume that there will be a strong demand for copper-based DS1 services in the local loop. More than half of the DS1 local loop installations in the next 10 years will be satisfied by fiber, but the demand for copper based DS1 systems will probably be in the hundreds of thousands of lines in the telephone industry.

ISDN Primary Rate Access

ISDN PRA uses up to the 1.536 mbps net payload capacity of the 1.544 mbps DS1 bandwidth. The typical channelization of PRA is 23B channels (B = 64 kbps), plus 1D channel (D = 64 kbps). Other Primary Rate channelization may include the use of HO (384 kbps) or HI (1536 kbps) channels with a D channel in a companion line. An HTU-C might also be integrated with other ISDN based network elements such as remote switch unit or a stand alone exchange termination that is used with an analog on non-ISDN switch.

Switched-DS1/Switched Fractional DS1 (SWF-DS1)

HDSL can provide access lines for transporting SWF-DS1 service. SWF-DS1 is a dialable, public, circuit switched service that provides digital connections from 128 kbps to 1536 kbps in 64 kbps increments. Clear channel capability is required for SWF-DS1, and all digital trunks used to support the SWF-DS1 must be B8ZS compatible. It needs to be provided on access lines synchronous with the network clock using Extended Superframe Format (ESF). HDSL would provide required access capabilities.

During the last few years, the term Fractional T1 (FT1) has often been misused. Interexchange Carriers (IECs) originally conceived of Fractional T1 as a means to provide scaled-down T1 access services to end-users who could not cost justify an entire T1 access line. The main idea was to hand-off a partially filled DS1 interface to the end user and charge them according to the number of DSOs in use. For Local Exchange Carriers (LECs) this has presented a problem. Since LECs have had to deploy a full T1 line for the Fractional T1 access, they have not been able to justify a reduced rate for FT1. The local access charges for T1 and FT1 are the same. With HDSL deployment coming soon, the ability to provide FT1 local loop access is practical.

LECs will be able to deploy FT1 services with one cable pair which will provide up to 12 DSOs using standard DS1 interfaces. This is important for end users because they can buy or lease industry standard T1 terminal gear such as multiplexers, PBXs, Channel Banks, CSUs etc., to use for FT1 service. If more DSOs are needed in the future, the equipment in use for FT1 would be reusable.

Switched Multi-Megabit Data Service (SMDS)

HDSL can provide DS1 based access lines for transporting SMDS. SMDS is a connectionless, public, packet-switched data transport service that extends Local Area Network-like (LAN-like) performance and features beyond the subscriber's premises across a metropolitan area. HDSL can provide a cost effective alternative to connect a Subscriber-Network Interface (SNI) to a MAN switching system (MSS).

DS1 Extension Access

An HDSL may provide DS1 rate access between fiber-based remote electronics in the distribution plant facilities and the customer premises. Multiplexers that provide DS1 level outputs might be co-located with or contain a line unit integrated version of the HTU, an HTU-LU. Here the HTU-C function is shown integrated into HTU-LU plug-ins in the terminal. Fiber feeding operating at the OC-3 SONET rate is shown connecting to the switch.

Digital Loop Carrier

DLC transport refers to the DS1 rate facilities that support a Digital Loop Carrier (DLC) system where both DS1 terminations are network-provided. An HDSL could support DS1 connections between DLC CO and Remote Digital Terminals (COT and RDT) that are widely deployed in the distribution plant.

Line Code

The line code for HDSL is 2B1Q (2 Binary, 1 Quarternary). This is a 4-level pulse amplitude modulation (PAM) code without redundancy. The 784 kbps data bit stream, comprised of 768 kbps DS1 payload plus 8 kbps DS1 framing plus 8 kbps of HDSL overhead, are grouped into pairs of bits for conversion to quarternary symbols that are also called quats. In each pair of bits so formed, the first bit is called the sign bit and the second is called the magnitude bit. The combination is used to create a Quarternary Symbol that represents the intelligence in the two bits. This is also called four level signaling. This coding is relatively immune to line impairments and allows the use of "dirty pairs" containing bridged taps and noise sources that are normally found in distribution loop cable.

First Bit (Sign)	Second Bit (Magnitude)	Quarternary Bit (Quat)
1	0	+3
1	1	+1
0	1	-1
1	0	-3

The four values in the table are to be understood as symbol names, not numerical values. At the receiver, each quarternary symbol is converted to a pair of bits by reversing the table above descrambled and finally formed into a data bit stream.

ADSL (Asymmetrical Digital Subscriber Loop)

The technology is essentially the same as HDSL with the exception that it is 1.544 mbps in one direction and 64 to 128 kbps in the other direction. The main intended use of this technology is to transport one way compressed video, with a control channel to send instruction in the other direction. The ADSL signal can be transmitted up to 18 Kft over a 24 gauge copper pair. The same 2B1Q line coding, echo suppression, filtering, and a simpler design of Transceiver as in HDSL, are used.

Telephone companies and CATV providers are very interested in this technology because of the potential of providing cable quality television channels on a copper pair. Tests by Bell Labs have thus far placed 10 ADSL type television channels in the same frequency slot as one analog video channel. The goal is to provide between 35 and 70 compressed video channels on a single existing television channel, to be made available one at a time on the ADSL loop at the direction of the customer.

The components to this system are as follows:

ADSL technology on the local loop to utilize existing copper pairs.

ADSL technology used for placing multiple channels into the existing analog TV bandwidth.

Improved compressed video providing CATV quality motion video.

Customer controlled 1.544 mbps switching capability.

With the exception of ADSL developments, which may be available in the 1992-93 timeframe, the other technologies are ready for deployment.

■ TELEPHONE INDUSTRY DIGITAL SERVICE OFFERINGS

The digital services offered today by the telephone industry are as follows:

1. Full time 56 kbps digital line. This is offered by local exchange carriers.

2. Switched 56 bkps digital service. This is available only in locations where the local digital switch has the capability to handle the 56 kbps line. Therefore this service is presently limited to larger cities. A combination of Switched 56 and dedicated 56 kbps service can be offered to extend the switched line further from the switch, but becomes very expensive.

3. Long distance carriers offer a 64 kbps dedicated service. Also available is a switched 64 kbps service. There may be some compatibility problems between the local telephone company and the long distance carrier, but the problems will be solved.

4. Most long distance carriers offer a 384 kbps service. This is 1/4 of a T1 (DS1) system, or a bundle of 6 digital voice channels. This offering is available in the larger cities.

5. Also offered by the long distance carriers, is a 768 kbps service, that is a bundle of 12 digital voice circuits. This service, as well as the 384 kbps service, is offered mainly for compressed video services.

6. The local telephone companies as of now do not offer a 384 kbps or a 768 kbps service. The main reason is that it would cost more for the telephone company to provision a local loop for these services than a full T1 (1.544 mbps, also called DS1), but they would have to charge less than the T1 price. What normally happens when a long distance carrier customer wants to extend their 384 kbps or 768 kbps service into the local loop, a full T1 service is connected and only a part of the T1 is used.

7. T1 service is offered by all telephone companies. The main use of this service has been for larger businesses that have large amounts of voice and data traffic. A growing market for T1 has been compressed video.

8. 45 mbps services are available in most telephone companies. The cost of 45 mbps (DS3) service is usually based on some factor of the 672 voice channel carrying capacity of the DS3 system.

9. Packet Switching is a service available in all telephone companies and mainly used for transporting low speed data between computers. It is a very efficient method of data transfer, and is priced per packet of information. The speed is usually below 64 kbps. The service is not practical at this time for video because the video signal requires a continuous signal, and Packet Switching is primarily designed for bursty type data. Paying for a continuous data stream by the packet would become prohibitively expensive.

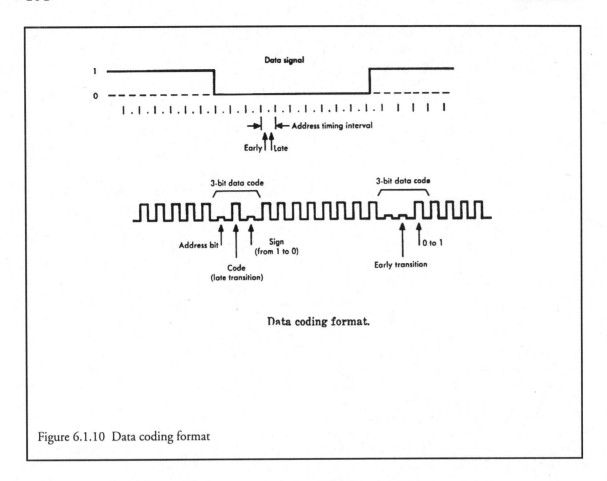

Figure 6.1.10 Data coding format

10. Video services that require from 56 kbps to 45 mbps, that are time and distance sensitive, are offered by a number of the long distance carriers. At least two of the local carriers are planning to add a dial-up time and distances sensitive video line service to their list of offerings.

11. One local carrier offers a full time 384 kbps and a T1 video service that provides the line and the video codec for a set price. This is an attempt to break through the price barriers that are based on the T1 ability to carry 24 voice circuits, to provide one circuit of video at a lower price.

12. One service that is planned, and in fact is in service on a limited basis is narrowband ISDN. The concept here is to provide two 64 kbps voice channels and one 16 kbps data channel on a single pair of copper wires in the local loop. The 144 kbps service has had a lot of attention for several years, and millions have been spent on the development of the technology. The islands of service do not require the ubiquitous switching capability of SS7 switching, which is an intelligent switch that is planned for deployment to larger cities over the next 5 to 10 years. There is a lot of criticism of the cost of providing universal ISDN. It will be slow coming to most cities and will be in competition with other technologies such as fiber to the curb or home, HDSL, Fast Packet etc.

13. HDSL is a new technology that is presently being developed and tested, that will provide up to 768 kbps on a copper pair up to 12,000 feet from the telephone central office. Two twisted pairs will be capable of providing 1.544 mbps. This technology holds much promise for providing sub-DS1 service on the local loop at a more reasonable cost. The same system technology that is used for ISDN is used for HDSL. This will be a very cost effective way to add bandwidth to the local loop, and may compete with ISDN.

14. ADSL is the same technology as HDSL except that it concentrates on the one way transmission of a full DS1 in one direction, on a single pair of copper wires. The main purpose for it's development is to provide one-way video and data services to the home.

15. Fiber to the home or business is another real possibility for the future. Local loop fiber has been deployed for trials to deliver full motion video, multiple voice and data circuits, with enough growth capability to add anything else that comes along.

16. SONET (Switched Optical Network) is an architecture that is planned to extend wideband capability to the home or business. It will require fiber to provide OC-3 (155 mbps) to a customer node that will distribute all or part of this bandwidth to the customer. A completely new telephone central office set of equipment will need to be deployed, such as multiplexers and wideband digital switching.

17. SMDS (Switched Multimegabit Digital

Service) is an architecture that is being planned to convert much of today's voice, data and video traffic to very fast packets that would lead to the ability to price by the amount of information being sent as opposed to time-distance-bandwidth pricing. Again, this technology may require fiber to provide at least a 45 mbps digital path to the customer.

18. Frame Relay is the same general concept as SMDS. It has been deployed for specific uses and seems to be cost effective for the telephone companies.

19. Cellular telephone is becoming more universal with the deployment of rural cellular systems that tie to the metropolitan cells. The message channels are digital in format which allows more channels to be placed in service.

20. Video in digital form can be sent over all of these technologies. Eventually it may be possible to send video by cellular telephone channels so that some form of video can be available almost everywhere.

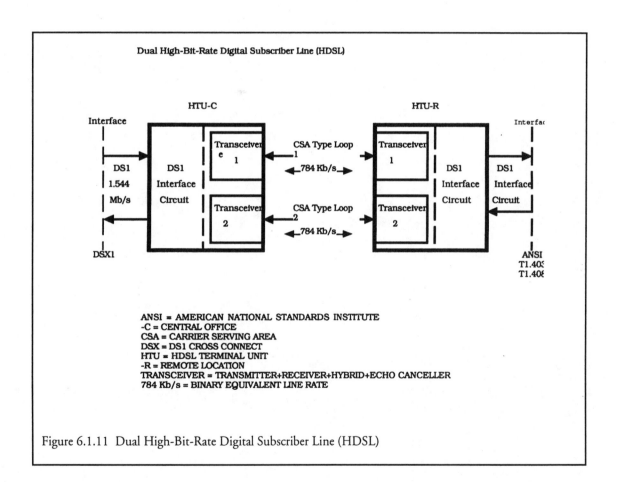

Figure 6.1.11 Dual High-Bit-Rate Digital Subscriber Line (HDSL)

Chapter 7

Richard A. Schaphorst is founder and president of Delta Information Systems, Horsham, PA. He obtained his BSEE degree from Lehigh University and worked at Philco Ford Aerospace in the areas of image communications prior to founding Delta. In recent years, he has been one of the key leaders in the development of standards for teleconferencing and videophone. He is a key member and contributor to the ANSI standards committee T1A1.5 and was one of three Core Members of the U.S. delegation to the Specialists Group for Coding for Visual Telephony which developed the H.261 Recommendation for the codec operating at Px64 kbps ISDN rates. Mr. Schaphorst was recently appointed to be one of two Coordinators for the U.S. delegation to the ITU Group for ATM Video Coding, and the ITU has designated him as Rapporteur for Very Low Bit Rate Videophone.

He has published articles and given numerous technical presentations on the subject of teleconferencing standards. Notable recent publications include an IEEE Press Publication entitled "Teleconferencing" and a book (co-authored) entitled "Digital Fax Technology and Applications."

Mr. Schaphorst was past chairman of the Facsimile Equipment and System Standards Committee (TR-29) of the EIA, and has been awarded three patents in the area of image processing.

VIDEOCONFERENCING STANDARDS (AUDIO/VISUAL COMMUNICATIONS STANDARDS)

■ INTRODUCTION

The facsimile marketplace was restrained for many years because fax units manufactured by different vendors were not compatible. The facsimile explosion we are experiencing today is traceable directly to the Group 3 standard developed by the ITU which makes it possible for fax units from different manufacturers to communicate with each other.

A similar scenario is in the process of unfolding for the video teleconferencing/video telephone market. The point-to-point teleconferencing market has been restrained by the lack of compatibility, and that restraint has been eliminated by the development of ITU Recommendation H.320 (and related standards) described herein. Table 7.1.1 is a list of all the recommendations related to audiovisual communications indicating their standardization status. There is little question that we will experience a rapid expansion in the teleconferencing area based upon the new standards.

There are three other developments which are occurring concurrently with the emergence of the standards which will also contribute to this growth. The first is the video compression technology upon which the standard is based. By combining the advantages of Predictive Coding, Discrete Cosine Transform (DCT), Motion Compensation and Variable Length Coding, the standard makes it possible to transmit TV pictures of acceptable quality at very low bit rates; rates which are reduced enough for low-cost communications over switched networks.

The second development is the advent of VLSI technology which reduces the cost of the video codecs. Chips are already on the market to implement the H.320, MPEG and JPEG Standards.

The third parallel development is ISDN (Integrated Services Data Network) which promises to provide low-cost, switched digital communication services to the user at a wide range of bit rates. The most basic ISDN communication element is the Basic Access which consists of two full-duplex 64-kbps B channels and a full duplex 16-kbps D channel. Thus the 64-kbps B channel is the basic building block on ISDN, and the H.261 standard is based upon this ISDN structure. This gives rise to the title of the H.261 Recommendation "Video Codec for Audiovisual Services at Px64 kbps." Although it will take several years for ISDN to be broadly available, the H.320 video terminals can readily operate over existing digital networks in the interim.

The purpose of this chapter is to provide a general background in the standards for audiovisual communications. The following topics are covered in the following sections.

■ STANDARDS BACKGROUND

The ITU is a part of the United Nations, and its purpose is to develop formal "Recommendations" to insure worldwide communications are accomplished efficiently and effectively.

In 1984, the first Recommendations for a teleconferencing codec (H.120 and H.130) were established. These recommendations were defined specifically for the European region (625 lines; 2.048 mbps primary rate) and for interconnection between Europe and other regions. Since no recommendation existed for non-European regions it lacked true international scope, and in 1984, the ITU established a "Specialists Group on Coding for Visual Telephony" to develop a truly international recommendation. The ITU established two objectives for the Specialists Group: 1. to develop a recommendation for a video codec for teleconferencing application operating at the bit rates of Nx384 kbps (N=1 through 5), and 2. to begin the standardization process for a video codec for teleconferencing/video telephone application operating at bit rates of Mx64 kbps (M=1,2).

The Specialists Group, chaired by S. Okubo from Japan, met 17 times from 1984 through 1989. Representatives from Canada, Finland, France, Germany, Italy, Japan, Korea, Netherlands, Norway, Sweden, the U.K. and the United States were members of the committee.

At the September meeting in 1988, it was determined that the compression algorithm chosen for Nx384 kbps was sufficiently flexible so that it could be extended, with good performance, down to 64 kbps. At that time, the Specialists Group shifted their focus to develop a single recommendation to code at all bit rates from 64 kbps to 2 mbps; i.e. to code at rates of Px64 kbps, where the key values of P are 1,2,6,24 and 30, and a final draft of a more complete H.261 was approved in December 1990.

Figure 7.1.1 also illustrates the domestic and international standards organizations which develop the U.S. technical positions on issues related to video

teleconferencing and video telephones. The T1 committee, which is accredited by ANSI (American National Standards Institute), works on two different aspects of teleconferencing: the coding algorithms and defining and measuring the quality of service to be provided by teleconference systems.

H.320: Px64 Overview

In December 1990, the ITU finalized a set of five recommendations (H.261, H.221, H.242, H.230, and H.320) which collectively define an audiovisual terminal to provide video teleconferencing (VTC) and video telephony (VT) services over the Integrated Services Data Network (ISDN). Since the basic building block of the ISDN is a basic channel operating at 64 kbps, the generic term "Px64 kbps" refers to operation of this terminal at integral values of P up to a maximum of 30. (Values of P which are of greatest interest are 1, 2, 6,12, 24 and 30.)

ITU Recommendation H.320 defines the interrelationship between all of the five Px64 recommendations as shown in Figure 7.1.2. The H.221 recommendation specifically makes provision for multiplexing data signals with video and audio signals.

One function of the H.320 Recommendation is to define the phases of establishing a visual telephone call as listed below.

Phase A: Call set-up, out-band signalling
Phase B1:   Mode initialization on initial channel
Phase CA: Call set-up of additional
               channel(s), if relevant
Phase CB1:   Initialization on additional channel(s)
Phase B2 (or CB2):   Establishment of com
               mon parameters
Phase C: Visual telephone communication
Phase D: Termination phase
Phase E: Call release

Another function of Recommendation H.320 is the definition of 16 different types of visual telephone terminals and their modes of operation.

Video Coding Standard

If the standard TV signal were to be encoded using conventional 8-bit PCM, a bit rate of approximately 90 mbps would be required for transmission. Video compression technology is employed to reduce this bit rate to the primary rates (1.544Mbps, 2.048 Mbps), half primary rates, 384 kbps and basic rates (64 kbps and multiples) which are employed for economical transmission. The compression function is performed by a video codec, and H.261 is the video codec ITU Recommendation for teleconferencing.

Figure 7.1.3 is a functional block diagram of the

video codec as defined in Recommendation H.261. The heart of the system is the source coder which compresses the incoming video signal by reducing redundancy inherent in the TV signal. The multiplexer combines the compressed data with various side information which indicates alternative modes of operation. A transmission buffer is employed to smooth the varying bit rate from the source encoder to adapt it for the fixed bit rate communication channel. A transmission coder includes functions such as forward error control to prepare the signal for the data link.

One of the most challenging problems to be solved by the codec was the reconciliation of the incompatibility between European TV standards (PAL, SECAM) and those in most other areas of the world — NTSC. PAL and SECAM employ 625 lines and a 50 Hz field rate while NTSC has 525 lines and a 60 Hz field rate. This conflict was resolved by adopting a Common Intermediate Format (CIF) and QCIF (Quarter CIF) as the picture structure which must be employed for any transmission adhering to H.261. The CIF and QCIF parameters are defined in Table 7.1.2.

The QCIF format, which employs half the CIF spatial resolution in both horizontal and vertical directions, is the mandatory H.261 format: full CIF is optional. It is anticipated that QCIF will be used for videophone applications where head-and-shoulders pictures are sent from desk to desk. Conversely, it is assumed that the full CIF format will be used for teleconferencing where several people must be viewed in a conference room.

Figure 7.1.4 is a functional block diagram outlining the H.261 source coder. Interframe prediction is first carried out in the pixel domain. The prediction errors are encoded by the Discrete Cosine Transform using blocks of 8 pels x 8 pels. The Transform coefficients are next quantized and fed to the multiplexer. Motion compensation is included in the prediction on an optional basis.

Picture Structure

In the encoding process, each picture is subdivided into Groups of Blocks (GOB). As shown in Figure 7.1.5, the CIF picture is divided into 12 GOB's while QCIF has only three GOB's. From the GOB level down, the structure of CIF and QCIF is identical. A header at the beginning of the GOB permits resynchronization and changing the coding accuracy.

Each GOB is further divided into 33 macroblocks, as shown in Figure 7.1.6. The macroblock header defines the location of the macroblock within the GOB, the type of coding to be performed, possible motion vectors, and which blocks within the mac-

roblock will actually be coded. There are two basic types of coding. In Intra coding, coding is performed without reference to previous pictures. This mode is relatively rare, but is required for forced updating, and every macroblock must occasionally be Intra coded to control the accumulation of inverse transform is match errors. The more common coding type is Inter, in which only the difference between the previous picture and the current one is coded. Of course, for picture areas without motion, the macroblock does not have to be coded at all.

Each macroblock is further divided into six blocks, as shown in Figure 7.1.7. Four of the blocks represent the luminance, or brightness, while the other two represent the red and blue color differences. Each block is 8x8 pixels, so it can be seen that the color resolution is half of the luminance resolution in both dimensions.

Example of Block Coding

Figure 7.1.8 shows a simple example of how each 8x8 block is coded. In this case, Intra coding is used, but the principle is the same for Inter coding. Figure 7.1.8a shows the original block to be coded. Without compression, this would take 8 bits to code each of the 64 pixels, or a total of 512 bits. First, the block is transformed, using the two-dimensional Discrete Cosine Transform (DCT), giving the coefficients of Figure 7.1.8b. Note that most of the energy is concentrated into the upper left-hand corner of the coefficient matrix. Next, the coefficients of Figure 7.1.8b are quantized with a step size of 6. (The first term {DC} always uses a step size of 8.) This produces the values of Figure 7.1.8c, which are much smaller in magnitude than the original coefficients and most of the coefficients become zero. The larger the step size, the smaller the values produced, resulting in more compression.

The coefficients are then reordered, using the Zig-Zag scanning order of Figure 7.1.9. All zero coefficients are replaced with a count of the number of zero's before each non-zero coefficient (RUN). Each combination of RUN and VALUE produces a Variable Length Code (VLC) that is sent to the decoder. The last non-zero VALUE is followed by an End of Block (EOB) code. The total number of bits used to describe the block is 25, a compression of 20:1.

At the decoder (and at the coder to produce the prediction picture), the step size and VALUE's are used to reconstruct the inverse quantized coefficients, which, as shown in Figure 7.1.8e are similar to, but not exactly equal to, the original coefficients. When these coefficients are inverse transformed, the result of Figure 7.1.8f is obtained. Note that the differences between this block and the original block are quite small.

Motion Compensation

The operation of motion compensation is shown in Figure 7.1.10. Block "A" is a block in the current picture that is to be coded. Block "B" is the block at the same position as "A" but in the picture that was previously stored in both coder and decoder. Because of image motion, block "A" more closely resembles the pixel data from block "C" than that from block "B". The displacement of block "C" from block "B", measured in pixels in x and y directions, is the motion vector. The pixel-by-pixel difference between blocks "A" and "C" is transformed and coded. The motion vector and code data are transmitted to the decoder, where the inverse transformed block data is added to the data in block "C" pointed to by the motion vector, and placed in the block "A" position.

The use of motion vectors is optional in the coder, where the calculation of the optimum motion vectors is complex, but required in the decoder, where the reconstruction of the motion is relatively simple.

The H.261 standard does not define all aspects of image coding and decoding. Rather it is just an interoperability specification, guaranteeing that any codecs manufactured according to the standard will be able to communicate with each other. This still allows considerable freedom for manufacturers to offer better performance, and new developments may be able to be incorporated. (This is in contrast with the G.722 audio standard, where the coding algorithm is rather precisely defined.) For example, the encoder strategy is not defined. Which blocks will be encoded, with what type of code, and with what accuracy is under control of the designer. While there is less freedom for the decoder, post processing, such as filtering or interpolation of the image is under control of the designer.

Furthermore, the H.261 permits two codecs to negotiate to an altogether different proprietary algorithm that they both incorporate, with the H.261 algorithm becoming a fall-back mode for codecs of different manufacture.

H.221: Frame Structure for a 64 to 1920 kbps Channel in Audiovisual Teleservices

The purpose of this recommendation is to define a frame structure for audiovisual teleservices in single or multiple B or HO channels or a single H11 or H12 channel which makes the best use of the characteristics and properties of the audio and video encoding algorithms, of the transmission frame structure, and of the existing ITU Recommendations. It offers several advantages:

• It is simple, economic and flexible. It may be implemented on a simple microprocessor using well-known hardware principles.

• It is a synchronous procedure. The exact time of a configuration change is the same in the transmitter and the receiver. Configurations can be changed at 20 ms intervals.

• It needs no return link for audiovisual signal transmission, since a configuration is signalled by repeatedly transmitted codewords.

• It is very secure in case of transmission errors, since the code controlling the multiplex is protected by a double-error correcting code.

• It allows the synchronization of multiple 64 kbps or 384 kbps connections and the control of the multiplexing of audio, video, data and other signals within the synchronized multiconnection structure in the case of multimedia services such as videoconference.

This recommendation provides for dynamically subdividing an overall transmission channel of 64 to 1920 kbps into lower rates suitable for audio, video, data and telematic purposes. The overall transmission channel is derived by synchronizing and ordering transmissions over from 1 to 6B connections, from 1 to 5HO connections, or an H11 or H12 connection.

A single 64 kbps channel is structured into octets transmitted at 8 kHz. Each bit position of the octets may be regarded as a sub-channel of 8 kbps. The eighth sub-channel is called the Service Channel (SC), containing these two critical parts.

• FAS (Frame Alignment Signal): This 8 bit code is used to frame the 80 octets of information in a B channel.

• BAS (Bit-rate Allocation Signal): This 8-bit code describes the capability of a terminal to structure the capacity of the channel or synchronized multiple channels in various ways, and to command a receiver to demultiplex and make use of the constituent signals in such structures. This signal is also used for controls and indications.

The video bit stream is carried in frames of data as shown in Figure 7.1.11. Each frame corresponds to a 64 kbps B channel in ISDN. Two frames are shown, one for the audio portion of the conference, and the other for the video portion. In each, there is an 8-bit Frame Alignment Signal (FAS) that permits synchronization of the frame and low-speed signalling of communication overhead. There is also an 8-bit Bitrate Allocation Signal (BAS) that defines how the H.221 channels and sub-channels are divided, and what type of service is used on each section. For example, one BAS code is used for "Standard Video to Rec. H.261", while another might indicate that two B channels are allocated to this service. The BAS codes can change from frame to frame to indicate complex protocols or changes of mode.

Each frame of 640 bits is transmitted in 10 msec, giving an overall rate of 64,000 bits per second. However, the FAS and BAS use 16 of the 640 bits, so the net rate available for video is only 62.4 kbps for a single B channel. The order of transmission is left to right across each row, and then the row below. Higher bit rates can be obtained by using multiple B channels (up to 2 for ISDN Basic access, up to 30 for Primary access).

H.242: System for Establishing Communication Between Audiovisual Terminals Using Digital Channels up to 2 mbps

Recommendation H.242 defines the detailed "handshake" protocol and procedures which are employed by H.320 terminals in the preliminary phases of a call. Major topics covered in this Recommendation are listed below.
- Basic sequences for in-channel procedures
- Mode initialization, dynamic mode switching and mode O forcing recovery from fault conditions
- Network consideration: Call connections, disconnection and call transfer
- Procedures for activation and deactivation of data channels
- Procedures for operation of terminals in restricted networks

H.230: Frame Synchronous control and Indication Signals for Audivisual Systems

Digital audiovisual services are provided by a transmission system in which the relevant signals are multiplexed onto a digital path. In addition to the audio, video, user data and telematic information, these signals include information for the proper functioning of the system. The additional information has been named 'control and indication" (C&I) to reflect the fact that while some bits are genuinely for "control", causing a state change somewhere else in the system, others provide for indications to the users as to the functioning of the system.

Recommendation H.230 has two primary elements. First, it defines the C&I symbols related to video, audio, maintenance and multipoint. Second, it contains a table of BAS escape codes which clarifies the circumstances under which some C&I functions are mandatory and others optional.

Audio Coding

The BAS codes of H.221 are used to signal a wide range of possible audio coding modes. The most prominent modes define existing ITU Recommendations G.711, G.722 and G.728. Recommendation G.711 (Pulse Code Modulation of Voice Frequencies) is used for narrowband speech since it samples only at 8,000 samples/sec. and

encodes to 8 bits/sample for a transmission rate of 64 kbps.

Recommendation G.722 (7kHz Audio-coding with 64 kbps) describes the characteristics of an audio (50 to 7,000 Hz) coding system which may be used for a variety of higher quality speech applications, The coding system uses sub-band adaptive differential pulse code modulation (SB-ADPCM) within a bit rate of 64 kbps. In the SB-ADPCM technique used, the frequency band is split into two sub-bands (higher and lower) and the signals in each sub-band are encoded using ADPCM. The system has three basic modes of operation corresponding to the bit rates used of 7kHz audio coding: 64, 56, and 48 kbps.

G.728 is a new (1992) ITU Recommendation used for the transmission of toll voice quality at 16 kbps. The coding algorithm is a Code Excited Linear Predictive Codes.

H.320: Data Channels

The H.320 set of recommendations defines two classes of data channels — [1] LSD/HSD, [2] MLP — which can be multiplexed for transmission with the video and audio signals. Each of these classes is, in turn, divided into high speed and low speed categories. The low speed channels — LSD and MLP — operate at rates less than 64 kbps. The high speed channels — HSD and H-MLP — operate at multiples of 64 kbps. While H.320 defines the means for establishing this wide range of transparent data channels, it does not define a communication protocol which is required to transmit data over these channels. The ITU is in the process of standardizing two different, but interrelated, protocols for this purpose. The most general protocol is the T.120 stack which is described later in the chapter under the heading of audiographic terminal. This protocol is designed for use on the MLP channels. An additional protocol (H.224) is being considered for the LSD/HSD channels for applications requiring low delay, simplex transmission such as far-end camera control. The structure of the H.320 data channels is illustrated in Figure 7.1.12.

Multipoint

The two standards for multipoint operation of H.320/Px64 terminals listed below were completed in 1993:

H.231 Multipoint Control Unit (MCU) for Audiovisual Services. H.243 System for establishing communication between three or more audiovisual terminals using digital channels up to 2 mbps.

The MCU provides optional means for conference control and add automatic control by voice activation, manual control by a chairman. Several manufacturers are now manufacturing MCU systems meeting the H.231 standard.

Privacy

The ITU is actively working on recommendations to provide for the privacy of transmission between audiovisual terminals. A privacy system consists of two parts; the confidentiality mechanism or encryption process for the data, and a key management subsystem.

Recommendation H.233, draft recommendation of confidentiality system for audio visual services, was finished in 1993, and H.234, Draft Recommendation on Authentication and Encryption Key Management System for Audiovisual Services. was completed in 1994.

H.233

This document describes the confidentiality part of a privacy system suitable for use in narrowband audiovisual services conforming to ITU Recs. H.221, H.230 and H.242. Although an encryption algorithm is required for such a privacy system, the specification of such an algorithm is not included here. The system provides for three specific algorithms [DES, FEAL, BCRYPT], and more are being added.

H.233 will be completed in two phases. Phase I for point-to-point encryption was completed in 1993 while multi-point encryption will be defined later.

H.234

This document described authentication and key management methods for a privacy system suitable for use in narrow band audiovisual services conforming to ITU Recs. H.221, H.230 and H.242. Privacy is achieved by the use of secret keys. The keys are loaded into the confidentiality part of the privacy system and control the way in which the transmitted data is encrypted and decrypted. If a third party gains access to the keys being used, then the privacy system is no longer secure. The maintenance of keys by users is thus an important part on any privacy system. The H.234 Draft Recommendations is considered stable at this time.

■ AUDIOGRAPHIC TERMINAL

Study Group 8, of the ITU, is developing a set of recommendations which defines an audiographic terminal used for teleconferencing. The series of standards is listed in Table 7.1.1, and is generally designated as the T.120 stack since that recommendation

is the overview for the entire set of standards.

As shown in the table, some of the recommendations (T.123, T.122) have been fully approved, while others (T.120, T.125) are not yet stable working drafts. Nevertheless, in a few years, it is anticipated that the T.120 series will be the dominant mode for transmitting graphics, screen sharing, collaborative computing, still pictures, facsimile, white boards, cursor, etc. between audiographic teleconferencing terminals. In addition, it is anticipated that T.120 will be the primary protocol used to transmit this type of data information in the H.320 terminal. One of the most important attributes of T.120 is that it inherently provides a robust multipoint capability. Figure 7.1.13 illustrates how the standards are configured for multipoint operation. Several companies have formed a consortium (CATS), the purpose of which is to exploit the potential advantages of the T.120 terminal.

Video Coding at Very Low Bit Rates

In the last couple of years, it has become apparent that there is a class of audiovisual terminal requirements which cannot be met by existing standards such as the MPEG1/2 and H.320 families. For example, only the conventional telephone network will continue to exist for many years in rural locations. Also, wireless communication is poised to be a critical technology, and will always be limited in bitrate. Recognizing this need, the ITU (International Telecommunications Union) and ISO (International Standards Organization) have initiated parallel projects to develop standards to code video at very low bit rates. Both organizations are working on long term standards to be completed around the 1998 time frame. The ITU has also begun a short term program to develop a family of videophone recommendations by 1995.

The purpose of the near term ITU Recommendations is to provide an international standard to insure interoperability between videophones connected via the Public Switched Telephone Network [PSTN]. One reason for the short schedule is that there are several PSTN videophones already in the marketplace none of which are interoperable. To achieve the 1995 objective it is necessary to use existing standards wherever possible or to make incremental improvements to existing technology.

Key objectives of the long term program are to significantly improve the video and speech quality relative to the near term standards, and to provide for transmission via mobile radio networks as well as the PSTN. Figure 7.1.14 is a list of the ITU Recommendations to be developed on both the near term and long term schedules. Figure 7.1.15 is a block diagram of the Very Low Bitrate Videophone illustrating the functions of the various Recommendations being developed. Two comments on the work which is underway are provided below.

H.DLP will define a data port used to transmit miscellaneous data in addition to speech and video. This is a new capability not available in most videophone products which are now in the market. The data channel makes possible a true multimedia terminal on a workstation platform to provide for a wide range of services such as collaborative computing, white board graphics, retrieval from a data base, etc.

The performance of existing videophones is significantly limited by their transmission bit rates. The proposed H.32P system will employ the V.34 modem which is in the final stages of standardization. The V.34 has a maximum bit rate of 28.8 kbps which will significantly improve the videophone performance.

Figure 7.1.16 is a chart which illustrates one example of a bitrate budget for the proposed H.32P videophone system operating at the full range of bit rates from 9.6 to 28.8 kbps. Four virtual channels are shown for speech, video, data, and overhead/supervision. It is not likely that the channel priorities will be standardized. Nevertheless, it is likely that the speech would have the highest priority followed by data and video.

To achieve the schedule requirement, the H.261 video coding algorithm will be an extension of H.261. However, to adapt H.261 for the videophone application, a number of significant changes are required. Examples of the changes being considered are listed below:

• H.261 transmits motion vectors with a quantization precision of one pixel. It is clear that more precision is required for the videophone. Accuracies of one-quarter and one-half pixel are being studied. Independent studies are underway for both luminance and chroma motion compensation.

• H.261 employs motion compensation on a macroblock basis [16x16 pixels]. Studies are underway to determine whether this also should be accomplished with more precision—8x8 or 16x8/8x16.

• VLC codes may be further optimized. As indicated earlier, the ITU work to develop Recommendation H.26P/L is being accomplished in close collaboration with the ISO MPEG4 activity. The objective is to achieve a video coding algorithm which significantly outperforms the H.26P technique. To accomplish this objective it is expected that a significant change in the video source model is required. Other models which may be considered are illustrated in Figure 7.1.17.

Broadband ISDN

Integrated Service Data Networks (ISDN) are divided into two parts - Narrowband and Broadband.

N-ISDN operates at rates equal to, or less than, the primary rates (e.g. 1.544 mbps) while the B-ISDN operates at rates above the primary rates.

Broadband aspects of the ISDN (B-ISDN) are being studied by ITU Study Group XVIII for a future customer-switched digital network. SGXVIII decided to standardize the Network Node Interface (NNI) by a worldwide unique synchronous Digital Hierarchy (SDH). This was achieved by Working Party 7 which is responsible for transmission aspects of digital networks. Figure 7.1.18 illustrates the world-wide unique NNI. The SDH specified 155.52 Mbps as the world-wide unique interface bit rate. The proposal of Study Group XVIII for B-ISDN as described in Recommendation I.121 is that the target transfer mode is the Asynchronous Transfer Mode (ATM), in which the data is transmitted in a series of fixed size blocks called cells. Packet-switched networks already exist for the transmission of digital data for non-real time services (for example, the exchange of information between computer databases). In this instance, if a packet is corrupted or lost, the receiving terminal can request that the particular packet be retransmitted. Recommendation I.121, however, envisages that the B-ISDN will carry all the telecommunications services provided in the future including real-time services such as telephony, videoconferencing and videophony, as well as television and sound contribution and distribution services. For these real-time services, if a cell is corrupted or lost, retransmission of cells is not possible and so degradation of the signal may occur.

The main advantage of ATM is that the network switches are no longer bit-rate and service specific; in the B-ISDN all services (including future new, and as yet unspecified services) are expected to be carried, and a common user-network interface will exist for all services. Many of the important parameters of B.ISDN have still to be specified. However, an ATM-based network will introduce some effects not experienced in synchronous networks, such as cell delay jitter and occasional cell loss.

An ATM-based network will, in principle, provide the user with whatever bit rate is required (within the constraints of the interface and the network), so that teleconference users, for example, could decide on the optimum picture quality required by sessions. Additionally, new television services at different bit rates could be transmitted over the network through the same user-network interface. With continuing improvement in picture coding algorithms, and with advances in technology allowing more complex algorithms to be implemented, service providers could, in the future, offer either an improved quality of service at the same average bit rate, or the same quality of service at a lower average bit rate. An ATM-based network will in principle have the flexibility to provide additional transmission capacity when required, and could allow the development of a variable bit rate/constant quality coding scheme.

Constant Bit Rate (CBR) and Variable Bit Rate (VBR) Coding

Restrictions of traditional circuit switched networks have meant that all commercial digital video codecs operate at a constant bit rate, despite the inherently varying information content of a motion video sequence (being dependent on changing image complexity, degree of motion, frequency of scene changes, etc.). The internally varying rate in these codecs is smoothed by buttering and dynamic control of codec parameters (sensitivity, quantizer stepsize, etc.) to ensure that the buffer neither empties nor overflows. Such codecs operate in a fixed rate, but variable quality, mode.

ATM Networks will support VBR coded video, allowing the transmitted bit rate to dynamically reflect the information content of the changing video signal, limited by the maximum channel capacity and parameters agreed with the network management system. AVBR codec can therefore (usually) maintain a fixed quality, variable bit rate mode of operation. The possible advantages to this are:

• Because data is not transmitted when the information content is low, and because high rates are only used when necessary, VBR codecs are expected to deliver a given overall quality at a lower average rate than a CBR codec.

• The reduction in buffer size and easing of constraints on rate control means that there could be savings in codec complexity and cost.

• Reduced buffering may mean that end-to-end delays will be reduced; this is an important consideration for communicative services such as videotelephony and videoconferencing.

• Working Party 1 of ITU Study Group 15 has established an "Experts Group on ATM Video Coding" to develop a number of Recommendations for a video transmission via the B-ISDN.

• The Experts Group is developing Recommendations for the three audiovisual terminals listed below.

• H.32X An audiovisual terminal for operation over the B-ISDN (see Figure 7.1.19).

• H.32Y Emulations of the H.320 terminal for operation over the B-ISDN.

• An audiovisual terminal for operation over Legacy LANs such as Ethernet, Token Ring, FDDI.

• ISO Standard for AV Storage and Retrieval (MPEG)

The ISO (International Standards Organization) standard organization has established a Working Group, (ISO/IECJTCI/SC2/WG11) known as MPEG (Motion Picture Experts Group), to develop two standards for coding AV signals for storage on digital storage media. The bit rates for the two standards (MPEG1, MPEG2) are up to about 1.5 and 40 mbps respectively. MPEG1 is essentially complete and is designated as International Standard 11172. The MPEG1 standard has three parts (Systems, Video, Audio) which are summarized here.

Systems Layer

An ISO 11172 bit stream is constructed in two layers; the outer most layer is the system layer and the inner most is the compression layer. The system layer provides the functions necessary for using one or more compressed data streams in a system. The video and audio parts of this specification define the compression encoding layer for audio and video data. Coding of other types of data are not defined by the specification, but are supported by the system layer providing that the other types of data adhere to the system constraints. The system layer supports four basic functions: The synchronization of multiple compressed streams on playback, the interleaving of multiple compressed streams into a single stream, the initialization of buffering for playback start up, and time identification.

Video Coding

The MPEG video coding standard specifies the coded representation of video for digital storage media and specifies the decoding process. The representation supports normal speed forward playback as well as special functions such as random access, fast play, fast reverse play, normal speed reverse playback, pause and still procedures. This international standard is compatible with standard 525- and 625-line television formats, and it provides flexibility for use with personal computer and workstation displays.

This international standard is primarily applicable to digital storage media supporting a continuous transfer rate up to about 1.5 mbps, such as Compact Disc, digital Audio Tape, and magnetic hard disks. The storage media may be directly connected to the decoder, or via communications means such as busses, LANs, or telecommunications links. This international standard is intended to non-interlaced video formats having approximately 288 lines of 352 pels and picture rates around 24 Hz to 30 Hz. The basic MPEG1 coding algorithm, i.e., the objective is to provide VCR picture quality, is similar to H.261; i.e. 8x8 DCT, interframe prediction, motion compensation.

Audio Coding

This standard specifies the coded representation of high quality audio for storage media and the method for decoding of high quality audio signals. It is compatible with the current formats (Compact Disc and Digital Audio Tape) for audio storage and playback. This representation supports normal speed playback. This standard is intended for application to digital storage media providing a total continuous transfer rate of about 1.5 mbps for both audio and video bit streams, such as CD, DAT and magnetic hard disc. The storage media may either be connected directly to the decoder, or via other means such as communication lines and the MPEG Systems layer. This standard is intended for sampling rates of 32kHz, 44.1kHz, 48kHz, and 16 bit PCM input/output to the encoder/decoder.

MPEG2

Work on the development of the MPEG2 standard is nearing completion. It has been decided that the MPEG2 coding algorithm will be an extension of the MPEG1 algorithm: 8x8 DCT, interframe prediction, motion compensation. It is anticipated that MPEG2 will provide picture quality fully equivalent to interlaced broadcast TV.

Like MPEG1, MPEG2 consists of three interrelated standards; Video Coding, Audio Coding, and System. The video coding standard is the most advanced of the three having reached the status of Draft International Standard [DIS]. Again like MPEG1, MPEG2 is a generic toolkit designed for a wide range of applications such as storage/retrieval, broadcast TV [contribution, distribution], satellite TV transmission, High Definition TV, etc. In many ways, MPEG2 is similar to MPEG1. The primary difference is that MPEG2 provides for the coding of an interlaced TV frame while MPEG1 is restricted to the coding of one field. This difference, and others, are summarized in Table 7.1.3. The table also compares MPEG operation with H.261 coding.

The MPEG2 toolkit standard is organized into levels and profiles as illustrated in Figure 7.1.20. In general, the Level axis refers to picture resolution, where the 1440 level refers to the number of pixels in an HDTV picture. The Profile axis describes the complexity of the set of tools required for the application. For example, the Simple Profile prohibits the use of B-frames. The Main Profile/Main Level is the most basic and fundamental application of MPEG2 employing a resolution and complexity appropriate for conventional broadcast TV.

JPEG Coding Algorithm

The Joint Photographic Experts Group (JPEG) is an ISO/ITU working group which developed an international standard ("Digital Compression and Coding of Continuous-tone Still Images") for general-purpose, continuous-tone (gray scale or color), still-image compression. The aim of the standard algorithm is to be general-purpose in scope to support a wide variety of image communications services. JPEG reports jointly to both the ISO group responsible for Coded Representation of Picture and Audio Information (ISO/IEC JTC1/SC2/WG8) and to the ITU Special Rapporteur group for Common Components for Image Communication (a subgroup of ITU SGX). This dual reporting structure is intended to ensure that the ISO and the ITU reduce compatible image compression standards.

The JPEG draft standard specifies two classes of encoding and decoding processes, lossy and lossless processes. Those based on the discrete cosine transform (DCT) are lossy, thereby allowing substantial compression to be achieved while producing a reconstructed image with high visual fidelity to the encoder's source image. The simplest DCT-based coding process is referred to as the baseline sequential process. It provides a capability which is sufficient for many applications. There are additional DCT-based processes which extend the baseline sequential process to a broader range of applications. In any application environment using extended DCT-based decoding processes, the baseline decoding process is required to be present in order to provide a default decoding capability. The second class of coding processes is not based upon the DCT and is provided to meet the needs of applications requiring lossless compression (e.g. medical x-ray imagery.) These lossless encoding and decoding processes are used independently of any of the DCT-based processes.

The Baseline System

Baseline System is the name given to the simplest image coding/decoding capability proposed for the JPEG standard. It consists of techniques well-known to the Image coding community, including 8x8 DCT, uniform quantization, and Huffman coding. Together these provide a lossy, high-compression image coding capability, which preserves good image fidelity at high compression rates. The Baseline System provides sequential build-up only.

The baseline system codes an image to full quality in one pass and is geared towards line-by-line scanners, printers, and Group 4 facsimile machines. Typically, the process starts at the top of the image and finishes at the bottom; allowing the recreated image to be built up on line by line basis. One advantage is that only a small part of the image is being buffered at any given moment. Another feature stipulates that the recreated image need not be an exact copy of the original. The idea being that an almost indistinguishable copy of the original is just as good as an exact copy for most purposes. By not requiring exact copies, higher compression, which translates into lower transmission times, can be realized. Together these features are know as lossy sequential coding or transmission.

Figure 7.1.21 shows the main procedures for all encoding processes based on the DCT. It illustrates the special case of single component image (as opposed to multiple component color images); this is an appropriate simplification for overview purposes, because all processes specified in this International Standard operate on each image component independently.

In the encoding process, the input component's samples are grouped into 8x8 blocks, and each block is transformed by the forward DCT (FDCT) into a set of 64 values referred to a DCT coefficients. One of these values is referred to as the DC coefficient and the other 63 as the AC coefficients.

Each of the 64 coefficients is then quantized using one of 64 corresponding values from a quantization table (determined by one of the table specifications shown in Figure 7.1.21). No default values for quantization tables are specified in this International Standard; applications may specify values which customize picture quality for their particular image characteristics, display devices, and viewing conditions.

After quantizations, the DC coefficient and the 63 AC coefficients are prepared for entropy encoding. The previous quantized DC coefficient is used to predict the current quantized DC coefficient, and the difference is encoded. The 63 quantized AC coefficients undergo no such differential encoding, but are converted into a one-dimensional zig-zag sequence which is common for DCT coding.

All of the quantized coefficients are then passed to an entropy encoding procedure, which compresses the data further. Since Huffman coding is used in the Baseline System, Huffman table specifications must be provided to the encoder, as indicated in Figure 7.1.21.

Huffman coding has two forms: fixed and adaptive. Fixed Huffman coding assumes that coding tables can be generated in advance from test images and then used for many images. In Adaptive Huffman coding, the encoder analyzes an image's statistics before coding and devises Huffman tables tailored to that image. These tables are then transmitted to the decoder. Then the image is coded and transmitted. Upon receipt, the decoder can reconstruct the image using the previously transmitted, tailor-made, Huffman tables.

Figure 7.1.22 shows the main procedures for all DCT-based decoding processes. Each step shown performs essentially the inverse of its corresponding main procedure within the encoder. The entropy decoder decodes the zig-zag sequence of quantized DCT coefficients. After dequantizations, the DCT coefficients are transformed to an 8x8 block of samples by the inverse DCT (IDCT). For DCT-based processes, two alternative sample precisions are specified: Either 8 bits or 12 bits per sample. The baseline process uses only 8-bit precision.

Extended System

Extended system is the name given to a set of additional capabilities not provided by the Baseline System. Each set is intended to work in conjunction with, and to build upon, the components internal to the Baseline system, in order to extend its modes of operation. These optional capabilities, which include Arithmetic coding, progressive build-up and "progressive lossless" coding, and others, may be implemented singly or in appropriate combinations.

Arithmetic coding is an optional, "modern" alternative to Huffman coding. Because the Arithmetic coding method chosen adapts to image statistics as it encodes, it generally provides 5-10 percent better compression than the Huffman method chosen by JPEG. This benefit is balanced by some increase in complexity.

Progressive build-up, the alternative to sequential build-up, is especially useful for human interactive with picture databases over low-band-width channels. For progressive coding: first, a coarse image is sent, then, refinements are sent, improving the coarse image's quality until the desired quality is achieved. This process is geared towards applications such as image data bases with multiple resolution and quality requirements, freeze-frame teleconferencing, photovideotex over low speed lines, and data-base browsing. There are three different, complementary, progressive, extensions: spectral selection, successive approximations, and hierarchical.

Progressive lossless refers to a lossless compression method which operates in conjunction with progressive build-up. In this mode of operation, the final stage of progressive build-up results in a received image which is bit-for-bit identical to the original (though at press time, there is the possibility that this mode may become a "pseudo-lossless" capability, which would guarantee only that output pixels would be within one level of the original pixels.)

The JPEG draft standard includes the requirement that the Baseline System be contained within every JPEG-standard codec which utilizes any of the Extended System capabilities. In this way, the Baseline System can serve as a default communications mode for services which allow encoders and decoder to negotiate. In such cases, image communicability between any JPEG sender and receiver which are not equipped with a common set of Extended system capabilities is assured.

Lossless Coding

Figure 7.1.23 shows the main procedures for the lossless encoding processes. A predictor combines the values of up to three neighborhood samples (A,B, and C) to form a prediction of the sample indicated by X in Figure 7.1.24. This prediction is then subtracted from the actual value of sample X, and the difference is losslessly entropy-coded by either Huffman or arithmetic coding.

The JBIG Coding Algorithm

In 1988, an experts group was formed to establish an international standard for the coding of bi-level images. The JBIG (Joint Bi-level Image Group) is sponsored by the ISO (IEC/JTC1/SC2/WG9) and the ITU (SG 8). In 1993 the JBIG finalized the standard entitled "Progressive Bi-level Image Compression Standard" and much of the material in this section is derived from this document.

The JBIG standard defines a method for compressing a bi-level image (that is, an image that, like a black-and-white image, has only two colors.) Because the method adapts to a wide range of image characteristics, it is a very robust coding technique. On scanned images of printed characters, observed compression ratios have been from 1.1 to 1.5 times as great as those achieved by the Modified READ (MMR) encoding of the ITU Recommendations T.4 (G3) and T.6 (G4). On computer generated images of printed characters, observed compression ratios have been as much as five times as great. On images with greyscale rendered by halftoning or dithering, observed compression ratios have been up to 30 times as great.

The method is bit-preserving, which means that it, like the ITU T.4 and T.6 recommendations, is distortion-less and that the final decoded image is identical to the original.

The JBIG standard provides for both sequential and progressive operation. When decoding a progressively coded image, a low-resolution rendition of the original image is made available first with subsequent doublings of resolution as more data is decoded. Progressive encoding has two distinct benefits. One is that one common data-base can efficiently serve output devices with widely different resolution capabilities. Only that information in the compressed image file that allows reconstruction to the resolution capability of the particular output device need be sent and decoded. Also, if additional resolution enhancement is desired, for say, a paper copy of something already on a CRT screen, only the needed updating information has to be sent.

The other benefit of progressive encoding is that it provides subjectively superior image browsing (on a CRT) over low-rate and medium-rate communication links. A low-resolution rendition is rapidly transmitted and displayed, with as much resolution enhancement as desired then following. Each stage of resolution enhancement builds on the image already available. Progressive encoding makes it easy for a user to quickly recognize the image being displayed, which makes it possible for the user to quickly interrupt the transmission of an unwanted image.

Although the primary aim of this recommendation is bi-level image encoding, it is possible to effectively use this standard for multi-level image encoding by simply encoding each bit plane independently as though it were itself a bi-level image. When the number of such bit planes is limited, such a scheme can be very efficient while at the same time providing for progressive buildup in both spatial refinement and greyscale refinement.

■ CONCLUSIONS

It is concluded from the above discussion that there are a large number of standards for audiovisual terminal equipment which have already been finalized, and there are many enhancements on the drawing boards. The explosive growth in videoconferencing, distance learning, videophone and multimedia services is underway.

It is generally acknowledged that the growth of these industries, which is still in the infancy, will become a fax-like revolution. This revolution will occur for the following reasons.

• The standards provide picture quality
which is at least competitive with propri etary systems in the marketplace.

• The picture quality will continually improve due to competitive market pressure.

• The systems are very flexible permitting a widerange of operational modes — videooperation, voice quality, still imagery, transmission bit rates, etc.

• Codec cost will rapidly drop due to the availability of low-cost VLSI chips and high-volume production.

• Communication costs will drop as the ISDN is introduced and traffic volume increases.

Table 7.1.1
Status of ITU Audiovisual Recommendatons

REC. NO.	TITLE	WORKING DRAFT (WD)	1) STABLE DRAFT, NOTICE TO APPLY RESOL. 1 PROCEDURE (SD)	2) SUBMIT DRAFT REC. TO RESOL. 1 BALLOT 9 MOS (R1)	3) APPROVED RECOMMENDA-TION 5 MOS. (REC)	REVISION STATUS
N-ISDN AUDIOVISUAL TERMINAL						
H.320	SYSTEM				12/90	WD
H.261	VIDEO CODEC				12/90	-
H.221	MULTIPLEX				12/90	SD
H.242	COMMUN. PROTOCOL				12/90	WD
H.230	CONTROL, INDICATION				12/90	SD
AUDIO						
G.711	BW-3KHz 64Kbps				1972	
G.726	BW-3KHz 32Kbps				1987	1992
G.728	BW-3KHz 16Kbps				1992	SD
-	BW-3KHz 8Kbps					
G.722	BW-7KHz 64Kbps				1988	
-	BW-7KHz 16Kbps					
DATA						
H.224	DATA PROTOCOL-LSD/HSD (DLL)		11/93	5/94		
H.280	FAR END CAMERA CONTROL		11/93	5/94		
MULTIPOINT						
H.231	MULTIPOINT CONTROL UNIT				3/93	WD
H.243	MULTIPOINT COMMUNICATION PROTOCOL				3/93	WD
ENCRYPTION						
H.233	ENCRYPTION ALGORITHIM				3/93	SD
H.234	KEY MANAGEMENT		11/93	5/94		
H.244	AGGREGATION		5/94	2/95		
AUDIOGRAPHIC TERMINAL						
T.120	AUDIOGRAPHIC SERIES OVERVIEW	X				
T.123	AUDIOGRAPHIC PROTOCOL STACK				3/93	
T.122	MULTIPOINT COMMUNICATION SERVICE				3/93	
T.124	GENERIC CONFERENCE CONTROL		6/94	3/95		
T.125	PROTOCOL FOR T.122	X		11/93		
T.SI	STILL IMAGE		1994			

CONTINUED ON NEXT PAGE

REC. NO.	TITLE	WORKING DRAFT (WD)	1) STABLE DRAFT, NOTICE TO APPLY RESOL. 1 PROCEDURE (SD)	2) SUBMIT DRAFT REC. TO RESOL. 1 BALLOT 9 MOS (R1)	3) APPROVED RECOMMENDATION 5 MOS. (REC)	REVISION STATUS
LOW BIT RATE VIDEOPHONE						
H.32P	SYSTEM	X	2/95			
H.26P	VIDEO CODEC	X	2/95			
AV.25Y	SPEECH CODEC	X	2/95			
H.22P	MULTIPLEX	X	2/95			
H.DLP	DATA INTERFACE	X	2/95			
H.24	COMMUNICATION PROTOCOL	X	2/95			
B-ISDN AUDIOVISUAL TERMINAL						
H.32X	SYSTEM	X	2/95			
H.262	VIDEO CODEC (MPEG2)					
H.22X	MULTIPLEX	X	2/95			
H.24X	COMMUNICATION PROTOCOL	X	2/95			
H.32Y	H.320 B-ISDN EMULATION	X	2/95			
H.32Z	ETHERNET, TOKEN RING	X	2/95			

STATUS OF ISO AUDIOVISUAL RECOMMENDATIONS

STANDARD NO.	TITLE	COMMITTEE DRAFT (CD)	DRAFT INTERNATIONAL STANDARD (DIS)	INTERNATIONAL STANDARD
11172-1	MPEG1 System			X
11172-2	MPEG1 Video Coding			X
11172-3	MPEG1 Audio Coding			X
11172-4	MPEG1 Compliance Test	X		
11172-5	MPEG1 Simulation	X		
13818-1	MPEG2 System	X		
13818-2	MPEG2 Video Coding	X		
13818-3	MPEG2 Audio Coding	X		

	CIF	QCIF
Coded Pictures per Second	29.97	(or integral submultiples)
Coded Luminance pixels per line	352	176
Coded Luminance lines per picture	288	144
Coded Color Pixels per line	176	88
Coded Color lines per picture	144	72

Table 7.1.2
CIF and CIF Parameters

Table 7.1.3
Comparison of Video Coding Standards

	H.261	MPEG1	MPEG2
General Standard Structure	Narrow Profile	Generic Tool Kit	
Picture Format	176x144 (Mandatory) 352x288 (Optional)	One Field	Field or Frame
Quantization Precision of Motion Vectors	One Pixel	One Half or One Pixel	
B-Frames	None	An Available Tool	
Intraframe Coding	Usually Distributed	Full Frame is Mandatory	
Color Coding	4:2:0	4:4:4, 4:2:2, 4:2:0	
Picture Structure	Group of Blocks	Slice	
Dual Prime (a special motion comprensation mode)	No	No	Yes
Nominal Bitrate	56 Kbps to 1,936 Kbps	1.5 mbps	4 mbps to 20 mbps

Applications	- Interactive Audiovisual Services
	
	- Videophone
	- Videoteleconferening
		- VCR	- Broadcast TV
			  • Contribution
			  • Distribution
			- DBS
			- HDTV

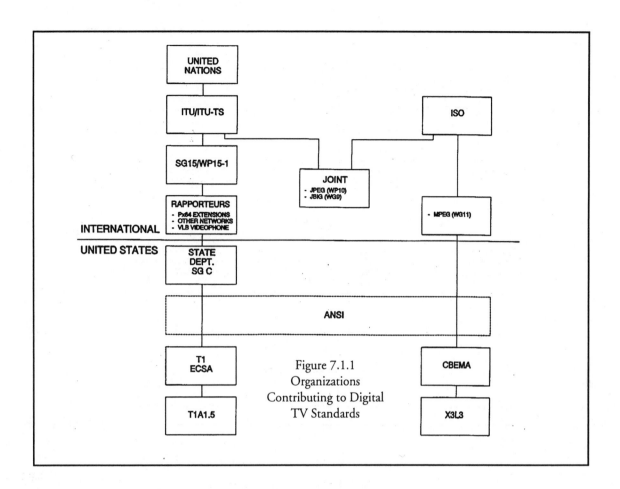

Figure 7.1.1
Organizations
Contributing to Digital
TV Standards

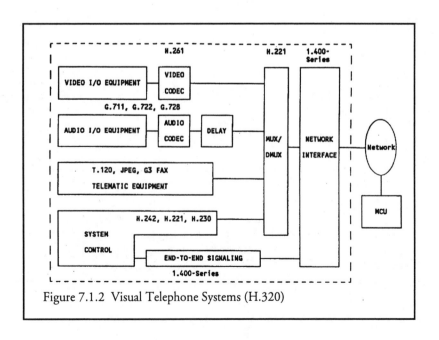

Figure 7.1.2 Visual Telephone Systems (H.320)

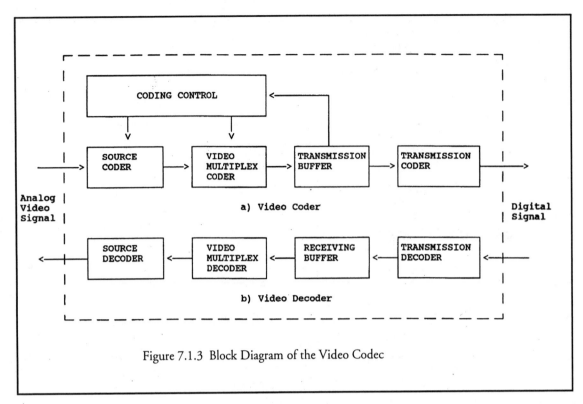

Figure 7.1.3 Block Diagram of the Video Codec

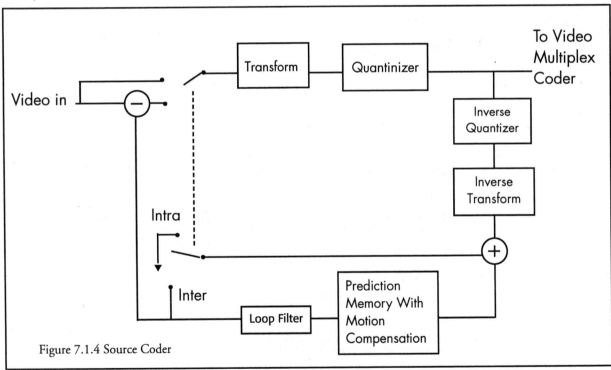

Figure 7.1.4 Source Coder

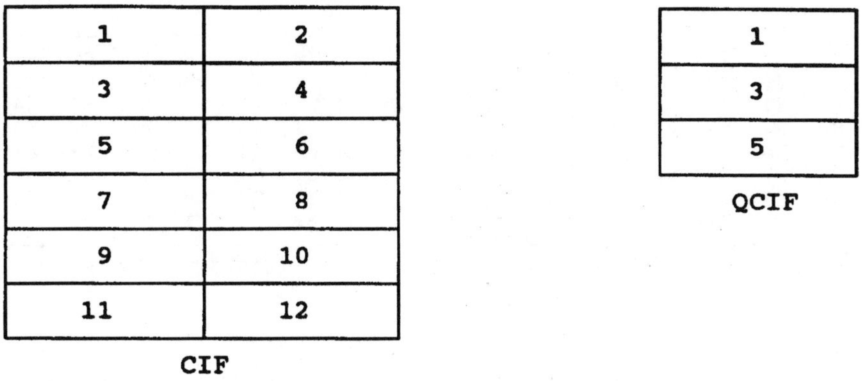

1
2
3
4
5
6
7
8
9
10
11
12
CIF
1
3
5
QCIF

Figure 7.1.5 Arrangement of GOBs in a Picture

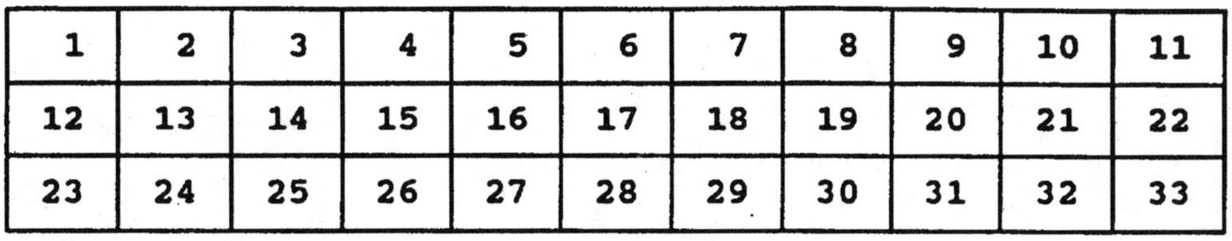

1 2 3 4 5 6 7 8 9 10 11
12 13 14 15 16 17 18 19 20 21 22
23 24 25 26 27 28 29 30 31 32 33

Figure 7.1.6 Arrangement of macroblocks in a GOB

1
2
3
4
Luminance
5
Blue
6
Red

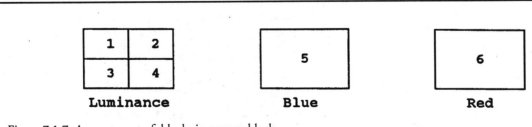

Figure 7.1.7 Arrangement of blocks in a macroblock

```
75  76  77  78  79  80  81  82          76  76  77  79  80  81  82  83
77  78  79  80  81  82  83  84          77  77  78  80  81  82  83  84
79  80  81  82  83  84  85  86          79  79  80  81  83  84  85  86
81  82  83  84  85  86  87  88          81  82  83  84  85  87  88  88
83  84  85  86  87  88  89  90          84  84  85  87  88  89  90  91
85  86  87  88  89  90  91  92          86  87  88  89  91  92  93  93
87  88  89  90  91  92  93  94          88  89  90  91  92  94  95  95
89  90  91  92  93  94  95  96          89  90  91  92  93  95  96  96
```

a) ORIGINAL BLOCK (8x8x8 = 512 BITS)   f) RECONSTITUTED BLOCK

```
684  -19  -1  -2   0  -1   0  -1        688  -21   0   0   0   0   0   0
-37    0  -1   0   0   0   0  -1        -39    0   0   0   0   0   0   0
  0    0   0   0   0   0   0   0          0    0   0   0   0   0   0   0
 -4   -1  -1  -1  -1   0  -1  -1          0    0   0   0   0   0   0   0
  0    0   0   0   0   0   0   0          0    0   0   0   0   0   0   0
 -2    0   0  -1   0  -1   0  -1          0    0   0   0   0   0   0   0
  0    0   0   0  -1  -1  -1  -1          0    0   0   0   0   0   0   0
 -1   -1  -1   0  -1   0  -1   0          0    0   0   0   0   0   0   0
```

b) TRANSFORMED BLOCK COEFFICIENTS   e) INVERSE QUANTIZED COEFFICIENTS

```
86  -3   0   0   0   0   0   0
-6   0   0   0   0   0   0   0
 0   0   0   0   0   0   0   0
 0   0   0   0   0   0   0   0
 0   0   0   0   0   0   0   0
 0   0   0   0   0   0   0   0
 0   0   0   0   0   0   0   0
 0   0   0   0   0   0   0   0
```

RUN   LEVEL   CODE

0   86   01010110
0   -3   001011
0   -6   001000011
      EOB   10

TOTAL CODE LENGTH = 25

c) QUANTIZED COEFFICIENT LEVELS   d) COEFFICIENTS IN ZIG-ZAG
                                              ORDER AND VARIABLE LENGTH
                                              CODED

Figure 7.1.8
Sample Intra Block Coding

1	2	6	7	15	16	28	29
3	5	8	14	17	27	30	43
4	9	13	18	26	31	42	44
10	12	19	25	32	41	45	54
11	20	24	33	40	46	53	55
21	23	34	39	47	52	56	61
22	35	38	48	51	57	60	62
36	37	49	50	58	59	63	64

Figure 7.1.9 Scanning order in a block

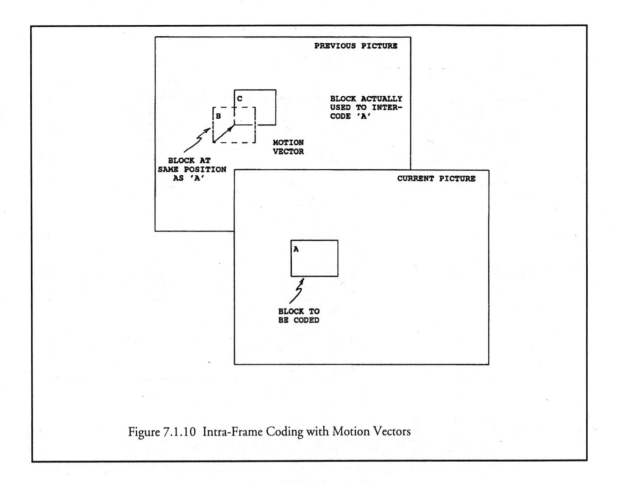

Figure 7.1.10 Intra-Frame Coding with Motion Vectors

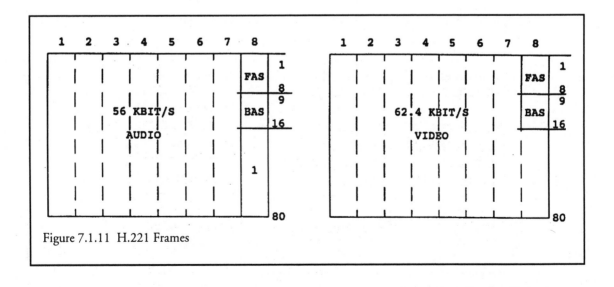

Figure 7.1.11 H.221 Frames

Figure 7.1.11
Data Transmission in H.320

DATA CHANNELS	Low Speed (<64 Kbps)	MLP (4, 6.4, VAR)	LSD (Low Speed Data)
	High Speed (Multiples of 64 Kbps)	H-MLP	HSD (High Speed Data)
COMMUNICATION PROTOCOL		T.120 Series - T.123 Protocol Stack - T.122 Multipoint Comm. Svc. - T.124 Conference Control - T.125 Multipoint Protocol	H.224 - Data Link Layer (DLL)
APPLICATIONS		T.si - Still Image T.res - Reservations T.avc - Audio & Video Control T.bwc - Bandwidth Control T.tdc - Transparent Data Channels T.fax - Multipoint Fax T.pro - Terminal Profiles T.bft - Binary File Transfer	H.280 - Far End Camera Control (Simplex, low delay)

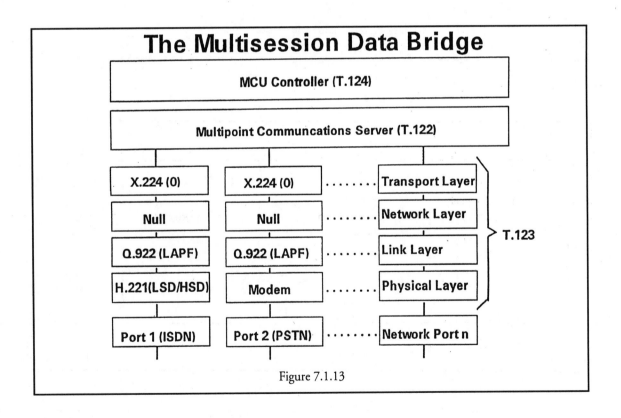

The Multisession Data Bridge

MCU Controller (T.124)

Multipoint Communcations Server (T.122)

X.224 (0)	X.224 (0)	· · · · · · ·	Transport Layer
Null	Null	· · · · · · ·	Network Layer
Q.922 (LAPF)	Q.922 (LAPF)	· · · · · · ·	Link Layer
H.221(LSD/HSD)	Modem	· · · · · · ·	Physical Layer
Port 1 (ISDN)	Port 2 (PSTN)	· · · · · · ·	Network Port n

T.123

Figure 7.1.13

Figure 7.1.14
ITU-T Recommendations for the Very Low Bit Rate Videophone

FUNCTIONAL ELEMENT		NEAR TERM (1995)	LONG TERM (1998)
SYSTEM			H.32P
VIDEO CODER		H.26P	H.26P/L
SPEECH CODER		AV.25Y	WP 15/2 (4Kbps)
DATA INTERFACE		Based On MLP or HDLL	
SUPERVISION CONTROL			H.24P
MULTIPLEX/ ERROR CONTROL			H.22P
MODEM	PSTN	V.34/V.8	
	MOBILE RADIO	--	FPLMTS

Recommendation will not be developed by WP 15/1

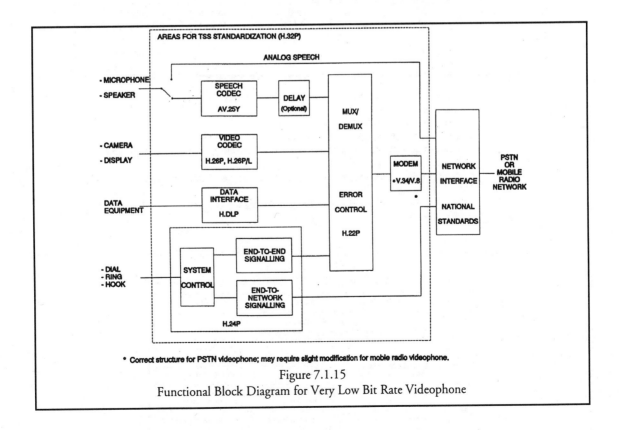

* Correct structure for PSTN videophone; may require slight modification for mobile radio videophone.

Figure 7.1.15
Functional Block Diagram for Very Low Bit Rate Videophone

Figure 7.1.16
Example of a Bitrate budget for Very Low Bitrate Visual
Telephony (kbit/s)

		Virtual Channel			
		Overhead/ Supervision (5%)	Speech	Video	Data
Overall Trans- mission Bit Rate	9.6 Kbps	0.5 Kbps	5.0 kbit/s	4.1 kbit/s[2]	Variable
			6.8	2.3[2]	
	14.4	0.7	5.0	8.7	Variable
			6.8	6.9	
	[3] ⋮	⋮	⋮	⋮	⋮
	21.6	1.1	5.0	16.5	Variable
			6.8	14.7	
	[3] ⋮	⋮	⋮	⋮	⋮
	28.8	1.4	5.0	22.4	Variable
			6.8	20.6	
Virtual Channel Bitrate Characteristic		Variable Bitrate	Dedicated, Fixed Bitrate [1]	Variable Bitrate	Variable Bitrate
Priority [4]		High Priority	High Priority	Lowest Priority	Higher Than Video, Lower Than Overhead/ Speech

(1)   The plan includes consideration of advanced speech codec technology such as...
- a dual bitrate speech codec
- reduced bitrate when voiced speech is not present.
(2)   Achievement of motion video is considered to be questionable at 4.1 kbit/s and inadequate at 2.3 kbit/s.
(3)   V.FAST operates at increments of 2.4 Kbps; i.e. 16.8, 19.2, 21.6, 24.0, 26.4, 28.8 Kbps.
(4)   The channel priorities will not be standardized; the priorities indicated are examples.

Figure 7.1.17
Classification of Video Coding Techniques Based on Source Models

LEVEL	SOURCE MODEL	CODED INFORMATION	CODING TECHNIQUE
1	Pels	Color of pels	PCM
2	Statistically dependent pels	Color of pels or block of pels	Predictive coding Transform coding
3	Translationally moving blocks	Color of blocks and motion vectors	Motion compensated hybrid DPCM/DCT coding
4	Moving structures	Mapping parameters or shape and motion	Fractal coding contour/texture coding
5	Moving unknown objects	Shape, motion and color of each object	Analysis/synthesis coding
6	Moving known object	Shape, motion and color of the known object	Knowledge based coding
7	Facial expressions	Action units	Semantic coding

Figure 7.1.18 World-wide Unique NNI

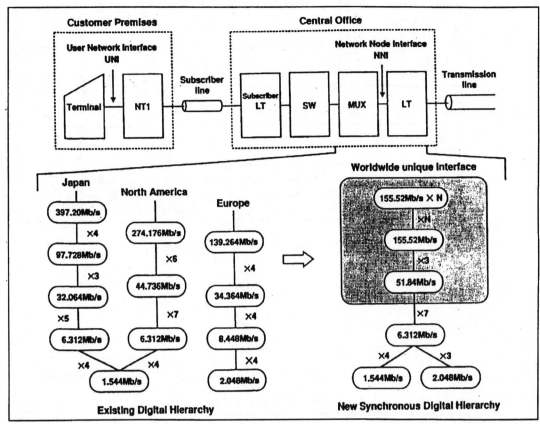

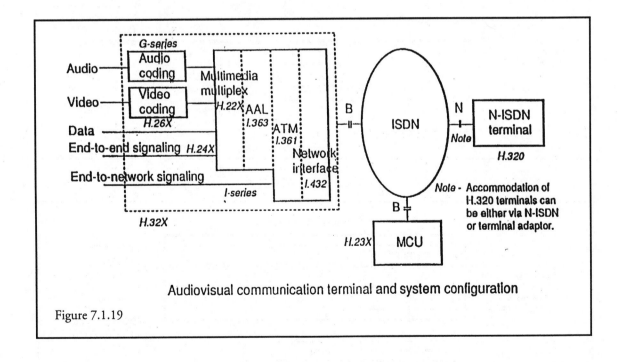

Audiovisual communication terminal and system configuration

Figure 7.1.19

Levels (Resolution)		Simple Profile	Main Profile	SNR Scalable Profile	Spatially Scalable Profile	High Profile
High Level			X			X
High - 1440 Level			X		X	X
Main Level		X	X	X		X
Low Level			X	X		

PROFILES (TOOL COMPLEXITY)

Figure 7.1.20
MPEG2 Structure

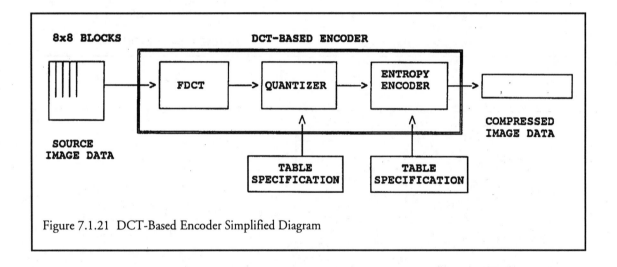

Figure 7.1.21 DCT-Based Encoder Simplified Diagram

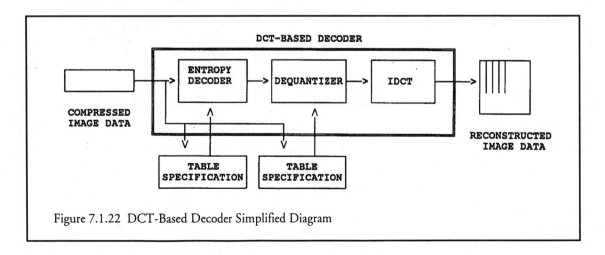

Figure 7.1.22 DCT-Based Decoder Simplified Diagram

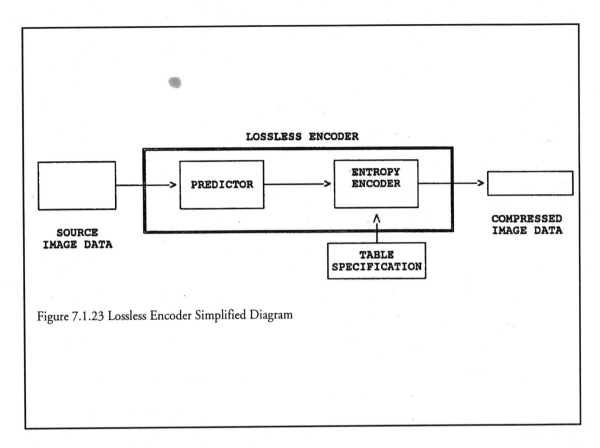

Figure 7.1.23 Lossless Encoder Simplified Diagram

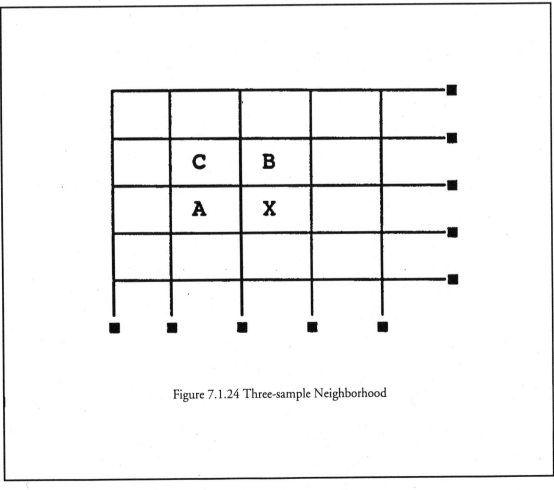

Figure 7.1.24 Three-sample Neighborhood

Chapter 8

Specializing in distance education, Dr. Carla Lane is the co-executive director of The Education Coalition (TEC) with affiliate agencies (schools of education, school districts, state departments of education and resource agencies) in 17 states producing pre- and in-service and K-12 programming in a range of media. She is the project director of the Distance Learning Resource Network (DLRN), a Star Schools Dissemination Project, and the evaluator for the TEAMS Star Schools Project (both funded by OERI, U.S. Department of Education and based at Far West Laboratory, San Francisco). She is a distance education consultant to Pacific Bell, the California Community College System, the California Department of Education, OTAN, Applied Business teleCommunications, University of Phoenix, and Chabot-Las Positas Community College District. She is a contributing editor of "ED," the official publication for the United States Distance Learning Association (USDLA), published for educators and trainers using electronically mediated instruction; she was the editor for four years. She holds one of the few doctorates in distance learning and has authored of many articles, manuals, and telecommunications guides, she has over 20 years experience in management, marketing, video production, and writing. Listed in the "Who's Who of Teleconferencing," she holds many Emmy and Addy Awards for her productions. Dr. Lane is an assistant professor in the telecommunications program at California State University-Hayward. She was the area chair for marketing/communications courses and has been an instructor with the Online Division of the University of Phoenix for four years.

Sheila Cassidy is the co-executive director of The Education Coalition (TEC) and is highly regarded for her role in TEAMS Distance Learning, one of the most rapidly growing Star Schools Projects. She was the main architect of the project and assisted in management of the project from 1990-93, when she also acted as executive producer of TEAMS programs. She won two awards for TEAMS student programming, and TEAM programs won seven national awards in those three years. Sheila has an in-depth background in school change and program evaluation as well as expertise in designing successful projects using strategies based on research and practice. Her background includes: telecommunications, systemic reform, mathematics education, programs for at risk students, bilingual education, learning styles, violence prevention and learner empowerment. Sheila has conducted research for the California State Legislature, California Department of Education, school districts in California and Far West Laboratory for Educational Research and Development. She was co-author of the Profile of Resources document published by Far West Laboratory which documented the services and programs available through Star Schools Projects in 1993. She has managed her consulting business since 1976.

TECHNOLOGY AND SYSTEMIC EDUCATIONAL REFORM

■ ABSTRACT

Systemic reform of education and training is built upon a number of converging events. This study reviewed the literature in the areas that directly affect systemic reform, and concludes with recommendations that will enable technology to play a significant role in this movement. Literature from the following areas were reviewed: paradigms, change strategies, systemic reform, Goals 2000, legislation, constructivism, student empowerment, andragogy, equitable access, current use of technology in education, problem-based learning, evaluation through performance based assessment, technology as a force for systemic reform, technology's potential as a partner and tool for systemic reform, new ways to think about the use of technology for education, competing in the global economy, total quality management in education, using the national information infrastructure for education and training and instructor in-service and pre-service.

When "A Nation at Risk" was released in 1983, it sounded an alarm for educators and the nation. It stated that the educational system of the United States was not serving the nation well. Approximately ten years after the release of "A Nation At Risk," the Goals 2000: Educate America Act was signed (March 31, 1994). The Information Infrastructure Task Force (1994) found that the way Americans teach, learn, transmit and access information remains largely unchanged from a century ago. They found the following conditions in American education and training:

• The textbook remains the basic unit of instruction. Absorption of its contents tends to be the measure of educational success.

• Teachers and instructors use "chalk and talk" to convey information. Students are often recipients of instruction rather than active participants in learning.

• School teachers work largely in isolation from their peers. Teachers interact with their colleagues only for a few moments each day. Most other professionals collaborate, exchange information and develop new skills on a daily basis.

• Although half of the nation's school teachers use passive video materials for instruction, only a small fraction have access to interactive video, computer networks or even telephones in the classroom.

• While computers are a frequent sight in America's classrooms and training sites, they are usually used simply as electronic workbooks. Interactive, high performance uses of technology, such as networked teams collaborating to solve real-world problems, retrieving information from electronic libraries and performing scientific experiments in simulated environments, are all too uncommon.

• "U.S. schooling is a conservative institution, which adopts new practice and technology slowly. Highly regulated and financed from a limited revenue base, schools serve many educational and social purposes, subject to local consent. The use of computer technology, with its demands on teacher professional development, physical space, time in the instructional day, and budget...has found a place in classroom practice and school organization slowly and tentatively" (Melmed, 1993).

In late June, 1994, the Carnegie Foundation for the Advancement of Teaching reported on the results of an international survey of 20,000 college professors in five European, four Western Hemisphere and four Asia-Pacific nations, plus Hong Kong. American college professors rated their students the lowest of the 14 countries participating. Only 15 percent said American high schools adequately prepared students for college-level math and quantitative reasoning. Asked if undergraduates are adequately prepared in writing and speaking skills, 20 percent or less of the faculty thought so in the United States (San Jose Mercury News (1994).

Since "A Nation at Risk," a number of forces came into play. As with any paradigm shift, it takes years and a number of events to reach a critical mass that moves theory and practice forward. Blanchard (1994) refers to "The Structure of Scientific Revolutions," Thomas S. Kuhn's (1970) book which describes how in the history of science, no major discovery ever came from those scientists who were vested (i.e., job, status, professional reputation) in the current predominant paradigm of thought. "Instead, what happened time after time was that a young person (a maverick or rebel to the scientific community) would discover some finding that could not be explained by known scientific law. He/She was scoffed at by mainstream scientists who clung tightly together in blasting the work of the upstart, and if need be the person him/herself. But the young maverick, if persistent, has an ally on his side: truth. And over time other newcomers to the field recognize that truth and a following emerges for a new paradigm that is better able to explain the current reality as it is now known to all."

Certainly, this describes how early adopters and users of technology in education and training have felt. Blanchard (1994) says, "The old command and control paradigm is failing because those who have been clinging to it (corporate America: GM, IBM, Sears, etc.) are now stumbling, and upstart companies that are representing a new era of assumptions about work and the nature of employment are thriving. The fact that Wal-Mart gave ownership interests to all its employees has a direct correlation with its rise in overtaking Sears and others on a national scale." This all underscores the "fact that the times they are a changin'," says Blanchard.

To unlock an epochal technology's power, Magnet (1994) says, companies have learned that they must restructure themselves and how they work as they weave computers into their most basic processes. "A technological revolution," he says, "is more than a merely technological matter; it entails an organizational transformation too. That's what U.S. business' recent frenzy of re-engineering has been all about, as companies flatten and decentralize along a unifying nervous system of the new information technology." As education and training has stretched to realign itself with the economy, it too has gone through a frenzy. Downsizing has occurred in teachers, resources and facilities; technology has not been added, or has been added sparingly. Criticism has resulted in changes in curriculum, most notably state frameworks and national standards. Instructional methods have been questioned and new ones put into place. Certainly, this is the basis of systemic reform where schools have decentralized, reorganized and rethought their production capabilities in the learning and teaching process and are now organizing along a unifying nervous system of the new information technology. Educators should become more comfortable with information technology as they become familiar with technology, as information has always been their business.

There are a number of elements that created the movement for systemic reform and enabled the passage of the Goals 2000: Educate America Act. The Goals 2000: Educate America Act covers all Americans; it includes the cradle to grave lifelong learning that has become necessary in a global economy. It should not be perceived to cover only K-12 education.

■ THE PARADIGM SHIFT

Baker (1992) provides the following definitions for paradigm shift:

Thomas S. Kuhn (1970): Scientific paradigms are "accepted examples of actual scientific practice, examples which include law, theory, application and instrumentation together—(that) provide models from which spring particular coherent traditions of scientific research."

Adam Smith (1975): A shared set of assumptions. The paradigm is the way we perceive the world; water to the fish. The paradigm explains the world to us and helps us to predict its behavior." Smith concludes that "when we are in the middle of the paradigm, it is hard to imagine any other paradigm."

Willis Harmon (1970): "The basic way of perceiving, thinking, valuing and doing associated with a particular vision of reality. A dominant paradigm is seldom if ever stated explicitly; it exists as unquestioned, tacit understanding that is transmitted through culture and to succeeding generations through direct experience rather than being taught."

Barker (1992): "A paradigm is a set of rules and regulations (written or unwritten) that does two things: (1) it establishes or defines boundaries; and (2) it tells you how to behave inside the boundaries in order to be successful.

"Barker (1992) identifies the following Paradigm Principles:

1. Our perceptions of the world are strongly influenced by paradigms.

2. Because we become so good at using our present paradigms, we resist changing them.

3. It is the outsider who usually creates the new paradigm.

4. Practitioners of the old paradigm who choose to change to the new paradigm early, must do so as an act of faith rather than as the result of factual proof, because there will never be enough proof to be convincing in the early stages.

5. Those who change to a successful new paradigm gain a new way of seeing the world and new approaches for solving problems as a result of the shift to the new rules.

6. A new paradigm puts everyone back to zero, so practitioners of the old paradigm, who may have had great advantage, lose much or all of their leverage.

Barker states that in times of crisis, people expect and demand great change. This willingness to accept great change generates two results: More people try to fine new ways, i.e., new paradigms, to resolve the crisis, thus increasing the likelihood of paradigm shifts. Because of the crises mentality, more people are willing to accept fundamentally new approaches to solve the crisis, thus increasing the opportunity to change paradigms. This sets the stage for radical change. Barker provides the following sequence for a paradigm shift:

1. The established paradigm begins to be less effective.

2. The affected community senses the situation, begins to lose trust in the old rules.

3. Turbulence grows as trust is reduced (the sense of crisis increases).

4. Creators or identifiers of the new paradigm step forward to offer their solutions (many of these solutions may have been around for decades waiting for this chance).

5. Turbulence increases even more as paradigm conflict becomes apparent.

6. The affected community is extremely upset and demands clear solutions.

7. One of the suggested new paradigms demonstrates ability to solve a small set of significant problems that the old paradigm could not.

8. Some of the affected community accepts the new paradigm as an act of faith.

9. With stronger support and funding, the new paradigm gains momentum.

10. Turbulence begins to wane as the new paradigm starts solving the problems and the affected community has a new way to deal with the world that seems successful.

■ CHANGE STRATEGIES

Even if we know exactly where we want to be in ten years and what the National Information Infrastructure will be...and even if we knew how much funding we could count on to get us there, we would still need to plan. There is a great body of literature on planning for technology, change, adoption of innovation and strategies of adoption. As you read through this literature review on systemic reform and the use of technology, it may be useful to refer to model change strategies. Blanchard (1994) suggests that we can "make the times change faster" through planning. His recommendation for a viable blueprint for the pending evolution (revolution?) is based on a study of six organizations (Beer, Eisenstat and Spector, 1990) on the process of change that leads to performance improvement." The six-step change strategy includes:

1. Mobilize commitment
2. Develop a shared vision
3. Foster consensus
4. Spread revitalization without directive
5. Institutionalize revitalization through formal policies
6. Monitor and adjust strategies.

Mojkowski (1990) suggests that a strategic approach to technology implementation should include the following:

1. Consider curriculum and learning outcomes first, then technology

2. Link the use of technology to organizational priorities

3. Develop a strategic sense guided by the organization's vision, mission and goals

4. Simultaneously transform and integrate technology in the learning and teaching process

5. Document and evaluate the implementation.

Farrell and Gring (1993) suggest another five-step model that is tied to a milestone timeline.

1. Needs assessment; gathering and analyzing data (where are we today)

2. Shared vision that leads to creating goals (where do we wish to arrive)

3. Select goals - clarify attainability, measurability and appropriateness

4. Prioritize goals and write a plan (how do we get from here to there and when)

5. Implement and evaluate the progress of the plan (how do we know when we arrived).

Pearson (1990) identified a model specifically for distance education programs. There were nine elements in the program and to be successful, all must be followed.

1. Decide to plan for change: Awareness
2. Recognize real vs. perceived need: Interest
3. Understand the real reason for implementation: Advantage
4. Mission of the organization: Evaluation
5. Plan the program: Trial
6. Review What the organization does now: Observability
7. The gap: Compatibility
8. Contingency: Pre-Adoption (Pilot)
9. Implementation: Adoption

■ WHAT IS SYSTEMIC REFORM?

During the last decade, it has become obvious that the contributions of teachers, administrators and the use of technology have made important changes in the lives of students. What became apparent is that the successes in the classroom needed to be viewed in the larger context of the educational system and curricula reform. The changes in the larger system were needed in order to enable wider spread changes in the classroom. Thus we began to use the words change, reform, restructure and finally systemic reform which includes K-12, higher education and training.

The National Commission on Time and Learning (1994) states that "Higher education needs to get involved. Colleges and universities, as institutions, have been bystanders for the most part in the school reform debate." They can do this by admission requirements that validate learning and not seat time, by aligning programs that educate teachers with the movement to higher standards (which will require changing offerings in schools of education and the design of undergraduate programs in core disciplines) and by becoming involved in the struggle to reinvent local schools.

"We really need to be more involved in collaboration with the public schools," says Dr. Thomas Schnell, associate dean of research at the University of Missouri, St. Louis (1994). "In the past we've had our differences. Public school teachers and administrators have viewed colleges as being ivory towers with no sympathy for life in the trenches, and colleges have viewed public schools (teachers) as being practitioners with no appreciation for innovative theory."

The goal is still to have an impact on the students, but impacting their learning depends on the teacher and the support that the teacher is receiving from the school's administration, district, higher education, state and federal government. As this definition broadened, other groups were identified to be included in systemic reform. Parents needed to be involved as first teachers and remain involved through their children's academic careers. Schools of education needed to be involved as they significantly impacted the pre-service and in-service of teachers and administrators. Curriculum groups began developing curriculum frameworks for all areas. Performance based assessment and authentic assessment became major movements. Teachers began to learn facilitation methods.

Textbook authors, media program producers, the Public Broadcasting System, educational associations, teachers' unions, communities, PTAs, county offices of education, state departments of education, U.S. Department of Education and the administrations of Presidents Bush and Clinton became involved. Each group defined the contributions it could make as well as the changes that needed to be made within its

structure in order to contribute to successful systemic reform. At the federal level, the use of technology was enabled through the Star Schools legislation. The movement became synergistic until finally, interest, understanding and change developed a critical mass that enabled the passage of the most sweeping educational legislation in the history of America, Goals 2000: Educate America Act.

Yet another layer of systemic reform deals with time — the time to learn, think and reorganize for teachers and students. "Unyielding and relentless, the time available in a uniform six-hour day and a 180-day year is the unacknowledged design flaw in American education," according to the National Education Committee on Time and Learning (1994). "By relying on time as the metric for school organization and curriculum, we have built a learning enterprise on a foundation of sand, on five premises educators know to be false." The first is the assumption that students arrive at school ready to learn in the same way, on the same schedule, all in rhythm with each other. The second is the notion that academic time can be used for nonacademic purposes with no effect on learning. Next is the pretense that because yesterday's calendar was good enough for us, it should be good enough for our children — despite major changes in the larger society. Fourth is the myth that schools can be transformed without giving teachers the time they need to retool themselves and reorganize their work. Finally, we find a new fiction: it is reasonable to expect 'world-class academic performance from our students within the time-bound system that is already failing them. These five assumptions are a recipe for a kind of slow-motion social suicide. The key to liberating learning lies in unlocking time."

Meyer, Brooks and Goes (1990) describe discontinuous or second order change as the type of strategies organizations employ when confronted by rapid change. "Discontinuous or second-order change transforms fundamental properties or states of the system." Fullan (1991) describes it as changes which "alter the fundamental ways in which organizations

Figure 8.1.1
Comparison of Conventional and Reform Approaches to Instruction (Means, et al, 1993)

CONVENTIONAL INSTRUCTION	REFORM INSTRUCTION
Teacher-directed	Student exploration
Didactic teaching	Interactive modes of instruction
Short blocks of instruction on single subject	Extended blocks of authentic and multi-disciplinary work
Individual work	Collaborative work
Teacher as knowledge dispenser	Teacher as facilitator
Ability groupings	Heterogeneous groupings
Assessment of fact knowledge and discrete skills	Performance-based assessment skills

are put together, including new goals, structures and roles.

"Meyer, Brooks and Goes (1990) describe two strategies which are used at individual work sites and at the industry level to react to sudden, jolting changes in the environment. The first is metamorphosis which is frame-breaking change within an individual organization; the second is revolution which is described as the emergence, transformation and decline of entire industries.

Conley (1993) distinguishes between three types of changes that schools undergo, sometimes simultaneously. They are renewal, reform and restructuring. He defines them as follows:

1. "Renewal activities are those that help the organization to do better and/or more efficiently that which it is already doing."

2. "Reform-driven activities are those that alter existing procedures, rules and requirements to enable the organization to adapt the way it functions to new circumstances or requirements."

3. "Restructuring activities change fundamental assumptions, practices and relationships, both within the organization and between the organization and the outside world."

David (1991) says that the central feature of restructuring is that it is a system wide process.

For the purposes of this article, systemic reform will be defined as second order change, or restructuring of educational agencies and processes, at the local, regional and national levels.

The dominant thinking about reform and its student level elements is to increase learning, especially advanced or higher-level skills and to enhance student motivation and self-concept. Reform stresses the elements in the second column of Figure 8.1.1 (Means, et al, 1993)

■ A FRAMEWORK FOR UNDERSTANDING SYSTEMIC REFORM AND TECHNOLOGY

Elements that have lead to and still are contributing to educational systemic reform and technology include a number of significant factors.

- **Global economy**
  - Chance that US may become a third world country due to a failure of education
  - Quality movement - Total Quality Management (TQM)
  - Industry
  - Education
  - Health industry
  - Re-engineering the corporation
  - The enabling role of information technology

- **Legislation**
  - Clinton Administration concentration on technology
  - Goals 2000: Educate America Act
  - School-to-Work Act
  - 1934 and 1994 Communications Act
  - Star Schools Legislation and Programs
  - Equitable access for all students
  - NTIA Programs
- **Research**
  - "A Nation At Risk"
  - Star Schools programs
  - Authentic Performance Assessment
  - Student Portfolios
  - Teacher Portfolios
- **Technological advances**
  - National Information Infrastructure
  - Digital fusion/convergence - movement of technologies to tele-computing
  - Ability to provide interaction
  - Easy to use machine interface - user friendliness
  - Interoperability through worldwide technology standards
  - Cable company advances in the use of interactive programming
  - Game technologies
  - Telephone companies wanting to provide programming
  - Cost reductions in technology
  - Computer Laboratories phased out
  - Computers installed in classrooms with access to information through modem
  - Computer drill and practice phased out in favor of problem solving
  - Movement from Industrial Age to Information Age to Communications Age
- **Understanding how to use technology**
  - Uses for education
  - Uses for business
  - Uses for home
  - Uses for entertainment — Nintendo
- **Educational System**
  - Collaborations between K-12 and post-secondary
  - Star Schools Grant RFP calls for teacher pre- and in-service
  - The system being seen as a system rather than separate entities of parents as first teachers, pre-school, elementary, middle school, high school, college, continuing education, training, adult basic education, GED, ESL
  - Learning
  - Understanding of learning styles
  - Understanding the mismatch of teaching styles and learning styles
  - Application of learning styles in teaching
  - Attitude change toward technology

Facilitation is used rather than "teaching"

Problem solving skills are used by all learners

Learners construct their own knowledge

Collaborative study groups are used by all learners

Experiential, hands-on learning is used by all learners

Seeing new teaching role models that do more than lecture from the book

Merging of constructivism and andragogy

- **The Millennium**

    The expectation that something dramatic occurs not only at the change of the century but at the end of the second one-thousand years on the planet.

As can be seen from the list, any one change or advance was not significant enough to cause a paradigm shift. However, when all of these factors are taken together, the critical mass is established that can move the country to a paradigm shift in education.

■ GOALS 2000: EDUCATE AMERICA ACT

At the kickoff meeting of Goals 2000: Educate America Act, Secretary of Education Richard Riley said that "technological and economic changes have occurred so fast that even the standards that we thought were high — just a few years ago — have to be higher today if we want all of our young people to be properly prepared. Technology has been growing exponentially. Education is even more important. We have to strive once again for higher standards. Here, it is important to recognize that in the next ten years this nation will have more children and young people in school than ever before. That's another change. The Baby Boomers filled up our classrooms in the 1960s and '70s, and some of us still remember the enormous amount of effort it took to teach all those children" (Riley, 1994b).

"Now, we have the same dynamic and something more," Secretary Riley continued. "By year 2004, data tells us we will have approximately 55.7 million students going to school, seven million more than we have today. The majority of these young people will not be suburban kids. The majority will be Hispanic, African-Americans, Asian and new immigrants — children who can learn if we teach them to high standards. But we have to expect them to learn and teach them high standards. If we ignore their education, if we continue to give them a watered-down curriculum and link inner city schools up last to the Information Superhighway, we will find this country in an economic pickle of the first order. We will have a work force that simply does not know how to work in today's economy."

"Like many of you, Ray Cortines, the very fine chancellor of education in New York city, is setting an example by raising academic standards for every student in New York city. He had it right, a few weeks ago, when he announced the new academic requirement and said that, 'the easy way out is the road to nowhere.' We can all find an easy way out. For that is the heart of the matter. If we accept 'the easy way out,' it will surely be the road to nowhere for a generation of young people who need our help. It will be a great injustice to their families and costly to this great nation of our's. We must do this for all of our students — our most talented students as well as all those young people who are in the middle, students from disadvantaged backgrounds, students whose first language is not English and students with disabilities. Even homeless students like those who are being taught by our 1994 National Teacher of the Year, Sandra McBrayer — San Diego children who have no one to speak up for them — who go to college like many of Sandra's students...if we believe in them. We need to remember that good teachers like Sandra are at the heart of it all. They make the difference. If we want our children to be 'living report cards,' we simply have to honor the work and effort of America's teachers — and listen to what they are telling us about our children. They simply have to be at the very center of our reform efforts. If you try to reform the schools in any other way, it will fail" (Riley, 1994b).

"Americans want the best for our children. They always have. They know our children need to learn more to get ahead...that the world they are growing up in is so different from the one that even all of us grew up in...more global, more knowledge-driven and certainly it's more competitive. This is why Goals 2000 is at the very center of an effort to redesign the American education system for the coming times — to create a strategy of lifelong learning that begins with Head Start and Even Start and many other preschool programs...that makes our schools safer...that raises standards and expectations across-the-board at the K-12 level...that helps young people make the transition from school-to-work...that prepares many more young people for college or other postsecondary learning. Goals 2000 is your act — your victory, your achievement...and our opportunity...together, to lift up American education (Riley, 1994b).

In March, 1994, Goals 2000: Educate America Act legislation was passed and signed into law. It's preamble states: "To improve learning and teaching by providing a national framework for education reform; to promote the research, consensus building and systemic changes needed to ensure equitable educational opportunities and high levels of educational achievement for all students; to provide a framework for reauthorization of all Federal education programs; to promote the development and adoption of a voluntary national system of skill standards and certifications; and for other purposes."

■ THE GOALS

1. School Readiness: By the year 2000, all children in America will start school ready to learn.

2. School Completion: By the year 2000, the high school graduation rate will increase to at least 90 percent.

3. Student Achievement and Citizenship: By the year 2000, all students will leave grades 4, 8 and 12 having demonstrated competency over challenging subject matter including English, mathematics, science, foreign languages, civics and government, economics, arts, history and geography and every school in America will ensure that all students learn to use their minds well, so they may be prepared for responsible citizenship, further learning and productive employment in our Nation's modern economy.

4. Teacher Education and Professional Development: By the year 2000, the Nation's teaching force will have access to programs for the continued improvement of their professional skills and the opportunity to acquire the knowledge and skills needed to instruct and prepare all American students for the next century.

5. Mathematics and Science: By the year 2000, United States students will be first in the world in mathematics and science achievement.

6. Adult Literacy and Lifelong Learning: By the year 2000, every adult American will be literate and will possess the knowledge and skills necessary to compete in a global economy and exercise the rights and responsibilities of citizenship.

7. Safe, Disciplined and Alcohol-and Drug-Free Schools: By the year 2000, every school in the United States will be free of drugs, violence and the unauthorized presence of firearms and alcohol and will offer a disciplined environment conducive to learning.

8. Parental Participation: By the year 2000, every school will promote partnerships that will increase parental involvement and participation in promoting the social, emotional and academic growth of children.

Technology is a vital part of Goals 2000 and is embodied as Title III. "Educational technology is a vital part of systemic reform. Electronic networks, for example, can facilitate the development and implementation of content and performance standards by allowing teachers and other educators to converse with one another across States. Instructional tools can be powerful resources in raising motivation and performance of all students. Furthermore, students must have technologies to enhance their learning and

to master the skills necessary for the workplace. States, districts and schools will benefit from examining technology issues as an integral part of systemic reform" (Guidance, 1994). Title III technology grants are awarded to state educational authorities (SEAs) to assist them in integrating the use of state-of-the-art technology as part of their State improvement plans.

In Goals 2000, we address the need for equitable access to information for all learners. You know the metaphor of providing an on-ramp to the information highway for school buses. You know that our natural resources are the same as other countries. The only thing that will set us apart in the global market is the ingenuity of our work force gained through an educational system that provides the tools for learning. In the past we provided buildings, teachers, blackboards, chalk and books. Now we must add daily access to information for all learners. These are major demands. If they are not met, the prediction is that we will create a new third world country called the United States. In the U.S., education has always been a third world country (Lane, 1994a).

When universal access was provided in the 1934 Communications Act, the only place that did not get access was the classroom. This can change if media connections are provided in all classrooms for voice, data and video. This is the only way to provide equitable access to information for all learners. Learners need access to resources that can be provided through telephone, cable, satellite, public broadcasting, wireless, commercial on-line services, Internet and any other technology. Learners need access whether they are in school, work or home. Providing one education cable drop and education rates for telephone service is a beginning...but it will not provide equitable access for all learners. We need more if we are to educate employees of the communication age (Lane, 1994a).

There are a number of problems in the use of technology that have been seen in the schools. Most schools are not wired for telephone, cable or television access. Most schools have only two to four telephone lines. One line is usually reserved for emergencies. Teachers must go to the school office to call parents. They do not have privacy. All of the teachers in the school share the same phone lines with the principal, office staff and school nurse. Many schools have computer labs but do not have computer modems to access information. Students go there several hours a week but may spend as little as ten minutes a week working on the computer. Computer labs do not enable integration of technology with curricula. They do not develop information gathering skills, or the application of newly found information to problems on which the children are working in class. Most schools do not have a satellite dish or access to programming through a satellite dish (Lane, 1994a).

A cable company may provide a cable drop and one channel for educational use, but the school has to wire the building, buy the equipment and buy or produce the programming. In a K-12 district, available programming overlaps for the age groups. Few get the programming that they want — even on tape. With limited funds, schools can usually get the cable signal to one classroom which they call a resource classroom. All of the classes at the school take turns using the room — usually only six classes per day can use the room. University campuses may have only one sophisticated distance learning classroom to be used by 20,000 students; it is rare for that classroom to be assigned to or built by the school of education. Teachers teaching in groundbreaking on-line programs such as the University of Phoenix On-line Program, have purchased the equipment in order to teach classes for students throughout the U.S. (Lane, 1994a).

Teachers do not have training in the use of information technologies. Funding is still too limited to provide this. Telecommunications operators seldom provide the training that is needed. Most states do not require information technology courses for students graduating from schools of education. Several require one three hour course. In K-12 schools that receive Star Schools programs, many do not have a satellite dish. Students view the programs on the cable access channel, public television station or on tape (Lane, 1994a).

The Star Schools programs are meant to be interactive, but most schools cannot afford the telephone line. Some teachers provide interaction by taking students to the phone — in the principal's office. Some teachers provide interaction by fax with the television teacher. Most schools do not have a fax machine. One teacher gives the fax to her husband who drops it off at the high school where there is a fax machine (Lane, 1994a).

In research on the TEAMS Star Schools Project, a new model of teacher re-education was identified (Lane, 1994 b). The research shows that teachers who watch a television teacher are presented with a role model for the new instructional methods. After using the programs for three years, the teachers report that they have changed their instructional methods so that the benefits to the children include constructing their own knowledge and becoming self-directed learners (Lane, 1994a).

What would help? To make the benefits of distance learning available, all classrooms must have access to information technologies – voice, data and video – as part of a network. Just as we believe that libraries should be open to the public so that everyone can share the information, we must extend the same provision to all information providers – whether it arrives on a telephone line, coaxial cable, fiber optic cable, satellite, wireless or broadcast. Low educational rates should be available from all telecommunications entities which come under the jurisdiction of the FCC. Specifically, there should be lower educational rates for cable, telephone, satellite, broadcast, commercial on-line services, Internet and provisions made for low educational rates for future information providers (Lane, 1994a).

In order to compete, we must build an educational system to match the needs of the Information Age. Being successful in the global economy is clearly linked to a strong education system. The critical natural resources of the Information Age are knowing how to learn, and access to education...which includes access to information. Because of this, it is imperative that we provide equitable access to all learners. If universal access is to mean equitable access for learners then legislation must state this and provide methods by which it can be implemented. Providing one line to a school is not enough. Providing ten lines to a school with 600 students is not enough. If we were expected to share access to information with 60 other people, little work would get done. The same goes for children. Their job is to learn. Most new and exciting learning resources that inspire and motivate them to construct their own knowledge and prepare them to work in the future are available through learning technologies. They can't do the work of childhood — they can't learn — without access to information. The same is true for adults (Lane, 1994a).

Distance learning facilitates high performance education by encouraging new instructional techniques and by allowing electronic access to information from any location. It is a driving force in the restructuring efforts of American education. The restructured school must bring these resources to the classroom and substantially supplement or replace the dated, non-interactive material used today if we are to implement Goals 2000. Working and learning are becoming synonymous. The work world for which students are being prepared requires learners who know how to learn and construct knowledge. These learners will become continuous lifelong learners. The real superhighway, is the highway of the mind (Lane, 1994a).

■ MODELS OF LEARNING: CONSTRUCTIVISM, STUDENT EMPOWERMENT AND ANDRAGOGY

Constructivism

The changes that are being advocated are based on a growing body of work and research that theorizes that learners actively construct their own understanding of the world around them by fitting their perceptions of the world into their existing knowledge and understanding. The positions represented by psychological, philosophical, pedagogical and thinking skills advocates are called constructivism. These theorists (von Glaserfeld, 1987; Shapiro, 1989; Confrey, 1990; Noddings, 1990; Davis, 1990; and Cobb et al, 1992) generally agree that learners must construct their own knowledge. Each learner has conceptions and skills "with which he or she must construct knowledge to solve problems presented by the environment. The role of the community — other learners and the teacher — is to provide the setting, pose the challenges and offer the support that will encourage construction."

Anderson, et al (1994) elaborates this perspective on learning and teaching.

1. Learning is dependent upon the unique prior conceptions that the learner brings to the experience. Old knowledge is the foundation into which the new knowledge must be integrated.

2. The learner must construct his or her own meaning. Students must organize and reorganize knowledge themselves until it fits with prior conceptions and has meaning within the learner's overall system. Learning is not memorizing or taking in knowledge in a form designed by someone else.

3. Learning is contextual and should be based on concrete experiences. The meaning that new knowledge has is highly dependent upon its context; it should not be presented in the abstract, independent of any meaningful context.

4. Learning is dependent upon the shared understandings that learners negotiate with others. Learners and teachers bring individually held knowledge, beliefs and feelings to the classroom and through daily interaction, they negotiate shared understandings of knowledge. These intersubjective meanings, "consensual domains," increase the importance of discussion and cooperative learning.

5. Constructivist teaching involves understanding students' existing cognitive structures and providing appropriate learning activities to assist them. Teachers need to attend to students' existing cognitive structures and provide learning activities accordingly.

Rote learning is often used because it is easier than dealing with the learners preconceptions and misconceptions. It is easier to memorize facts because it does not build on prior learning, and therefore, is not influenced by misconceptions. Students when exposed to rote learning for several years come not only to accept it but to actually prefer it. Learners must be made aware that meaning is something they construct, not something given to them by the teacher.

6. Teaching can utilize one or more of several key strategies to facilitate conceptual change depending upon the congruence of the concepts with student understanding and conceptualization. Models include some variation of awareness, disequilibrium and reformation. Awareness is based on the student working with information sources which link prior knowledge and construct meaning. In disequilibrium, learners evaluate the new constructs for consistency (agreement) or dissonance (disagreement) with prior knowledge. To reformulate their thinking, during the reformation phrase, students may be presented with formal concepts that lead to the resolution of anomalies and to the dissipation of cognitive dissonance.

7. The key elements of conceptual change can be addressed by specific teaching methods which address awareness, disequilibrium and reformation.

8. Constructivism leads to new conceptions of what constitutes excellence in teaching and learning and in the roles of both teachers and students. The classroom is no longer teacher-centered but is student centered. The teacher serves as a facilitator, pathfinder, guide, clarifier and maintains the environment. Students shift from that of a passive receptacle to that of an active participant, exploring, investigating, discussing and constructing his/her own knowledge. These role changes are among the most difficult to attain.

9. In constructivist teaching and learning, more emphasis is placed on learning how to learn than on an accumulation of facts, creating a philosophy of content in which "less is more."

■ STUDENT EMPOWERMENT MODEL OF LEARNING

Cummins (1989), in a synthesis of research and theory on successfully learning by language, cultural and ethnic minority students, provides the model of student empowerment. At its core are four components, each with a major shift from the traditional educational paradigm to an empowered paradigm. The four core components are:

1. Cultural — Linguistic Incorporation: moving from additive rather than subtractive; where students' languages and cultures are incorporated into the school program rather than being seen as hindrances to their learning

2. Pedagogy: interactive/experiential pedagogy, rather than one of transmission of information; promotes intrinsic motivation on the part of student to use language in order to formulate meaningful questions and generate knowledge; based on the work of Paolo Freire and others

3. Community Relations: inclusive rather than exclusive; full community participation encouraged as

an integral component of children's education

4. Assessment: advocacy oriented rather than legitimization oriented; professionals involved in assessment become advocates for students by focusing primarily on the ways in which students' academic difficulty is a function of the interactions within the school context rather than legitimizing the location of the "problem' within students

ANDRAGOGY

Malcolm Knowles identified a model for adult education (1975) which he called andragogy and defined as "the art and science of helping adults learn." According to Knowles (1983) the media have not been used effectively for adult education because they have been seen as one-way transmissions of teacher-controlled instruction which does not result in optimal learning; they are based upon the pedagogical model of education and the entertainment model of media use. Knowles recommends the andragogical model of learning and the educational model of media.

Knowles makes the distinction among the andragogical and pedagogical models of teaching based upon sets of assumptions about learners which teachers make. The teacher who makes one set of assumptions will teach pedagogically whether he or she is teaching children or adults, whereas the teacher who makes the other set of assumptions will teach andragogically whether the learners are children or adults (Knowles, 1975). The key features of the model are interaction, task-centeredness, individualization and self-directedness. Knowles states that learning is most effective when learners engage interactively in the inquiry process. Interaction can be introduced between the learner and the program using interactive video disc, computers and interactive reading materials. There are many striking similarities between andragogy and constructivism.

1. Concept of the learner: The learner is self-directing. The psychological definition of adult is "one who has arrived at a self-concept of being responsible for one's own life, of being self-directing." Adults who have arrived at that point develop a deep psychological need to be perceived by others, and treated by others, as capable of taking responsibility for themselves. In situations where others impose their wills on the adult without allowing the adult to participate in making decisions which affect the adult, he/she will often experience a feeling of resentment and resistance. Adults entering a situation labeled "education" or "training" hark back to their conditioning in school, assume role of dependency and demand to be taught. However, if they really are treated like children, this conditioned expectation conflicts with their much deeper psychological need to be self-directing, and their energy is diverted away from learning to dealing with this internal conflict.

2. The role of the learner's experience: Adults enter into an educational activity with both a greater volume and a different quality of experience from youth. For many kinds of learning, adults are the richest resources for one another, and hence the greater emphasis on group discussion, simulation exercises, laboratory experiences, field experiences and problem-solving projects that make use of the experiences of the learners. Because of the vast difference in learners' experiences, emphasis is placed on individualized learning plans through self-directed learning contracts. Adults derive their self-identity from their experience; if this is ignored, not valued or not made use of, it is not just the experience that is being rejected - it is the person.

3. Readiness to learn: Adults become ready to learn when they experience a need to know or do something in order to perform more effectively in some aspect of their lives. Chief sources of readiness are the developmental tasks associated with moving from one stage of development to another; but any change — birth of children, loss of job, divorce, death of a friend or relative or change of residence, is likely to trigger a readiness to learn. To induce a readiness to learn, learners can be exposed to more effective role models, engaged in career planning or provided with diagnostic experiences in which they can assess the gaps between where they are now and where they want and need to be.

4. Orientation to learning: Because adults are motivated to learn after they experience a need in their life situation, they enter an educational activity with a life-centered, task-centered or problem-centered orientation to learning. For the most part, adults do not learn for the sake of learning; they learn in order to be able to perform a task, solve a problem or live in a more satisfying way. The chief implication is the importance of organizing learning experience (curricula) around life situations rather than according to subject matter units. Another implication is the importance of making clear at the outset of a learning experience what its relevance is to the learner's life tasks or problems. One of the first tasks of a facilitator of learning is to develop "the need to know" what will be learned.

5. Motivation to learn: The most potent motivators are internal — self-esteem, recognition, better quality of life, greater self-confidence, self-actualization and the like. These intrinsic motivators are superior to external motivators such as a better job, a salary increase and the like.

6. The basic format of the andragogical model is a process design which assigns a dual role to the facilitator of learning; first and primarily, the role of

designer and manager of processes or procedures that will facilitate the acquisition of content by the learners; and only secondarily, the role of content resource. Besides the facilitator, other resources include peers, experts, media, experiential learning and field experiences. It is the facilitators job to link the resources and the learners.

7. Climate setting create a climate that is informal and conducive to learning. The physical environment may be one large circle or several small circles of chairs. The psychological climate includes mutual respect, collaborativeness, mutual trust, supportiveness, openness and authenticity, pleasure and humanness.

8. Involve learners in mutual planning as people tend to feel committed to any decision in proportion to the extent to which they have participated in making it.

9. Involve participants in diagnosing their own needs for learning. This involves meshing the needs of which the learners are aware (felt needs) with the needs their organizations or society has for them (ascribed needs). Using a model of competencies allows learners to identify the gaps between where they are now and where they need to be.

10. Involve learners in formulating their learning objectives. Learning contracts provide structure for this. Goals are set by mutual negotiation.

11. Involve learners in designing learning plans which help learners identify resources and devise strategies for using the resources to accomplish their objectives.

12. Help learners carry out their learning plan. Knowles sees the model as being a process design rather than a content plan so that there is no attempt to cover particular content areas; instead the student samples content in relevant problem situations. It is useless to have a stockpile of content information without having a process or method by which to handle it.

13. Involve learners in evaluating their learning by judging the quality and worth of the total program and their learning outcomes. The evaluation of the learning which has occurred is done through mutual assessment of the evidence which is prepared by the learner.

For media programs to be effective with adult learners, Knowles states that they must be organized around the acquisition of the knowledge, skills, understandings, attitudes and values that are applicable to performing the life tasks with which adults are concerned. Knowles (1975) states that one of the most significant findings from research (Tough, 1979) about adult learning is that when adults go about learning something naturally, rather than being taught, they are highly self-directing. He finds that evidence is accumulating to support that what adults learn on their own initiative – through planning and

constructing their own learning – they learn more deeply and permanently than what they learn by being taught.

■ CURRENT USE OF TECHNOLOGY IN EDUCATION

"In the 1980s, reform efforts tried to improve student performance by increasing course requirements," according to Means and Olson (1994). "Reformers did not, however, examine the way that teaching and learning unfold." Technology applications "exist primarily on the periphery of a school's instructional program" (Farrell and Gring, 1993). Many technology initiatives are unconnected to the school's improvement activities and not part of any long-range plan.

Means and Olson (1994) believe that earlier attempts to introduce technology into schools failed because "the attempts were based on the wrong model of teaching with technology. Product developers believed in their content knowledge, pedagogical techniques and in the power of technology to transmit knowledge to students. With satisfaction, the developers touted the so-called 'teacher-proof' instructional programs." Means and Olson observe that the vendors must have been surprised that their applications were never used for very long. Another primary problem is that the applications were an "imperfect and incomplete match with the bulk of the core curriculum" (Means and Olson (1994). They believe that education reform and the use of technology is a basis for optimism.

Ohler (1991) lists a number of ways that distance education has been used.

1. To overcome geographic isolation in order to receive a state-sanctioned education
2. To avoid or reinforce particular content
3. Because of incarceration
4. To avoid social influences
5. To experience or avoid certain learning dynamics
6. Because of a disability
7. To avoid having to abandon a life-style or culture
8. To avoid a schedule conflict
9. Because the student is not learning in school
10. To escape tracking
11. To learn in a more global context
12. To learn information-economy skills
13. Remediation
14. Because schools are too expensive for the state to provide
15. To improve local communication under certain conditions

16. To reduce anxiety and improve face-to-face communication
17. Because the media are motivational
18. To associate with a particular segment of society, or conversely, to become diffused within a heterogeneous population
19. To take advantage of a world of experts and resources that only media can provide.

"Schools are not moving to integrate technology, nor are they keeping up with the latest development; in fact, they are falling farther and farther behind as the equipment they purchased in the 1980s becomes obsolete and they are unable to purchase new equipment" (Elkmer-Dewitt 1991). "Schools are out of step with the times," according to David (1991). "Inside and out, schools today look very much the way they did a hundred years ago: the buildings, the size and shape of classrooms, the divisions based on age, and the ways of delivering instruction have changed very little. Yet the world has changed remarkably. Families, jobs, social organizations and entertainment look nothing like they did at the turn of the century. From inside a school, however, one would hardly know that visual images, rapid motion, technology and change are pervasive in the world outside."

"Schools are neither organized nor funded in a way that enables them to keep up with changes in knowledge or changes in technology used to store and present such knowledge" (Elkmer-Dewitt 1991; Levinson 1990). "Schools have defined technology as computers. There are many types of technology in addition to computers that will have an equal or greater impact on learning" (Conley, 1991). "Textbooks are an obsolete technology, yet they continue to be central to the way schools conceive of teaching and learning" (Conley, 1993). "Statistics such as the number of computers per teacher are worse than useless as a measure of progress to determine effective use of technology in schools; careful examination of schools' attempts to use computers yields results that are dismaying and disheartening" (Borrell 1992).

Conley (1993) says that the most striking observation one reaches about technology in education over the past dozen years is not its impact but its lack of impact." Informational technologies have been adopted in the central offices, but "technology has not revolutionized learning in the classroom, nor led to higher productivity in schools. While telecommunications may prove to be a powerful tool for restructuring, its use at this point is primarily to expand, not to change, the existing curriculum by offering courses such as physics or French to schools not otherwise able to offer them and by employing traditional instructional strategies." Certain technologies have definitely found niches in education, but Smith and

O'Day (1990) say that the technology of the last two decades has changed schools far less than it has the worlds of work, entertainment and communication. On the whole, they say, teachers have simply closed their classroom doors and gone right on teaching just as they were taught.

According to the Information Infrastructure Task Force (1994) (IITF), while computers and some communications capabilities are present in American schools, high speed communications technology is limited to very few classrooms. Substantial local infrastructure investments will be necessary to realize the promise of National Information Infrastructure applications. The installed base of computers, modems, networks and video technology indicates that growth has been, at best, uneven. Most schools, communities and state and local governing bodies have neither recognized nor acted on the need to build the technological capability to access the information superhighway. A key, but not well understood requirement, is for technical expertise to install and maintain high speed connections to the NII. Once the high speed communications linkages of the NII are brought to the schoolhouse door, the challenge is to build the internal high speed linkages within the building to connect the user hardware.

Instructional video has made the most notable inroads into the schools. Seventy-five percent of America's schools have cable television, and half of its teachers use video material in their courses (CPB, 1991). The Stars Schools programs are reaching over 200,000 students in 48 states with advanced placement courses in mathematics, science and foreign language instruction using fiber optics, computers and satellites. Cassette videotapes for instruction are widely used in schools and work places, and the development of these videotapes for both education and training has become a vigorous industry.

Hundreds of thousands of students in schools, community colleges and universities now take courses via one-and two-way video and two-way audio communication. In South Carolina, high school students across the state study with a teacher of Russian based in Columbia through South Carolina Educational Television. Boise State University offers a masters degree program conducted entirely over networked computers to students all over the country. The University of Phoenix Online Division offers an undergraduate and graduate degree in business. The Education Coalition (TEC) has been formed to collaboratively produce programming for teacher in-service and pre-service, and K-12 social studies programming. The California State University System has distance learning classrooms on each campus which can be linked. The University of Missouri, St. Louis and St. Louis Community College offer courses over the Higher Education Channel (cable) that reaches over three million homes. The TEAMS Star Schools program produced by the Los Angeles County Office of Education has provided mathematics and science programs to over 60,000 fourth through sixth graders throughout the U.S. in one year.

The Department of Defense is investing well over one billion dollars in the development and implementation of networked distributed interactive simulation. This technology, which allows dispersed learners to engage in collaborative problem solving activities in real time, is now ready for transfer to schools and workplaces outside of the defense sector.

The IITF states that the installed base of computers in American elementary and secondary schools is largely incapable of supporting multimedia graphical applications because of obsolete or obsolescent hardware. Eighty percent of the base includes Apple IIs (about 55 percent) and IBM PCs, XTs, ATs or similar class machines (about 24 percent), with limited modern graphic or multimedia capabilities; the remaining part of the base is made up of 10 percent Apple Macintoshes and eight percent IBM compatible 386s or 486s and is capable of supporting high level applications. The number of computers in the schools, 2.5 million, is equivalent to one per classroom (Malmed, 1993).

In a 1993 survey of NEA members, only four percent of teachers reported having a modem in their classroom, while 38 percent reported having access to a modem somewhere in the school building (Princeton, 1993). Another survey found that among 550 educators who are actively involved in using telecommunications, less than half have access to the Internet. They use the Internet services twice as often for professional activities as for student learning activities (Honey and Henriquez, 1993).

In 1994, 3Com Corporation based in Santa Clara, CA committed $1.9 million in cash, personnel and networking equipment to link every high school classroom in San Jose and eventually all 35 schools in the county to each other, and the Internet (Bank, 1994). A four week summer institute at San Jose State University trained 450 teachers, 100 students and 50 administrators to master the technology, and develop teaching methods and curricula that take advantage of the new capabilities. By the end of the second year of the project each teacher will have a computer workstation linked to the network, and four student workstations with access to the Internet through the teacher's computer.

3Com was attracted to Santa Clara because they "did not look at this as a technology issue but as a fundamental change in curriculum development and the way teachers collaborate among themselves and with administrators. Networking is just a piece, but it's a fundamental catalyst" according to Eric Benhamou, 3Com's chief executive. When the network is done, the 33-district network will be one of the largest of any kind in the country. David Katz, director of the San Jose Education Network stressed that this is not a project to teach students to use computers, but "rather to use computers as a tool for learning in general. In a world where al the information is available with a keystroke, the teacher has to change to the role of a facilitator, to help kids learn to find information, to think critically, to decide what's important. In the workplace of the 21st century, that's how value is going to be determined, by your ability to add value to information" (Bank, 1994).

Laser video disc-based programs enable the user to interact with still or moving images and print. Texas, Florida and West Virginia have approved laser video discs for instruction. In 1990, Texas approved Optical Data Corporation's "Windows on Science" program for use in lieu of textbooks. The corporation's studies strongly suggest that the medium improves a variety of educational outcomes (Hancock and Betts, 1994).

Information collection, according to the IITF (1994) includes location and retrieval of documents such as lesson plans and research reports, but it also includes newer data sources such as CAD databases for workplace technologies and equipment and multimedia information retrieval from digital libraries that can be accessed by students, workers or people in homes, libraries and museums. Over 60,000 electronic bulletin boards are used by more than 12 million Americans every day (Investor's Business Daily, 1994). The annual rate of Gopher traffic on the Internet, which directly represents an effort to use NII facilities to gather information, is growing at an annual rate of approximately 1000 percent (Treese, 1993). The Department of Education has a Gopher server which points to or contains educational research information, such as the AskERIC service and information from sources such as CNN, Academy One and the Educational Testing Service. NASA Spacelink makes lesson plans on space flight and related science topics available on the Internet. Until compelling applications are available, education will not realize the potential of the NII.

Two-way communication includes communication via electronic mail and conferencing among teachers, students, workers, mentors, technicians and subject matter experts of every sort (IITF, 1994). Approximately one-quarter of the teachers in Texas regularly sign on to the Texas Education Network, or TENET, to share information, exchange mail and find resources. A professor at Virginia Polytechnic Institute and State University teaches a writing course entirely on-line. Students swap writing projects and discuss their assignments on-line. In the workplace, electronic mail is used by more than 12 million workers, increasing to over 27 million workers by 1995.

Just less than a sixth of U.S. homes now have at least one computer connected to a modem, and this percentage is growing rapidly. As of July, 1993, there were four Internet hosts for every 1000 people in the United States. There are now 60 countries on the Internet. About 137 countries can now be reached by electronic mail (Treese, 1993).

Current application of NII capabilities (IITF, 1994) to work place training is more extensive and technologically advanced than educational applications, yet it lags well behind what is needed and available. Workplace training seems to be a case of the haves receiving more and the have-nots remaining neglected. Small firms, those with 100 employees or less, provide about 35 percent of total U.S. employment, but they lack the expertise to provide in-house training, the resources to pay for outside training and sufficient numbers of people who need training at any one time to justify focused training efforts. Larger firms are more likely to provide training than smaller ones, but the training they provide is mostly limited to college-educated technicians and managers. The lower the level of skills possessed, the less likely the worker is to receive training from any source. Transportable, quality controlled training and lifelong learning could be made readily and inexpensively accessible using the NII and will have a major impact on improving worker skills and workplace productivity.

In Senate testimony, NEA's Kenneth Malley said, "It should be of little surprise then that teachers and others do not have access to new information technologies and telecommunications services. Last year (1993), NEA conducted a nationwide survey to determine classroom teachers' access to computing and telecommunications technologies. Only 12 percent of U.S. classrooms have a telephone. Only four percent of teachers have a modem in their classroom and these are concentrated in affluent districts. Further, only four percent report having any access to Internet. The consequences of this resource scarcity are obvious. A mere six percent of the teachers surveyed reported that they and their students had ever used electronic networks to collaborate with other teachers. Numbers like these would be astounding in a business environment, yet we continue to expect teachers to work educational miracles without even the most basic communication tools. We do not have any choice but to make this investment. What is at stake is the future of these children, the future of these communities, indeed, the future of the nation itself" (Malley, 1994).

Malley (1994) contends that "we will be unable to attain these technological advances without recognizing that we have a larger problem. Many of our schools are simply incapable of accepting the promise of telecommunications. The physical infrastructure of our nation's schools is in need of massive renovation and repair. Three-quarters of the schools in the nation are at or near the end of their useful life. While we can bring the Information Superhighway to their door, many are not prepared to receive it in their classrooms. Many teachers work in classrooms with only one electrical outlet and when workers attempt to install additional electrical wiring, they often encounter asbestos and other hazards brought on by years of neglect. Computers sit in cartons because they cannot be safely installed in rooms with leaky ceilings. Schools may have telephone lines, but they most often extend to the principal's office — not to the teacher's desk, let alone a student's computer."

"The structure of knowledge is rapidly evolving. The division of academic disciplines is no longer appropriate for understanding or solving the problems that exist in the world, yet schools cling to the old structure" (Conley, 1993). "Information is seen less as an end in itself than as a means to an end, an essential ingredient in problem-solving. Curricula that focus on information as an end in itself (fact-based rote learning) can be counterproductive, extinguishing the curiosity and inquisitiveness of the learner and providing little practice in applying information to solve problems" (Conley 1993).

In 1991, then U.S. Labor Secretary Lynn Martin established a Commission on Achieving Necessary Skills (SCANS) which reported five learning areas of increasing importance in the workplace and foundation skills (Whetzel, 1992). Workers use foundation skills — academic and behavioral characteristics — to build competencies on. Foundation skills fall into three domains:

1. Basic skills — reading, writing, speaking, listening and knowing arithmetic and mathematical concepts;

2. Thinking skills — reasoning, making decisions, thinking creatively, solving problems, seeing things in the mind's eye and knowing how to learn; and

3. Personal qualities — responsibility, self-esteem, sociability, self-management, integrity and honesty.

Competencies more closely relate to what people actually do at work. The competencies that SCANS has identified fall into five domains:

1. Resources — identifying, organizing, planning and allocating time, money, materials and workers;

2. Interpersonal skills — negotiating, exercising leadership, working with diversity, teaching others new skills, serving clients and customers and participating as a team member;

3. Information skills — using computers to process information and acquiring and evaluating, organizing and maintaining and interpreting and communicating information;

4. Systems skills — understanding systems, monitoring and correcting system performance and improving and designing systems; and technology utilization skills — selecting technology, applying technology to a task and maintaining and troubleshooting technology.

"Economic survival in an information society requires being able to work with media, by using computers to navigate and move information effectively" (Wodtke, 1993). It requires being able to relate to and build computer models that address the realities you are working with. It means being able to work with methods that involve computers and a wide range of electronic devices that enable you to communicate and collaborate effectively using electronic media. These thinking skills coupled with hardware and software application skills are becoming the computer literacy essential for survival on many levels — essential for individuals to survive in an information society, for individuals to compete in the job market, for the enterprises of work groups to be competitive in the marketplace and for countries to compete in the global community" says Wodtke.

Implementation issues center on the fact that technologies "can be used to deliver instruction," but that "does not mean that they will be" (Means, et al, 1993). What has happened in the past is that the technology is adopted by a school but is adapted to traditional school structures and teaching styles if it is flexible, or is discarded if it cannot be adapted (Cohen 1988; Cuban 1986). Cohen contends that the dominant use of distance learning is wider dissemination of a traditional mode of teaching — the lecture. Microcomputers have provided an on-line version of drill and practice seat work. Piele (1989) suggests that computer labs have failed to transform schools because they are typically not supervised by the classroom teacher so teachers "can ignore them altogether." Cohen (1988) contends that the central instructional program remains much as it was 50 years ago, untouched by the technological revolution going on around it.

Technology has not made a real difference in teaching and learning primarily because of the "imperviousness of the education system to any kind of fundamental change; the barriers that are specific to technology-based changes are very real, but a lesser impediment (Means, et al, 1993). Sheingold (1991b) asserts that it is now understood that the "challenge of integrating technology into schools and classrooms is much more human than it is technological. What's more, it is not fundamentally about helping people to operate machines. Rather, it is about helping people, primarily teachers, integrate these technologies into their teaching as tools of a profession that is being redefined through the incorporation process (p. 1).

Media literacy also plays a part in the use of technology for learning and teaching. To date media literacy skills are seldom taught to teachers or to students. Four principles have been identified by groups working for media literacy (Thoman, 1993)

Media Construct Reality. Media are constructions made through editing in writing or production edits. What is created becomes a version of reality. Understanding the construction process and how the media shape what we know and understand about the world we live in is an important way of helping them navigate their lives in a "global and technological society."

Media Use Identifiable Techniques. Camera angles, music, special effects, layouts and lighting heighten responses to media and vie for attention. Each medium has production codes that are used to construct media — and can be used to de-construct media so that people are less susceptible to manipulative uses.

Media are Businesses with Commercial Interests. Corporations spent $130 billion in advertising in 1991 — the equivalent of $6 per week for every man, woman and child in the U.S. Young people are susceptible to advertisers; media literacy provides a grounding to help students gain perspective on what's really important and how to make decisions about commercial messages and the program content that is designed to make sure the right audience is watching when the commercials appear.

Media Contain Ideologies/Value Messages. All media — TV, movies, news, sports, game shows, video games, even supposedly objective newspapers — contain points of view. Media literacy aids in recognizing the point of view and the entertainment. The job for educators is to teach critical thinking and critical reading of all media so that people, young and old, can recognize what values are embedded — and accept or reject them.

Interactive instruction systems, especially those combining video disc and computer technologies, are gaining widespread acceptance within educational and training communities according to Rockley L. Miller (date unknown). These systems have been available for over ten years, with hundreds of off-the-shelf programs and custom applications produced to date.

Johnstone (1991) reported that one of the assumptions "inherent in the design of most contemporary distance learning systems is the need for interaction between students and their teachers. Do teachers need to see one another for effective learning and interaction to take place? Is real-time student-teacher or student-student interaction really the best or only model?" Johnstone reports two instances where teachers said that instructors for computer mediated classes reported higher levels of critical thinking than do traditional class discussions and papers.

Harasim (1989) and Quinn, Mehan, Levin and Black (1983) found that most of the verbal exchanges in face-to-face classrooms come from the instructor, while the reverse is true on-line. Harasims' research indicated that in an active on-line learner-to-learner exchange, between 60 -80 percent of the verbal exchange in an on-line class comes from the students which indicates a high level of interaction and collaboration. Harasim found that communication in an interactive on-line class is more equitably distributed among class members, whereas a conventional classroom frequently has one or two students dominating the discussion while the majority remain silent.

Quinn, et al (1983) noted that students in their electronic classroom system produced longer and more complex responses than in the classroom groups they compared. The researchers observed that the time-delay in asynchronous communication contributes to the quality and quantity of student interaction. They report that the delay between receiving a message and sending a response allows reflection and the time to compose a substantive answer.

Students using audio conferencing for interaction in conjunction with video programs delivered by satellite, produced longer and more in-depth interaction with a facilitator. Smaller groups produced longer and more in-depth responses than did larger groups (Lane, 1990).

A survey of studies on the effectiveness of technology in schools concluded that "courses for which computer-based networks were used increased student-student and student-teacher interaction, increased student-teacher interaction with lower-performing students and did not decrease the traditional forms of communications used" (Interactive Educational Systems Design, 1993).

Miller reports that corporate, institutional and governmental users state the following benefits for educational technologies.

1. Reduced learning time. Well over thirty studies compiled have found that interactive technologies reduce learning time requirements by an average of fifty percent. This reduction is attributed to a variety of factors: (a) self-paced instruction encourages students to take the most efficient path to content mastery — skipping areas of existing strength while investing more time in areas of weakness; (b) the combination of visual presentation with audio delivers information in an easily understood format; (c) immediate interaction and feedback provides constant, highly-effective reinforcement of concepts and content; and (d) personalized instruction accommodates different learning styles to maximize student learning efficiency. The IBM Principles of the Alphabet Literacy System, an interactive video-based course, is achieving increases of over two years in reading and writing skills, with only 100 hours of instruction.

2. Reduced cost. The primary costs of interactive instruction lie in design and in production, not replication, distribution and delivery. Thus, the cost per student is reduced as more students use the same program. With traditional instructional methods, the costs of instruction lie primarily in the delivery (i.e. teacher salaries, overhead) and remain constant or even increase as more students place demands on fixed resources. A typical, cost-per-student break-even point for interactive instruction might occur when from 100 to 200 students are using a program. Beyond that number, savings build dramatically. Federal Express projects saving over $100 million by using interactive systems for employee training.

3. Instructional consistency. Technology-based instructional systems do not have bad days or tire at the end of a long day. Instruction is delivered in a consistently reliable fashion that does not vary from class to class or school to school.

4. Privacy. With one-on-one systems, students are free to ask questions and explore areas that might cause embarrassment in group situations. Because instructional systems never lose patience, they encourage learners to persist in asking questions and reviewing materials until real mastery is achieved or natural curiosity is appeased. In one example, an interactive program called "TeenScope" allows teenagers to prepare for the transition to living on their own. Covered topics such as finding a job, pregnancy and parenting, sexuality, building self-esteem, what to do with a paycheck and feelings about families are difficult to openly explore in group situations.

5. Mastery learning. Unlike a normal classroom situation, the interactive system will not move on to new material until current material is mastered. This ensures that students have strong foundations for continued learning. In one example, at-risk students in Everett, WA, achieved a 53-point gain, from 38 percent pre-test scores to 91 percent post-test scores; using interactive mathematics instruction from Systems Impact. Similarly, remedial and Chapter I students in Bethel Park, PA, achieved 300 percent improvement, jumping from 21 percent pre-test scores to 88 percent post-test scores using the same program.

6. Increased retention. The process of interaction with material being studied provides a strong learning reinforcement that significantly increases content retention over time. Spectrum Interactive (a division of National Education Corporation) reports over 25 percent improvement in retention with interactive video courses.

7. Increased safety. With interactive systems students can explore potentially dangerous subjects without risk. These dangers might be in academic areas (chemistry explosions, burns) or social areas (drugs, sexually transmitted diseases, pregnancy). In one example, the TARGET system allows students to learn about drugs and alcohol and consequences of

substance abuse without the dangers of experimentation. In another example, a course on basic electronics and maintenance allows the student to accidentally touch the wrong parts without risking electrocution.

8. Reduced behavior problems. Interactive systems focus attention and increase individual involvement, thereby reducing the potential for misbehavior.

9. Increased motivation. Interactive systems provide a level of responsive feedback and individual involvement that has proven to be highly motivating in both individual and classroom learning environments.

10. Increased access. Interactive systems can provide greater and more equal access to quality instruction. Systems can deliver peripheral subjects in schools where student populations are insufficient to support full time teachers for such subjects or where qualified teachers are otherwise unavailable. Further, interactive systems can be used to simulate laboratory equipment that would be too expensive to actually acquire.

The Institute for Defense Analyses conducted a quantitative analytical review (meta analysis) in defense training and in the related settings of industrial training and higher education (Fletcher, 1994) in which by comparison with over all instructional settings and applications, interactive video disc instruction was found to improve achievement by about 0.50 standard deviations over less interactive, more conventional approaches to instruction. This improvement is roughly equivalent to increasing the achievement of students at the fiftieth percentile to that of students currently at the 69th percentile. An improvement of 0.38 standard deviations was observed across 24 studies in military training (roughly an increase from fiftieth to 65th percentile achievement). An improvement of 0.69 was observed across 24 studies in higher education (roughly an increase from fiftieth to 75th percentile achievement). Interactive video disc instruction was more effective, the more the interactive features of the medium were used. It was equally effective for knowledge and performance outcomes. It was less costly than more conventional instruction. Overall, interactive video disc instruction demonstrated sufficient utility in terms of effectiveness, cost and acceptance to recommend that it now be routinely considered and used in Defense training and education.

A review of computer-based instruction used in military training found that students reach similar levels of achievement in 30 percent less time than they need using more standard approaches to training (Orlansky and String, 1979).

Hawkins (1991) describes five issues that have evolved in the development of distance learning projects that have spanned the available mediating technologies. The five overall categories have emerged

from a synthesis of research. They include the practical issues of technology functioning, issues of community creation and definition, quality of discourse, activity definition for the distance learning work and quality control. While most of these are obvious topics of research, the issues of community creation have grown out of "a real reason to collaborate or communicate with others across distances," according to Hawkins.

Activities that structure the communicative exchange or information retrieval around carefully designed problems, activities that support collaborative work through the customized design of the software and/or the materials that coordinate the work, and a coordinator responsible for assuring the smooth flow of information throughout the community and who helps with problems helps to build the sense of community according to Hawkins (1991). "Expectations for what it means to participate in the community, and coordinating meaningful collective work or exchange are part of the ongoing definition of a social fabric for these media in education. It is a social world that is emergent, whose norms for interaction and purpose for communications becomes defined in practice," says Hawkins. She contends that as we move into the second phase of research in distance learning, the goal is to understand what designs for distance learning are effective for cognitive and social development.

At the New Jersey Institute of Technology, Hsu (1990) compared the achievement of students in an on-line class and face-to-face class by evaluating their success in learning to make business decisions. When both groups were given a business simulation project to complete, the on-line group fared better by 50 percent. Hsu noted that the most significant difference between the two groups was in their cohesion. The on-line students were better coordinated and more collaborative when working on the project.

University of Phoenix (1992) faculty who taught courses in both traditional classroom and on-line programs compared a number of factors of both programs. The results indicated that faculty were very satisfied with the overall educational environment. They were especially satisfied with on-line students' academic and professional preparation. Compared to the classroom environment, faculty were somewhat more satisfied with the students' mastery of the course content in the electronic environment. Faculty also rated the written communications skills of on-line students higher than the classroom students they teach. Research on the costs of instruction delivered via distance learning, videotape, teleconferencing and computer software indicates that savings are often achieved with no loss of effectiveness. Distance learning vastly broadens the learning environment, often providing teaching resources simply not available heretofore. Technology-based methods have a positive impact on learner motivation and frequently save instructional time. Savings in training time produce benefits both by reducing training costs and by shortening the time required to become and remain productive in the workplace. A Congressionally mandated review covering 47 comparisons of multimedia instruction with more conventional approaches to instruction found time savings of 30 percent, improved achievement, cost savings of 30-40 percent and a direct, positive link between amount of interactivity provided and instructional effectiveness (Fletcher, 1991).

A comparison of peer tutoring, adult tutoring, reducing class size, increasing the length of the school day and computer-based instruction found computer-based instruction to be the least expensive instructional approach for raising mathematics scores by a given amount (Fletcher, Hawley and Piele, 1990).

A landmark study of the use of technology for persons with disabilities found that "almost three-quarters of school-age children were able to remain in a classroom, and 45 percent were able to reduce school-related services" (National Council on Disability, 1993).

Most computer mediated on-line education programs are based upon asynchronous (not real time) communication. Computer conferencing messages are text files that are stored in the central computer database awaiting access by the students and instructor. Students participate at a time and a pace convenient to them. This attribute impacts upon the group dynamics and the learning process. Hegegaard (1994) assessed the similarities or differences that might exist in the affective domain, due to the educational delivery system selected by the student. The 15 affective constructs are as follows:

Self-efficacy — Professional self-confidence and esteem

Teamwork — Value toward working together to accomplish common organizational objectives

Strategic Thinking — Value toward systematic, long-range planning

Education — importance placed on life-long learning

Developmental Path — Belief in current career path as a means to professional progress

Risk-Taking — Value assigned to taking risks in order to achieve goals

Applied Orientation — Value of action/experience in the learning process

Communication — Importance of communication skills in contributing to professional success

Anti-Education — Negative value assigned to formal education as a meaningless barrier to progress.

Proactive Education — Value assigned to long-term appreciation and benefits of higher education

for working adults

Practical Education — Importance assigned to short-term, immediate value and benefits of education for working adults

Ethics — Importance of having standards of fairness in dealing with employees and customers

Civic Action — Importance in being an active and responsible member of the community

Long-term Orientation — Value toward focusing on innovation and efficiency in the long versus the short-term

Cooperation — Importance of establishing and rewarding cooperative relations in organizations.

The results indicated that of the 15 affective constructs, six were significant as a result of time (from beginning to end of program): teamwork, strategic thinking, developmental path, proactive education, applied orientation and practical education. Nine affective constructs were shown to be significant as a result of the delivery system (classroom or on-line): self-efficacy, risk-taking, applied orientation, communication, practical education, ethics, civic action, long-term education and cooperation. Both groups placed their highest values on ethics, civic action, teamwork, long-term orientation and cooperation, on their pre and post assessments (Hedegaard, 1994). Figure 8.1.2 shows the factors in which each group experienced significant change from the time they started a program to the time they graduated. While both the on-line and campus students experienced some changes in values over time, only one of the areas of change is common to both groups: teamwork. This indicates that the method of delivery may play a role in fostering certain values over others, but that both delivery methods are equal in their reinforcement of teamwork (Hedegaard, 1994).

The on-line group rated the following factors higher than the campus group: Self-efficacy, Education, Risk-taking, Communication, Ethics, Civic Action and Cooperation. They manifested higher self-concept and esteem, placed a greater value on education to enhance their professional life and also had a greater propensity toward risk-taking. They rated the importance of communication to their professional development higher than the campus group did. Civic Action was a value which the on-

| Figure 8.1.2 |
| Significant Change Over Time for UOP Students |

Computer-Mediated	Campus
Teamwork	Teamwork
Strategic Thinking	Applied Orientation
Developmental Path	Practical Education
Pro-active education	

line group not only valued higher than the campus group, but their ratings on this value increased over the program, while the campus group's value toward Civic action declined. The value assigned to having standards of fairness in dealing with employees or customers (Ethics construct) was significantly higher for the online group, over the campus group on both pre- and post-tests (Hedegaard, 1994). The on-line program does not have face-to-face interaction or instantaneous feedback. It relies on processing ideas through the written word, and lends itself to strategic-thinking approaches. As noted by both Harasim (1989) and Quinn, et al. (1983), the on-line program also reinforces a greater quantity and quality of peer interaction. Every member must participate in the discussion and all ideas are first thought through, then put into written form. Finally, they are reviewed and available for comment and archiving by all members of the group. The very mechanics of learning on-line requires a systematic orientation, the benefits of which may not be felt except over time, and this may be why the computer-mediated group experienced the most change in these three areas, as compared to the campus group (Hedegaard, 1994).

Cooperation was valued higher by the on-line group. While it may be characterized as a physically isolated learning experience, it still contains all of the qualities of a cooperative learning environment; one where ideas are shared, scholarly collaboration takes place and group support is strongly evident. Mason (1989) asserted that collaboration leads to higher-order learning through the process of cognitive restructuring and conflict resolution. Through these collaborative activities, new ways of understanding the material emerge because of contact with new or different perspectives (Hedegaard, 1994). Development and monitoring of evaluation and assessment standards will influence the effectiveness of on-line delivery. This may involve a restructuring of the traditional university as it is presently perceived. "Virtual classrooms" might give way to the proliferation of "virtual universities" in the future. Universities are strongly traditional, but reflective of the history and culture they serve. Many are slow to change, and may be reluctant to do so even when provided with empirical evidence that new teaching-learning models and new technologies can be as effective as old methods (Hedegaard, 1994). Honey and Henriquez study (1993) showed that penpal exchanges, scientific data collection and social awareness and opinion exchanges represent the telecommunications activities most frequently done as classroom exchange projects. When students conduct research projects, encyclopedias, news retrieval services, weather information and educational databases are the resources they use most frequently. Penpal exchanges were not rated as highly effective learning tools. They prefer science and social awareness projects. The most useful information resources to use with students were news retrieval services, scientific databases, encyclopedias, ERIC and social studies databases. In contrast to the frequency with which telecommunications is used as a professional resource, student learning activities happen with much less regularity. Much of the activity done with students takes place in the classroom, but teachers also telecommunicate from computer labs, library media centers and their own homes.

The benefits to using telecommunications technology with students include expanding students' awareness about the world in general, accessing information that would be difficult to get otherwise, enabling students to gain familiarity with basic computer applications, helping students to feel successful and allowing students to undertake more collaborative group-based activities. One of the most important benefits is its impact on their students' higher order thinking skills, suggesting that inquiry-based analytical skills-like critical thinking, data analysis, problem solving and independent thinking-develop when students use a technology that supports research, communication and analysis. In contrast, these educators report that students' involvement with telecommunications does not directly help to improve their performances on state- or city-mandated tests. This finding suggests that there is a gap between what teachers know the creative use of telecommunications can do for their students, and what traditional measures of assessment actually account for (Honey and Henriquez,1993).

There are a number of factors that these educators believe influence the success of student-based telecommunications activities. When teachers are using networks to carry out classroom exchange projects, advanced planning and full cooperation of all participating teachers is viewed as important to the project's success. The scope and content of the activity also need to be well designed to support and enhance the curriculum, the relevance of the telecommunications activity to the teacher's ongoing curriculum is important. In addition, timelines that specify when data will be collected and transmitted or when stories will be written and exchanged are viewed by these educators as critical to the success of classroom exchange projects, as is ongoing technical support to ensure that the project runs smoothly. While important, preparing participating students in the use of telecommunications skills and having students perform the mechanics of telecommunications by logging-on, uploading and downloading information are factors that received a lower rating of importance than those mentioned above. These findings suggest that central factors that influence the success of any shared learning activity are important to the success of a telecommunications project: planning, cooperation and well-defined and relevant project goals (Honey and Henriquez,1993)

The research continues to validate that the use of technology and distance learning is a viable means of delivering education and training.

■ TECHNOLOGY AS A FORCE FOR SYSTEMIC REFORM

Many critics of American schools see technology as an important tool in bringing about the kind of revolutionary changes called for in these new reform efforts (Means, et al, 1993). "Having seen the ways in which technology has transformed the workplace, and indeed, most of our communications and commercial activities, the business community and the public in general are exerting pressure for comparable changes within schools."

David and Shields (1991) state that today's reform efforts "strive to change the education system by fostering a different style of learning." The efforts seek to move classrooms away from conventional didactic instructional approaches where teachers lecture and students listen and complete exercises on well-defined, subject-area-specific material (Means and Olson, 1994). Instead, students are challenged with complex, authentic tasks and reformers are pushing for lengthy multidisciplinary projects, cooperative learning groups, flexible scheduling and authentic assessments. "In such a setting, technology is a valuable tool," say Means and Olson (1994). "It has the power to support students and teachers in obtaining, organizing, manipulating and displaying information. These uses of technology will, we believe, become an integral feature of schooling."

"When technology is used as a tool for accomplishing complex tasks, the issue of mismatch between technology content and curriculum disappears altogether. Technological tools can be used to organize and present any kind of information. Moreover, it is not necessary for the teacher to know everything about the tools that students use; students and teachers can acquire whatever technology skills they need for specific projects. In fact, one of the best things that teachers can do with respect to technology is to model what to do when one doesn't know what to do" (Means and Olson, 1994).

"The primary motivation for using technologies in education is the belief that they will support superior forms of learning. Advances in cognitive psychology have sharpened our understanding of the nature of skilled intellectual performance and provide a basis for designing environments conducive to learning. There is now a widespread agreement among educators and psychologists (Collins, Brown and Newman, 1989; Resnick, 1987) that advanced skills of comprehension, reasoning, composition and experimentation are acquired not through the transmission of facts but through the learner's interaction with content. This constructivist view of learning, with its emphasis for teaching basic skills within authentic contexts (hence more complex problems), calls for modeling expert thought processes, and for providing for collaboration and external supports to permit students to achieve intellectual accomplishment they could not do on their own. It also provides the wellspring of ideas for many of this decade's curriculum and instruction reform efforts" (Means, et al, 1993). Sterns (et al, 1991) notes that from the successful uses of technology "we have learned that technology often produces unexpected benefits for students and teachers. From the failures we have learned that implementation without thoughtful planning or sustained support is nearly always futile."

"Thus, support for the use of technology to promote fundamental school reform appears to be reaching a new high. At the same time, we have the opportunity to profit from the experiences of those educational institutions that already have implemented various technological innovations within the context of serious reform efforts. In these cases, technology is viewed as a means of supporting goals related to increased student involvement with complex, authentic tasks and new organizational structures within classrooms and schools (Sheingold, 1990).

David (1991) makes a case for technology and restructuring efforts as partners in educational change. She says: "The concepts behind restructuring the education system and the technology that can contribute to that effort are both part of the Information Age. Together they reinforce a new viewpoint that magnifies their potential to change education. To the extent that restructuring and technology are twisted to fit the Industrial Age of the past, they will not affect educational practice. To the extent that restructuring and technology are driven by challenging goals for students and supported by long-term commitments to change and investment in human resources, they will increase the productivity of our schools — and ultimately of our society.

"Dede (1991) observes that technology is not a silver bullet to resolve the schools' crises, but "advanced information technology is a crucial component in implementing a more effective paradigm for education." He contends that individualized learning and decentralized institutional structures require "complex organizational practices that necessitate sophisticated computational and communications capabilities.

"Farrell and Gring (1993) state that a planning foundation for technology could be based on the concepts that technology is not a panacea for the current problems facing education or a substitute for the basic tenets of a good learning environment; it serves as a tool for the learning and teaching processes. They state that there is no single best use of technology and the potential power of technology is in 'how it meets the needs of the learner."

Educators "have not seen that the information technologies are as central to the operations of education as they are to business, research and the arts," according to White (1990). She contends that without this understanding, "there can be no conceptual framework for the role of technologies in education. Without the conceptual framework, there has been no commitment to invest in the technologies as a rational educational investment.

"Stepping into a classroom, says Dede, "should be like entering a time machine hurtling forward; today's educational system should foreshadow the intelligent tools and interactive media that will pervade future workplaces and communities." Dede projects that a critical mass for large-scale school restructuring can be built through an "iterated process" of the following:

• define basic assumptions about learning, instructional technology and organizational development;

• imagine ideal teaching/learning environments based on design principles stemming from the basic assumptions;

• delineate action in the present to initiate an evolutionary process to shape the desired future; and

• assess the strengths and weakness of the emerging paradigm to minimize unwanted side-effects from technology.

Conley (1993) presents a view for a need for second order change because the worlds in which schools exist are changing rapidly, and in many ways. He says "Social, political and economic systems are evolving (and in some cases imploding) at an ever-increasing rate. Old institutions, beliefs, assumptions and behaviors no longer seem adequate to explain and cope with the problems and issues that present themselves to citizens in complex societies." Conley makes the point that educators may find themselves in a time of sudden, unpredictable jolts — changes in the environment for which incremental change could be disastrous. One of these forces, technological advances "create an ever-changing environment in which human behavior and relationships are altered, and new skills are needed to prosper and survive."

■ THE POTENTIAL OF TECHNOLOGY AS A PARTNER OF AND A TOOL FOR SYSTEMIC REFORM

"Reform can only succeed if it is broad and comprehensive, attacking many problems simultaneously," according to the National Education Commission on Time and Learning (1994). The Commission offers eight recommendations:

1. Reinvent schools around learning, not time.

2. Fix the design flaw; use time in new and better ways.

3. Establish an academic day.

4. Keep schools open longer to meet the needs of children and communities.

5. Give teachers the time they need.

6. Invest in technology.

7. Develop local action plans to transform schools.

8. Share the responsibility: Finger pointing and evasion must end.

"Technology is a great unrealized hope in education reform. It can transform learning by improving both the effectiveness of existing time and making more time available though self-guided instruction, both in school and out," states the National Education Commission on Time and Learning (1994). But the true promise of technology lies in the classroom. Technology makes it possible for today's schools to escape the assembly-line mentality of the "factory model" school. With emerging hardware and software, educators can personalize learning. Instead of the lock-step of lecture and laboratory, computers and other new telecommunications technologies make it possible for students to move at their own pace. Effective learning technologies have already demonstrated their ability to pique student interests and increase motivation, encouraging students not only to spend more of their own time in learning but also to be more deeply involved in what they are doing.

The Commission also notes that the NII can "reshape education." The revolution of schools "depends both on a concerted investment strategy to help educators obtain these technologies and on educators confronting their reluctance to supplement the techniques of the 19th century (textbooks, chalk and blackboards) with the technologies of the 21st (CD-ROMs, modems and fiber optics). They must do so.

"David (1991) say that there are essential conditions for making systemic reform: an invitation to change, authority and flexibility, access to knowledge and time. "Technology alone cannot provide any of these conditions, but it can contribute in a variety of ways to each one of them. Following are potential contributions of technology.

Technology invites change and can act as a catalyst for change in several ways. Technology can "provide an occasion for change — a necessary step in restructuring. The presence of technology not only provides an opportunity for change; it also symbolizes that change" (David, 1991). Using technologies, teachers practice learning how to learn — technologies change so rapidly they realize there is no right answer, but that learning is a process. Teachers act as colleagues and decision makers as technology foster interactions among teachers. Technology provides problem solving opportunities. And, new relationships in the classroom result from the presence of technology, as collaboration among students and between students and teachers occurs" (David, 1991).

The potential for technology to reallocate and extend existing resources is only beginning to be tapped. Technology can also bring into the classroom resources that facilitate active, problem-based learning and can access information otherwise unavailable or prohibitively expensive. Teachers and administrators, as well as parents and students, can avail themselves of a variety of workshops and courses by means of telecommunications. Teachers no longer need to rely exclusively on what is available in their districts. Through networks, teachers and students can exchange ideas and expertise with their peers around the world (David, 1991).

Conley (1993) sees technology bringing "a broad array of new techniques for organizing, communicating and disseminating information." David (1991) does not believe that the use of technology will simplify teaching. In fact, introducing technology into schools as currently organized "vastly increases the complexity of teachers' jobs because it makes possible more complex — though more effective — approaches to teaching." She notes that under the best of current circumstances where teachers have access to the latest technology and sufficient training and support, "the presence of technology complicates teachers' jobs enormously. They are learning not only how to use the technology but also how to teach differently, how to relate in new ways to their students, and how to assume new role as learners, researchers and equipment technicians. Technology offers the potential to undertake more complex tasks in the classroom. Ultimately, when organizational changes — team teaching, flexible grouping and scheduling, time for learning — and the larger culture of the school and district support restructuring, the potential of technology to simultaneously increase and manage complexity will be exploited."

Bugliarello (1990) and Sheingold (1991a) say that because knowledge is becoming more accessible to more of the population, the teacher's role as gatekeeper must change. Information need no longer be stored in memory for it to be useful; the ability to access information will be as or more important than the ability to store information in one's memory.

Thornburg (Betts, 1994) says that the Information Age is now over. He says we're entering a new Communication Age. "Of course, a lot of schools still have not taken advantage of the tools of the Information Age" and others are barely entering the Information Age. As an example of this, he cited schools which use word processors as glorified typewriters. He says the challenge "is not to use them to do the old job better but to do something new." Does it matter what we call the age in which we live? Probably not, but Thornburg's observation signifies another paradigm shift — another way of thinking about our lives and purpose, before we had become accustomed to the last one. The Information Age is generally accepted to have begun in 1985. In less than nine years, another age has begun.

"Now we're seeing tools of the Communication Age starting to change the face of the Information Age," says Thornburg (Betts, 1994). "Students will integrate technologies to create their own multimedia projects. At this point, schools will shift from focusing on Information processing to an emphasis on communication. When we take a look at all the research on learning styles — Gardner's Theory of Multiple Intelligence (1985), McCarthy's 4-MAT process (1981), for example — we find that students learn best when they learn in the way best suited to them." He feels that by acknowledging that each of us has components of at least seven different intelligences — different pathways to learning — we can activate more of those pathways each day. The more we do, the more effective the learning environment will be, he says. "If you create rich learning environments where multiple intelligences are addressed simultaneously, kids really thrive. Just by mixing things up and making the classroom a more multi-sensory environment, you take advantage of these multiple pathways to learning and benefit all of the students. And multimedia is a great tool because it combines images, text, animation — all sorts of sensory experiences.

"Information Age technology can empower learners in different ways. Kids will acquire information themselves in ways that are congruent with their natural styles of learning and that's exciting." In the Information Age, "we had our computers bolted to the desk and chained to the wall by two cords, one for power, and the other, when we had it, for telecommunications. Now we're moving toward transportable equipment and wireless communication. That means that the informational tools and communications tools merge into one and become transportable; you can carry them and use them wherever you are." Thornburg observes that information technology has become personalized" so that "students will be taking more control over their learning, taking control away from the educator. The whole role of the educator has changed." New social factors will drive curriculum changes. Technology is one way we respond to changes in our environment," say Thornburg (Betts, 1994). The idea that learning can occur only within four walls when twenty-five young people interact with one certified teacher is rapidly being replaced with models in which varying combinations of adults and children interact both inside and outside of school (Ratzki and Fisher 1989/1990). The world around the school is becoming a source for curriculum according to Beane (1991). Local issues, problems and resources are being integrated. Information from around the world, available to teachers and students via technology, serves as the framework within which local issues can be understood and examined, creating curriculum that allows students to understand global events in relation to the world in which they live.

Technology is almost an icon in some school restructuring plans says Collins (1991). In other settings, he continues, technology is emerging as an extension of the interaction between teacher and student. In almost all visions for restructured schools, it holds an important, if still indeterminate, place. Conley believes that In the new vision of education, technology is an integral component (1992). Technology is used to provide basic skills support, interface with information sources outside of the school, support individual student creativity, manage information about student performance and achievement, assist teachers in their dual roles as instructors and clerks, and provide students with greater control over their own learning (Conley, 1992).

Technology now makes it possible to "replicate critical elements of the teaching-learning process in live time or asynchronously," according to James Hall, president of Empire State College and SUNY's vice-chancellor for educational technology (Jacobson, 1994b). That "was not able to be done easily by distance learning in the past," which means that "the quality of the campus experience is approached in distance learning, and that's new." He believes that the latest technology can be used to raise the quality of education even as it increases the number of students served.

Dede (1991) observes that the two most common errors in impact assessment are to overestimate how soon a new technology will change society and to underestimate the magnitude of its eventual effects. Typically, he says, communications devices have their impact on institutions in four sequential stages:

Stage 1: The new technology is adopted by an institution to carry out existing functions more effectively;

Stage 2: The institution changes internally —

work roles, organizational structure — to take better advantage of these new efficiencies;

Stage 3: Institutions develop new functions and activities enabled by additional capabilities of the technology; as the roles of different types of institutions expand, new competitive relationships emerge; and,

Stage 4: The original form of the institution may become obsolete, be displaced or be radically transformed as new goals dominate the institution's activities (Coates, 1977).

According to Naisbitt (1982), new technologies pass through three stages. In the first stage, the new technology follows the least resistance into a ready market. In the second stage, users improve or replace previous technologies with the new technology. In the third stage, users discover new functions for the technology and, based on its attributes ask, "What can we do now that was not possible before?" This is the stage that many educational technology users are in now, but there are others in the earlier stages. For example, educators using computers to create puzzles, or assess student progress are in a stage two usage pattern. "In contrast, educators who have moved to stage three are asking, 'How can these new tools contribute to a more powerful educational experience?' These educators are searching for a paradigm shift, not just a way to squeeze technological tools between the existing bricks of yesterday's educational practices" (Peck and Dorricott, 1994).

Academic leaders should begin planning now for the time, perhaps as little as five years away, when distance learning will give most students virtually unlimited course options from institutions all over the world, regardless of where they may formally enroll, according to Rich Gross, dean of telecommunications at Kirkwood Community College, Iowa (Jacobson, 1994). Gross suggests that by then many colleges will need to become local "gate keepers," helping students choose courses from different networks, showing them how to obtain instructional materials and providing counseling and other traditional student services. Gross projects an increasingly competitive market in distance learning and thinks colleges may need to follow a model in which they strive to be "net exporters" of education no matter what their students may be using from other places.

For those who use technology as an integral component of learning, Peck and Dorricott (1994) developed a "top ten list" of reasons to use technology:

1.   Students learn and develop at different rates. Technology can individualize instruction with integrated learning systems which offer thousands of lessons covering the same basic skills now taught in a lock-step way through textbooks to groups of students with different backgrounds, interests and motivation. Students can move at an appropriate pace in

a nonthreatening environment, developing a solid foundation of basic skills rather than the shaky foundation a calendar-based progression often creates.

2. Graduates must be proficient at accessing, evaluating and communicating information. Educational technologies can, by design, provoke students to raise searching questions, enter debates, formulate opinions, engage in problem solving and critical thinking and test their views of reality. On-line tools and resources allow students to efficiently gather and evaluate information, then communicate their thoughts and findings. This communication may require reading, thinking, creating charts, graphs and other images; or the organization and production of information using spreadsheets and databases.

3. Technology can foster an increase in the quantity and quality of students' thinking and writing. Perhaps one of the best documented successes with computers in education is in developing students' writing. Word processors reduce writing phobia and the temporary positive feeling makes it easier to take creative risks. Difficulty with handwriting usually does not transfer to the keyboard, so frustration is reduced. Editing and revising are quick, spelling and grammar check programs teach to weaknesses; printed papers generate a sense of accomplishment.

4. Graduates must solve complex problems. Higher-level process skills cannot be "taught" in the traditional sense, or transferred directly from the teacher to the learner. They need to struggle with questions they have posed and search out their own answers. Computer productivity tools could revolutionize the way students work and think. Databases, spreadsheets, computer assisted design, graphics programs and multimedia authoring programs allow students to independently organize, analyze, interpret, develop and evaluate their own work. These tools engage students in focused problem solving, allowing them to think through what they want to accomplish, quickly test and retest solution strategies and immediately display the results.

5. Technology can nurture artistic expression. Modern technology-based art forms (video production, digital photography, computer based animation) have great appeal, encourage artistic expression among a diverse student population. These tools provide forms of artistic communication for those students who have been constrained by the traditional options of verbal and written communication, and they increase motivation and foster creative problem-solving skills as students evaluate the many possible ways to communicate ideas.

6. Graduates must be globally aware and able to use resources that exist outside the school. Children's domains of discovery at school are limited to the classroom and the school. Technological tools allow students to inexpensively and instantly reach around the world, learning firsthand about other cultures. Various technologies can provide up-to-date maps and demographic data, and computer-based wire services can bring a newsroom quality stream of current events into the school.

7. Technology creates opportunities for students to do meaningful work. Students need to produce products that have value outside school, receive feedback on their work, and experience the rewards of publication or exhibition. Technology can provide a widespread audience for students' work. Computers link students to the world, provide new reasons to write and offer new sources of feedback. Students' video products shown on local cable can produce high levels of motivation and accomplishment.

8. All students need access to high level and high-interest courses. Electronic media can bring experiences and information previously unimagined by students into the classroom. Through instructional television, students can view and discuss events they otherwise could not experience. Laser discs and CD-ROMs put thousands of images and topics at students' fingertips. Distance education technologies can bring important learning experiences to students, even in districts where small student populations have made some courses impossible to offer.

9. Students must feel comfortable with the tools of the Information Age. Computers and other technologies are an increasingly important part of the world in which students live. Many of today's information producers are converting their knowledge bases to digital format and are constructing new technologies to increase speed, capacity and reliability of dissemination. As telephone, computer, television and other media merge, incredible resources will become available. An "I tell you, you tell me, and I'll grade you" model of education will not prepare students to take advantage of these resources.

10. Schools must increase their productivity and efficiency. Technology can re-place (not replace) the teacher. When stage three educators determine what students should do and how teachers and technologies can support students, many of the routine tasks done by teachers can be reassigned to technology, elevating the role of teacher. Some things only teachers can do. Teachers can build strong, productive relationships with students. Technologies can't. Teachers can motivate students to love learning. Technologies can't. Teachers can identify and meet students' emotional needs. Technologies can't. Technology-based solutions in education can, and must free the teacher to do the important work that requires human interaction, continuous evaluations and improvement of the learning environment (Peck and Dorricott, 1994).

Hancock and Betts (1994) observe that identifying "promising advanced technologies and planning how to integrate them into instruction are two very different issues. To rise to the latter challenge, planners need to know what to expect when technology is used intensively and effectively.

1. Learning experiences proliferate. Teachers expect more of their students and present more complex material. The range of learning experiences extends far beyond those offered in traditional classrooms.

2. More individual attention is possible. Time-consuming paperwork for teachers is reduced, permitting them to focus their attention elsewhere. Teachers can better meet the demands of individual students, give them more attention, allow more independent work, and accommodate different learning styles.

3. Roles shift. Teacher-centered classrooms tend to evolve into student-centered ones. The teacher acts more as a coach than an information dispenser. More collaboration and small-group work occurs. To realize any vision of smarter schooling by using technology, school districts and colleges of education must prepare teachers to use the technology. Apart from funding considerations, adequate teacher preparation is probably the most important determinant of success.

Another essential according to Hancock and Betts (1994) is to put technology into teachers' hands. Ways to do so include rent-to-own agreements in cooperation with local businesses; professional contract revisions to recognize that the ability to do productive work is not restricted by time or place; teachers-only electronic tools provided in classrooms, teachers lounges or library/media centers; technology loan programs for teachers' home use; technological competency requirements in all teacher education programs; state-of-the-art electronic tools for professors of education; and a telephone line in every classroom.

Research by Sheingold and Hadley (1990) suggests that seven years of administrative support, staff development and planning time are required before teachers fully integrate technologies into their repertoires. To move this process along, teachers must have timely opportunities to use technology for classroom and personal productivity. Hancock and Betts (1994) suggest that putting a laptop and telephone into teachers hands will provide the proper impetus. Distance learning and other applications of technology hold great promise for contributing to educational reform efforts. Exploiting that promise, however, requires that activities such as those sponsored by Star Schools be more closely tied to other reform efforts than currently is the case. "Over the long term, both distance learning and educational reform will benefit from a close relationship" (Tushnet, et al, 1993). To facilitate the integration of distance learning technology into educational reform, the U.S. Department of Education could provide grants to projects which effectively show the contribution to systemic reform that distance learning and educational technologies make.

■ EQUITABLE ACCESS THROUGH TECHNOLOGY

Educational technologies can enable change. Milken (1994) writes that, "In approaching the millennium, Americans may justifiably argue that millennial ideas are in short supply. Faced with a multitude of challenges from within, our society is seeking the visionary concepts — from worlds as diverse as economics, multiculturalism and even particle physics — that will lead us into the 21st Century. However, such direction will not be found in financial innovation, social service or scientific discovery. Rather, it will come from the frontiers of education."

Milken values the people as a society's most valued resource, rather than the factories, building or natural resources. Comparatively, our natural resources are the same as our competitors in a global economy. The only significantly different resource that we have is people. If education does not improve the resource, we run the very real risk of creating another third world country in the United States. Certainly, the gap is widening between the technological haves and have-nots.

Equity is a second major concern according to the NEA (Malley, 1994). "Rural areas face special challenges in modernizing their infrastructure to support advanced technologies. Similar challenges are found in many of our most urban districts, where critics have charged that telecommunications providers are 'redlining' certain communities out of their modernization plans." Malley asserts that "without addressing this issue head-on, we will assure a divided country unprepared to face the challenges of the information age. We will condemn whole segments of the population to dead-end jobs and menial existence. We cannot have any particular segment of our student population falling behind, because their community has not been connected to the Infrastructure or because advanced services are not being offered there. Our whole social fabric relies increasingly on education to offset the social and economic disparities that exist in our society. Schools, then, need to be one place where any student, anywhere, can have access to the best learning tools available. Deficits in education lead fairly directly to difficulties in the job market and to other social problems."

U.S. Department of Education Deputy Secretary

Madeleine Kunin asks "will technology narrow the gap between the haves and have-nots or if we don't do the right thing, does it have the potential to widen the gap even further? I think this is such a fundamental question, not only for technology's sake and what that means, but because it goes right to the heart of education itself. Access to a quality education is what our whole education reform effort is about. And now that we have some marvelous new tools such as Goals 2000 and the School-to-Work Legislation, we have an opportunity that has not been given us before to bring excellence to scale" (Kunin, 1994).

Kunin provided figures that showed "where we are in terms of equity right now. We have an average of about 2.4 computers per 30 students. Most of the computers — about 55 percent — are Apple. They're usually at the low end of the scale. There are some ways in which these statistics are not as bad as you might expect; 41 percent of the computers are in classrooms, 46 percent are in labs; 64 percent of schools with access to outside networks, 12 percent of schools have some telephones in the classroom. That's probably the cheapest investment — to simply get a telephone — and that's the one where we have the furthest to go. This can be the great equalizer. It can be our great opportunity to really put muscle behind the rhetoric of equal opportunity. That is exactly what we must achieve because the stakes are extraordinarily high. I don't think we'll have a chance quite like this again. We won't have a chance, first of all, to really keep the faith in public education itself" (Kunin, 1994).

"Tomorrow's promise," according to Milken (1994), "is grounded in the marriage of education and technology, in interactive networks that will bring ideas, knowledge and new ways of thinking to people young and old, in schools, homes and workplaces. Satellites, data compression technology, CD-ROM and speedier computers have all brought new worlds of information to ever-larger audiences. The prospects are limitless for students of any age, for training, retraining and pure knowledge enhancement..." He concludes that, "A combination of education and technology, made readily available to our school children, teachers and adults at home or in the workplace, represents the brightest opportunity for achievement and for the knitting together of an America too often divided by economic and cultural differences. Access to knowledge is the most lasting millennial idea of them all."

Levin (1993) discusses the idea of undereducating a significant proportion of our students which has serious economic consequences. Two criteria define undereducation: the level of attainment (years) that youngsters make in the educational system and their level of knowledge, values and behavior. "If young people do not have enough schooling for the labor force, if they do not have enough schooling to meet certain standards that the society sets, then we believe they are undereducated. If, at the same time, we find that they do not have the skills, the attitudes, the values and the knowledge that we expect, then we also say that they are undereducated even if they have many years of schooling. So the two criteria overlap to a substantial degree, but they are two, somewhat different, criteria." High school is a minimum requirement in years, but "there is considerably less agreement on what we mean by undereducation in terms of knowledge and skills, except when we get to an extreme," according to Levin.

Levin defines "at-risk" students as those who are unlikely to succeed in schools as schools are currently constituted because they do not have the experiences in the home and family on which school success is premised. "We know, simply based on children's origins and family background," Levin says, "which children are likely to succeed in schools, which children are not. Schools are not neutral with respect to the types of student backgrounds for which the existing curriculum and instructional practices offer success. Students who come from middle class and non-minority backgrounds with both parents present and who speak a standard version of English are much more likely to succeed educationally than those from poverty, minority, immigrant, non-standard English and single-parent backgrounds. At-risk students are caught in a mismatch between their home situations and what schools require for success. This is very important because the kinds of results that many school systems are getting today with at-risk students are not inevitable. They are heavily premised on the way that we think about schools and on a curriculum that builds on some children's experiences and either ignores, or even negates, the experiences of other children."

Levin believes that probably more than 35 percent of elementary and secondary students are in the at risk population, with California and Texas reaching over 50 percent. He estimates that by 2020, more than half of the elementary and secondary children will be in at-risk categories. The economic consequence is that these are the future employees who tend "to be very much undereducated relative to modern work force standards," and they "will not have the foundation typically required for regular employment or for further training. Unless we are able to intervene," cautions Levin, "unless we are able to get the resources, employers are going to face lagging productivity, higher training costs and competitive disadvantages at a time when there is a general concern in this nation about just such things."

There is also a forewarning for training as "there is a growing gap between the competencies technology professionals currently have and those they will need in the future" (Wendt, 1994). The pace of change "exceeds the capacity of their current skills acquisition process. To close the gap, management must balance the rates of change across people processes and technology so people develop new skills just in time to perform new processes and use new technologies."

Wakin (1994) suggests that the reasons for the gap are "more easily identified than corrected. Significant changes are overtaking people, processes and technology faster than ever before," Wakin says. "Employees in flattened organizations are becoming empowered, taking on new roles and straining to develop new skills. Work processes are being re-engineered and technology is evolving at breakneck speed." Not only does this put more pressure on information technology (IT) managers to develop appropriate training and development programs, but it adds another level to the gap between the haves and have nots, and the education that the haves are currently getting.

Watkin (1994) says that what is needed is a "continuous learning process — one that can include delivery of training to desktop PCs, in-house seminars, performance support systems and coaching." He recommends that managers must stay in close touch with the progress of each employee and provide an environment that encourages and rewards professional development. Finding the time to train is a problem shared by IT managers and educators as the workload overwhelms training plans. This is further compounded by trying to anticipate changes in technology — in existing technology and that which is still in the planning stage so that short-term and long-term planning can be revised continuously.

The warning is clear for education and training. We are at-risk of creating haves and have-nots in technological skills and even the haves are now at risk of having too few competencies. Certainly, this is yet another powerful argument for providing equitable access for all. Distance learning activities that are designed to improve equal educational opportunities can be funded in two ways: seed money or formula base (Tushnet, et al, 1993). Schools can receive seed money grants that allow them to modify facilities and purchase equipment to implement distance learning technologies. Seed money is particularly important for schools that serve low-income students because there is a relationship between the quality of the facilities and the income level of students. Schools should receive funding to support distance learning based on the needs of students. In rural areas, a measure of curriculum isolation can be derived from the size and geographic location of schools. It is more difficult to derive a formula for urban students, but viewing distance learning as an approach to supplementing instruction for educationally disadvantaged students provides one way of considering the problem. Money also can be allocated for gifted, low-income students. Congress should engage in broad consultation in order to develop an equitable formula (Tushnet, et al, 1993).

■ PROBLEM-BASED LEARNING

In problem-based learning students meet an "ill-structured problem" before they receive any instruction. "In the place of covering the curriculum, learners probe deeply into issues searching for connections, grappling with complexity and using knowledge to fashion solutions" according to Stephen and Gallagher (1993). As with real problems, students will not have most of the relevant information needed to solve the problem or know what actions are required for resolution. After they tackle the problem, the definition of the problem may change. And even after they propose a solution, the students will never be sure they have made the right decision. They will have had the experience of having to make the best possible decision based on the information at hand (Stephen and Gallagher 1993). Information access through computer modem, working with other students at a distance, or collaborating with experts by computer conferencing can bring more reality to problem-based learning. It also helps students weigh the relative value of information found on the Internet, other resources and the opinions of others. Not only do students find the information and process it, but they apply it. Through this process they develop information gathering skills, writing and communication skills and learn to work in collaborative teams.

They will also have had a stake in the problem (Stephen and Gallagher 1993). "In problem-based learning students assume the roles of scientists, historians, doctors or others who have a real stake in the proposed problem. Motivation soars because students realize it's their problem. By having a stake, they come to realize that no real-world problem is objective, that every point of view comes with a bias toward interpreting data in a certain way."

Teachers take on new roles in problem-based learning, too. First they act as models, thinking aloud with students and practicing behavior they want their students to use (Stephen and Gallagher 1993). "They familiarize students with metacognitive questions such as, What's going on here? What do we need to know more about? What did we do during the problem that was effective? Then they coax and prompt students to use the questions and take on the responsibility for the problem. As time goes on, students become self-directed learners. To encourage the students' independence, the teachers then fade into the background and assume the role of colleagues on the problem-solving team.

Students using technology in problem solving should have access to a variety of media. A variety of media ensure that all learning styles are met — visual, verbal, written and hands-on — so that every type of intellectual skill is exercised in solving the problem (Lane, 1992).

In the process of problem solving, students crisscross a variety of disciplines. They build substantial knowledge bases through increasingly self-directed study. Through collaboration with their classmates, students refine and enlarge what they know, storing their new knowledge in long-term memory in such a way as to promote transfer to new problems. As they move toward solutions, they identify conflicting ethical appeals. And when it is time for resolution, they present, justify and debate solutions, looking for the "best fit." Problem-based learning is apprenticeship for real-life problem solving. (Stephen and Gallagher 1993).

Electronic media make it possible for students to work with information interactively and proactively. Instead of passive viewing, good instructional design will engage the learner in proactive ways. Video may provide a general introduction and demonstrate or show information through electronic field trips or hypermedia and engage users in the creation of new media. The highest participation is gained through interaction with the work group where problems are solved.

Reganick (1994) reports that students with behavior problems 'overwhelmingly prefer the individual and immediate reinforcement that computers provide, rather than being taught by traditional methods." Technology can serve as the bridge as school reform moves the educational paradigm from passive to active learning where students are engaged in a curriculum that connects them to the real world. Our goal "is to facilitate significant changes in student achievement and behavior. Using computers in the classroom will help our students realize three objectives: significant improvement in academic achievement; improvement in behavior to a marked degree; and a reduction in interpersonal deficits." Means and Olson (1994) list five features of reformed classrooms. They are presented in the context of a project called Local Heroes which was created by the students.

1. An authentic, challenging task is the starting point. Authentic tasks are completed for reasons beyond earning a grade. Students also see the activity as worthwhile in its own right.Students needed to earn money for a trip to camp. The students wanted to find more material about contemporary Hispanic leaders as their textbooks and libraries were limited to old material and the reading level was too high for ESL students. The students become convinced that there was a market for such materials in other schools. The Local Heroes multimedia project was born. Students needed technology to assemble their materials and produce copies of salable quality. They identified local heroes, conducted and videotaped interviews with them and composed written highlights from the interviews.

2. All student practice advanced skills. Complex tasks involve both basic and advanced skills. Heroes involved student in a wide range of tasks, some of which called for high-level thinking. They prepared for interviews, wrote questions to elicit information and interesting responses, learned presentation techniques for the interview and taped and edited the interviews. The students learned and practiced complex skills in a variety of domains (cognitive, social, and technical).

3. Work takes place in heterogeneous, collaborative groups. Three-member students conducted each interview (one asked questions, one videotaped and one took notes.) Afterwards, they discussed how to improve their technique for future interviews. They transcribed the interviews and notes and entered the text on a computer and each took responsibility for aspects of the work (typing, spelling).

4. The teacher is a coach. This doesn't mean fading into the background, but providing structure and actively supporting students' performances and reflections. Moving among small groups to check on progress, monitoring students' practice, suggesting questions and helping groups improve their interviewing technique are part of the coaching job.

5. Work occurs over extended blocks of time. Serious intellectual activity doesn't usually fall neatly into 50-minute periods.

Means and Olson (1994) observe that technology itself is not the driving force behind the learning in the Local Heroes Project, but rather the technology "amplifies what teachers are able to do and what they expect from students." Technology has this positive effect because:

• teachers see complex assignments as feasible

• technology appears to provide an entry point to content areas and inquiries that might otherwise be inaccessible until much later in an academic career

• technology can extend and enhance what students are able to produce

• selection and manipulation of appropriate tools appear to stimulate problem solving and other thinking skills

• technology lends authenticity to school tasks because the professional quality seems to make schoolwork real and important, thus students take pride in using the same tools as practicing professionals

• technology gives teachers the opportunity to become learners' again through the challenge of planning and implementing technology-supported activities

• technology provides a context in which an initial lack of knowledge is not regarded as cause for embarrassment – teachers are eager to share their developing expertise and to learn from one another

• as teachers search out the links among their instructional goals, the curriculum and technology's possibilities, they collaborate more, reflect more and engage in more dialogue.

Means and Olson (1994) say that technology will not make the teacher's life simple because it requires teachers with multiple skills to deal with subject matter that is inherently challenging. Because technology is evolving and open-ended, it can never be totally mastered. "New roles pose many challenges, too. The teacher must be able to launch and orchestrate multiple groups of students, intervene at critical points, diagnose individual learning problems and provide feedback. Nevertheless, in classrooms where teachers have risen to this challenge, a profound change is occurring in the learning environment.

■ EVALUATION — PERFORMANCE BASED ASSESSMENT

During the last several years, there has been a national trend toward the use of authentic performance assessment — the actual demonstration of learned abilities or outcomes. Teachers are moving away from multiple-choice, pencil-and-paper tests. While most educators feel that authentic assessment will create greater learning opportunities for students, it is very time consuming. Classroom teachers have little formal training in student testing and assessment. They are required, in many cases, to invent as they implement new assessments.

The key features of performance based assessment according to Smith and Cohen (1991) are that it requires examining the purposes of education, identifying skills we want students to master and empowering teachers. "Testing that requires a student to create an answer or a product that demonstrates his or her knowledge or skills" is the definition of performance assessment provided by the Office of Technology Assessment (OTA) of the U.S. Congress (1992). Rudner and Boston (1994) observe that a key feature of all performance assessments is that they require students to be active participants. "Rather than choosing from presented options, as in traditional multiple-choice tests, students are responsible for creating or constructing their responses. These may vary in complexity from writing short answers or essays to designing and conducting experiments and demonstrations or creating comprehensive portfolios."

Rudner and Boston (1994) and Herman, Aschbacher and Winters (1992) insist that to imple-

ment performance assessment fully, administrators and teachers must have a clear picture of the skills they want students to master and a coherent plan for how students are going to master the skills. They need to consider how students learn and what instructional strategies are most likely to be effective. They need to be flexible in using assessment information for diagnostic purposes to help individual students achieve. (Figure 8.1.3)

"No longer is learning thought to be a one-way transmission from teacher to students, with the teacher as lecturer and the students as passive receptacles. Rather, meaningful instruction engages students actively in the learning process. Good teachers draw on and synthesize discipline-based knowledge, knowledge of student learning and knowledge of child development. They use a variety of instructional strategies from direct instruction to coaching, to involve their students in meaningful activities...and to achieve specific learning goals" (Herman, Aschbacher, and Winters, 1992).

Dede (1991) observes that, "The act of learning is always constrained by the characteristics of the communications channel between the student and the content to be mastered. For example, the wider the bandwidth of a communications medium, the more immediate and rich a learning experience can be: seeing a videotape on how to ride a bicycle conveys more information than reading an article about that subject. The greater the interactivity of a medium, the more feedback can be communicated to motivate and individualize learning: having a friend teach you to ride a bike is more effective than is watching videotape on the topic."

Performance assessment techniques include projects, group projects, interviews/oral presentation, constructed-response questions, essays, experiments, demonstrations and portfolios (Rudner, 1991). Proponents believe that performance assessment can measures skills that have not traditionally been measured in large groups of students — skills such as integrating knowledge across disciplines, contributing to the work of a group and developing a plan of action when confronted with a novel situation (Rudner and Boston, 1994). Performance based assessment is not new (Rudner and Boston, 1994) as ERIC has used "performance tests" as a descriptor since the birth of the ERIC system in 1966. What is new is the widespread interest in the potential. Proponents of "authentic assessment" make distinctions among the various types of performance assessments, preferring those that have meaning and value in themselves to those that are meaningful primarily in an academic context. "In a chemistry class, students might be asked to identify the chemical composition of a premixed solution by applying tests for various properties, or they might take samples from local

Figure 8.1.3
Aligning Instruction and Assessment: Implications from Cognitive Learning Theory (CLT) (Herman, et `al, 1992).

CLT: Knowledge is constructed. Learning is a process of creating personal meaning from new information and prior knowledge.

Implications for Instruction/Assessment:

- Encourage discussion of new ideas.
- Encourage divergent thinking, multiple links and solutions, not just one right answer.
- Encourage multiple modes of expression, for example, role play, simulations, debates and explanations to others.
- Emphasize critical thinking skills: analyze, compare, generalize, predict, hypothesize.
- Relate new information to personal experience, prior knowledge.
- Apply information to a new situation.

CLT: Learning isn't necessarily a linear progression of discrete skills.

Implications for Instruction/Assessment:

- Engage all students in problem solving.
- Don't make problem solving, critical thinking or discussion of concepts contingent on mastery of routine basic skills.

CLT: There is great variety in learning styles, attention spans, memory, developmental paces and intelligences.

Implications for Instruction/Assessment:

- Provide choices in tasks (not all reading and writing).
- Provide choices in how to show mastery/competence.
- Provide time to think about and do assignments.
- Don't overuse timed tests.
- Provide opportunity to revise, rethink.
- Include concrete experiences (manipulatives, links to prior personal experience).

CLT: People perform better when they know the goal, see models and know how their performance compares to the standard.

Implications for Instruction/Assessment:

- Discuss goals; let students help define them (personal and class).
- Provide a range of examples of student work; discuss characteristics.
- Provide students with opportunities for self-evaluation and peer review.
- Allow students to have input into standards.

CLT: It's important to know when to use knowledge, how to adapt it, how to manage one's own learning.

Implications for Instruction/Assessment:

- Give real-world opportunities (or simulations) to apply/adapt new knowledge.
- Have students self-evaluate; think about how they learn well/poorly; set new goals; why they like certain work.

CLT: Motivation, effort, and self-esteem affect learning and performance.

Implications for Instruction/Assessment:

- Motivate students with real-life tasks and connections to personal experiences.
- Encourage students to see connections between effort and results.

CLT: Learning has social components. Group work is valuable.

Implications for Instruction/Assessment:

- Provide group work.
- Incorporate heterogeneous groups.
- Enable students to take on a variety of roles.
- Consider group products and group processes.

lakes and rivers and identify pollutants. Both assessments would be performance-based, but the one involving the real-world problem would be considered more authentic" (Rudner and Boston, 1994.

"Linking curriculum, instructional strategies and performance-based assessment encourages teachers to focus on higher order, integrated skills, communicate goals and standards and experiment with approaches to help students achieve them. An aligned curriculum that features meaningful learning and offers students choice in demonstrating their knowledge empowers them to be more responsible for their own education and increases their motivation," according to Rudner and Boston (1994). "Assessments can be used to provide diagnostic information about what individual students know and can do and where they need additional assistance. They can also alert teachers to necessary changes in classroom instructional strategies. Rudner and Boston contend that the act of assessment is a learning opportunity for students and use portfolio assessment as a case in point. "Most versions of portfolio assessment call for student self-reflection either in selecting pieces or in evaluating progress over the course of a semester or a year. Students are thus responsible for monitoring their own learning and for assessing the implications of their progress," applying thinking skills, understanding the nature of quality performance and providing feedback to themselves and others. "Students and teachers alike are empowered through the experience" (Rudner and Boston, 1994).

There has been a fear that students would become frustrated by the technical demands of using technology. While using the technology is not a major problem for most students (Means, et al, 1993), students face other "kinds of challenges when they use technology to support themselves in activities, in inquiry learning which include: understanding their responsibilities as active learners; getting help with individual learning needs; and integrating their technology-supported inquiry learning with their larger school experience. Developing their own questions is the reversal of the role that students have traditionally played in schools. Because "learning is no longer a process in which a teacher who knows all passes on knowledge and students passively take it in," (Means et al, 1993), students need information about how the role of the teacher has changed and how their role must change.

Blanchard (1994) observes that putting learning into practice is a matter of focus. "We need to put ten times the amount of follow up on the application of training as we do on the training itself. In the past, trainers have been rewarded for the number of sessions they have done and the 'happiness' ratings those sessions have immediately produced. Increasingly, we need trainers to be focusing on the impact that man-

agers will have in applying principles that are discussed." Blanchard says that for his own firm which provides seminars for others, they have moved to an assessment which demonstrates an "impact on business practices — and the resultant savings in time and/or money."

■ NEW WAYS TO THINK ABOUT THE USE OF TECHNOLOGY FOR EDUCATION

That technology has been used successfully in education, is no longer the question. How to use it to support educational systemic reform is now the question. A framework for thinking about technology has evolved in the business sector. As business sought to reinvent itself, those that wanted to make dramatic changes were found to be asking themselves a different question. Hammer and Champy, (1993) observed that those who were succeeding were asking "Why do we do what we do at all?" rather than "How can we do what we do better?" or "How can we do what we do at a lower cost?" They reported that they found many tasks that "employees performed had nothing at all to do with meeting customer needs — that is, creating a product high in quality, supplying that product at a fair price and providing excellent service. Many tasks were done simply to satisfy the internal demands of the company's own organization" (p.4).

Certainly educational systemic reform requires the reinvention of education. As it has been for business, reinvention is not carried out in small steps. Hammer and Champy strongly suggest that "It is an all-or-nothing proposition that produces dramatically impressive results." Like companies, education has no choice but to gather our courage and do it. To paraphrase Hammer and Champy, "systemic reform is the "only hope for breaking away from the ineffective, antiquated ways of conducting" education "that will otherwise inevitably destroy" education.

If educational technologies and information technologies are to play a role in reinventing education, equating it with automation is not the answer. To date, we have looked for problems and then identified a technology solution. This has been called the "silo mentality" (West, 1994) where pockets of technology have been installed to meet the needs of one or two problems. The extension of the silo mentality is the situation which we have today. There are pockets of technology which are not connected to networks, few people have access to the technology and therefore few people have developed the skills to use the technology or the interest in using it because of the inconvenience.

If what we've been doing has not been productive, we need to think differently about the use of technology. It requires that we use a form of thinking which we teach — inductive thinking — where we recognize the powerful solution that technology can provide and then identify problems that it might solve. To date, we have used deductive thinking where we defined the problem, identified solutions and evaluated the relative merits of the solutions. We've looked at the technology and asked how it can help us to do what we do better? How can it teach more quickly? How can it help with drill and practice? How can it help take a teaching load off the teacher? We looked at the list of duties that went with teaching and administering education, and asked

how technology could do it for us? Yes, technology can do those thing, but we should also be asking "How can we use technology to allow us to do things that we are not already doing? (Hammer and Champy, 1993). One of the hardest things to do is to recognize the "new, unfamiliar capabilities of technology instead of its familiar ones."

The larger point that Hammer and Champy (1993, p. 88) make is that "needs, as well as aspirations, are shaped by people's understanding of what is possible. Breakthrough technology makes feasible activities and actions of which people have not yet dreamed."

Take, for example, the technology of two-way

Figure 8.1.4

Old Rule: **Information can appear in only one place at one time**
Disruptive Technology: Shared databases
New Rule: Information can appear simultaneously in as many places as it is needed
Educational Application: Shared lesson plans
Shared on-line texts
Shared assessment ideas
Shared student work
Access to information - Internet and commercial services
Access to daily newspapers and magazines
Access to student information
Saves trees and environment
Any number of students, teachers or staff can be using the same information

Old Rule: **Only experts can perform complex work**
Disruptive Technology: Expert systems
New Rule: A generalist can do the work of an expert
Educational Application: Students and teachers can learn material together
Teachers have access to information that increases their expertise
Substitute teachers have access to information about students

Old Rule: **Organizations must choose between centralization and decentralization**
Disruptive Technology: Telecommunications networks
New Rule: Organizations can simultaneously reap the benefits of centralization and decentralization
Educational Application:
Shared decision making
Information dissemination and sharing
Student articulation packages
Collective purchasing from a range of products for group discounts
Interaction between decentralized units
Access to information — shared libraries and interlibrary loans
Maintain flexibility and responsiveness

(Figure 8.1.4 Continued)

Old Rule:	**Managers make all decisions**
Disruptive Technology:	Decision support tools (database access, modeling software)
New Rule:	Decision-making is part of everyone's job
Educational Application:	Teachers, when properly trained, have sophisticated decision-making capability
	Access to information improves decision making at all levels
	Decisions can be made at the front line rather than kicking them up the ladder
	Decisions can be made more quickly and problems are resolved
	Empowered teachers enable empowered students
	Teachers make their own decisions based on the needs of their students
	Experts are available to provide information for making decisions
Old Rule:	**Field personnel need offices where they can receive, store, retrieve and transmit information**
Disruptive Technology:	Wireless data communication and portable computers
New Rule:	Field personnel can send and receive information wherever they are
Educational Application:	Resource teachers can be in several classrooms at once
	Classroom storage moves to electronic storage
	Students and teachers can be anywhere and still communicate
	Resource teachers can provide information from anywhere
	The virtual classroom can be created through computer conferencing
	Roving personnel can provide technical support from anywhere
Old Rule:	**The best contact with an information provider is personal contact with a teacher**
Disruptive Technology:	Interactive video disc, CD-ROM, on-line networks, distance learning
New Rule:	The best contact with an information provider is effective contact
Educational Application:	Information can be provided by anyone
	Information can be provided by any type of technology
	Effective learning is more important than who or what provides it
Old Rule:	**You have to find out where things are**
Disruptive Technology:	Automatic identification and tracking technology
New Rule:	Things tell you where they are
Educational Application:	Book counts and locations are easily available
	Supplies are located easily
Old Rule:	**Plans get revised periodically**
Disruptive Technology:	High performance computing
New Rule:	Plans get revised instantaneously
Educational Application:	Everyone participates in changes and notification

video conferencing for education. The prime reason that two-way video conferencing has been successful to date is that it seems to recreate the traditional classroom. Because of that, it seems to make users more comfortable with the use of technology. But, the traditional classroom is not successful in moving learners from the industrial age, to the information age and into the Communications Age. The traditional classroom and its instructional methods have been roundly criticized as it does not provide a facilitated learning environment or meet learning expectations. To recreate a situation that is not working through a costly technological solution seems pointless at best.

Two-way teleconferencing has also been valued because it provides a method of interaction that is valued in the traditional classroom. True interaction in the traditional classroom is seldom seen. Immediate interaction is a perception by the student and the teacher that the student can ask a question at any time and the teacher will be able to answer it. Teachers can also ask questions of the students; there has always been the perception that students could answer the questions. Education is using two-way video conferencing so that distance learning can be cost efficient; it allows one teacher to "teach" to several classes. Two-way video conferencing has also been utilized because it seems to provide equitable access to education for all students in classrooms capable of receiving the transmission. While this use may increase equitable access, the question of the quality of the learning still remains.

Those are some of the current reasons for using two-way video conferencing, and, basically, they are technological solutions to the old problems. They do not expand our ability to use technology in new ways.

Some of the new questions might be as follows. How might two-way video conferencing be used to invent new avenues for learning, to bring new resources into classrooms, to provide learning opportunities in settings other than the classroom? What are the advantages for students in using two-way video conferencing? How can we prevent students from becoming embarrassed because of the cameras? How can we increase their verbal communication skills through two-way video conferencing? Is there an advantage in having students together in the video conferencing classroom? How can the teachers learn from one another? What are the advantages of collaboration for students in distant classrooms? Are there other technologies that can augment two-way video conferencing? If students need time to think about new information, how can we provide that in the instructional design? How can we make hands-on demonstrations more realistic on video and how can they be recreated through other technologies or at the site perhaps through computer simulations? How do student created presentations increase learning? Is there something that makes two-way teleconferencing more valuable because it allows us to do something we have not done before? The real power of technology is not that it can make the old processes work better, but that it enables organizations to break old rules and create new ways of working" (Hammer and Champy, 1993, p 90).

Hammer and Champy also discuss "the disruptive power of technology, its ability to break the rules that limit how we conduct our work" that will give us an advantage. Breaking rules is the method that they recommend for people to learn to think inductively about technology. For example, two-way video conferencing breaks the rule that teachers can only work with a certain number of students. Using computer conferencing to turn in homework in an open class electronic forum breaks the rule that students can't see the work of other students and it results in students learning from one another and all having access to creating at the same standard level. Students writing for other students and sharing their work through computer conferencing breaks the rule that the students write only for the eyes of the teacher and that only the teacher can critique the work.

Hammer and Champy (1993, pp. 92-100) developed additional rules about how the organization of work can be broken by various information technologies. They are valuable in helping educators rethink the use of educational and informational technologies for teaching and learning. (Figure 8.1.4)

Gordon Aubaugh, executive director of the Council of Chief State School Officers (CCSSO) suggests keeping the focus on student and teacher needs. Work from what they need rather than from what the technologies can do. Aubaugh (1994) suggests that we think about what it will "look like five or six years from now when perhaps, we would have every child and teacher going back and forth from school with a PowerBook, with a laptop, or box X, or whatever it's going to be, in a knapsack, rather than something else like a pile of books. What is it going to be when the learning can come just as well and equally from outside the school as it can go on inside the school. How is it that we think about it from the perspective of the student and teacher need, and then design out. How is it that we come together to shape market and to put the muscle in, in order to get the instruments of learning designed and built which service education," (Aubaugh, 1994).

■ COMPETING IN THE GLOBAL ECONOMY

Much of the past argument for changes in education has been based on the needs of companies trying to survive in a global economy. While educators would not want to be called unpatriotic, they have resisted and resented the idea that the only worth of the educational services they provide is fodder for industry. The idea that humanities, arts, poetry, literature and music are worthless may be personally reprehensible to them and that worthlessness is underscored by reduced support or deletion of these content areas. The idea that they should teach only those skills which have been emblazoned through the SCANS Report and the ASTD Report is disagreeable. Certainly these views should be validated, but we cannot ignore the needs of the global economy if that means we lose our place in the world. Contemplating the effect on that in terms of democracy and human rights may help teachers reevaluate their position.

It seems that the reasonable course is to meet the needs of a global economy and the higher needs of education as well. The perception that the two are mutually exclusive is not based in the reality of what education needs to do to regain its previous reputation or to go on to greater excellence.

"Much of the current economy is there because of a lack of information," (Davidow and Malone, 1992). Information equals the things that it can replace economically. Those can include stores, retail clerks, consumption of gasoline. "If all the information you need is on your PC, you don't come into the office. That allows you to cut back on office space, secretaries and file cabinets which displaces the janitor, the heating bill, the construction worker."

Distance education is "more than just another attempt by the education community to respond to the rising chorus of criticism leveled at it by citizens, school boards and government," according to Ohler (1991). He asserts that "It is a discernible step in social evolution. It is an imaginative and yet practical attempt by society to invest itself with the survival skills needed in a highly competitive world that increasingly values the educated, cooperative, technologically competent citizen." Distance education provides a sorely needed fresh perspective. "It forces our thinking beyond the confines of the campus and out into the changing world about which we are supposed to be teaching and for which we are supposed to be preparing our students."

There are points to "accept about the march of digital technology and its effects on our lives in the new economy," says Huey (1994). "Nothing is going to stop it. You can't ignore its impact any more than an early 20th-century horse lover could ignore the intrusive onset of the automobile. Technology never advances without social consequences. So believe it: Life, especially work, will be different in the world emerging — as it already is for many." The "job" itself may "disappear and be remembered only as an artifact of the industrial age, to be replaced by meaningful, market-driven work assignments in post-job organizations." Paul Saffo of the Institute for the Future notes that, "The Nineties are not a good time to work in a large organization;" the average size of the "effective organization is plummeting." Saffo contends that in this era, the model organization mirrors our networked information structure. It's a web, not a hierarchy. The big difference: In a hierarchy, your title determines your power; in a web, it's who you know."

Huey points to these other indicators of the entrenchment of the Information Society as part of the new economy: The old economy rewarded hierarchical organizations; the new economy rewards webs; the microchip drives the new economy as powerfully as the internal combustion engine drove the old one. Since W.W.II the military has driven the advancement of technology — now the role is moving toward the entertainment industry where there is money, shot cycle time and people want to do things no one has done before. "Microsoft's near equaling of IBM in market value in 1993 was another indicator of the coming of the new economy in which companies that are low on physical capital but intense in intellectual capital — pure thought stuff — can blow by those burdened with the role of stamping out machines. Software is America's fastest-growing industry; for four years US industry has spent more on computer and communications equipment than on all other capital equipment combined. The computer has evolved into a device that will be used as much for communication as computation (Huey, 1994).

New tools for using electronic multimedia are rapidly evolving. "Your mind can help shape these tools," says Wodtke (1993). "At the same time, these tools change the way you think. A new Renaissance is emerging that is, once again, relating the arts and the sciences and influencing design. There are exciting possibilities for using computers to learn the knowledge base of a discipline more easily, as well as to enhance your creative capacity." Wodtke observes that most of us realize that the way in which we gather and work with information is changing dramatically and that we're working through some far-reaching changes. "Presently, a shift from verbal to more visual modes of thinking and expression is occurring. Children around the world today are being exposed more and more to visual images. Perception of real-life experience is highly visual. This is now augmented by passively viewing TV. Through computers, however, both the audio and visual tracks are becoming interactive, tapping into a whole range of visual capabilities. Although verbal capabilities seem to be declining, visual capabilities seem to be increasing." What Wodtke finds interesting as an educator who teaches visual thinking and design communication, is how education will respond to the shift that is occurring. The challenge, he thinks, "might be to teach both verbal and visual literacy. This would provide children with a better base for effectively using the tools that are emerging."

Wodtke (1993) calls for a "mind primer" for computer literacy courses in primary and secondary schools so that computer literacy goes beyond teaching only which button to push. He wants students to learn creative and critical thinking skills that they can draw upon when they use electronic media. He concludes that "Consummate realism came with photography and video. Ironically, by dealing with realism people also became more aware of the power of abstraction — working with simpler images. The mind will continue to evolve to make use of new technology. As tools and toys change, they reshape the way people work and play. This in turn changes the tools and toys people create."

Until recently, the industrial age model had us using resources such as materials and energy, rather than information. The information age model is to use information as our resources are essentially equal with the rest of the world. The factor that will contribute to the continuation of America's leadership during the information age is our ability to think and use information (Fitzsimmons, 1994). Fitzsimmons chronicles the changes saying that in the past "Americans were the highest paid workers in the world. We had more natural resources per capita by a factor of ten and better technology than the rest of the world. Through mass public universal education, workers could read, write and count. Today, Fitzsimmons says, all of that has changed. Raw materials are no longer part of the equation as they are 40 percent less costly today than in 1970. The new telecommunications/computer revolution has spawned global competition, but technology is no longer a guarantee of success as Japan is the market leader in technologies developed by U.S. companies. "The variables we can impact are the skills and education of our work force" asserts Fitzsimmons, but the development of U.S. skills and education are not globally competitive and 75 percent of Americans do not graduate from college. He warns that unless there is a dramatic change, we are in "danger of producing a third world economy with falling wages for the American work force."

Fitzsimmons (1994) points out that the industrial age learning system had three dominant elements which presented a consistent message to students that the character of the individual and the criteria for future success were substantially agreed to by society and were clearly defined and communicated. In the communications age learning system, increasingly, learning takes place randomly and learners receive conflicting messages from home, church, formal education, cable TV, video games, personal computers, movies and other information sources. According to Fitzsimmons, a new society is evolving from radical new forms of learning that compose a new societal learning system. As there is no dominant element of learning, all elements can have equal influence on the individual learner. While Fitzsimmons does not call this the Communication Age, this description seems to embody a way of thinking about how humans are bombarded with communication.

Fitzsimmons (1994) describes the problem with the following group of factors that form a basis for making significant changes quickly through system reform;

Literacy: Today, 90 million adults cannot figure out a bus schedule or write a simple letter explaining a credit card error.

Drop Out: A student drops-out every eight second during the school day.

Investments: Public school's annual budget nationwide estimated at $200-$215 billion

Annually, the U.S. spends an average of $6,300 per student for education. In 1993-94, schools spent only $40-$45 per student on information technology

Education's investment in R&D comes to only .025 percent of total annual revenues.

Training: U.S. industry spends about $210 billion per year on training and education. One third of the major U.S. corporation must provide basic skills training for employees. Employers invest 300 times more of their total budgets in computer-based

instruction than does public education. Small firms provide 35 percent of U.S. employment, but have no means to provide training. At the Fortune 2000 corporations, about 12 million workers use e-mail, increasing to about 27 million by 1995.

Implications for Education and Training and the U.S.: Although the current educational system is now a recognizable subsystem of the new society learning system, it is still expected to perform in a dominant role. Fitzsimmons believes that the cost to education is uneasy taxpayer support and wavering prestige of the teaching profession. He believes that because education is not functioning in real-time, up-to-date elements of the learning system are having a greater impact on the individual. He estimates that by 1995, 14 million Americans will be unprepared for the jobs that are available.

The implication for U.S. business and industry is to face the most serious shortage of qualified entry level workers by increasing the investment in training because workers are not prepared for the new workplace environment and do not possess life-long learning skills. It will take more time to find qualified job applicants, more time to train them and more time for new-hires to become fully productive. Fitzsimmons also observes that the concepts of quality and continuous improvements are foreign to the graduates of U.S. formal learning systems.

"We must take advantage of the technology revolution," says Fitzsimmons (1994). "Skills and education of the work force are the only sources of sustainable competitive advantage. There is no premium for natural resources. There is, however, a premium for a highly skilled work force of life-long learners." To support his contention that learning technologies work, Fitzsimmons sites a Hudson Institute review of twenty years of research on computer based instruction which indicates that there is 30 percent more learning, 40 percent less time, and at 30 percent less cost. He also chronicles the American Society of Training and Development (ASTD) survey which reports that business considers training technology to be the most cost effective method, that Federal Express trains data processors with a 60 percent time savings, and that the Armed Services are heavily invested in training technology.

Fitzsimmons concludes that the threat we face is the cultural lag between industry and education...and information is widening the gap. "What is in jeopardy are not the institutions of industry and education. What is in jeopardy is the American society that depends on them."

A possible solution is a computer system that provides just-in-time-learning. Also called an electronic performance support (EPS) system, a number of these sophisticated systems are in place in companies such as AT&T, Aetna Life & Casualty Inc. and

Delotte & Touche, an accounting firm. Hequet (1994) describes EPS as a system that can answer questions by electronically bringing up the correct page in a manual, provides computer-based training to build employee confidence, offer guidance on ticklish decisions and is an "all-purpose reference" that workers can access as they do their jobs. Workers in the field can also access the systems as they work with sales questions or request other information. U S West's Learning Systems uses a system that helps trainers design and create training materials and analyze training trends. Hequet says that companies with EPS systems report very favorable reactions from workers using them.

EPS systems have been credited with remarkable changes. Aetna reports that the total development cost was $20,000 and it paid for itself in just weeks. IDS Financial Services Inc. of Minneapolis reported that an EPS system replaced 15 hours of classroom training with one hour on the EPS system, reduced "time to competency" from two-three months to four weeks, and error rates are down 50 percent (Hequet, 1994). "Enthusiasts say that, in certain areas, EPS has the potential to render training all but obsolete," according to Hequet. The goal is to stop filling people up with knowledge and skills and putting them back to work. With performance support we can give them real work to do and provide them with the support to learn while they work. Hequet warns against over reliance on EPS because it may not include critical thinking for problem solving. What will happen if the EPS system isn't right at hand, Hequet asks. He concludes that EPS systems have their place with other training.

Cohen (1994) observes that traditional training design and delivery cannot "continue to concentrate solely on training as a one-time event that takes participants away from the job where the information will be used. He sees instead, a learning process initiated in short segments, delivered on the job and facilitated by those closest to the participants; their business unit (or team) leaders." Training professionals will "initiate training that can be delivered by line personnel in brief meeting formats, in the workplace, within the work day, and in ways that allow and encourage participants to immediately apply the skills and tools involved — with the line manager in an optimal position to evaluate progress and provide coaching."

During an on-line computer conference with Dr. Kenneth Blanchard (1994) and the faculty and students of the University of Phoenix Online Division, Blanchard drew a significant comparison between situational leadership and situational learning. "The comparisons are great," said Blanchard. For example in the learning environment at hand that we are using, all four styles of managing are being used: S1, Directing occurs when I gave you my initial perspec-

tive; S2, Coaching commences when we start to dialogue about the topic; S3, Supporting happens when you all carry on the conversation without me; and S4, Delegating occurs when each of you apply and use something we've discussed to make your life or job a bit more manageable."

When asked for help on how to make the transition to a new style of management in organizations when an old paradigm of management is dominant, Blanchard said, "Please do not fall victim in assuming that nothing can be done until the folks on top get enlightened and decide to empower us below. Life will pass you by." You have to remember that we are all groping together about what's happening in business today. And many of those upper managers are the most scared because they seemingly have the most to lose. The attention then must be placed on what can be done to help them — not to convince them that they are wrong. You need to catch your boss doing something right if you are lacking positional authority and build a personal power base from which you can then build to the point in which you can provide useful personal feedback to the manager as to how they can be more effective. Managers from the top down need to feel they are in a safe environment so that they can be open to change" (1994). Magnet (1994) agrees saying "managers have had a way of fighting hard to hold on to the information on which their power rested." Obviously, the same thing applies to educators.

Richman (1994) asserts that what it will take to provide workers with the attitudes and technical skills they need for the new economy include schools that are linked with employers to integrate classroom instruction with practical on-the-job experience. It will also take community colleges that offer the specialized training their local businesses demand, companies that find new ways to teach and motivate employees and government programs to steer dislocated people back into productive new careers. Magnet (1994) observes that there will be fewer people doing more work. "Those who survive must learn to master new information technologies...and then figure out how they fit into a new, more chaotic organizational structure. For productivity, growth directly affects the national standard of living, and the way we play the game may well determine the extent to which America can remain the dominate economic superpower in the years ahead."

■ TOTAL QUALITY MANAGEMENT IN EDUCATION

As Total Quality Management is adopted by schools, educators are discovering the natural fit that quality principles and practices have with their own aspirations for the continuous improvement of education (Bonstingl, 1992). Bonstingl has taken Deming's 14 points, Juran's Trilogy, Kaoru Ishikawa's Thought Revolution and has adapted the theories to education.

Bonstingl observes that teacher-students teams are the equivalent of industry's front-line workers. "The product of their successful work together is the development of the student's capabilities, interests and character. In one sense, the student is the teacher's customer, as the recipient of educational services provided for the student's growth and improvement. Viewed in this way, the teacher and the school are suppliers of effective learning tools, environments and systems to the student, who is the school's "primary customer." The school is responsible for providing for the long-term educational welfare of students by teaching them how to learn and communicate in high-quality ways, how to assess quality in their own work and in that of others, and how to invest in their own lifelong and 'life-wide' earning processes by maximizing opportunities for growth in every aspect of daily life."

In another sense, says Bonstingl, "the student is also a worker, whose product is essentially his or her own continuous improvement and personal growth. The school's stakeholders and "secondary" customers — including parents and family, businesses, members of the community and other taxpayers — have a legitimate right to expect progress in students' competencies, characters and capabilities for compassionate and responsible citizenship — not for the direct and immediate gain of the stakeholders but, rather, for the long-term benefit of the next generation and of generations to come. Total quality in education, as in life, is essentially generative. Within a total quality school setting, administrators work collaboratively with their customers: teachers. Everyone in the organization must be dedicated to continuous improvement, personally and collectively. Deming suggests that we "abolish grades (A, B, C, D) in school, from toddlers up through the university. When graded, pupils put emphasis on the grade, not on learning" (Bonstingl, 1992).

Bonstingl (1992) contends that if schools are to be true learning organizations, "they must be afforded the resources, especially time and money, needed for training, quality circles, research and communication with the school's stakeholders: parents, students, businesses, colleges, community residents, taxpayers and others. Schools must also rethink practices that focus narrowly on students' limitations rather than their range of innate strengths. Howard Gardner has pointed out the self-defeating nature of a narrow academic focus, encouraging educators to acknowledge the existence of multiple intelligences and potentials within each student and to help students develop

their many intelligences more fully day by day."

Barker (1992) identifies Total Quality Management as the most important paradigm shift of the twentieth century. He says that the paradigm "has created an epidemic of quality throughout the world" so that any organization that "doesn't catch this disease may have a very difficult time surviving the next twenty years." It is, he says, "a revolution of the human spirit" as it " brings back spirit to the workplace" and it " creates an attitude of constant innovation." Innovation takes us into "territories we have never been to before; and therefore, to be responsible to the future and to the things we value, we must develop a sense of anticipation of the implications of our innovations. This will allow us to pick from the many potential solutions to our problems and find the few that best support those values we wish to carry into the future."

Barker (1992) contends that with increased productivity and innovation comes a growing self-esteem in the workers which "often leads to the request to self-manage" as they realize that they can "be in charge of themselves far more effectively than a manager can." The result is a flattening of the organization and the disappearance of the classic middle manager. That leads to middle management resistance to this paradigm. Logical. But, in the long term, useless," says Barker. "Self-management is the most democratic, most efficient and most powerful way to get things done. And it frees up those who are middle managers to use their intelligence for more productive and innovative purposes. No more pushing papers, protecting turf, building empires."

■ USING THE NATIONAL INFORMATION INFRASTRUCTURE FOR EDUCATION AND TRAINING

"I knew we were on a solid rocket booster when I saw the growth rates in the Internet. It took off at 15 percent a month and it just kept going," says Douglas Van Houweling, vice-provost for information technology at the University of Michigan (Jacobson, 1994a). "When I saw that, I said to myself: Universities are about sharing knowledge and information, and what we're seeing now (with personal computing and networking) is about the same thing. The convergence here is going to be something we can't imagine." Jacobson sees signs that a growing number of institutions will soon be competing electronically in each other's back yards — and that more and more students will be taking courses and full degree programs over computer networks without setting foot on a campus. Stanford University's President Gerhard Casper agrees that the latest technology will transform both the content and delivery of higher education to an extent not yet fully understood or appreciated on most campuses (Jacobson, 1994a). Casper says he has no idea what the impending changes will be like, but "we'd better start thinking about it. I am amazed about how little discussion seems to be taking place."

Some college officials say a technology-driven restructuring of academe is only five to ten years away (Jacobson, 1994a). Van Houweling says that the potential of the technology is moving much more rapidly than we are able to take advantage of. Nor have many organizations begun the process of "realigning the way they teach — the way they create a learning environment, the way they reach out to students to take advantage of these new capabilities." Some of the questions are how will colleges respond when other institutions start offering electronic classes to their students? Are campuses necessary? How many faculty members will be needed? Will large lecture halls survive (Jacobson, 1994a)?

Carnegie Mellon is asking what might happen if technology made it routinely possible for anyone to send or receive anything electronically from or to any place at any time? The Fantasy and the Reality subcommittees provide the check and the balance. The Fantasy group is imagining a wireless world of ubiquitous, hand-held computers, communications devices and super-user-friendly software which might vastly expand the university's efforts in distance learning or lead to new collaborations with outside businesses and other organizations. "We see ourselves on the verge of potentially major technological advances," says Stephen W. Director, Dean of the School of Engineering. "We can see that the way we do computing, communicate with each other and interact with data and each other is changing. We want to "get in a position so that ten years from now we have what we need" (Jacobson, 1994a).

Stanford's Casper who proposed a three-year degree as a way of controlling education costs, has found affirmation for the idea in distance learning. He thinks some of the more quantitative courses may be delivered by distance learning in the future. John Bravman, chairman of a Stanford subcommittee focusing on technique and technology in teaching and learning, says that with two-way video and other options "one can imagine rethinking the whole purpose and structure and function of a university like Stanford." The committee recommended that Stanford establish a permanent group to coordinate efforts to bring technology to the classroom. Jacobson (1994a) observes that as the planning begins, universities know that two problems of the past decade will continue to inhibit the spread of technology: high costs and limited participation by many faculty members. Bravman says many institu-

tions will soon experience nothing less than culture shock. "This is going to affect academic publishing, teaching, the way we do research — everything that we do. People are either going to jump into this very fast moving stream or they're going to get left behind."

In 1994, the first (ever) technology conference was held by the U.S. Department of Education. For the first time the chair of the Federal Communications Commission (FCC) addressed a conference sponsored by the Department of Education. In his remarks, FCC Chair Reed Hundt said there are many ways in which all adult Americans already have contact with the Information Highway. "There is only one group, of course, that does not have access. Forty-five million people a day from the beginning of the day to the end of the day who basically are outside the scope of these networks — who basically do not participate in the Information Superhighway — the way that almost all adult Americans already do. I'm speaking, of course, of our children in the classrooms" (Hundt, 1994).

"Now the potential of the Information Highway is what led to the President of the United States mentioning it and our classrooms in the same sentence in the State of the Union speech. The first time that education and communications were ever mentioned by a president in the State of the Union together — and I hope not the last time — because I hope it marks the continuing attention to what the communications world brings to education and what education can, through communications, mean in terms of changing our country for the better. That potential, I think, consists of at least three things. It consists of the educators using the Information Highway to have access to information. It consists of educators using the Information Highway as a source of new techniques for teaching. And third, it consists of the Information Highway facilitating communication itself so that in the act of communication, all children, can, in fact, learn" (Hundt, 1994).

"It is so much easier to pierce the walls of the classroom with interactive communication," Hundt said. "It is so much easier, with fingertips to show the children how to go to every art exhibit in the world to see every painting of every painter brought to life in perfectly clear pixels on the computer screen. This is available, even today. For anyone who has ever used Mosaic, a computer program to navigate the Internet. There is literally no reason why that tool cannot be available to all the children in all the classrooms, except we don't have the lines into the rooms yet " (Hundt, 1994).

Hundt said that he had run into some people who said, " This sounds like the latest gee whiz gismo that won't work out and will cost lots of money." Hundt doesn't think so. "I believe those who tell me that putting the networks into the classrooms will free teachers to work with students who need individualized attention. I believe those who tell me that putting the networks into the classrooms will permit students and teachers to learn from anyone, anywhere, anytime. I believe those who tell me that putting the networks into the classrooms will allow students from different backgrounds and even different countries to learn from, and to teach each other, and to build community across our country. I believe those who tell me that putting the networks into the classroom will allow for the creation of new communities of learning among students. Any parent with a television, any teacher, knows the power of visual images in their building to create a community of interest. The power of those visual images has yet to be fully tapped and will not be until we can extend our networks into our classrooms" (Hundt, 1994).

In testimony before the Senate Commerce Committee, U.S. Department of Education Secretary Richard Riley testified that "it will be absolutely impossible to educate the coming generation of young people to high standards of excellence — if their access and use of the NII is seen as a secondary consideration to broad based commercial purposes." Riley said it was his "very strong belief that free connections to the NII may not be enough. If we want young people to actively use the technology of the future so it becomes second nature to them then we must go a step further and provide our schools with free usage of the telecommunications lines that will connect school children and young people to new sources of knowledge. The principle of 'free' public education for all children is the bedrock of our democracy. Not cheap, inexpensive or available for a fee, but in its very essence 'free.' We believe in this basic American principle because we know its long term value for society as a whole. This is something every business person in this room should understand. Every year millions, if not billions of dollars are being spent by business, our community colleges and our public universities on remedial education" (Riley, 1994a).

Riley said, "If we want to get out of the business of remedial education, if we want to create a well-educated and world-class work force, this is the time to get it right...to raise our standards and give our young people the access and the tools they need to get a world-class education. While nearly every school has computers, school traditions frown on teachers having telephone lines in their classrooms. Opportunities for teachers and school staff to learn how to use new technology and integrate it into instruction are all too scarce. Few schools budget adequately for ongoing technical support for these new tools. Only when these and other steps are taken will we truly be able to end the isolation of our nation's classrooms" (Riley, 1994).

The National Coordinating Committee on Technology in Education and Training (NCC-TET) is the largest, most diverse coalition of organizations to take up the task of defining the requirements to be met if the National Information Infrastructure (NII) is to support education and training. The National Education Association believes that education and training will provide people with the skills they need to use and develop the NII, but will do this only if the NII also serves their needs (Yrchik, 1994).

Yrchik asserts that we must ensure that all Americans have affordable access to the NII. "Accessing the best information to do a job or perform a task must become a cultural norm by the end of the century. Given the fact that at the present time, about 90 percent of K-12 classrooms lack even basic access to telephone service, this is a formidable challenge. The goal of connecting every classroom and home to the NII should be set for the year 2000. Rural and poor populations which have traditionally been underserved must have special attention given to them to make sure they have access to the network. This is the point at which we as a society choose whether we want to grow together or grow apart into the information rich and the information poor. The issue is equity. The issue is the vitality of our democracy. It is therefore imperative that we ensure that the NII is accessible in a variety of learning environments. The vision of the NII is one in which learning occurs in a variety of environments throughout the course of one's life — in homes, workplaces, schools, universities, libraries, museums and community centers. People should have access to information they want where they want it and when they want it. This sensibility should guide the design of all federally funded NII-related education and training programs" (Yrchik, 1994).

For another goal, to develop and disseminate NII guidelines for education and training applications, (Yrchik, 1994) says that instructional standards would ensure that educational and training applications of the NII help us attain the National Education Goals. He adds that to promote a teaching profession experienced in the effective use of technology, national teacher certification standards and credentialing requirements should be expanded to include applications of educational technology. Yrchik notes that many schools of education do not require training in the use of education technologies. "Without effective training programs for educators, the information highway will quickly become a dead end. Educators must be given training on how to use the equipment. But, as importantly, training should include teaching strategies that incorporate a wide variety of technologies. If educators are interested in collaborative learning, for example, they should be given the training to create collaborative learning environments that use telecommunications technologies to extend teaching and learning across district, state and national boundaries."

Another goal calls for emphasizing interactive, broadband transmission of voice, video and data for education and training. The need for a broadband NII is a critical need for education and training is critical for the widespread use of interactive video in education according to Yrchik (1994). He contends that this will open new dimensions in the learning process that are not possible with voice and data transmission alone. Electronic field trips, simulations and sensory immersion in new kinds of learning environments could become commonplace. "The wall that separates schools from the world outside — a wall that exists because of the technological deprivation of schools — must be dismantled" said Yrchik. "The ability to transmit voice, video and data with relative ease across networks will extend teaching and learning beyond traditional school walls, opening classrooms to the world outside. Students should have the opportunity to interact directly with the world's experts on a variety of subjects, to share their ideas and experiences with their peers from other districts or countries, to produce their own multimedia products and distribute them to their peers around the country or world. Teachers should be able to do the same. This will irrevocably change the way education is experienced and organized."

Ease of use is another goal recommended by the NCC-TET. Yrchik believes that the user interfaces in education must be easy to use and as consistent as possible across computer platforms, individual databases, information services and other applications. "The enormous complexity of such a system must be hidden from the user. The user should be encouraged to build rapport with it," Yrchik says. He believes that the use of intelligent software agents, knowledge robots (knowbots) represent one of the most promising directions because in response to a query, the software agent would enter numerous computers and databases and provide a single easy-to-understand response.

The NCC-TET also recommends that the NII support user collaboration for training and educational organizations of the future. Effective people-to-people communication is open-ended Yrchik says, and it requires a wide variety of data types including voice, video, audio, graphics, still pictures, text and animation. "It takes place in many different configurations and contexts — between two people or two groups, between an individual and a group, in work, at home and on the road. It's both synchronous and asynchronous. It occurs at the same time and at different times. Managing this communication in a way that simulates actual physical proximity is a challenging task, but an important one. Increasingly, education and training will take place in geographically distributed environments. Of all the potential uses of the NII, its use in education and training is the most important. Education and training applications alone will provide individuals with the skills and knowledge to make use of and further develop our National Information Infrastructure. In time, the National Information Infrastructure may become one of our national treasures" (Yrchik, 1994).

In a working paper, the Information Infrastructure Task Force (IITF) (1994) discusses the objectives for the future uses of technology and education. Among them are instructional delivery which will provide workers with a "Ph.D. in a pocket." Instruction and job performance aiding will be delivered on a device that resembles a pocket calculator. Every complex device will include sufficient embedded training and user assistance to make it easily usable. Instructional intelligence will support integrated individualized tutoring that integrates goal setting, instruction, job performance aiding and decision aiding into a single package. Natural language interaction will be an essential feature of this capability. The IITF believes that institutional integration will be the most difficult challenge to meet as the new instructional capabilities will first have to be integrated into the routine, daily practice of our current instructional and workplace institutions. "Just-in-time and just-enough training that is universally available will not only change the way people are treated in the workplace but the workplace itself."

IITF notes (1994) that instructional programs, simulations, materials and databases can all be accessed over the NII and delivered to schools, homes, libraries and workplaces wherever and whenever it is desirable to do so. Currently, there are massive exchanges of software, databases and files using the Internet, but relatively little of this activity occurs in the service of education, training and lifelong learning.

The NII will provide the backbone for a lifelong learning society (IITF, 1994) "Education and training communities will better accommodate an enormous diversity of learners in an equally diverse variety of settings. In addition to schools and work places, interconnected, high-performance applications will extend interactive learning to community centers, libraries and homes. Education, training and lifelong learning applications available from the NII may include:

• Multimedia interactive learning programs delivered to homes to immigrant children and their parents to learn English as a second language.

• Comprehensive interconnectivity for students that allows them to receive and complete assignments, collaborate with students in distant locations on school projects and interact with teachers and outside experts to receive help, hints and critiques.

• Simulated learning activities such as laboratory experiments and archeological digs.

• Universal access interfaces for computers and telecommunications devices for students, workers and others with disabilities to allow access to the NII.

• Affordable, portable personal learning assistance that tap into the NII from any location at any time and provide multimedia access to any NII information resource.

The NII, will be the vehicle for improving education and lifelong learning throughout America in ways we now know are critically important. Our nation will become a place where students of all ages and abilities reach the highest standards of academic achievement. Teachers, engineers, business managers and all knowledge workers will constantly be exposed to new methods, and will collaborate and share ideas with one another (IITF, 1994).

The NII will give teachers, students, workers and instructors access to a great variety of instructional resources and to each other. It will give educators and managers new tools for improving the operations and productivity of their institutions. The NII will remove school walls as barriers to learning in several ways. It will provide access to the world beyond the classroom. It will also permit both teachers and students access to the tools of learning and their peers — outside the classroom and outside the typical nine to three school day. It will enable family members to stay in contact with their children's schools. The NII will permit students, workers and instructors to converse with scientists, scholars and experts around the globe (IITF, 1994).

Workplaces will become lifelong learning environments, supporting larger numbers of high skill, high wage jobs. Printed books made the content of great instruction widely and inexpensively available in the 18th Century. The interactive capabilities of the NII will make both the content and interactions of great teaching universally and inexpensively available in the 21st Century (IITF, 1994).

The NII will provide a powerful tool to address many of the learning needs the country is facing according to the Computer Systems Policy Project (CSPP) (1994), an affiliation of chief executive officers of American computer companies that develop, build and market information processing systems, software and services. These include tailoring curriculum and instruction methods to meet the needs of individual students; providing teachers with the resources they need to improve their skills and update their knowledge; providing a means for Americans to continually acquire the knowledge to adapt to new or changed career objectives; providing better access to information that affects our quality of life and cultural awareness.

CSPP (1994) believes that the enhanced NII will provide many opportunities to better prepare the population for the work place of the future. For example, they cite the following contributions:

1. Computers will make learning complex ideas easier by providing learning environments that closely approximate real work environments or experimental apparatus.

2. Students interconnected via networks will be able to collaborate as teams, even though members of the team are geographically separated. Teachers and school and college administrators from around the country will be able to communicate electronically to exchange ideas and build a sense of community.

3. Instruction will be tailored to the specific learning needs of individuals, particularly adults re-entering a training environment, minorities, women people with disabilities and others that may benefit from customized approaches to instruction.

4. People who do not have direct access to a wide variety of opportunities, either because they live in remote areas or because of the demands of work or family responsibilities, will be able to access information from other locations in a variety of formats and media, obtain degrees from distant colleges and access facilities with unique learning resources.

5. Information technologies, such as visualization, will provide new ways to learn difficult concepts and data.

6. Information technologies will reduce the burden of record keeping and other paper work that consumes so much of teachers' and administrators' time.

CCSP (1994) suggests ways that the NII has the potential to move learning at any age, from "beyond the four walls of a classroom to a broad community of learners and to help meet Goals 2000." It can assist with helping children start school ready to learn (Goal 1, 8). Through links to the home, the NII can provide parents with the necessary tools to expand their roles in their children's education and better prepare their children to enter schools ready to learn. Through computers linked to social agencies, assistance with specific problems can be provided directly and confidentially to parents, or to schools, to help teachers and administrators deal more effectively with learning difficulties caused by social problems. Parents will be able to access high-quality and developmentally appropriate preschool education materials to use with their children, which promotes inter-generational literacy and strengthens communications between parents and educators very early on. Parents will be able to access nutritional and health care information and services, find model programs for early childhood care, and find customized information for children with special problems or circumstances. An interactive "Ask the Experts" electronic bulletin board could provide insights on all types of developmentally related issues.

The NII can play a crucial role in enhancing the

linkages to make learning more exciting and hands-one and relevant which increase the high school graduate rate (Goal 2). The NII can allow student access to scientists, researchers, journalists, entertainers and members of Congress. Students will see through firsthand interaction, skills and education being applied in relevant situations, which will help them gain an understanding of the benefits of staying in school. Career counselors and mentors can provide interactive information about career options and education requirements. Instruction will be more exciting and better tailored to students' interests, and needs, through the use of networked multimedia, speech recognition, intelligent tutors and tools and curricula that integrate voice, video, data and text. This will help enhance retention and provide the flexibility to meet the varying learning styles of students.

"In a global, knowledge-based economy, skills in generating, accessing, manipulating and managing information will be paramount" (CCSP, 1994). "Information and the base of knowledge about any particular subject are growing so quickly that a person can no longer expect to 'know' all there is to know about a subject. Instead, workers and students will need to know how to find the information they need quickly and in the right format. While tools and software will help make the NII easier to use and navigate, information technology-based education techniques will still require learners to conceptualize a problem, use critical thinking skills, access distant resources and collaborate with fellow students, all while using basic reading and writing skills" (Goal 3). Integrated curricula supported by information technology will move learning beyond memorizing "facts" like the periodic table, to inquiry-driven, on-line, interactive sessions, such as discussions with NASA researchers, collaborations with oceanographers or writing reports using images downloaded from the Voyager satellite. Global networking capabilities will enable students to interact with students and teachers around the world, bringing current events into the classroom. Projects like the National Geographer's "Kids Net" move lessons from static textbooks and outdated maps to powerful, interactive exchanges with students from cultures and countries around the globe and the resources those countries provide. "Understanding and appreciating diversity, whether different countries, races, cultures or languages, will be essential for American students and workers operating in an increasingly interconnected, global environment."

Through the NII, information technologies have a huge potential to reinvigorate how science and mathematics are taught, used and learned by students (Goal 5). With network-supported simulated experiments, project-based learning and authentic problem solving exercises, students can generate their own scientific data, which they can share with researchers

which makes students active participants in their education rather than passive observers. With electronic access to resources, teachers will be able to develop exciting, up-to-date curricula. Students will be able to collaboratively participate in student-conducted work that will provide hands-on experience in statistics, architecture, sample gathering, testing, design, assessment and forecasting. Tools and resources such as super computers, visualization and statistics software will bring difficult mathematics functions, including modeling, simulation, probability and finite analysis alive and within the reach of many students.

The global economy is moving to one driven by information and knowledge. High-wage jobs, quality of life and competitive advantage will be based on the ability to create, manipulate and deliver information quickly to the right person and place. Literacy, or technological fluency, will require the skills to acquire knowledge and to adapt to emerging technologies and work methods (Goal 6). Networked services and resources drawn from academia, business and the community will provide lifelong learning opportunities for workers, students and teachers. Distance learning technologies will enable innovative retraining initiatives, designed and adapted to meet the needs of mid-career and other non-traditional students. Expanded citizen access to federal, state and local information and the ability to learn at home, at work or other convenient places will make learning literacy skills easier (CCSP, 1994).

Technology will not cure societal problems but it can help create an environment in which the preconditions for drugs and violence are minimized (Goal 7). Students who are "excited, engaged and challenged by the education system are less likely to turn to destructive behavior" (CCSP, 1994). Innovative student assistance programs implemented with electronic support can provide anonymous and accurate information about specific problem areas to curious students or students-at-risk. An array of drug prevention and counseling techniques can be made available electronically to parents and teachers. Linking schools with community organizations, the police and other public safety agencies will facilitate communication among these agencies and assist in identifying trouble spots or issues before they erupt into classroom violence.

■ INSTRUCTOR IN-SERVICE AND PRE-SERVICE

"Researchers studying teacher development have found that beginning teachers progress through a series of stages: survival, mastery and impact" (Fuller, 1969; Hall and Loucks, 1979). Teachers focus on themselves first, concentrating on issues such as controlling student behavior. They become better able to anticipate and solve problems after they gain confidence, and gradually their focus shifts to their impact on students' achievement and attitude. Little is known about classroom management techniques in mediated environments. The research literature offers few suggestions for teachers with classrooms filled with computers, networks, laser discs, printers and other technological tools. Most studies have concentrated on computer laboratories or classrooms with one or two computers (Amarel, 1983; Ragsdale, 1983; Hoffman, 1984).

As teachers encounter new instructional methods, facilitation, systemic reform, Goals 2000 and technology, the three-stage model is useful in understanding the development of experienced teachers who are implementing educational innovations. Data from Apple Classrooms of Tomorrow (Dwyer, et al, 1990) support the assumption that experienced teachers entering high-access-to-technology classrooms also move through these stages.

One of the outcomes of systemic reform has been the recognition that teacher education — both in-service and pre-service — is not addressing the need to provide teachers with professional development that allows in-depth study of facilitation, constructivism, andragogy, technology and develop personal skill in instructional methods which support these theories. "As a result, reformers have sought to restructure schools to produce conditions that address these concerns. These restructured schools will make new demands on teachers, as well as provide them with new opportunities" (Abdal-Haqq, 1989).

Only now have people begun to realize that just putting computers in classrooms, even plugging them into the Internet, won't improve education unless the teachers understand how the computer revolution can be fully exploited, according to Dolores Gore at Austin Peay State University, TN (Wilson, 1993). She says she has embraced the technology as the only way to solve the nation's ills such as illiteracy, joblessness and crime. Many of these problems can be traced to inadequate and antiquated teaching methods starting with kindergarten and running through high school says Gore. Instead of rote memorization, Gore says students must be taught how to locate and retrieve information and to navigate through the vast quantities of data that the computer will make available to them. "We need to teach children how to learn. The technology gives us an opportunity to address all these problems" (Wilson, 1993).

When we think of teachers using technology, we tend to "focus primarily on their need to learn how to operate hardware and software" (Means, et al, 1993). Certainly this is a critical component, but teachers also need to develop skills in developing curriculum, allocating resources among students, managing instruction within the classroom using technology,

keeping abreast of new technologies and finding out about the "potential power each technology application has with respect to inquiry-based teaching and learning. "Any technology integration requires that teachers engage in rethinking, reshifting and reshaping their curriculum."

Restructuring schools not only changes the character of school culture but also creates a need for a nontraditional approach to in-service teacher education. Ongoing professional development replaces the sporadic, short-term staff development activities that constitute typical in-service education at present (Holmes Group, 1990). Traditional teacher pre-service programs have done little to prepare teachers for the demands or opportunities of restructured schools (Levine, 1988; Mahlios et al., 1987).

The U.S. Senate Committee on Labor and Human Resources has requested the Office of Technology (OTA) to investigate the "availability and quality of staff-development programs offered at the pre-service and in-service teaching level which encourage the use of technology in an integrated fashion across the curriculum" (Charp, 1994). The report which is due at the end of 1994, includes questions regarding pre-service teaching programs such as: Do programs concentrate primarily on operational procedures or are creative applications discussed? Are pre-service training programs a part of the school of education's courses or are the programs offered as electives as part of the computer science department? Is there increasing interest among schools of education in offering more technology-based instruction training? Are there model technology education courses in place at the nation's colleges and universities that could be emulated by others? Do any states license exams measure a prospective teacher's knowledge of technology? Are most in-service technology training courses offered at the school or at another site, and who provides the training? Are states developing statewide plans for coordinated integration of technology into the curriculum and if so, are teacher training programs crucial to the plans? What is the proper federal role in providing professional development to teachers regarding the role of technology in the classroom?

Abdal-Haqq, (1989), identified a number of promising emerging trends in teacher education. These included (1) research based, reflecting a reform trend that roots school improvement efforts in theoretical soil; (2) preparing teachers to examine and assess their own practice, to become inquiring, reflective practitioners; (3) emphasizing collegiality; (4) preparing teachers to participate in decision making on varied school issues; and (5) helping teachers to qualify for professional advancement through differentiated staffing programs. Abdal-Haqq did not reference any use of technology in this review.

Monk (1989) reviewed the use of technology and distance learning in small rural schools and concluded that teacher training was the most substantive problem in the use of technology to improve curriculum.

Means et al (1993) observed that there are a number of challenges for teachers using technology as a critical part of an inquiry-oriented learning-teaching process. "Challenges include learning how to use a variety of technology applications; using, adapting, and designing technology-enhanced curricula to meet students' needs; expanding content knowledge; taking on new roles; and responding to individual students. None of theses challenges stand alone; they are tightly interrelated." They (Means, et al ,1993) recommend that technology use should force teachers to pose questions such as: What does the technology offer my students in terms of developing concepts and content? How does it help them to carry out inquiry processes? How will they work together collaboratively or cooperatively? What is the relationship between the technology and other instructional materials? What knowledge, processes and skills do students need before using the technology? What new knowledge of my content or discipline, of teaching or of technology do I need in order to foster new learning in my students?"

When teachers become facilitators and develop inquiry-based curricula that integrate technology, "their role in the classroom becomes more that of a coach, or facilitator of student learning (Means, et al, 1993). In inquiry-based learning, teachers set the context, help students pose questions to explore, stimulate problem solving and give students tools and resources to use so that the students can construct knowledge. The knowledge construction process takes place within an individual student; it is highly individualistic because of the knowledge maker's prior knowledge, experience, skills and talent. Teachers cannot — and should not expect to — have a total grasp of the content related to every topic. What they do need to know is how to help guide students through the meaning-making process: how to ask probing questions, how to connect students to relevant resources, how to organize students into cooperative learning groups, and how to give them tools to store, manipulate and analyze information."

Teachers may feel awkward in this new situation. One does not move from being a teacher to a facilitator overnight, although this is often the expectation. "They feel vulnerable as they take the risk of shifting from a more comfortable knowledge transmission mode of teaching to inquiry-based teaching" (Means, et al, 1993). The conclusion (Means, et al, 1993) is that our initial enthusiasm for technology (especially computers) and the prediction that teacher's jobs would be easier so that they could spend time with students was naive. "Teachers are nearly unanimous

in concluding that in the early stages of technology implementation, at least, their job becomes harder." Yet teachers continue to use technology "because they sense that their students are learning more and approaching their classroom activities with a heightened level of motivation. The skills that teachers acquire and the "satisfaction of facing a challenge and overcoming it, add to teachers' sense of professional growth."

Russell, Sorge and Brickner (1994) contend that their experience has clearly delineated the steps necessary to successfully change the way technology is used in schools. Mandates have "often forced teachers into premature use of technology for the sake of using it rather than because it performed a genuine instructional functions." The steps include:

1.   Workshops distributed throughout the academic year that demonstrate integration of emerging technologies in education, cooperative learning techniques and strategies for implementing technology in classrooms;

2.   On-site visits between workshops, which include interaction with teachers, classroom observations, assistance with instruction and modeling of technology implementation;

3.   Proposal development to obtain funds for further implementing technology;

4.   Workshop goals are to equip teachers to select, implement and assess the effectiveness of technology in teaching; assist schools in the grassroots integration of technology into their curricula and to develop a plan for technology implementation; discuss the process of implementing technology; demonstrate strategies for successful technology implementation; present the effective uses of various types of technology; and describe student-assessment techniques that are appropriate for use with various technology initiatives.

5.   Teachers are taught to implement instructional technology using the ASSURE Model (Heinrich et al, 1993), "A Model to Help ASSURE Learning" — a mimetic device that means:

A   Analyze Learners
S   State Objectives
S   Select Media and Materials
U   Utilize Media and Materials
R   Require Learner Participation
E   Evaluate and Revise

Sammons (1994) reports on a program at Wright State University that began in 1991, but still had only about four percent utilization by faculty in early 1994. Sammons asserts that, "Higher education has lagged behind primary and secondary education in incorporating multimedia into both teaching and learning. The literature suggests that college faculty in general are slow to integrate new technology into the instructional process. Studies also indicate that the traditional faculty lecture mainly relies on the blackboard and overheads with occasional slides." In a survey, faculty reported that they were reluctant to use multimedia because they perceived they did not have equipment, lack of time to develop materials, lack of knowledge about multimedia or how it will help in the teaching and learning process and uncertainty about which material to incorporate into a multimedia lecture. Sammons recommends making the equipment more visible, providing hands-on seminars that last about two hours, provide time to produce multimedia...and start with simple computer presentations that allow faculty to grow into their use of multimedia.

Beatty and Fissel (1993) reported that after four years of operation of the Ball State University Video Information System which provides classrooms with a voice, video and data system, the faculty "have adapted to the new technology and many have added visual components to class material traditionally taught by straight lecture. System use has grown each year as faculty discovered new ways to employ technology in the classroom." They also say that on the whole, teachers were less dependent on linear forms of pedagogy and searched for cooperative learning strategies that took advantage of the information system. Beatty and Fissel believe that when educational technology is supported by far-sighted administration, it can "empower faculty to achieve classroom successes that no one anticipated." Faculty have determined curriculum development and the technology evolved as dictated by their needs. "The impetus has come from the teachers" and "student evaluation of the system has helped assess the pedagogical value of faculty innovations" so that "teachers and students are, together, creating the campus of the future."

Hirschbuhl and Faseyitan (1994) assert that faculty "should be trained in the use of computers and demonstrate a willingness to adopt computers for instructional purposes before the university launches a technology project. Instructor training should include a specific focus on how to design instructional content for the intended media; this is especially true for multimedia projects as the requirements for integrating video, audio, animations and graphics are fairly technical. Once having successfully finished one project, an educator's confidence and enthusiasm will increase exponentially and less support will be required. Decision makers should heed this 'train first' principle."

The Center for Professional Development and Technology at Southwest Texas State University has assessed the teacher training projects conducted there. Various evaluative studies have assessed the impact of the center on teacher preparation, professional growth and student learning. They report (Curtin, et al, 1994) that technology alters teaching. Project teachers report more independent student work, a transition to more student-centered classrooms, and more cooperative efforts among students. The evaluation showed that technology can serve teachers. Regardless of current skill levels, teachers see the technological skills that they have learned as very applicable to their job requirements. They report that technology invigorates learning. Elementary students believe that after technology was incorporated into their classes, the classes were different and more fun assignments were done faster and better, and more resources were available. One teacher said that students "were especially excited about coming up with their own things to do." Another teacher said that "the computer especially helps the slower kids, getting them excited about learning and introducing other avenues for them." The evaluation also showed that parents support the use of technology and were very aware of the technological advances that their children were using. Parents believed that the children were achieving more academically than in past years. One parent said, "My child cannot be more enthusiastic about school and learning than she is now."

In 1994 (DeLoughry) reported on the American Association for Higher Education's (AAHE) new interest in determining how computer and other technologies can help bring about the teaching reforms it has long sought. DeLoughry says the Association's new thinking is significant because "it promises to bring advocates for technology who have operated on the periphery of higher education together with the faculty members, provosts and presidents who have influence over teaching and who are involved in the organization's conferences and other activities." It is the first large-scale effort on the part of a leading higher-education group to deal with issues related to technology. DeLoughry cites CAUSE and EDUCOM as other organizations involved with these issues. Stephen Gilbert, AAHE director of technology projects, said he believes that info tech will be embraced "eventually by higher ed, but that we need to facilitate that process by focusing first on the educational and scholarly needs of faculty." According to information collected by AAHE, only about one-third of all college campuses have teaching improvement centers; many were not established until after 1988.

DeLoughry (1994) observes that other organizations are "clamoring behind" AAHE as technology becomes more of a mainstream topic in higher education. "The Association of American Colleges and Universities had several sessions related to technology at its annual meeting in January. The National Association of State Universities and land-Grant Colleges has an information-technology council that has been involved in monitoring federal legislation that could affect the growth of computer networks in academe. The Association of American Universities has been working with the Association of Research Libraries to explore the potential impact of technology on scholarly publishing. And the American Association of State Colleges and Universities is conducting a poll of its members to determine what kinds of technology they have and how they are using them. Honey and Henriquez (1993) studied the impact of telecommunications on teaching. Slightly more than two thirds of the educators felt that integrating telecommunications activities into their teaching has made a real difference in how they teach. However, when compared to the difference that integrating computers into teaching made for educators in the "Accomplished Teachers" study (Sheingold and Hadley, 1990), the impact of telecommunications on how teachers teach was less pronounced. In the earlier study, 88 percent of the sample indicated that computers made a difference in their teaching, compared to 68 percent in the telecommunications survey.

In the earlier study (Sheingold and Hadley, 1990) , most respondents reported that computer technology had an impact on multiple aspects of their teaching. Teachers' expectations of their students' ability to pursue independent work increased; they spent more time working with individual students; and they were more comfortable with students' working independently. They reported that computers allowed them to present more complex material to students and tailor students' work to individual needs. When these same questions were posed to educators in the telecommunications survey, the impact of this technology on their teaching practices was significantly different from the Accomplished Teachers study (Sheingold and Hadley, 1990). There are at least two possible explanations for this difference (Honey and Henriquez, 1993).

One explanation is that educators' use of telecommunications technology directly affects what students learn as well as the quality of teachers' professional lives, and does not affect as directly teachers' pedagogical practices. And indeed, the most highly rated incentives for using telecommunications for student learning and professional development support this assertion. Telecommunications broadens students' perspectives on the world, and provides access to information that would not otherwise be available. Telecommunications has an impact on what teachers teach, not necessarily how they teach. One teacher wrote that, "Topics are of a more global significance. I require students to apply higher level thinking skills of analysis and synthesis." A second explanation centers on their sophistication with computer technology. The majority of these respondents have been using computers in teaching for years and may have already undergone significant changes in the way they teach. To the extent that they have taken place,

changes in these educators' pedagogical practices came with the integration of computers into their teaching (Honey and Henriquez, 1993).

Current approaches to general staff development implicitly use a model of experts giving teachers information (Tushnet, et al, 1993). Distance learning technologies allow all teachers to have access to expensive consultants and lower costs to each. At the same time, as now used, the technology facilitates imparting information to teachers as passive recipients. The effectiveness of staff development is likely to increase if projects provide intermediate support to participating schools and provide for more interaction among teachers and other school staff. Such support can come in the form of technical assistance from distance learning providers who can use the interactive aspects of the technologies to foster "learning communities." Computer networks may support teleconferencing and other approaches to staff development.

Within the Star Schools Program, at least three Funding Cycle One and Two projects have worked to demonstrate the uses of varieties of distance learning technologies to reform education (Tushnet, et al, 1993). This focus is even more evident in Cycle Three. One finding from the first year of the Star Schools study is that using technology to support educational reform requires a different approach from using technology to equalize educational opportunity. In the latter instance, personnel at the receiving school need moderate amounts of technical support, which all Star Schools projects provided with a high degree of professionalism and attention to the field. In contrast, using technology to reform education requires greater amount of support at the school site. The approach requires collaboration with teachers so they become comfortable with the technology, understand the cognitive and pedagogical demands of the reform and are able to use the curriculum and instructional methods to advance student learning.

Projects working on reform require time to develop educational applications of technology (Tushnet, et al, 1993). When they bring innovative technology to teachers, it should be as "bug free" as possible, which entails fairly extensive field tests. The materials and approaches also must meet high standards, which rely on rigorous quality control that includes content experts. In addition, because educational reform rests on teachers' approaches to curriculum and instruction, they should be supported in their efforts to use technology and change educational practice. Regular and intensive staff development provides such support. Indeed, among the Star Schools-sponsored activities that aimed at educational reform, the most successful projects used well-developed technology and provided fairly intensive ongoing support at the site level.

Just as we expect students to go through a

process of learning, we should also allow teachers the same privilege. The reality is that teachers receive limited in-service that is counted in days per year, and their pre-service experiences in a school of education were provided by faculty members who were role models for the traditional stand-up and lecture role model. Providing meaningful in-service in the new instructional methods and uses of technology has been a problem. However, research emanating from the TEAMS Star Schools Program in Los Angeles County has presented a significant new model (Lane, Cassidy, and Lake, 1994). This Three-Tier Distance Learning Staff Development Model developed for TEAMS Distance Learning by Sheila Cassidy in 1990 includes:

1. Theoretical Training: information, theory, demonstration and two-way communication about the theoretical basis of the instruction and training.

2. Implementation Training: theory, demonstration, practice and peer discussion of curriculum and instructional methods involved in the student programming, providing training to implement the student programs.

3. Simultaneous Teacher Training and Student Instruction: teacher training through in-class experience, practice and support from the studio team-teacher, through live, interactive student instructional programs.

The pattern that emerged during the evaluation has the potential to create a new model for teacher pre-service and in-service, because the model actually creates change in the classroom, as can be seen from a summary of the results.

TEAMS teachers reported in the survey that they viewed the TEAMS television teacher as a role model (on a scale of one to four where four was high, first year mean 3.9; second year mean 3.6; third year mean 3.7).

First year TEAMS teachers reported that there was a great deal of preparation for TEAMS. First year TEAMS teachers who used the program on videotape usually previewed the tape. First year TEAMS teachers reported that they felt that the TEAMS programs required a lot of work on their part to learn the new instructional methods, but they felt it was worthwhile because their students were learning so much more.

Second year TEAMS teachers reported that they had to prepare less for TEAMS programs as they now knew what the programming contained and understood the instructional methods. During this year, they reported a higher comfort level with the instructional methods, so much so that they used the same methods — collaborative learning, hands-on, and discovery methods — in the other content that they presented to students for math or science. Many teachers reported that the TEAMS television teacher was a role model who provided step by step guidance in presenting material to students. Teachers reported that they received

more usable information on new instructional methods through TEAMS programming than through in-service seminars.

Third year TEAMS teachers reported that they were very comfortable with TEAMS programming and instructional methods. They spent very little time gathering the materials for the class for TEAMS programs, and felt that the instructional methods had become natural components of their teaching style. They had become so immersed in the new methods that they used the methods in all content areas that they taught.

Using TEAMS has effectively provided teachers with new methods which they use because they have watched the TEAMS television teacher demonstrating the methods and have had opportunities to practice them in their own classrooms with their own students before, during and after the student programs. These results were reported across the United States at all evaluation sites as well as in the surveys. Principals also noted these changes in TEAMS teachers saying that TEAMS teachers showed more enthusiasm for math/science, a higher use of interactive and hands-on methods and that teachers were more confident of their ability to teach math and science. The survey question that dealt with planning and preparation for TEAMS also showed that teachers were increasingly comfortable with TEAMS.

The survey included one question which asked how well prepared teaches felt to use a variety of methods due to their TEAMS experiences and prior to their TEAMS use. Teachers reported an increase in the ability to teach heterogeneous groups, teach math/science in an active learning environment, manage a class of students who are using manipulatives, use cooperative learning in math/science instruction, involve parents in their child's math/science education, use the textbook as a resource rather than as the primary instructional tool, use a variety of alternative assessment strategies and follow national mathematics standards/science recommendations.

TEAMS was chosen by districts, principals and teachers for a variety of reasons including the fact that it was based on the mathematics and science standards/recommendations, hands-on procedures and distance delivery, which would enhance teaching and learning. Schools and teachers continued to use TEAMS in the second and third years because it fulfilled its original promise.

Students are learning from TEAMS. There are increases in skills in math and science content that TEAMS teachers can directly attribute to students viewing TEAMS programming and using TEAMS materials. Teachers reported that students who had difficulty learning about mathematics and science through other methods, were now learning from the TEAMS hands-on methods and manipulatives. Students revealed in student focus groups that it was fun to learn with TEAMS as opposed to the "other" way which seemed to be the "hard" way. Teachers reported a positive change in student behavior even with normally disruptive students. Teachers reported increased self-esteem, increased attendance and an increased interest by girls in math/science.

Teachers reported that students became comfortable in using scientific inquiry, increased participation in science fairs and many selected a TEAMS topic for their science fair projects. Teachers reported that students are more interested and motivated to do math, including students who were lower achievers in math. They felt that there was more retention of math skills.

TEAMS has effectively provided teachers with new instructional methods by viewing the TEAMS television teacher during the student programming. The TEAMS model has changed the teaching styles and the instructional methods of TEAMS teachers by the time teachers have used TEAMS three years. The most significant changes in TEAMS teachers were achieved by those who used TEAMS on regular basis. Based on the information emerging from the TEAMS evaluation, it is possible to identify how the TEAMS program can be most successfully adopted by a district and its school. Teachers' ability increased in a variety of ways. Because of the TEAMS teaching model they reported increased skills in teaching heterogeneous groups, teaching math/science in an active learning environment, managing the student use of manipulatives, using cooperative learning in math/science instruction, involving parents in their child's math/science education, using the textbook as a resource rather than as the primary instructional tool, using a variety of alternative assessment strategies and following national mathematics standards/science recommendations.

Teachers, students, principals and TEAMS site coordinators reported that they liked TEAMS programming and that it was increasing the time allocated to math and science in the classroom. Teachers increased their class time in math and science by an average of four hours per week.

TEAMS motivates students to learn math and science because they enjoy it and because it maintains their enthusiasm through interaction with the TEAMS television teachers and the use of hands-on manipulatives for learning. TEAMS is also used as a taped program and the student learning in these classes is equivalent to that of the students who view the program live.

In a research study on interactive technologies for Apple Classrooms of Tomorrow (ACOT), students and teachers were provided with technology to use for a year in the classroom and at home. Part of the study focused on the evolution of classroom management in ACOT high-tech classrooms. Analysis of data suggested a three-stage model of development that reflects

teacher concerns about classroom management: survival, mastery and impact (Dwyer, et al, 1990).

In the survival stage, according to the ACOT study, teachers are "preoccupied with their own adequacy. The concerns center on their ability to control the class and they spend considerable time reacting to problems instead of anticipating and avoiding them. In the mastery stage, they begin to anticipate problems and develop strategies for solving them. Finally, in the impact stage, teachers focus on the effects of their teaching on students' achievement and attitudes, and begin to use the technology to their advantage" (Dwyer, et al, 1990).

"Evidence of moving from stage to stage is not always clear cut, however, as individuals may vacillate between phases. For instance, in this study, by the second year, most teachers had learned to expect occasional technological problems, such as disk failures or network bombs, and planned accordingly. Yet, when new software, hardware or students arrived on the scene, many teachers temporarily reverted back to the survival stage" (Dwyer, et al, 1990).

The Education Coalition (TEC) Staff Development Model: The TEC Staff Development Model is a unique and unprecedented collaboration between the TEC Schools of Education, K-12, broadcast, cultural and historical Affiliate Agencies. This collaborative model provides an environment for these Affiliate Agencies to work together to design distance learning program series and courses for preservice and in-service teachers, students, parents and community (Lane and Cassidy, 1994).

The series and courses integrate live-interactive teleconferences with computer-based conferencing and information access and audio conferencing. This model provides for national resources of many types to be used for local training and development. The TEC model, incorporating inter-agency collaboration and multiple technologies, is an expansion of the Three-Tier Distance Learning Staff Development Model used by TEAMS Distance Learning, including: theoretical training; implementation training; and simultaneous teacher training and student instruction.

Overall, this approach answers many of the problems related to traditional staff development design, in that it:

- is long term, sequential training
- fosters immediate transfer of learning, with skills becoming a part of the teacher's repertoire of instructional methods
- is conducted mostly in the teacher's own classroom during the school day
- creates immediate changes in the roles of the teacher and student
- provides opportunities for teachers to see their own students being successful with a rich and

challenging curriculum, allowing them to change their attitudes and behaviors related to instruction and expectations of their students

- provides motivation for teachers to participate in other staff development after the regular school day because it is directly related to their classroom program

This model is based on research and practice in the fields of staff development and adult learning, as well as national and state standards and guidelines. The basis of the staff development research is formed by the work of Joyce and Showers (1988), Cassidy and Taira (1988, 1989) and the Rand Corporation (Berman and McLaughlin, 1978). The adult learning principles are summarized in work by Jones and Woodcock (1984) and Knowles (1975, 1984).

The staff development research (Joyce & Showers, 1988) provides insights on the relationship between training outcomes and specific training components. They analyzed the training outcomes of knowledge, skill and transfer of training for participants engaged in training programs options providing:

1. information
2. theory
3. demonstration
4. theory and demonstration
5. theory and practice
6. theory, demonstration and practice
7. theory, demonstration, practice and feedback.
8. theory, demonstration, practice, feedback and coaching in participants.

Their research clearly shows that training which provides only information and theory produces only increased knowledge. That by encompassing any of options four through eight shows greater knowledge and skill outcomes. Option eight provides the greatest outcomes in knowledge, skills and transfer of training. Practice, feedback and coaching can be considered an in-classroom, on the job, experiential and support component.

This model provides a distance learning alternative to option eight. It clearly provides theory, demonstration and practice. Although distance learning cannot provide a full face-to-face feedback and coaching component, part of what feedback and coaching provides is an in-class support system. That is provided through the in-class team teaching with the studio instructor.

In retraining of teachers, Cassidy and Taira (1988, 1989) found that teachers reported the factors which contributed to their success were: a sound theoretical basis; experience and practice with the particular curriculum and instruction being adopted/adapted; a support system designed specifically to their needs; convenience, with training during the school day and at their site when possible; training with no expense to teachers. The simultaneous in-class training component meets all of these criteria. The Rand Corporation found that successful projects had these common characteristics for staff development (Berman and McLaughlin, 1978):

1. training is concrete, continual, and tied to the world of the teacher
2. local resource personnel provide direct follow-up assistance
3. peer observation and discussion provide teachers with reinforcement and encouragement
4. school leader participates in staff development
5. regular meetings held with teachers for problem solving and adapting techniques and skills of the innovation
6. release time used for teacher staff development
7. staff development planned with teachers prior to and during the project.

Cassidy (1985) reviewed programs with findings similar to the Rand study but with additional information.

1. individualized staff development activities are more effective than large-group activities
2. programs incorporating demonstrations, trials and feedback of ideas are more effective than lecturing and reading of ideas
3. staff development programs are more successful when teachers are active planners and help each other.

During an investigation by Lane (1993), the Three-Tier approach was shown to be highly successful with changing attitudes and behaviors of students and teachers. In an earlier national investigation, Lane (1989) found that distance learning educators were not using self-directed learning, facilitation, hands-on or other elements of constructivism or andragogy in either selecting distance learning programs or in developing such programs. A second national study (Lane, 1992, p 211-215) ultimately led to national standards which won the Teleconference Magazine "Most Significant Advance in Distance Learning Overall 1991." Based upon the results of that research, program providers would need to meet at least 85 percent of the criteria in each section to meet national quality standards. The educational objectives included providing specific learning experiences and skill transfer, small group work, learning outcomes, andragogy and a variety of presentation techniques to reach varied learning styles (visual, auditory, experiential, tactile). In research on audio conferencing for instruction (Lane, 1992, p 224-231) it was shown that this medium provides a useful method of interaction with the facilitator in structuring national programming.

■ CONCLUSIONS

Systemic reform of education and training is built upon a number of converging events. This study reviewed the literature in the areas that directly affect systemic reform, and concludes with recommendations that will enable technology to play a significant role in this movement. Literature from the following areas were reviewed: paradigms, change strategies, systemic reform, constructivism, student empowerment, andragogy, Goals 2000, legislation, equitable access, current use of technology in education, technology as a force for systemic reform, technology's potential as a partner and tool for systemic reform, new ways to think about the use of technology for education, problem-based learning, competing in the global economy, total quality management in education, evaluation through performance based assessment, using the national information infrastructure for education and training and instructor in-service and pre-service.

Paradigm Shift: Teaching and learning in American schools looks very much as it did a century ago, leaving students and society economically and socially at risk. We have entered a time of great societal and technological change which indicates a paradigm shift. We have moved into the Information age and may have moved into the Communications Age. These changes will continue to impact our educational institutions in ways of ever increasing magnitude, making systemic reform inevitable. Total Quality management and theory are a part of systemic reform.

Change Models and Change Agents: There are a variety of change models which can be used to plan for systemic reform. Just as young scientist, "outsiders," are instrumental in scientific paradigm shifts, conceptualization of educational systemic reform will be influenced by those inside as well as those outside of our traditional educational institutions. Change agents looking at education in non-traditional ways will play a vital role in the conceptualization of a vision for second order change.

Learning Models: Constructivism, student empowerment and andragogy models for learning provide a basis for systemic reform. Each of these models views the learner as responsible and able to learn, capitalizing on strengths, rather than working from deficits. Even though andragogy is an adult learning theory, younger learners should be thought of as developing on a continuum, coming closer and closer to adult learning. Therefore, the principles of andragogy may be applied to younger learners to the extent that is developmentally appropriate. Students should be taught about the learning models so that they understand their responsibilities in becoming self-directed, proactive learners. Students who are suddenly moved from a teaching style to a facilitation style will need time to process this information, learn new behaviors and understand what is expected of them. Cognitive dissonance is common during the transition period as they have come to expect certain things from teachers that will not be done with facilitation. This is true for all learners regardless of their age. Problem based learning, authentic learning and authentic assessment are integral parts of these learning models. Educational technology is an excellent way to support and enhance these models.

Learning Styles: It has become accepted that students learn in different ways and that teachers must provide instructional methods that reach the learning styles of all students. Students should be taught about their learning style preferences and guided in selecting materials that will help them learn. Students should also be encouraged to develop methods that will help them learn in their non-preferred learning style. Teachers should know what their personal learning style is and what their teaching style is so that they can avoid a mismatch in working with students. Parents should also have information about their child's learning style and how they can guide the child in learning. Educational technology is an excellent way to support and enhance students' learning styles.

Policy and Legislation: Goals 2000 and The Goals 2000: Educate America Act provide a framework for movement toward national systemic reform. Passage of the legislation has brought greater understanding to the problems education has faced and the importance it carries in continuing to transform national fiscal goals into reality. The legislation carries with it significant funding that will enable education to meet goals in the short term. Long term funding for education is in question. The 1994 Communications Act provides a significant quantity of information access, but funding to pay for the equipment and access charges will come from local and state legislation.

Equitable Access: Equity of access and opportunities to learn will continue to be the driving force in providing technology to learners so that a nation of haves and have-nots does not further burden the tax system. The uses of educational technologies for adult literacy programs, immigrants, workplace skills for those on welfare and education for the prison population are promising, but these populations are virtually untouched by educational technologies. Once educational technology is accepted and the research continues to mount in its favor, it is anticipated that these populations will be the next recipients of the benefits of educational technologies.

Educational Technology: Technology is both a force pushing us toward systemic reform and a partner and tool in reform. The environment has been set to create the paradigm shift, but to maintain the forward thrust, other factors must be in place. There must be a clear understanding that technology changes the context in which education takes places. It allows educators to think differently about how and where learning takes place and what "basic education" needs to be in a world driven by technology but where terms such as distance education and computer mediated learning still have to be explained to otherwise enlightened members of society.

Systemic reform necessitates:

1. Redefinition of the educational community, the roles and relationships between the partners in that community

2. Restructuring of curriculum, instruction and assessment

3. Redefinition of the structures and technologies of "school," recognizing that it is one player in the educational process

4. Redefinition of where learning takes place and what it means to be "educated"

5. Time for teachers to learn to use technology, to experiment with its use and to create effective lesson plans that contribute to the learning needs of students.

6. Consistent access to a range of similar technologies at all levels of education for teachers and students, in schools, individual classrooms, libraries, home and the workplace

7. Ongoing needs assessment and evaluation of technology use, with recommendations for continuing improvement to meet the needs of the community and the workplace.

National Information Infrastructure (NII): The National Information Infrastructure will provide vast opportunities for learning and collaboration for learners when they gain access. Continuing improvement and innovation will lead to a convergence of media — video, data, audio and document manipulation, simulations and virtual environments will become dominant. Ease of use through user interfaces will attract more users.

Competing in the Global Economy: Educators need to fully understand the needs of employers. Collaborative planning for local needs between employers and educational institutions will enhance understanding, support planning and lead to curricula that support these needs as well as academic pursuits. Students should have access to the same types of technologies as those used in the work place. Ongoing analyses and evaluation of graduates should lead to curricula that meet the changing needs of the workplace. Trainers should use adult education instructional methods, and focus on programs that develop employees ability to learn (learning how to learn), self-directed and proactive learning with intrinsic rather than extrinsic reward systems.

Instructor Pre-Service and In-Service: K-12 teach-ers and faculty at colleges, universities and other adult education programs need to receive education about the academic theory of distance education, effective uses of educational technologies, learning styles, learning models and systemic reform. These should be provided by schools of education. Just as teachers perform a practicum in the real classroom, they should also perform a practicum where they work with all forms of educational technologies. Educational technologies and their effective use should be required core courses provided by schools of education. Certification should be required in educational technology in all certification areas. Colleges and universities should include effective use of educational technologies as a requirement for tenure and continuing employment for full-time and adjunct faculty. College and university faculty should act as role models for new instructional methods that support systemic reform and the use of educational technologies.

Systemic Reform: The model that has emerged for technology to have a significant impact on systemic reform is multi-faceted. A number of forces must continue, and others must be successfully operationalized for technology to impact systemic reform. Without operationalizing all of these factors, educational technologies will continue on the periphery of education as alternative education. If the factors move into place and begin to develop a synergy, educational technologies will move into the mainstream of education and training and will be a powerful force and partner in how we learn in the future. In order to use technology wisely for systemic reform, we need to project answers to the following questions.

Technology

1. Will the technology continue to improve?

2. Will the cost of technology continue to drop?

3. How can the pace of adoption of technology by industry and education be brought more closely together?

4. How fast will technology change society?

5. Have we underestimated the magnitude of technologies' eventual effects?

Policy/Legislation and Funding

6. Will the National Information Infrastructure be built to include education as a major partner?

7. Will legislation for systemic reform continue to be supported by sufficient funding?

8. Will the administration and government continue to support technology in general and educational technology in particular?

9. Will education be treated as a system which includes parents as first teachers, pre-school programs, college and continuing education for lifelong learning?

10. How can we ensure that the media are not used to shape the instructional message in unwanted ways?

Teaching:

11. Will schools of education embrace new instructional methods and technology and act as the role model for faculty and students?

12. Will the use of technology receive grass roots support from educators?

13. Will education successfully adopt the innovation of technology?

14. Will educators receive sufficient training, time and continuing support in the use of technology for education?

Learning — Technology Applications:

15. Will research continue to support the use of technology as an effective educational method?

16. Will educational technologies change the place where education is delivered from the school to the home and workplace?

Futuring:

17. What other factors should be included to sustain the paradigm shift?

18. Are we encountering a significantly quickened pace between Ages?

19. How can change be easier to accept?

20. Will the Millennium have the expected impact?

21. What other questions should be asked?

It is obvious that the technology will continue to improve and the costs will continue to drop as more units move to the end user. It is likely that the use of technology for education will continue to be sustained through research that shows learning does take place as well through technology as it has through traditional methods.

There is no current evidence to support the hope that all schools of education will embrace the use of technology. The Goals 2000 legislation will need to be re-authorized in the future. Goals 2000 includes a technology component, but there is not enough legislated funding to support full use of technology in all classrooms. The in-service for educators is minimally supported now and new legislation does not provide substantial amounts of funding to provide sufficient in-service for all educators. As such, it is doubtful that a true grass roots movement for the use of technology will be supported by rank and file faculty members. The problem here is that the use of technology represents a personal change (re-learning), an increase in personal productivity through technology, insufficient release time to learn the technology and new instructional methods and a personal threat to the importance of the job of teaching. Because few teachers have made the change to becoming facilitators of learning (sage on the stage, to the coach on the side), they perceive the use of technology and facilitation as a loss of personal empowerment — rather than the more positive side — where students are empowered to construct their own knowledge.

While future administrations and government will continue to support technology and build a National Information Infrastructure, it is doubtful that the 1994 Communications Act will enable the use of the NII to the extent that it could be used for education. Congress will not impose more than a token amount of wiring for education on the telecommunications companies. Lower rates for telecommunications services will enable more use of technology for education. The true problem for most schools will be wiring the buildings with sufficient bandwidth to provide access for learners and teachers. Most schools have only two telephone lines at this writing and it is unlikely that this will increase to ten telephone lines in the near future. The question of universal access comes into play here. The 1934 Communications Act, required that universal access be provided throughout the United States — except to schools. As a result, sixty years later, schools do not have the same universal access that the rest of the country enjoys.

It does seem likely that education will be perceived and treated as a system rather than the fragmented components that make up the system now. Pre-school, elementary, middle school, junior high, high school, community colleges, vocational education, universities, continuing education and industry training make up the largest pieces of the pie now. There is a jumble of administration offices to deal with education including the local school, district, county, region and state department of education. Many collaborations have been put into place because they "are rooted in the belief that the two entities need each other to achieve the goals they hold in common" (Ishler, 1994). Ishler states that if students are not given the proper foundation in the first twelve years, there is no prospect that they can be successful when they enter college.

Each higher education system is fragmented as well. While a local community college may be autonomous, part of a community college district, or part of a statewide system, a state's university system is also fragmented between the land-grant university system, and the state college system with each vying for the minimal funds distributed by the state for higher education. Vocational education is also a part of the mix, as are the schools managed by the Bureau of Indian Affairs and the defense schools for children of military personnel. Special education plays a significant part in the mix along with adult basic education (ABE), general equivalency degree programs (GED) and English as a second language (ESL) programs. Private educational institutions are involved at all levels.

At the corporate and industry level, training departments have characterized their operations as being the trainer of last resort in providing basic skills programs and re-training programs. The downsizing of the military has also heaped additional pressures on an already overworked system.

Poverty programs operate within the educational system with notable programs such as Chapter I and Chapter II programs, Pell grants provide funding within the system for students or provide funding to pay into the system for tuition.

Arguably, all elements of education are not included in the previous paragraphs. The question is, how will all of these groups be treated equally so as to provide equitable access and how will they begin to work together? How will each of these groups use technology to promote and advance their movement through systemic reform? As students move into and out of each system, what guarantee is there that the technology will be consistently available so that students will be able to move easily through the system?

Within the next ten years technological advances will be monumental. Four factors which should impact the learning environment are:

1. voice, data and video delivered into the home, community learning centers, workplaces via cable television

2. access to worldwide information via telecommunications

3. access to worldwide dialogue via telecommunications

4. learning through entertainment-like devices.

To date, it is clear that technology has played a minor role in changing schools in what Conley calls renewal activities — helping schools to do what they already do better or more efficiently. It is also evident that technology has played an even lessor role in the restructuring of schools. It is time to let the genie out of the bottle — to transform education as it is transforming the world in which we live. It is time for the broad American educational community to be creative in how it wishes to reinvent the educational system, and to specify how technology can be used as a powerful force to create it! The technological genie provides a need, opportunities and resources for truly "break the mold" systemic reform. In the larger sense, the genie is creating a global learning environment in which "schools" of some type will play an important role, but one which will take on different dimensions, new structures, new relationships and new practices. Schools are but one of the places and ways in which learning occurs. Soon — as the genie advances — homes, libraries, community centers, workplaces and recreational areas will provide opportunities for greater educational access — if we meet the challenges to use them for truly educational purposes. Schools will change as technology changes the world in which they exist. It's even possible that schools won't exist as they do today, and that we may call that a cause for celebration.

Under the command of effective instructors, the technological genie can lessen the isolation and unidimensionality of schools as we now know them. It can provide opportunities for access, dialogue and resources for learners of all ages from rural to urban America as well as global study groups. It can help to create a seamless, lifelong process of education, with public and private agencies working together to provide an array of learning opportunities and a multitude of learning options. Teachers and students will learn together through sharing of resources and interacting with peers and experts in a variety of fields. Individuals from a variety of agencies will be able to work together on issues of common concern. K-12 educators need not be separated from post-secondary educators — all can work to better meet the needs of a growing and diverse student population. Education will become an open road map for learners to explore rather than a one lane road that all must travel.

Through multiple technologies, learners will have access to real-world problems thus providing them with an environment rich for exploration and will use problem-based learning. Students and teachers will work as colleagues on problem-solving teams. Learners will assume the role of investigators — bringing to bear interdisciplinary content information to the solution of the problems. They will contribute their ideas for solutions to those formulated by others throughout the country and world.

Distance learning brings a new model for teacher training, teacher support and systemic reform. It can now provide easy access through television and computer conferencing to course work to prepare teachers in the use of instructional applications of technology, in moving into systemic reform, in developing new expertise in problem-based learning, in developing authentic instruction and assessment moving students toward lifelong learning outcomes, and to provide a support structure of and for teachers throughout the country.

All of these are at hand. Now is the time to envision what is possible, unleash the power of the genie, and direct the use of technology toward the accomplishment of these ends. Distance learning has changed. It is being used more extensively by more organizations and learners. It is using more technologies.

Dede (1991) said that "Advanced information technologies are transforming society, altering our conceptions of quality, effectiveness and intelligence. These shifts require profound changes in every aspect of our current educational model. Because new functionalities are altering the characteristics of the communications channel between student and subject, distance learning is particularly affected." He suggests that even our "paradigm for distance learning must evolve so that we can replicate the workplaces and communities of the future" in schools to help students master, filter and interpret the "complex, pervasive informational environments that sophisticated media are creating in society." He contends that distance learning is not only a "method of delivering instructional services" but students familiarity with technology-permeated experience is vital for coping with the world of tomorrow."

Starting is the hardest part, but we have over thirty years of demonstration projects and pilot programs behind us. The excitement that we wanted to bring to education through the use of technology and new instructional methods that met all learning styles has now eclipsed us with its own synergy. The excitement is truly systemic reform, and its goals and those of educational technologists are well aligned. The old basic beliefs have been challenged and found to be lacking. Technology is a vehicle that helps us continue to challenge our beliefs and do things differently because it empowers the student and enables success for all learners.ue to challenge our beliefs and do things differently because it empowers the student and enables success for all learners.

■ REFERENCES

Abdal-Haqq, Ismat (1989). The Influence of Reform on In-service Teacher Education. ERIC Digest. Clearinghouse Number: SP032647, Accession Number: ED322147.

Amarel, Marianne (1983). Classrooms and computers as instructional settings.Theory into Practice, 22 (4), 260-266.Anderson, R. D.,

Anderson, B. L., Varanka-Martin, M. A, Romagnano, L., Bielenberg, J., Flory, M., Mieras, B., and Whitworth, J. (1994). "Issues of Curriculum Reform in Science, Mathematics and Higher Order Thinking Across the Disciplines." Studies of Education Reform Program, US Department of Education, Office of Educational Research and Improvement, Office of Research. Washington, DC.

Aubaugh, Gordon (May, 1994) Address to the U.S. Department of Education, Secretary's Technology Conference, Washington, DC.

Bank, David (June 20, 1994). "Schools Look to Future," San Jose Mercury News. San Jose, CA., 1a, 15a.

Barker, Joel (1992). "Paradigms: The Business of Discovering the Future." New York, Harper.

Beane, James. (October 1991) "The Middle School: The Natural Home of the Integrated Curriculum." Educational Leadership 49,2, 9-13.

Berman, Paul and McLaughlin, M. (1978). The Rand Corporation.

Beatty, Thomas R. and Fissel, Mark C. (October, 1993). "Teaching With a Fiber-Optic Media Network: How Faculty Adapt to New Technology." T.H.E. Journal, 82-84.

Betts, Frank (April, 1994). "On the Birth of the Communication Age: A Conversation with David Thornburg." Educational Leadership.

Blanchard, Kenneth (February, 1994). Personal communication, Seminar, University of Phoenix, Online Division, San Francisco, CA.

Bonstingl, John Jay (1992). "The Quality Revolution in Education." Columbia, MD.

Borrell, Jerry. (September 1992). "America's Shame: How We've Abandoned Our Children's Future." MacWorld 9,9.

Bugliarello, George. (1990). "Hyperintelligence: Humankind's Next Evolutionary Step." In Rethinking the Curriculum: Toward an Integrated, Interdisciplinary College Education, edited by Mary E. Clark and Sandra A. Wawrytko. New York: Greenwood Press.

Cassidy, Sheila (1985). Unpublished manuscript.

Cassidy, Sheila and Taira, Susan. (1988). "Study of Bilingual Teacher Training Programs in California." Rancho Palos Verdes, CA; Educational Development Network.

Cassidy, Sheila and Taira, Susan. (1989) "Follow-Up to the 1988 Study," Rancho Palos Verdes, CA; Educational Development Network.

Cassidy, Sheila. (1990). "Three-Tier Distance Learning Staff Development Model for Teachers." TEAMS Handbook 1993-94.

Coates, Joseph F. (1977). "Aspects of Innovation: Public Policy Issues in Telecommunications

Development," Telecommunications Policy, 1(3):11-13.

Cobb, P., Wood, T., Yackel, E. and McNeal, B. (1992). "Characteristics of Classroom Mathematics Traditions: An interactional Analysis. American Educational Research Journal, 29(3), 573-604.

Cohen, D. K. (1988). "Educational technology and school organization. In R. S. Nickerson and P. P. Zodhiates (Eds.), Technology in education: Looking toward 2020 (pp. 231-264). Hillsdale, NJ: Erlbaum.

Cohen, Herb (May 1994). "State of the Art: 1999," Training, S10.

Collins, A., Brown, J. S., and Newman, S. E. (1989. Cognitive apprenticeship: Teaching the craft of reading, writing, and mathematics. In L. B. Resnick (Ed.), Knowing, learning, and instruction: Essays in honor of Robert Glaser (pp. 453-494). Hillsdale, NJ: Erlbaum.

Collins, Allan. (September 1991). "The Role of Computer Technology in Restructuring Schools." Phi Delta Kappan 73,1 : 28-36.

Computer Systems Policy Project (March, 1994). "Information Technology's Contribution to Lifelong Learning." Computer Systems Policy Project, Washington, D.C.

Confrey, J. (1990). "What Constructivism Implies for Teaching." In R. B. Davis, C.A. Maher and N. Noddings (Eds.), Constructivist Views on the Teaching and Learning of Mathematics: Journal for Research in Mathematics Education, Monograph No. 4 (107-122). Reston, VA. NCTM.

Conley, David T. (1992). "Some Emerging Trends in School Restructuring. ERIC Digest, Number 67. ERIC Clearinghouse on Educational Management, Eugene, OR. Office of Educational Research and Improvement (ED), Washington, DC. EDO-EA-91-9, ED343196.

Conley, David T. (1993). Roadmap to Restructuring, ERIC Clearinghouse on Educational Management, University of Oregon.

Corporation for Public Broadcasting Spring (1991). " Study of School Uses of Television and Video." Alexandria, VA.

Cuban, L. (1986). "Teachers and machines: The classroom use of technology since 1920. New York: Teachers College Press.

Cummins, Jim (1989). "Empowering Minority Students." California Association for Bilingual Education. Sacramento, CA.

Curtin, Pat, Cochrane,Lucy, Avila, Linda, Adams, Laura, Kasper, Susan and Wubbena, Curtis (April, 1994). "A Quiet Revolution in Teacher Training." Educational Leadership.

David, Jane L. (September, 1991). "Restructuring and Technology: Partners in Change." Phi Delta Kappan 73, 1.

Davidow, William H. and Malone, Michael S. (1992). The Virtual Corporation, New York, Harper Colins.

Davis, R. B., Maher, C.A. and Noddings N. (1990) (Eds.), "Constructivist Views on the Teaching and Learning of Mathematics: Journal for Research in Mathematics Education, Monograph No. 4. Reston, VA. NCTM.

Dede, Christopher J. (March, 1991). "Emerging Technologies: Impacts on Distance Learning." In The Annals of the American Academy of Political and Social Science: Electronic Links for Learning. 146-158.

DeLoughry, Thomas J. (May 11, 1994). "Stamp of Approval." The Chronicle of Higher Education.

Dwyer, David C., Ringstaff, Cathy, Sandholtz, Judith Haymore (1990). "Teachers Beliefs and Practices Research Summary #8, Part I: Patterns of Change," Cupertino, CA, Apple Computer, Inc.

Elkmer-Dewitt, Philip. (May 20, 1991). "The Revolution That Fizzled." Time 137, 20.

Farrell, Rod and Gring, Stephen (November, 1993). "Technology Strategically Planned: A Dismal or Bright Future?" T.H.E. Journal, 119-122.

Fitzsimmons, Ed (April 21, 1994). Presentation to TEAMS Distance Learning, Office of Science and Technology Policy, Executive Office of the President, Washington, DC.

Fletcher, Dexter J. (1994). "Effectiveness and Cost of Interactive Videodisc Instruction in Defense Training and Education." Institute for Defense Analyses, Alexandria VA.

Fletcher, J. D. (1991). "Effectiveness and cost of interactive videodisc instruction. Machine Mediated Learning," 3, 361-385. Fletcher, J. D., Hawley, D. E., and Piele, P. K. (1990). "Costs, effects, and utility of microcomputer assisted instruction in the classroom." American Educational Research Journal, 27, 783-806.

Fuller, Frances F. (1969). "Concerns of teachers: A developmental conceptualization." American Educational Research Journal, 6 (2), 207-226.

Hall, Gene E., & Loucks, Susan (1979). "Teacher concerns as a basis for facilitating and personalizing staff development." In A. Lieberman and L. Miller (Eds.). Staff development: New demands, new realities, new perspectives. New York: Teachers Press.

Hancock, Vicki, and Betts, Frank (April 1994). "From the Lagging to the Leading Edge." Educational

Leadership.

Harasim, L. M. (1989). "Online Education: A New Domain." in R. Mason and A. Kaye (Eds). Mindweave, Communication, Computers and Distance Education. New York: Pergamon.

Harmon, Willis (1970). "An Incomplete Guide to the Future." New York, W. W. Norton.

Hawkins, Jan (March, 1991). "Technology-Mediated Communities for Learning: Designs and Consequences." In The Annals of the American Academy of Political and Social Science: Electronic Links for Learning. 159-174.

Hedegaard, Terri (1994). Learning online and on campus: A comparison of adult students' professional attitudes and values. Thesis, University of Phoenix, San Francisco, CA.

Heinrich, R., Molenda, M. and Russell, J.D. (1993). "Instructional media and the New Technologies of Instruction (4th Ed.), New York, NY, Macmillan.

Hequet, Marc (May, 1994). "Should Every Worker Have a Line in the Information Stream." Training, 99-102.

Herman, J. L., Aschbacher, P. R., and Winters, L. (1992). "A Practical Guide to Alternative Assessment, Alexandria, VA. Association for Supervision and Curriculum Development, ED 352 389.

Hirschbuhl, John J. and Faseyitan, Sunday O. (April, 1994). "Faculty Uses of Computers: Fears, Facts and Perceptions." T.H.E. Journal.

Hoffman, Ruth I. (1984). "Recommended changes in the teacher preservice program to reflect computer technology and its impact on education." Educational Perspectives, 22 (4), 21-23.

Holmes Group. (1990). "Tomorrow's schools: Principles for the design of professional development schools." East Lansing, MI: Author.

Honey, M. and Henriquez, A. (1993). "Telecommunications and K-12 Educators: Findings from a National Survey." New York, NY: Bank Street College of Education.

Hsu, E. Y. F. (1990). "Running Management Game in a Computer Mediated Conferencing System: A Case of Collaborative Learning." Proceedings of the Third Symposium on Computer Mediated Communication. Guelph, Canada: University of Guelph.

Huey, John (June 27, 1994). "Waking Up to the New Economy." Fortune, 36-46.

Hundt, Reed (May 10, 1994). Speech to U.S. Department of Education, Secretary's Conference on Technology, Washington, DC.

Information Infrastructure Task Force (1994). "A Transformation of Learning: Use of the NII for Education and Lifelong Learning." Washington, DC.

Interactive Educational Systems Design (1993). "Report on the Effectiveness of Technology in Schools 1990-1992," Software Publishers Association, p. 2.

International Society for Technology in Education (October, 1990). "Vision: Test. Final Report, Eugene OR: ISTE.

Investor's Business Daily (February 17, 1994)

Ishler, Richard E. (1994). "Together We Can make A Difference: Collaboration Between Schools and Universities," National Forum, The Phi Kappa Phi Journal, Spring 1994, Vol. Lxxiv, No. 2.

Jacobson (April 27, 1994a). "The Coming Revolution." The Chronicle of Higher Education.

Jacobson (July 6, 1994b) "Extending the Reach of 'Virtual' Classrooms." The Chronicle of Higher Education.

Johnstone, Sally (March 1991) . "Research on Telecommunicated Learning: Past, Present, and Future." In The Annals of the American Academy of Political and Social Science: Electronic Links for Learning. 49-57.

Jones, John E. and Woodcock, Michael. (1984). "A Manual of Management Development"; Aldershot, Hampshire, England. Gower Publ. Ltd.

Joyce, Bruce & Showers, Beverly. (1988). "Student Achievement through Staff Development." New York: Longman, Inc.

Knowles, Malcolm (1975). "Self-directed learning: A guide for learners and teachers. "New York, Cambridge.

Knowles, Malcolm (1984). "Andragogy in Action." San Francisco, Jossey-Bass.

Kuhn, Thomas S. (1970). "The Structure of Scientific Revolutions." Chicago: University of Chicago Press.

Kunin, Madeleine (May 10, 1994). Address at the U. S. Department of Education, Secretary's Conference on Technology, Washington, DC.

Lane, Carla (1989). "Development of a pre-adoption evaluation instrument for distance education tele-courses."

Lane, Carla (1990). "Use of Audio Interaction in a Telecourse Offered by Satellite: Foundations of Adult

Basic Education." Proceedings of the Midwest Research to Practice Conference.

Lane, Carla (1992) In "A Technical Guide to Tele-conferencing & Distance Learning." Portway, P. and Lane. C (Eds), Applied Business teleCommunications, San Ramon, CA.

Lane, Carla (1993). "1993 Evaluation of TEAMS (APOLLO 2000)."

Lane, Carla (May 25, 1994a). Testimony before the Senate Commerce Committee on S.1822, 1994 Communications Act.

Lane, Carla (1994b). First Year Evaluation of the TEAMS Program, 1992-93. Far West Laboratory, San Francisco, CA.

Lane, Carla and Cassidy, Sheila (June, 1994). "The Education Coalition (TEC) Staff Development Model." The Education Coalition Quarterly, McLean, VA.

Levin, Henry (May 1993). "The Economic Consequences of Undereducation": address to the second annual NEA Education Finance Workshop, Washington, DC.

Levine, M. (July-August, 1988). Professional practice schools: Teacher education reform or school restructuring, or both? Radius, 1(2), 1-7.

Levinson, Eliot. (October 1990). "Will Technology Transform Education or Will the Schools Co-Opt Technology?" Phi Delta Kappa 72, 2.

Magnet, Myron (June 27, 1994). "The Productivity Payoff Arrives" Fortune, 79-84.

Mahlios, M., Moore, C., Barnes, D., Bell, H., Kachur, D., and Bettis, N. (1987, February). Master teacher model development program. Paper presented at the AACTE Annual Meeting, Washington, DC. ED 277 699.

Majkowski, Charles (1990). "Developing Technology Applications for Transforming Curriculum and Instruction," Technology in Today's Schools, Association for Supervision and Curriculum.

Malley, Kenneth (May 25, 1994). Testimony before the Senate Commerce Committee, Washington, DC.

Martin, Lynn. (1991). "On Achieving Necessary Skills." U.S. Labor Secretary's Commission Report, Washington, DC.

Mason, Robin, (1989). Mindweave. Ed. R. Mason and A. Kaye. New York: Pergamon Press

Melmed, Arthur (June 1993). "A Learning Infrastructure for All Americans." Fairfax, VA: Institute of Public Policy, George Mason University.

Meyer, Alan; Geoffrey Brooks; and James Goes. Environmental Jolts and Industry Revolutions: Organizational Responses to Discontinuous Change. Strategic Management Journal 11 (1990).

Means, Barbara; Blando, John; Olson, Kerry; Middleton, Teresa; Morocco, Catherine Cobb; Remz, Arlene R; Zorfass, Judith. (1993). Using Technology to Support Education Reform, Office of Research, U.S. Department of Education, Washington, DC.

Means, Barbara and Olson, Kerry (April, 1994). "The Link Between Technology and Authentic Learning." Educational Leadership.

Milken, Michael (May 11, 1994). "Perspectives on Education Technology Married to Learning: a Millennial Wedding." Los Angeles Times.

Miller, Rockley L. (Date unknown). "Learning benefits of interactive technologies." The Videodisc Monitor.

Monk, David H. (1989). Using Technology To Improve the Curriculum of Small Rural Schools. ERIC Digest. Clearinghouse Number: RC017156. Accession Number: ED308 056.

Naisbitt, J. (1982). "Megatrends." New York: Warner Books.

National Council on Disability (March 4, 1993). "Study on the Financing of Assistive Technology Devices and Services for Individuals with Disabilities: A Report to the President and the Congress of the United States," Washington, DC.

National Education Commission on Time and Learning (April, 1994). "Prisoners of Time" Report of the National Education Commission on Time and Learning, Washington, DC.

Noddings, N. (1990). "Constructivism in mathematics education." In R. B. Davis, C.A. Maher and N. Noddings (Eds.), Constructivist Views on the Teaching and Learning of Mathematics: Journal for Research in Mathematics Education, Monograph No. 4. Reston, VA. NCTM.

Office of Technology Assessment, Congress of the United States (1992). Testing in American Schools: Asking the Right Questions. Washington, DC.: Government Printing Office. ED 340 770.

Ohler, Jason (March, 1991). "Why Distance Education?" In The Annals of the American Academy of Political and Social Science: Electronic Links for Learning. 22-34.

Orlansky, J., & String, J. (1979). "Cost-Effectiveness of Computer Based Instruction in Military Training," (IDA Paper P-1375). Institute for Defense Analyses, Alexandria, Virginia.

Pearson, Virginia (1990). "Strategic Planning for Distance Education Programs." Dissertation, Oklahoma

State University.

Peck, Kyle, L. and Dorricott, Denise (April, 1994). "Why Use Technology?" Educational Leadership, 11-14.

Piele, P. K. (1989). The politics of technology utilization. In D. E. Mitchell & M. E. Goetz (Eds.), "Education politics for the new century: The twentieth anniversary yearbook of the Politics of Education Association (pp. 93-106). London: Falmer Press. Princeton Survey Research Associates. (Spring, 1993) NEA Communications Survey.

Quinn, C. N., Mehan, H., Levin, J. A., and Black, S. D. (1983). "Real Education in Non-Real Time: The Use of Electronic Message Systems for Instruction." Instructional Science, 11, 322-324.

Ragsdale, Ronald G. (1983). Integrating computers into the curriculum. Paper presented at the Canadian Symposium on Instructional Technology, Winnipeg, Canada.

Ratzki, Anne, and Angela Fisher. (December-January, 1989-90). "Life in a Restructured School." Educational Leadership; 46,4 :46-51. EJ 400 500.

Reganick (June, 1994). "Using Computers to Initiate Active Learning for Students with Severe Behavior Problems." T.H.E. Journal, 72-74.

Resnick, L. B. (1987). Education and learning to think. Washington, DC: National Academy Press.

Richman, Louis S. (June 27, 1994). "The New Work Force Builds Itself." Fortune, 68-76.

Riley, Richard (May 25, 1994a). Testimony before the Senate Commerce Committee, Washington, DC.

Riley, Richard (May 25, 1994b). Address to Goals 2000 Conference, Washington Hilton, Washington, DC.

Rudner, L. M and Boston, C. (Winter, 1994). "Performance Assessment." Eric Review, Vol. 2 Issue 1.

Russell, James D, Sorge, Dennis, and Brickner, Dianna (April, 1994). "Improving Technology Implementation in Grades 5-12 With the ASSURE Model." T.H.E. Journal.

Sammons, Martha C. (February, 1994). "Motivating Faculty to Use Multimedia as a Lecture Tool." T.H.E. Journal, 88-90.

San Jose Mercury News (June 20, 1994). "Professors call students unprepared."

Schnell, Thomas (Summer, 1994). "A New Lesson Plan for Teachers." UM St. Louis.

Shapiro, B. L. (1989). What children bring to light: Giving high status to learners' views and actions in science. Science Education 73(6), 711-733.

Sharp, Sylvia (April,1994). "Editorial." T.H.E. Journal.

Sheingold, Karen. (1990). "Restructuring for Learning with Technology: The Potential for Synergy." Restructuring for Learning with Technology, pp. 9-27.

Sheingold, Karen and Hadley, M. (September, 1990). "Accomplished Teachers: Integrating Computers into Classroom Practice." New York: Center for Technology in Education, Bank Street College of Education.

Sheingold, Karen. (September 1991a). "Restructuring for Learning with Technology: The Potential for Synergy." Phi Delta Kappan 73, 1.

Sheingold, Karen (October, 1991b). "Toward an alternative teaching environment (Draft for Technology Leadership Conference, Council of Chief State School Officers, Dallas).

Smith, Adam (1975). "Power of the Mind." New York: Ballantine Books.

Smith, M., and Cohen, M. (September, 1991). "A National Curriculum in the United States?" Educational Leadership, 49 (1): 74-81.

Smith, M. S., and O'Day, J. (1990). "Systemic School Reform." In Politics of Education Association Yearbook, edited by R. S. Nickerson and P. P. Zodhintes. London: Taylor and Francis.

Stephen, William and Gallagher, Shelagh (April, 1993). "Problem-Based Learning: As Authentic as it Gets." Educational Leadership, Volume 50, Number 7 Association for Supervision and Curriculum Development.

Stearns, M. S., David. J. L., Hanson, S. G., Ringstaff, C., and Schneider, S. A. (January, 1991). "Cupertino-Fremont Model Technology Schools Project research findings: Executive summary (Teacher-centered model of technology integration: End of year 3). Menlo Park, CA: SRI International.

Thoman, Elizabeth (Spring, 1993). "Educating for Today — and Tomorrow." ASCD Curriculum/Technology Quarterly, Vol. 2, No 3 , Association for Supervision and Curriculum Development, Alexandria, VA.

Treese, Win (December ,1993). "Internet Index."

Tough, Allen. (1979). "The adult's learning projects." Toronto: Ontario Institute for Studies in Education.

Tushnet, N. C., Bodinger-de Uriarte, C., Manuel, D., van Broekhuizen, D., Millsap, M. A., Chase, A. (1993). "Star Schools Evaluation Report One," Southwest Regional Laboratory (SWRL) and Abt Associates.

University of Phoenix Institutional Research Department (1992). "Management of Academic Quality in Distance Learning Environments: Report on Education Processes and Outcomes of the Online Campus. Report IR-SR92.10.3. Phoenix, AZ.

UM St. Louis (Summer, 1994). "A New Lesson Plan for Teachers." The Magazine of the University of

Missouri, St. Louis St. Louis, MO.

von Glaserfeld, E. (1987). Learning as a Constructive Activity. In C. Janvier (Ed.). Problems of Representation in the Teaching and Learning of Mathematics, 3-18. Hillsdale, NJ: Lawrence Erlbaum Associates, Publishers.

Wakin, Edward (July/August, 1994). "Training for Tomorrow Today," Beyond Computing. P 55-56.

Wendt, Nancy (July/August, 1994). "Training for Tomorrow Today," Beyond Computing. P 55-56.

West, Tom (May, 1994). Presentation at the Alliance for Distance Education in California Conference, Long Beach, CA.

Whetzel, Deborah (1992). "The Secretary of Labor's Commission on Achieving Necessary Skills. ERIC Clearinghouse on Tests, Measurement and Evaluation, Washington, DC. ED339749.

White, Mary Alice (1990). "A Curriculum for the Information Age," Technology in Today's Schools, Association for Supervision and Curriculum Development.

Wilson, David L. (December 1, 1993). "Producing Computer-Literate Teachers." The Chronicle of Higher Education.

Wodtke, Mark von (1993). "Mind Over Media: Creative Thinking Skills for Electronic Media." New York, McGraw-Hill.

Yrchik, John (1994). "The National Information Infrastructure & Education." Address — exact date and place unknown. National Education Association, Washington, DC.

Chapter 9

Distance Education

by Dr. Carla Lane

■ DEFINITION OF TERMS

The term "distance education" refers to teaching and learning situations in which the instructor and the learner or learners are geographically separated, and therefore, rely on electronic devices and print materials for instructional delivery (Keegan, 1983; Holmberg, 1981; Sewert, 1982). Distance education includes distance teaching — the instructor's role in the process — and distance learning — the student's role in the process (Keegan 1982, 1983).

"In the images we hold of American education, none is as prominent as the self-contained classroom. The classroom is an island on which a teacher, a group of students, standardized textbooks and other limited resources determine the educational process. From time to time, the teachers and students make forays into the world outside to do research or take field trips to relevant sites. On the whole, though, the educational process is focused inward on the resources that exist within the classroom and the activities that occur there (NEA, 1991).

"The use of telecommunications technology in classrooms literally inverts the typical locus of educational activities. Classrooms face the world outside rather than the world inside. Instead of islands, classrooms have become links in communication highways transmitting data, video and voice to thousands of other sites. Teachers and students have easy access to vast databases and participate in joint activities that involve classes in other states and countries by traveling on these highways," (NEA, 1991).

A major movement in K-12 and higher education in the United States today is the use of telecommunications technologies to teach students at many sites or at a distance from the campus. The theme that runs through all of the literature about distance education is the contribution that it will make to educational reform. The current crises in funding for

education is a long-term change. This calls for a fundamental change in the dominant model and mode of operations of our educational institutions. "Fundamental adaptation is needed, as well, in our modes of instruction which, at this time, are neither pedagogically nor financially effective," (Lynton, 1992). Coupled with a responsibility to help practitioners acquire and maintain their competence is the need to change the balance between didactic instruction, self-directed learning and collaborative learning.

The electronically mediated courses which have evolved to serve the student are composed of live and taped video programming which varies in time up to 48 hours and replaces the traditional classroom lecture. The video program is augmented by textbooks, study guides, anthologies, audio tapes, computer programs, computer and audio conferencing, multimedia and other instructional material required by course content. Instructors are generally assigned to a course and may require other meetings with the students including laboratories and seminars which may be conducted in traditional ways or by audio, video or computer teleconference.

Delivery technologies for the video program include broadcast television (including public television stations), cable, satellite, fiber optic cable, computers, CD-ROM, CDI/DVI, video disc and radio. Through these technologies, institutions reach learners who are at other sites or are unable to attend campus classes due to distance time, or disability constraints, and make education accessible to them.

An accompanying development is the electronic campus, classroom and dormitory. These rooms and buildings have a wiring infrastructure connected to a central hub delivery system for audio-visual materials, access to learning resources, access to local and national on-line databases, on-line library catalogsand e-mail. Faculty, students and administrators frequently share these systems which can be accessed by computer, modem, telephone and video equipment.

Another development is the delivery of instruction entirely through computer conferencing. Accredited universities and colleges are offering undergraduate and graduate degrees, including the doctorate through computer conferencing. These classes are also augmented by textbooks, study guides, additional readings, audio conferences with the instructor, and in some cases, videoconferencing or face to face meetings. The use of computers may be bringing about a fundamental change in teaching methods at the institutions and in the homes of students.

The establishment and acceptance of the validity and effectiveness of distance education courses, together with the production of more and better course materials, will increase student demand and institutional interest in offering them. The ability of educational institutions to reach more students, wide though it already is, will be multiplied almost beyond imagination by the proliferation of relay technologies, the growth in regional and national consortia, digital fusion and the technical merger of computers to television.

An additional factor in the growth of distance education is the cost of building new campuses and maintaining existing campuses. Distance delivery allows the institution to continue to meet its mission cost effectively. Institutions have decided that there is a need to share courses, enrollment and instructors between campuses, and are able to increase the number of courses being offered without increasing the number of instructors. This allows the delivery of education to the home and the workplace. Without distance education, low enrollments at one site would force the cancellation of the course. With distance learning, that rarely happens as courses can be broadcast to a variety of campus, home and workplace sites, and students can study independently.

During the early stages of distance education planning on some campuses, there was a perception that distance education would replace instructors. This is not the reality. In fact, through distance education, instructor positions have been saved because once students perceive that the course will be taught, the demand for the course will rise. Distance education adds additional avenues to the education process as it enables the expansion of the campus outreach. Through distance education, many people are taking courses who simply would not have enrolled for a traditional course because of their work, home or distance circumstances.

Once distance learning systems are installed, they also save money on in-service training and administrative conferences by reducing travel time. Systems can be used for instructor certification and a wide variety of other continuing education courses such as real estate certification, IRS training, regional, state and national meetings, fire and police personnel training, and a variety of other community and state needs.

Unwin (1969) suggests that through these tech-

nologies we communicate in the idiom of the age and argues that if the development of an educational system is to be in line with the technologies and truths from which it draws its reason for existence, then teachers must reconcile traditional methods of instruction with new ideas by integrating new methods.

Electronically mediated instruction (EMI) is a new domain. The primary use of EMI has been to duplicate the traditional face-to-face classroom. Some institutions have made or are in the process of a paradigm shift from thinking of EMI as a replacement for the traditional classroom, to thinking about distance education and EMI as a new educational domain. The enabling tools are the technology, a mix of media, and a focus on learner centered classes rather than teacher centered classes. Through the use of self-directed learning methods, the responsibility for learning is shifted to the student and the instructor facilitates the learning by acting as a coach, resource guide, and companion in learning. Through this attitude shift, the student becomes proactive rather than reactive.

In higher education, we are experiencing dramatic shifts to notably a move toward lifelong learning as a result of the need to retrain individuals whose skills are no longer marketable. Adult students now constitute over 83 percent, or 10 million of the nation's 12 million college students (U.S. Department of Education, 1987). The stereotypical 18-22 year-old, full time, residential college student is greatly in the minority at 17 percent (2 million) of this population. In 1970, older students constituted only 28 percent. United States institutions primarily use distance education to reach the same adult audience that is returning to the campus to complete course work (Daniel, et al., 1982; Frankel & Gerald, 1982; Lewis, 1983). The adult population increase indicates a continued growth in the demand for distance higher education as it better meets the needs of adults (Mayor & Dirr, 1986). Another chapter contains case studies and a discussion of higher education utilization of distance education.

In kindergarten through twelfth grade, distance learning has been embraced by public school systems which can no longer fund the luxury of specialized teachers at all schools. The shortage of math and science teachers has fueled the utilization of distance education technologies for K-12. Other courses are offered nationally, such as German by Satellite and Calculus by Satellite which originate at Oklahoma State University. It is not unusual for K-12 distance education instructors to have 900-1200 students. By year 2004, we will have approximately 55.7 million students going to school, seven million more than in 1994. Teachers will be in very short supply with only one million available for two million teaching positions. The shortage will be most acute in science and foreign languages with rural schools suffering more. Well placed distance education technology will be needed to fill the void opened by the teacher shortage. The Star Schools program has lead the K-12 field nationally in the use of distance technologies.

Corporate and business educational programs are also using distance learning to make their programs more accessible and cost effective, and to increase the timeliness of delivering new information. Where traveling trainers used to take six months to a year to deliver training sessions to their sites, one trainer can deliver the same information to all sites (national and international) in one day. Utilization of distance education technologies by business for training is an enabling factor in maintaining global competitiveness. Institutions are confronted with the need to deliver more educational activities despite shrinking resources and the increasing cost of delivering services with traditional methods (Meierhenry, 1981). Economic considerations will continue to act as a force on post-secondary institutions to find ways to use telecommunications. In the mid '80s, the Carnegie Commission stated that if higher education does not integrate telecourses, the private sector will. Galagan (1989) and Bowsher (1989) liken this to the situation in industry where economic considerations forced companies to cut training costs and utilize distance education techniques. Designing courses with advanced technology is more cost effective than traditional courses. Distance education is an economically feasible way for post secondary institutions to confront shrinking resources and the increased cost of delivery of educational services by traditional methods.

Media and technology can provide the packaging and delivery of educational programs. Knowles (1983), Galagan (1989) and Bowsher (1989) conclude that by the end of the 1990s most education will be delivered electronically. It is likely that through the National Information Infrastructure (NII) these predictions may become reality, but at a later date.

Adult educators (Moore & Shannon, 1982) reported that their interest in video was due to programming availability, video's ability to expand the service area, reusable videotapes, and its being less expensive than traditional classes. In 1989, 66 percent of American homes had at least one VCR and industry projections forecast 90 percent by the late 1990s (Ladd, 1989).

Fiber optic cable installed in homes will vastly expand phone company services to include information, video, education, and other developments we cannot even imagine. With the FCC ruling in July 1994, approval was given to Bell Atlantic to offer interactive television service for 38,000 homes in New Jersey. Video transmission to homes is the big promise of the next two decades. At press time, the FCC was expected to act on a backlog of 21 similar requests from other regional Bell companies that would cover millions of homes nationwide. Cable companies have filed a blizzard of objections against their well-financed would-be rivals.

Flanigan (1989) predicts that fiber optic cables will be the industrial highways of the information age. Brey's (1988) study showed that broadcast television is the most important delivery system but video tape and cable will soon overtake it. New services are most likely to include access to local, state, national and international libraries and databases, pay per view on demand, full news text, community information, banking by mail, global e-mail, and video dial tone which will allow the wide use of video telephones and/or desktop videoconferencing and document sharing. All new telecommunications plans include the ability to provide voice, data and video. Education will be a part of the new communications systems, but of the 500 cable channels that may ultimately be offered, it is unclear whether there will still be only one educational access channel or if there will be many.

The 1994 Communications Act (still in Congress at press time) if passed, may also significantly impact education. Proponents have lobbied intensively to include schools in providing universal access. The 1934 Communications Act did not include schools in the provisions for universal access; 60 years later, this has had a major impact on providing telecommunications to educational organizations.

Because of the move toward distance learning, research has led to findings on the nature of learning. Experts in education agree that there is a need to reinvent education in America. The reasons are twofold: Academic performance is inadequate, and our understanding of how people learn has shifted dramatically. We now perceive learning as a transformational process where the student is changed as opposed to mimetic where information is merely transferred or distributed to the student.

To a great extent, the Goals 2000: Educate America Act passed in 1994 will add to the impetus to use new instructional methods and new educational technology tools. The Act is the most sweeping educational reform legislation that has ever been passed in America.

The thrust in this paradigm shift in learning theory is that education means two-way communication, both inside and outside of the classroom. The typical classroom today is cut off from other academic environments. It does not have access to knowledge centers such as museums, universities, libraries or other databases and it is isolated from the working world. The majority of classrooms are not equipped with a telephone, computer, television or any other means to learn about the outside world. The challenge in transformation learning is to help students make intellectual, social, emotional, and physical connections to make knowledge meaningful.

During the last two decades, theorists and researchers have developed a new understanding of the nature of learning. Here are some of their insights (Gomez, 1992):

1. Learning is transformational. New knowledge is built on previous knowledge and on intuitive, informal experiences. Students need to acquire facts, principles and theories as conceptual tools for reasoning and problem solving in meaningful contexts.

2. Learning is enhanced when students can interact and perform authentic tasks. Meanings, the building blocks of knowledge, are best learned through two-way communication supported by props which include gestures, models, sketches, white boards, computer screens, and other vehicles for expression.

3. The classroom is an opportunity to expose students to people who apply knowledge in practical, professional contexts. Learning is a process that should facilitate the eventual transition of the student into a professional community and into the community at large. Two-way contact with representatives from these communities can give students a clearer understanding of the mapping between classroom work and real-world applications of that work.

4. The central mission of teaching is to support learning, not simply to deliver information. Conversations are the means by which people construct a common ground of beliefs, meanings, and understandings. Teachers should create classroom communities in which thinking and problem solving are supported by extensive interaction with people and information.

Other research has indicated that interaction may be either synchronous (in real-time) or asynchronous and still be successful.

Audioconferencing and audio technologies are greatly underutilized for delivery of instruction. Audio conferencing is an efficient way for instructors to communicate with students in real time and at a distance. By having a set agenda, encouraging interaction, sharing of information and experiences, it provides a class situation without having the students and instructor physically in one place. It is also cost efficient to use audioconferencing instead of the more costly videoconferencing when the course content does not require visual contact or visual demonstrations.

Voice mail (also called voice messaging), has become one of the most important business productivity tools and can increase productivity by as much as 30 percent because it lets people communicate with other people in their absence. Industry studies indicate that 60 percent of all calls can be completed without a two-way conversation. Voice messaging allows information to get to the recipient quickly, accurately, and with the nuances of voice inflection intact.

Interactive voice response (IVR) technology allows people to talk with a computer via a touch tone telephone. The IVR unit answers the call, greets the caller, and guides the caller through possible responses with a series of voice prompts. The desired information is provided via prerecorded voice fragments (words) or computer-generated speech.

Computer conferencing is another grossly underutilized distance education technology. It will support a number of students asynchronously and can be quite interactive when appropriate methods are utilized in the instructional design process. It is an easy and convenient method to use for messaging, homework transmission, and grading. For adult students who must travel, laptops have been a boon which enables them to interact with classmates and communicate with their instructor from any part of the world. Instructors who have taught in the medium feel that the level of interaction is superior to traditional classroom interaction for two reasons; students are required to participate and cannot hide behind the class stars who easily assimilate information and whose verbal skills enable them to shine. Computer conferencing is a significant way to orchestrate class discussion where all are able to participate equitably. Teachers often fail to plan the use of video, and effective ways to support instructional objectives with video (Gueulette,1988), audio or computer. Since the decision to use distance education usually rests with administrators and faculty who are unlikely to have media expertise, there is a need to train them in media selection and utilization.

Bates argues that institutions should define overall objectives for integrated media at a program level including how programs affect students and how students can easily integrate programs into their mode of learning.

Improved media selection procedures can change the current situation (Sive, 1978, 1983; Niemi, 1971; Teague, 1981). An adoption process which includes an evaluation instrument for distance education based on media selection methods would ensure that adoption personnel evaluate materials using the best available media selection methods to help ensure the selection of resources that will make genuine contributions to student learning (Teague,1981). Many educators maintain that we are just beginning to learn how to use media for educational purposes (Knowles, 1983; Hewitt,1982; Lane 1989).

In its 1979 report on the future of public broadcasting, the Carnegie Commission stated, "Television and radio have great unused potential for learning, and new technologies are on the verge of greatly enhancing this potential. We believe it is time to launch new efforts to tap the power of broadcasting and the new telecommunications media for learning" (Carnegie, 1979, pp. 255-256). The report concluded that, "It is clear that with careful planning, skillful execution, and thorough evaluation, telecommunications will play an increasingly fundamental role in the learning processes of Americans of all ages" (p. 273).

Two national studies (Lane, 1989, 1990) developed media selection and evaluation models.

■ HISTORY OF TELECOURSES

American Experience

Distance education is often viewed as a recent development when in fact, correspondence courses were established in the 1870s. By 1882, the University of Chicago had established a home study division. In 1915 the National University Extension Association established a Correspondence Study Division. By 1923 over ten percent of all broadcast radio stations were owned by educational institutions which delivered educational programming. In 1926, the National Home Study Council was established. Over 55 million students have studied at home.

In 1934, Congress established the Federal Communications Commission. The Association of College and University Broadcasting Stations and other associations were organized at that time and pushed to keep frequencies open for educational uses. During the 1940s and 1950s, these same groups applied for television station licenses.

In 1947, the Truman Commission articulated a strong position on universal education; this action was followed by even stronger pronouncements by the Eisenhower Commission. In 1952 when the Federal Communications Commission (FCC) assigned frequencies to establish public broadcasting, one of the objectives was the provision for instructional television.

Efforts to produce educational materials for television broadcast are almost as old as the medium, but early efforts bear little resemblance to the soundly designed, sophisticated telecourses available to today's students. In the 1950s the first educational television programs were created for open broadcast. In 1951, the City Colleges of Chicago pioneered the first large-scale instructional television programs for credit by organizing an institution through which students could obtain a degree by taking only television courses. It has served over 200,000 students. The soldiers returning from World War II wanted to utilize their educational benefits with distance education programs, but the Veterans Administration prevented this arguing that off-campus programs would be abused. Disabled veterans were able to attend classes via telecommunications courses when their counselors approved it.

WOI-TV at Iowa State University went on the air in 1950. It was the world's first non-experimental, educationally owned television station. Following an early and fairly enthusiastic acceptance of educational television in the early 1950s, more producers entered the field and used a variety of methods to teach via television. As there were more failures than successes, disenchantment followed in the 1960s as it became apparent that television could not solve all of education's problems. Early programs tended to use the medium as an electronic blackboard for elementary and secondary teachers, and televised lectures at the college level. Educators regarded television as an extension of the classroom, not as a medium with its own enormous advantages and capacities and this is largely, still the case today. The capacities and strengths of the medium were not recognized for a long time and early efforts to teach by television were largely disappointing. Yet the telecourse evolved from these blackboard and talking-head approaches as well as from the older independent study models long familiar to higher education, and recognition of television's unique potential came with this evolution.

Many have recognized and criticized the failure of educators to use the medium to its best advantage, noting that taking pictures of a talking head or what is done in a regular classroom and televising it was not using television for the unique medium that it is. Television must involve careful design, scripting, and production that provide a high quality that could never be replicated in a regular classroom presentation.

Use of the community cable television (CCTV) facility to prepare telecourses was one of television's potentials. A CCTV system enabled an institution to tailor its telecourse to fit the local needs. Videotape and kinescope made packaging and storing educational programs possible. The University of Denver reported programming telecourses in accounting and zoology. At Iowa State University, sixteen classrooms in a new building were equipped with two receivers each to receive taped programs. The University of Akron (Ohio) used CCTV to telecast seven required courses and students had no alternative as CCTV was the only way the courses could be taken. In 1960, the University of Missouri presented 27 taped television courses; 19 were presented on the University CCTV channel and the others were split between CCTV and broadcast stations including four on St. Louis' PBS station KETC-TV. Institutions continued to perceive television as a partial solution to burgeoning enrollments and instructor shortages.

In 1963, Instructional Television Fixed Service (ITFS) was created by the FCC which mandated that the microwave spectrum channels be used for educational purposes. The first university to apply for licensing was the California State University (CSU) System.

During the 1970s and 1980s, there was a renewed acceptance of educational television based on an understanding of the medium's potential, strengths and limitations, and an increasing sophistication in the development of a system of learning elements which were integrated to reinforce mutually the learning experience. In the 1970s several new United States organizations began to produce and offer telecourses. In 1970 the Maryland Center for Public Broadcasting and the Southern California Consortium for Community College Television produced and offered telecourses regionally and nationally. The Consortium makes college credit telecourses available to its member colleges and usually has three or more new telecourses in development. In 1972, three community college districts began producing and offering telecourses; Miami-Dade Community College District in Florida, Coast Community College District in Costa Mesa, CA, and Dallas County Community College District.

Since the early 1970s numerous organizations have produced and offered telecourses. Chief among these was the now defunct University of Mid-America, a consortium that consisted of nine state universities. Telecourses produced by this group are now available through the Great Plains Network (GPN).

The rush of institutions and their students to take advantage of instructional television began suddenly toward the end of the 1970s and accelerated rapidly thereafter (Hewitt, 1982). Purdy(1980) and Grossman (1982) refer to the revolutionary nature of the swift increase and the extent to which telecourses were used in the 1980s. Likely catalysts for this increase were the refinement and sophistication of telecourses and the technological means to deliver them (Munshi, 1980). These concurrent events have had strong impact, spawning several other developments of national significance. These include: establishment of the PBS Adult Learning Service — a public programming service which is devoted to national delivery of educational programs; the Annenberg/CPB Project, a $150 million fund to encourage the development of innovative television and radio courses; establishment of the National University Consortium and the University of Mid-America; organization of large and small consortia representing hundreds of institutions which share production and licensing costs; and the emergence of several multi-campus community colleges as leaders in the production and use of telecourses.

Since the mid-1970s, immense improvement has been made in telecourses through application of sound principles of academic design and the participation of professionals in the fields of television, writing, editing, and publishing. Both the concept of the telecourse and the use of telecourses are still changing and evolving, and it would be incorrect to suggest that all the prob-

lems of this form of education have been solved. There is still room for improvement in the quality of telecourses.

When ordinary broadcast delivery or closed-circuit channel is not possible, telecourses are being relayed by cable, satellite, telephone, videotape, and videodisc to hundreds of adult learners who probably never could — or would — attend courses offered on campus. The use of television in higher education today is widespread and growing. Establishment of the Annenberg/CPB Project continues to stimulate the production of superior courseware and the growing number of consortia, task forces, and commissions will encourage and expand the use and production of telecourses.

The Public Broadcasting Service has identified adult learning as one of its primary objectives. Colleges, universities and public broadcasting stations are working together to make education available to individuals who would not have this opportunity without the intervention of telecommunications. As cable becomes more available and new technologies offer additional avenues, more opportunities will become possible.

Problems continue to beset distance education programs but despite this, growth continues. Accrediting agencies are still dealing with how to accredit courses that cross state lines as well as accrediting agency borders. High front end costs for equipment and production discourage many entrants as their funding shrivels, but educators continue to find imaginative new solutions such as wireless cable, computer conferencing, audiographics, and audioconferencing to augment the more expensive television programming. Unfortunately, educators continue to try to capture the essence of the traditional classroom in telecommunications classrooms, but are steadily moving toward a new model which is a paradigm shift brought on by the use of electronically mediated courses and a new understanding of learning and learning styles. This new learner centered model includes components which utilize the concepts of instructor facilitation, student learning styles, interaction, collaborative and self-directed learning, electronic access to resources, hands-on experiential learning, authentic learning or problem based learning which is based on reality, authentic assessment, and a mix of media. Ultimately, when a national infrastructure is in place, the mix of media will be delivered as multimedia on a telecomputer.

British Experience

A major advance in instructional television and telecourses was made by the establishment in 1969 of the British-Open University. It was designed to offer students non-traditional opportunities for education and placed particular emphasis on instruction by television. Probably no institution has had such a dramatic impact on the use of television in higher education as has the Open University of Great Britain. Perry (1977) writes that the Open University evolved from the convergence of three major educational trends: adult education, educational broadcasting, and the spread of educational egalitarianism. In the United States, the success of the Open University rekindled interest in the use of educational television (Hewitt, 1982).

The Open University enrolled its first students in 1971 and continues to enroll about 40,000 students each year, many of whom earn regular degrees. Some telecourses produced by the Open University are used by American institutions. Several dozen distance learning institutions now exist in many countries around the world.

The Open University sees satellites as an important development in the next few years to make telecourses available and to extend its work with industry and commerce in the field of professional and technological training.

The end point of what can be done when television is combined with other media for education has not been reached; rather, this is probably just the beginning of a revolution in education which will involve many forms of telecommunications.

■ DISTANCE EDUCATION PLANNING

There is little variance of opinion about the value of coordinated telecommunications planning. Hezel's (1987) study showed that most distance educators recognize economies of scale in the development and installation of services for multiple institutions. Even though the use is extensive there is a growing feeling that telecommunications is not being used to its full capacity. As a result, educators have strong inclinations to develop uniform systems that can equitably provide education to dispersed populations (Hezel, 1987; Ladd, 1989). Because of the high front-end costs of telecommunications, the cost of building new campus buildings, and the reduction of faculty and staff, there is a renewed interest in forming regional and state consortia. It is the community development model raised to an expanded geographic area so that educational communities can share their limited resources.

Brey (1991) observed that one of the most important consequences of distance learning during the 1990s will be the accelerated removal of the traditional barriers to competition among postsecondary institutions for students and institutional resources. Most states are confronted with conflicts between institutions that want to limit this new competition,

and hence prevent the growth of distance learning programs, and those institutions that want the removal of all barriers. This will be a problem at the interstate and national levels, where the power of state agencies to regulate the delivery of distance education programs into their states may not exist, and will undoubtedly lead to calls for intervention by the federal government and regional accrediting associations. There are already a number of national programs including Mind Extension University (ME/U), the National Technological Network (NTU), the University of Phoenix Online and Access Divisions, Nova University, and the Fielding Institute to name only a few. The accrediting agencies are still grappling with courses that cross state borders as well how to accredit distance learning programs. Which accrediting agency has jurisdiction over the program will also pose a problem.

The technological concept of digital fusion is driving the installation of a wide bandwidth infrastructure. Digital fusion describes the merger of telecommunication technologies through computer control and the ability of laymen to use them more easily. The components are wideband transmission services; fiber optic or coaxial cable in homes and offices to deliver audio, data, and video educational programming; computer desktop video to produce programming; and high definition television (HDTV) which is digitized video. Through merged technologies, video, audio, and data can be delivered by fiber optic cable to the computer, stored on disc, and utilized to produce educational programming. At this writing, we have moved technologically to the telecommunications dream — audio, video and data — anytime and anyplace. Many high level examples of this telecommunications ideal exist. Since it is most likely that more video will be used, it must be used judiciously and correctly. To date, most educators have not learned how to use media, and this has resulted in media not being used effectively as a learning resource (Knowles, 1983; Lane, 1989)). We do not know enough about media and how to use them in an educational context; educators are not technology literate, and worse, very often, are afraid to admit it. Historically and currently, there is little emphasis on how to plan, prepare, and utilize media in education. If the use of media and technology is to increase, educators must learn how to reach educational goals and objectives through electronically mediated instruction.

Brey's 1991 study found that community colleges and universities may double the average number of telecommunications technologies used for live instruction. The total number of technologies used by community colleges will increase 51 percent between 1991 and 1994; for universities it will increase 79 percent. Importantly, this illustrates that educators are not focusing on only one media such as video. This will allow components of a program to better address the varying learning styles of students. By using a mix of media now, educators will provide themselves with an understanding of each individual medium while the national infrastructure is being built. This will enable them to fully utilize multimedia when it is easily available.

The use of telecommunications has increased and hundreds of telecourses augmented by print materials now exist and are offered for graduate and undergraduate credit. Of the 3,000 United States colleges and universities, user institutions increased from 25 percent in 1978 to 32 percent in 1986. A total of 902 (32 percent) colleges and universities offered one or more telecourses during 1984-85; 10,594 telecourses, an average of 12 per institution, were offered to 399,212 students by 1986. Courses are produced by at least 56 institutions and video production houses and are offered in departments which range from business to computer science. Faculty in these areas seldom have media expertise.

Brey's 1991 study indicates that the number of colleges and universities with distance education programs will increase during the 1990s. Approximately two thirds of higher education institutions have distance education programs now. By 1994, 80 percent of community colleges and 78 percent of universities will have distance learning programs.

Since the mid-1970s, improvement has been made in distance education but the concept and use is still evolving; all of the problems have not been solved (Hewitt, 1982). In the face of growing trends in electronic education, institutions will expect quality distance education programs, however, the literature does not show that all are of equal quality. There has been an ongoing demand for quality since telecourses appeared. In 1952, Newsom stated that programming must be first-rate or instructional television will fail. Eash (1972) evaluated 1960s materials and notes that he became painfully aware of the shortcomings of many glossy, highly advertised materials. Evaluation is important because of the lack of quality programming (Berkman, 1976) and, unfavorable student attitudes, and thus the success of the learning experience rides on it (Berkman, 1976; Curtis, 1989). Bates (1974) contends that the wrong criteria are applied to judge the value of a program.

In 1984, the Center for Learning and Telecommunications reviewed over 900 telecourses for possible inclusion in their Telecourse Inventory (1984). Out of the 900 submissions, they were able to recommend only 139. The 1985 Annenberg study (Lewis, 1985) showed that faculty valued technology's potential but were highly critical of the quality of most video and computer software. Kressel (1986) notes that the quality and evaluation of telecourses continues to plague educators and policy makers;

material is being "cranked out" (pp. 4-6) everywhere from obscure garage-top attics to high-tech production facilities. Kressel asks if the issues of educational quality will be addressed so that distance education will thrive? She warns that without evaluation and quality control, distance education will fail; failure is preventable if good practice is ensured by dissemination of effective models, quality criteria, evaluation methods, and assistance to state planners.

More than 95 percent of the nation's public schools now have one or more computers, according to a report by the Office of Technology Assessment. School reform movements emphasize the importance of technology in instruction and computers are common in a growing number of homes. Despite this, many teacher-training programs produce graduates who are less proficient with technology than their future students. While many schools of education offer media courses, most did not require media courses for graduation.

David G. Imig, executive director of the American Association of Colleges for Teacher Education says that roughly 20 percent of the nation's teacher-training programs are on the cutting edge of technology. About 60 percent offer one or two courses which introduce students to technology or concentrate its use in a few areas, while the remaining 20 percent have not taken the first steps.

Ideally, schools of education should try to weave technology throughout a teacher's education through modeling the use of technology in courses, research and administration. If technology is not used throughout, future teachers will perceive technology as isolated and not relevant to a teaching career or a valid instructional method. To date, schools of education have had varying degrees of success in integrating technology into their curricula. Many professors are reluctant to use technology in their classes and are equally reluctant to use computers in their work. Many schools cannot afford the equipment that would let them make technology a priority. As the cost of equipment drops, this situation should change. However, the case can be made now for the cost effectiveness of delivering instruction through distance education technologies.

Stages of Acceptance or Adoption

Rogers (1962) analyzed the stages of acceptance and adoption used to accept things which are new and different. He identified five states which included awareness, interest, evaluation, trial, and adoption. Acker (1985) analyzed Rogers' work and its application to education technology. To Rogers' process he added six other conditions which included the perceived and actual relative advantage of the innovation, complexity, observability, ability to hold a trial, compatibility, and other issues which include cost, type of equipment, equality, difficulty of use, other uses, evaluations, comfort, and culture.

Conditions for the success of educational technology include a recognized existence of a need for the programming, articulation of purpose, identification of structure, leadership of the innovation, teacher participation and support, appropriate technology, and evaluation mechanism, and continuing adequate resources (Armsey and Dahl, 1981).

Human Factors

In 1983, Olgren and Parker established that dealing with human factors required as much or more time for planning than technical design and order to build user acceptance and the sustained use of the new applications. Park (1983) suggested that the innovation should be introduced by an influential person prior to use. He noted that teleconferencing users must "clear a hurdle" to get from their first teleconferencing experience to familiarity and acceptance of current teleconferencing technology. That is, you must lose the mystique and fear that they feel for teleconferencing. Those users who don't get across the hurdle tend to drop their teleconferencing commitment and become poor ambassadors to others contemplating adoption of the technology.

Bell and Weady (1984) suggest that human factors and systems should be developed simultaneously. People will not use a technology simply because it seems like a good idea, or because it will save them time and money. To accept and adopt teleconferencing and distance education, the technical structure and the human interface must be both initially and lastingly rewarding. Much remains to be done before education defines its objectives and the world of communicators, in turn, opens its mind to the problem of education (Souchon, 1986). Imaginative planning and vigorous action are necessary to maintain a viable educational system. The educational system of the future will be shaped by planners in purposive fashion, or it will by default, be shaped by accident, tradition and the senseless forces of environment (Irvine, 1983).

■ GENERAL BARRIERS TO THE USE OF EDUCATIONAL TECHNOLOGY

A number of barriers to the use of educational technology have been identified in recent years. They include: lack of information about technology (Baer ,1978), length of time for widespread use (Baer, 1978), inappropriate match between technology and service (Lucas, 1978), and an approach where a technology solution is seen as a Panacea (Benne, 1975). Pacey (1983) identified a machine mysticism where there was a misperception that technical advances lead unalterably to progress. This was based on a myth that a cultural lag occurs everywhere as we try to keep with progressive technology. A better solution is to use technology to answer new patterns of problems. Dirr (in Barron, 1987) identified barriers of lack of money, lack of faculty commitment, and a lack of trained support staff. Barron (1987) identified faculty concerns as barriers to adoption which included the class size, the lack of support for faculty from their peers, and the lack of discussion and face-to-face involvement between faculty and students. There is a perception that benefits are assumed to accrue to students from face-to-face interaction with the instructor which has not been validated in the research. There is also a perception that distance education prevents students from having hands-on experience in subjects such as chemistry when the campus reality is that more students are performing chemistry and physics experiments on computer keyboards instead of at laboratory benches.

Holt (1992) suggests that "many who wish to discredit the use of telecommunications claim that there is no student-teacher interaction, as if face-to-face contact is the only kind of interaction. Such criticism ignores the potential of available sophisticated computer hardware and software or the utility of the telephone line. It also assumes, erroneously, that significant one-on-one interaction occurs in a classroom of 25 students or more in a 50-minute period." He also validates the "ain't made here" syndrome saying that "Local control of the curriculum is highly cherished at all levels in the educational establishment. Rather than relinquish any control over either the subject matter or the teacher to outsiders, preference is often given to local staff, even if they are poor instructors or teaching outside of their field."

Holt suggests an astute selection of the on-camera teacher backed up by an invitation (rather than a demand) to teach and a reward for participation will start off the collaboration effectively. He states that because some programs treat distance education as an extension of the traditional classroom (lectures in front of the camera), there is a perception that television-taught classes take the same amount of instructor time as a traditional class. In programs that make fuller use of the medium, time should be allowed for a quality program to develop.

The need to organize the program, collect material, and script telecasts increases the workload. Holt recommends that instructors in such classes should be held responsible for no more than two class preparations per day. He also believes that the educational philosophy should be to use the medium to its fullest extent and make extensive use of preproduced footage of illustrative material, music videos, pretaped demonstrations, as well as the computer and video effects. This brings a visual richness to the program that students expect of commercial television.

Class size may demand that other faculty members and non-faculty assistants be employed. This is the case for Dr. Harry Wohlert, the ASTS German by Satellite teacher who may typically have 1,500-2000 students. For interaction an 800 number is provided for off-air hours. Holt says that two-way video with six to eight class sites will lose an interactive advantage. According to data currently available, student performance is the same whether interaction is accomplished using two-way or one-way video, but is strongly affected by other factors, such as the quality of classroom management and the commitment of the on-site teachers and administrators.

The barrier of teacher certification borders on the "absurd" according to Holt. Some K-12 program providers report that to obtain certification in certain states, the teachers are required to take a physical examination, even though they will never stand in front of a class in those states. ASTS has had it's courses accredited in 48 states, without faculty holding certification in those states. He points out that they are viewed differently because they are university based and all hold a doctorate.

Teaching partners at the distance sites can ensure success or failure; "Appropriate training and a positive attitude toward distance learning virtually ensure success; lack of training and a negative attitude almost always ensure failure" according to Holt. He recommends training in management of instruction and operation of the equipment which may vary in length from one day to a week depending on the complexity of the receive system. Training for the on-camera instructor should emphasize a tightly organized and well-paced program and working in the studio environment to develop a sense of confidence in handling the medium.

Faculty support for the distance education teacher can be surprisingly negative. To avoid this, Holt suggests that highly vulnerable non-tenured faculty should not be hired. Older tenured faculty are often receptive to new methods and their acceptance of distance learning by virtue of teaching in the program may increase acceptance by associate and assistant professors.

Holt concludes that "Convincing people of the efficacy of the medium, working with accrediting agencies, cajoling good faculty into participating as teachers, convincing site coordinators that their role is crucial — all these and more are crucial if a program is to be a success. It takes tact, determination, commitment, and good humor, but given these, no barrier is too high to overcome.

■ PSYCHOLOGICAL BARRIERS TO DISTANCE EDUCATION

A number of psychological barriers to the use of educational technology have been identified. In addition to "It's never been done that way before," other psychological barriers include suspicion and fear of change as well as telephobia which is a suspicion of change which involves television. Others fear that they will make a fool of themselves in front of their peers. "People who have watched TV for 20 years have built up all kinds of cultural expectations about people ... on the screen. They expect to see a polished performer reading a script without a hair out of place. In contrast, executives or managers on a videoconference tend to have their ties askew, don't always look at the camera ... and seem unsure of what to say" (E.C. Gottschalk, Jr., Wall Street Journal). Goldstein (1991) says that he is certain that the "move from the tutorial to the lecture that accompanied the rise of the modern university was greeted with similar outcries.

I am equally certain that the differences in learning outcomes are as overstated today as they were then." Educational television and videoconferencing have been categorized as only hype or show biz and there is still an unsubstantiated fear that television may only entertain rather than inform. "As long as that attitude exists, teleconferencing will be limited to that use ... there must be a recognition that teleconferencing is used not in a show biz environment but in a day-to-day environment, married to applications" (Jack Fox, Western Union).

Distance learning is perceived as being somehow fundamentally different from traditional instruction. What is the difference between a live lecture delivered to 600 students in a campus lecture hall and the same lecture delivered over a telecommunications system? This is an "intellectual trap" that leads us to believe that distance learning is so inherently different from what we have come to define as traditional instruction that it demands entirely different rules or it cannot possibly meet the established standards and therefore it is not worth fixing (Goldstein, 1991).

While distance education has become well established, there is still skepticism within the academic community about whether this form of education is of comparable quality to the more familiar classroom-based learning, as well as opposition from those who regard it as a threat to traditional faculty roles and classroom enrollments (Reilly and Gulliver, 1992). As long as regulators and accreditors continue to apply measures intended for classroom-based instruction to distance education, the skepticism will be reinforced. This uneasiness with distance education is heightened by the sense of a "competitive threat from the entry of an "outside" institution into a state. The "Not Invented Here" syndrome reflected in this response is one of the greatest potential barriers to the national expansion of distance education" according to Reilly and Gulliver.

Others have noted that television does not transmit a personal high touch environment, but is rather a cold, high tech medium which looses body language, chemistry, electricity, does not maintain a lengthy audience attention span, is not interactive and is known for low quality. In addition to those problems, educators have noted that it lacks central grading, testing and measurement elements.

The advantages of educational technology have been noted as being cost efficient, providing access to programming and having the ability to enrich education (Seidman, 1986; Wilson, 1987; Lewis, 1985). Changes have occurred. In 1994, the Goals 2000: Educate America Act and the School to Work Bill were passed; it appeared likely that the 1994 Communications Act would also pass. Each bill provided major new educational funding. Equally important was the embodiment of education's system reform in the first two bills, and the recognition by Congress that education needs continuing support for educational telecommunications if we are to avoid creating a nation of technologically literate haves and have-nots.

◼ STRATEGIC PLANNING FOR THE IMPLEMENTATION OF DISTANCE EDUCATION PROGRAMS

The literature lists major barriers to distance education program implementation (Pearson, 1990). Lack of successful institutional planning for the delivery of distance education programs at educational institutions represents a major barrier to implementation and success. The problem is that there was no validated process for planning for implementation of successful distance education programs. Pearson's study determined what critical factors leaders of successful distance education programs considered to be important prior to, during, and following implementation of the program at their institution.

Thirty administrators in education, distance

education specialists and program providers were invited to participate in a three round Delphi to determine the 20 critical factors that should be considered in the planning process to implement a distance education program at an educational institution. The 30 key leaders were asked with each Delphi round to refine and rank those critical factors that they listed. The final round produced 20 critical factors in rank order.

Panelists also indicated that the factors were dependent upon each other for the ultimate success of the implementation of the program. The critical factors they generated contained a planning model which included the steps of purpose, philosophy, organizational structure, people, finances, equipment and facilities. The experts indicated that successful implementation depended upon the completion and thorough investigation of each of these critical factors.

The model set a high priority on human and fiscal resources that can serve as a model for the strategic planning of administrators of new programs in long distance instruction. Planning for the implementation of the program requires a major investment in time, people and funding. Serious consideration should be given to the number one critical factor: "identification of the need for the program." All the experts agree that without this identified need, an institution should not move ahead to purchase equipment, hire people, or even think about delivering a long distance program. Faculty involvement, incentives, motivation and training were ranked as serious issues for these successful institutions. According to these experts, the educator is a high priority in the delivery of long distance coursework. While the fear of teachers being replaced by the technology appears to be an overriding concern and barrier for many institutions, the importance of the teachers remains critically high in the electronic classroom.

◼ DISTANCE EDUCATION PROGRAMS

Telecommunications technologies support a variety of educational activities. Students have used teleconferencing to speak with scientists, interview government officials and talk rock stars and link with one another. Electronic field trips have taken them to China, Germany, France, Russia, and the ocean floor. We are witnessing the birth of new learning communities, which are defined by telecommunications links and not by the space in which they exist. They permit a level of geographic and cultural diversity in education that could only have been imagined previously. In these new communities, the roles of teachers and students are undergoing dramatic redefinition. Teachers have moved from centerstage to concentrate more and more on overall coordination, planning, the organization of learning opportunities, and the mentoring of individuals and small groups of students. At the same time, students have considerably more freedom to define their activities, choose the direction of their research, and collaborate with their peers in their own and other classrooms. The use of telecommunications technology redefines teacher and student roles and plays a major role in efforts to restructure American schools.

Courses delivered at a distance over a number of telecommunication platforms serve as important resources for the curriculum of many school districts. Growth in distance education has occurred in both K-12 and higher education. In the K-12 setting, there has been a special emphasis on courses in foreign languages, the sciences, and mathematics, but courses in the humanities, social sciences, business, and vocational education are also broadcast.

Networks range in size from relatively small district-wide networks to TI-IN, a national satellite-based educational network, which broadcasts whole courses, student enrichment programs, and professional development programs into 1,000 school districts in 29 states.

Distance education has been of particular benefit to rural communities that have a low demand for particular courses, difficulty in meeting state-mandated curriculum requirements, and problems in attracting teachers in specific areas of the curriculum. It has also been a solution for both insufficient enrollment and the lack of specialized teachers.

In addition to restructuring, telecommunications technology makes vast improvements in available educational resources. The inclusion of on-line databases such as Dialog, CompuServe, Prodigy, America Online and e-World in media centers and classrooms puts enormous research capacity at the fingertips of teachers and student. Comprehensive databases on scientific research, the social sciences, finance and current affairs can be accessed almost instantaneously. SCOLA, a satellite-based news service brings foreign TV new programming in 20 languages from 30 countries into schools.

The number of college and university students enrolled in distance education is in the high six figures nationally. Courses might be offered through interactive satellite courses with two-way audio hook-ups, compressed video with two-way audio and two-way video, interactive audiographics courses which employ two-way audio hook-ups, prerecorded video telecourses, cable, closed circuit, ITFS, computer and audioconferencing. For example 2,000 institutions of higher education use the Annenberg/CPB telecourses.

Perhaps one of the biggest reasons for the growth of distance education in higher education is the changing nature of work and society. Many students are working adults who have schedules that prevent taking classes at regular times, or they live so far away from any college or university that they would have to move to attend school. Table 9.1.1 and 9.1.2

Table 9.1.1

Distance Education Implementation
Critical Factors in Rank Order (Pearson, 1990)

1. Identified need (perceived or real) for the program.
2. Faculty and teachers supportive and given incentives for motivation.
3. Funds for capital costs; production, equipment, facilities.
4. Availability of on-going money for operations and expenses.
5. Quality of the educational content of the program (evaluation).
6. Adequate support staff to produce the program.
7. Ensuring equivalent learning experience to remote students.
8. Enthusiasm and belief by the institution in the overall distance education project.
9. Identification of a visible, spirited key leader/administrator initiating program.
10. Adequate receive sites, facilities, and staff.
11. Availability of appropriate and specialized equipment to deliver the programming.
12. Sufficient time for careful needs analysis; Identify the range of services and programmatic needs of students.
    Example: Number of people, type of courses, ages served, location.
13. Ensuring equivalent status for remote students: i.e., credit, degree.
14. Instructional design and TV production: the interactive   components, length, frequency and number.
15. Identification of a marketing plan for the network, system or program. Public relations with the public.
16. Cost effectiveness: feasibility and justification for delivery system to students and institution.
17. Identified or gathered support/partners for the program: industry, corporate, legislative, institutional.
18. Ensure continued credibility of the program with the public, faculty, students, and supporters.
19. Knowledge of educational administrators, teachers and staff at educational institutions on what
    distance education is and how to teach and use it effectively.
20. Ability to accredit courses, offer credit or transfer credit across states or institutions.

Table 9.1.2

The Strategic Plan — Steps for Change (Pearson, 1990)

I. Decide to Plan for Change: Awareness
    1. Key Administrators
    2. Super Leader
    3. Understand Elements of Change
    a) Flexible Environment
    b) Policy
    c) Philosophy
    d) Leadership

II. Recognize a Real Need vs. Perceived Need: Interest
    1. Identify the Recipient
    2. Why Have the Program? Who wants and who needs the program?
    3. The Competition: Who Else Is Doing It?
    4. Is the Program Really Needed?

III. Understand the Real Reason for Implementation: Advantage
    1. Value to the Organization
    2. Political Issues Involved
    3. Technology or Need Driven
    4. Competition Driven for Competition's Sake
    5. Philosophy of the Program
    6. Culture of the Organization Affects the programs: Political issues involved

IV. Mission of the Organization: Evaluation
    1. Does the Programming Fit the Organization's ... Goals, Objectives, Quality Standards
    2. How Will This Help the Organization? If it won't, don't!
    3. What is the Driving Force to Market the Program?
    4. Will it Make Money?
    5. Will It Be Self Sufficient?
    6. How Large Do We Want It to Become?
    7. What Is the Return on the Investment?

V. Plan the Program: Trial
    1. Time - Take the Time to Plan
    2. People - Faculty/Staff
    3. Space, Facilities, Equipment
    4. Production Capability
    5. Money - Now & Later

VI. Review What the Organization Does Now: Observability
    1. Will Distance Learning Duplicate Services? Classes, Staff, Departments
    2. Is the Organization Working Well In Training & Education
    3. Does the Organization Support Education & Training, Change, Technology
    4. Do We Have Enough People and Support to Add Change?
    5. What Are the Organization's Strengths and Weaknesses

VII. The Gap: Compatibility
    1. How Far to Go to Have a Successful Program
    2. Will the Organization Be Able to Change
    3. Subtract the Difference Between ...
    Where We Want to Be
    <u>Where We Are Now</u>
    = The Gap
    4. Can We Do It?

VIII. Contingency: Pre-adoption
    1. Trial & Pilot
    2. Flexibility
    3. Client Needs
    4. Institutional Perceptions
    5. Success vs. Failure. What happens if ... it won't, doesn't, can't, or if it is better or different?

IX. Implementation: Adoption
    Commit to the Ongoing Process
    1. Lead People
    2. Design Programming
    3. Train in ... Production Techniques and Technology
    4. Faculty Support
    5. Dollar Support
    6. Continued Resources — Finance the Program
    7. Plan for Change, Growth and On-going Growth
    8. Believe in the Program
    9. Garnish Support Again and Again
    10. Evaluate the Program

Conclusion
Follow: All Steps of Change and All Conditions of Success

■ DISTANCE LEARNING PROGRAMMING AND RESOURCES

Listed here are a number of distance learning programs for K-12, higher education and continuing education for multiple technologies including video, audio, and computer conferencing. Where many providers began by offering programming for only one market segment, many now offer a range of programs which include courses for students, teachers, parents, industry/business, and the community. Many users subscribe to several program providers to meet their needs.

America Online

AOL's strong point has always been its intuitive user interface which run from menus and icons, letting you easily open and flip between multiple windows and more. The AOL offerings have increased dramatically particularly in the area of on-line magazines and newspapers and the Internet. The San Jose Mercury News, Chicago Tribune, Time Magazine and the New York Times are just some of the on-line resources. Users get both full-text editorial on-line and interactive access to the news organization's editors and writers as well as information on how to use the resources for classes. The San Jose Mercury News is especially valuable, since it offers the full text of each day's edition, plus the ability to search back issues of the Mercury and 16 other national newspapers as far back as 1978 in some cases. The Mercury News is probably the most active newspaper forum as it conducts nightly chat sessions where guests, editors, writers, and on-line coordinators hold forth on such subjects as multimedia, education and issues in the news. During these chat sessions, you can participate live by typing your comments. You can also establish a private chat session with anyone you meet on-line. It's not uncommon to participate in a group session and carry out two or three private chats at the same time. AOL offers a number of education areas, ranging from on-line courses offered by the Electronic University Network, to resources for teachers and parents.

Annenberg/Corporation for Public Broadcasting (CPB) Project

The Annenberg/CPB initiative, "New Pathways to a Degree: Using Technologies to Open the College," funded seven model academic programs that are demonstrating how colleges can use technologies to offer richer and more accessible degree programs (Ehrmann et al., 1992).

Apple Classrooms of Tomorrow

Apple Classrooms of Tomorrow is a research project that explores learning when children and teachers have immediate access to interactive technologies. ACOT's extensive research portfolio includes R&D and longitudinal studies that examine the impact of technology on teaching and learning and create more powerful applications. ACOT R&D collaborates with university and research lab scientists in the development of technology tools that strengthen and accelerate learning. Currently there are over 20 such projects, including integrated media, simulations, intelligent applications, research tools, and other innovations. ACOT longitudinal research, coordinated by UCLA and Ohio State University, examines traditional and non-traditional outcomes of participants in the innovative program. The research team initially applied conventional measures to assess student achievement, and is currently developing more appropriate measures to assess problem-solving skills, process writing, and deep understanding.

Arts and Sciences Teleconferencing Service (ASTS), Oklahoma State University

When the OSU College of Arts and Sciences voted in 1983 to raise the college graduation requirements, they included a foreign language requirement. Superintendents in the three-hundred-plus high schools who had no foreign language teacher felt that such a requirement would place their graduates at a distinct disadvantage for entry into OSU. Responding to this concern and a need for course offerings in a broad range of subjects, OSU created ASTS. In 1984, it began delivering via satellite enrichment programming to ten public schools in western Oklahoma. In 1985, German I was included in the offerings and the network was expanded to 50 schools. In 1992, over 1000 schools in the U.S. received some form of programming from ASTS (Holt, 1992).

AT&T Learning Network

In the AT&T Learning Network, groups of seven to ten classrooms constitute learning circles, which form the basic units of collaborative research and interstate and international information sharing. A teacher from Lake Charles, Louisiana, sent the following message about the learning Network to the NEA Special Committee on Telecommunications via electronic mail: "In our efforts to move toward total school restructuring at Fairview Elementary, telecommunications particularly the AT&T Learning

Network, has become an integral part of the curriculum. It has served as the foundation for the development of interdisciplinary units. Electronic messages from Coober Pedy, South Australia, have unveiled explorations into aborigine caveart, homes built underground, and studios of the Southern Hemisphere. The interest of a class in Hilton, New York, in MardiGras has spurred in-depth research of our own Louisiana heritage"(NEA, 1991).

Big Sky Telegraph

Big Sky Telegraph was designed especially for Montanans. It went on-line January 1, 1988, funded inpart by US West through Western Montana College, to create an information exchange network for educators, students, business people, communities, and organizations. It features educational on-line resources such as a lending library of software, electronic newsletters, educational databases, technical and educational support, a children's literature library, and public domain software. It also offers people the chance to create their own on-line courses.

BITNET

BITNET is an electronic communication network linking institutional and departmental computers at 550 participating Corporation for Research and Educational Networking (CREN) members and affiliates in the U.S., including universities, colleges, and collaborating research centers. With its cooperating networks in other countries, BITNET is part of a single logical network connecting almost 3,500 mini- and mainframe computers in about 1,400 organizations spanning 47 countries, for the electronic exchange of non-commercial information among its participants in support of research and education. Gateways allow the exchange of electronic mail between this network and CSNET (a CREN network originally serving computer science research departments and research institutions), the Internet (NSFNET and its associated regional, state, and campus networks), and many other networks worldwide. BITNET users share information via: electronic mail to individuals and shared-interest groups; transfer of documents, programs, and data; access to BITNET server machines and associated data services; and brief, nearly-interactive messages. Nearly 3,000 discussion groups active on BITNET cover most topics of academic interest and may have from five participants to several thousand.

Black College Satellite Network (BCSN) (See USEN listing)

Blacksburg Electronic Village   The Blacksburg (VA)

Electronic Village is envisioned to link together virtually all individuals and groups in the community; elementary and secondary students and teachers; people in businesses and professional services; Virginia Tech students, staff and faculty; civic groups and those they seek to serve; and individual citizens. The network would provide access to Internet, electronic mail, worldwide electronic discussion groups, information databases, and a wide range of educational, financial, business, and general communications services. The Blacksburg network would serve as a national model for future "electronic villages" expected to emerge around the nation in the decade ahead. Many Virginia Tech students, staff and faculty already own personal computers. The university has installed its own voice, data, and video network throughout the campus and residence halls. Computers are needed to efficiently participate in the electronic village network.

Buddy System Indiana

The Buddy System project provides in-home computers, modems, and access lines to more than 2,000 fourth, fifth, and sixth grade students at 19 Indiana Schools. The project's goals are to improve the educational process, boost student motivation, and stimulate the state's economy through a better educated work force. It has achieved success in its three years of operation. A 1990 evaluation showed 90 percent of educators agreed student work was more creative and of higher quality because of the computers, and 40 percent of the project parents increased contact with teachers about their child's education.

Bunker Hill Community College, Boston, MA

Bunker Hill Community College runs an analog fiber based network which links the college to three inner city schools. Plans call for links to additional schools, a senior citizen center, and a correctional facility.

Cable in the Classroom

Cable in the Classroom is the cable television industry's $86 million public service initiative to enrich education. Each month, schools can use over 525 hours of commercial-free, high quality educational programming. Teachers can use as little or as much of the cable programs as they need to motivate their students and enrich their lesson plans. The programs may be used when teachers want, not just because of the technology, but also because they have "extended" copyright clearances. Cable in the Classroom supports teachers with study guides, other educational materials and workshops.

The goals of Cable in the Classroom are to provide curriculum-related support materials and help expand and improve teaching resources and to contribute free installation and free basic cable service to all schools which are passed by Cable in the Classroom members. As of February, 1994, Cable in the Classroom served over 34.5 students in 63,327 schools across the country. Over 25 cable networks donate programming as part of their commitment that learning will not be interfered with by advertising.

Network members provide high quality, award-winning programs that cover all disciplines and issues. Among them are programs about environment, weather, interaction between countries, current events, presidential campaigns and elections, money management, health issues, plus cultural and dramatic productions. Members work directly with school administrators and classroom teachers to assess the particular needs of their area. The national Cable in the Classroom organization is a clearinghouse for information on educational programs and coordinating teacher workshops and other activities to foster cable's use in the classroom.

In a national teacher survey of 816 teachers commissioned by Cable in the Classroom, 65 percent of those surveyed reported using at least one of the Cable in the Classroom program services. "Assignment Discovery " garnered 36 percent, "CNN Newsroom" claimed 24 percent, "C-SPAN I/II" had 18 percent, and "A&E Classroom" was chosen by 17 percent of the respondents.

Cable programs such as The Discovery Channel's "Assignment Discovery" bring broadcasts on topics ranging from science and technology to social studies and contemporary issues into classrooms and media centers. The Cable News Network's Newsroom is a 15-minute, commercial-free cable-delivered new program offered free of charge to middle an secondary schools each day.

Classmate Instruction Program

Dialog Information Services' Classmate Instruction Program allows students to do sophisticated research on-line. Beginning with special teacher training workshops to help educators log onto electronic databases, to special flat rates for schools, the Classmate program provides schools easy access to references from over 100 sources including popular magazines, research journals, and specialized databases. Students can retrieve information in such subjects as agriculture, the sciences, news, environment, medicine, psychology, and education. Teachers can get curriculum materials and guides for directing student activities on-line.

Community Learning Network — CLN

CLN of Indiana University and Purdue University at Indianapolis (IUPUI) uses technology to extend regular university courses and student services into an urban community to attract historically underrepresented groups (Ehrmann, et al., 1992). It has adapted a model of student interaction developed by Professor Uri Treisman at the University of California at Berkeley. Treisman and colleagues at other universities have demonstrated that learning can be dramatically enhanced when students collaborate and learn from each other, especially when the students' learning is reinforced by others of their community. The Treisman strategy also depends on giving students material that is more difficult, open-ended, and more interesting. Student mentors keep the student focused on learning objectives. The courses incorporate four forms of interaction with faculty and others students:

1. Didactic: televised lectures, recorded lectures available in cassette form on site, supplemental self-paced materials, a library of print and electronic materials available at learning centers and occasional lecture/discussion by faculty or peer mentors.

2. Real time: telephone, interaction among students and with the peer mentor, office hours on campus, faculty visits to the learning centers, and collaboration among students in person or through one of the media.

3. Time-delayed: electronic mail and computer conferencing, voice mail, Fax mail, bulletin boards (with thumb-tacks), and video playback.

4. Informal: through the involvement of community-based individuals, especially community-center personnel, family members, and peer mentors, a variety of informal interactions will be used to increase the sense that learning is a vehicle for personal satisfaction, growth, and empowerment. Guest speakers, field trips, experiments, and special assignments involve the students in their own learning.

Community Telecommunications Network - CTN Wayne State University Detroit, MI

Wayne State University has had a partnership with twenty local educational institutions for a number of years called South Eastern Michigan Television Education Consortium (SEMTEC). In return they helped to arrange for channel space with their local cable operators. To deal with the cost, video signals generated at the university were broadcast on two of the university's Instructional Television Fixed Service (ITFS) channels. A new wireless system went on-line in 1990 and the new CTN consortium was created. The 18 channel system has all the benefits of the earlier system so that the signal can be picked up and rebroadcast on cable systems. The new systems provides a high-quality cable signal within a 50-mile radius of Detroit. Schools, factories, government buildings, churches and other facilities with the right equipment can receive all the channels. CTN was created primarily because each of the seven institutions involved in it had submitted competing requests for ITFS channels that were offered in Detroit in 1985. When the FCC devised a point system to select the institutions to receive the channels, the competitors filed a joint proposal with the FCC which allowed the CTN members individually to acquire two or three channels each, but at the same time specified joint transmission of all channels because of interference. The CTN members decided to operate the television transmission facilities jointly. (Rahimi, 1992).

CompuServe

The CompuServe on-line service has been around since the 1970s and offers thousands of information sources. Interface software for Mac and Windows is available. It offers night-time access to Knowledge Index, a feature that provides full text of numerous newspapers, magazines, newsletters, and professional journals. Many software and hardware companies maintain technical support forums where you can receive advice, exchange messages with company staff and other users, and get the latest software updates. There's also Support On-Site, a new extra-cost service that contains answers to questions about thousands of software and hardware products.

CSUNET, CSU California Technology Project

CSUNet, the California State University Network, run by the California Technology Project, offers conferences and access to teachers and universities. There are collaborative projects such as the Kids2Kids writing project, which is run out a classroom in Costa Mesa.

Distance Learning Resource Network (DLRN)

DLRN provides a national information clearing-house for educators and agencies to support planning and implementation of distance leanring programming. By calling the 800 number educators can talk to a DLRN staff member or be forwarded to someone who has the desired information. A database can be accessed through the Internet for a wide array of information and plans are being made to make it accessible through other commercial services. Resources include information about distance learning programs, Star Schools programs, planning and evaluation resources, funding, staff development, research, policy reports, legislation, computer conferencing, instructional techniques, model programs, classroom design, Goals 2000 and other information.

DLRN focuses on how reform and restructure can be facilitated through the applications-based use of technology and telecommunications. DLRN can also provide consulting services for up to one-hour with no fee to discuss setting up distance leanring programs strategies of adoption and other topics. DLRN can also provide training services to potential users of distance learning and the resources to provide extensive background and in-depth understanding of how to operate a successful distance learning program.

DLRN provides a 60-page resource guide and an Implementation Kit which includes video produced by the Pacific Mountain Network as part of the Star Schools Farview grant and other resources for educators, decision makers, and policy makers. DLRN is funded through a Star Schools dissemination grant from the Office Educational Research and Improvement (OERI), U.S. Department of Education.

World, Apple Computer

World is an on-line service provider for both PC and Mac platforms. Windows software should be available in early 1995. e-World features an easy-to-use interface, which is based on the same software that drives America Online. Using a town-center metaphor, e-World helps users develop a spatial understanding of a large and complex service. When you start the service, you see a community of buildings, a business and finance plaza, computer center, learning center, community center, newsstand, shops, information booth, and various other locations that can be visited electronically. Eventually it will offer true Internet access, enhanced e-mail services (outgoing fax and paper mail), and will offer international access.

Education Network of Maine, University of Maine at Augusta

The University of Maine at Augusta operates the Community College of Maine, which is in fact a virtual electronic college. In 1992, the name was changed to The Education Network of Maine. It has responsibility for off-campus associate degree programs and administrative responsibility for several of the off-campus centers. It will also lay the groundwork for articulation agreements for UMA's associate degrees in business administration, general studies, liberal arts, and social services with appropriate baccalaureate programs among the system's seven campuses. The distance learning network has a backbone that uses fiber optic (45 Mb) transport to link the state colleges in Maine and also to feed instructional television fixed service (ITFS) transmitters that broadcast programming directly into other learning sites. The fiber network is two-way, video and audio. The ITFS network uses broadcast video with return audio.

New England Telephone (NET) has provided special contract pricing for the fiber-based network and has cooperated on providing conference bridging for the audio return on the broadcast side as well as in making other college resources available to distance learners, including the University's on-line card catalog. Computer conferencing is being added to the mix of technologies.

Education Satellite Network, ESN Education SATLINK

ESN was established by the Missouri School Boards Association in 1987 to ensure that all Missouri school districts, regardless of size or location, have equal access to instructional enrichment opportunities through high quality, satellite-delivered educational television programs. ESN received a grant from NTIA in 1987 which assisted schools with the installation of satellite receiving equipment. In 1988 and 1989, a Star Schools grant allowed significant expansion of ESN's programming and equipment. ESN serves as both a producer of educational programs and as a clearinghouse of program information.

"Education SATLINK," a monthly satellite program guide, lists more than 40 networks and independent program providers each month (September through June). It is also available as SATLINK OnLine — a searchable database of programming. Access is free but you must pay for the long distance charges. Call 1-800-221-6722 to receive a password. The Online modem number is 314-446-8139. ESN develops programming to help education professionals meet specifically targeted needs in local schools. Through production of original teleconferences and programs for students, teachers, administrators and school board members, ESN provides schools with cost-effective opportunities to enhance the way students are taught. ESN also provides support services to schools in Missouri.

EDUCOM K-12 Networking Project

Aims to link practitioners in primary and secondary education through computer-mediated communications networks, and, with this connectivity, to develop networked resources to support curriculum reform and institutional restructuring; develops directories of K-12 people and resources, including primers and guides to training resources; seeks opportunities for business and industry collaboration; helps to build a leadership organization for K-12 educators; conducts outreach to key practitioners and policy makers.

Eisenhower National Clearinghouse for Science and Mathematics Education

Provides free information about mathematics and science curriculum materials through electronic and print media; allows users to access a variety of databases.

Electronic Schoolhouse, America Online

The Electronic Schoolhouse links teachers and students for collaborations. An area has been established to assist teachers in setting up school to school and class to class connections using America Online. Using the Electronic Schoolhouse teachers can create learning experiences that go beyond the limits of the classroom's four walls. They can have an ongoing weather information exchange with a class in Florida, exchange students' writing projects with school newspapers in California, run a chess tournament with schools in Nebraska, Maine, and North Dakota, and display computerized art work for students in Connecticut. Up to 23 classes can meet at an agreed upon time for an electronic connection to enhance the standard classroom curriculum as well as to offer special events for students and staff.

FrEdMail - Free Educational Electronic Mail Network

Initiated in 1986, FrEdMail is the oldest and largest education network in America, linking, through the Internet, more than 150 electronic bulletin boards (called electronic mail centers) operated by individuals and institutions. Each bulletin board represents a "node" on the system and delivers Internet e-mail to as many as 300 teachers and students. In 1991, FrEdMail helped approximately 5,000 teachers and their students participate in a wide variety of learning experiences designed to motivate students to become better learners and writers. Recent projects include "Acid Rain," for which students from around the country collected rain samples, plotted national data, and shared research, conclusions, and essays on the causes and effects of acid rain; and "Experts Speak," which involved one group of students assuming the personalities of various historical figures and another group interviewing them to determine their identities. FrEdMail is also intended to promote the sharing of resources and experiences among teachers.

GALAXY Classroom

The GALAXY Classroom is a $24 million interactive satellite communications learning network that combines the best ideas about teaching and learning with the best in modern telecommunications technology. Using a satellite telecommunications network that aims to reach more than 10 million students in 20,000 schools by the end of the decade, GALAXY Classroom supplements elementary school curriculum by integrating commercial-free programs and curricula materials in the classroom. It offers a new way to better prepare children early when they are developing their attitudes toward learning. Programming and curricula help children at an early age build the cognitive, creative and literacy skills to become enthusiastic, lifetime learners and productive workers. The interactive curriculum engages students and teachers in challenging instruction that builds on every child's strengths.

GALAXY begins by training teachers and principals to integrate educational technology with the best curriculum first. Then by using satellite, fax, video, and computer networks, they combine high-quality, commercial-free programming with hands-on activities and materials including science kits, teachers' guides, student magazines and literature books. The network offers continuing professional development, classroom assessment and technical advice. Thirty-nine demonstration schools in 21 states and the District of Columbia, and one school in Mexico receive GALAXY Classroom programming.

There are three curricula series in language arts and science. Materials are in English and Spanish, and telecasts are closed-captioned. Each series is built upon a child-centered approach that encourages experimentation, collaboration and open-ended tasks, and recognizes that each child has something of value to contribute. GALAXY uses those strengths to create authentic situations that motivate students to read and write on their own and to participate in real-life problem-solving situations. The goal is not to recite irrelevant information, but to enable a child to construct meaning by instilling new ways of observing and thinking about the word.

GALAXY is an initiative of Hughes Aircraft operated by the non-profit GALAXY Institute for Education.

Global Schoolhouse

In the spring of 1993, the Global Schoolhouse (GSH) project put the most current Internet tools in the hands of classroom teachers and their students. During GSH activities, classrooms interact with each other over the Internet using multi-dimensional environments and videoconferences conducted with Cornell's CU-SeeMe software for Macintosh and Windows. Schools are grouped in clusters of four or five according to the topic of study. The Clearinghouse for Networked Information, Discovery, and Retrieval (CNIDR) provides teachers, students, and principals Internet tools training via desktop videoconferencing. The activities cluster around four themes: alternative energy sources, solid waste management, space exploration, and weather and natural disasters. Sponsors who support this classic example of government, industry, and education cooperation are: Arlington Cable Partners, California State University, CERFnet, Farallon, Internet World, InterNIC, Microsoft, McGraw Hill, Media General, Metropolitan Fiber Systems, Midnet, National Science Foundation, Network Solutions, O'Reilly and Associates, Palomar Software, Scripps, Howard Cable, Sim J. Harris Company, Sprint, SuperMac, SURAnet, QualComm, and Zenith.

Golden Gate University, San Francisco, CA

Golden Gate University in San Francisco is offering a telecommunications management course over the Sprint network of public rooms. The innovative use of the Sprint rooms was initiated by Patrick Portway. Six sites were included in the pilot project and a seventh site was used by different sites to include industry experts as guest lecturers.

Great Lakes Collaborative

The Great Lakes Collaborative uses multiple telecommunication technologies to allow teachers to match a wealth of worldwide nationally-approved science and math curriculum resources to the needs of students in grades pre-K through 9. It provides an instructional support system for teachers making it easy for them to gain access to a full spectrum of worldwide innovative classroom resources. Educators gain access to an interactive network to exchange ideas and to maximize effective uses of the resources. "Explorer" software for the Macintosh allows educators and students to easily navigate their way through the maze of broad-based resources to find the right material based on curricula, learning objectives and grade level. The materials are available on-line immediately (or the software gives instructions or how to obtain them) and are interactive. Teachers and students can input their experience with the resource, add their own materials or create new materials using the database. Users will also be able to access Internet. Curricula listed are learner driven noting different learning styles, levels and readiness addressed through multi-media activities and is matched to the national mathematics and science curricula standards. GLC provides staff development and technical assistance to teachers involved in system reform of math and science instruction, integrating curricula with technology and support to implement interactive multi-media activities., Michigan, New York and Pennsylvania participate through their state boards of education and there are nine partners from state and national telecommunication and education organizations. GLC produces "Around the Lakes," a newsletter for professional development in telecommunications. It has models of how telecommunications is used in K-12 schools. Each issue focuses on a thematic topic for telecommunications or multimedia instruction, and contains a skill module and related activities for teachers. On-line resources, related to a given issue's theme, are included. ECNet (Education Connection Network) teacher sites are featured in each issue to encourage information exchange between schools. 16 issues per year. The 16 issues of the newsletter each year can be received on-line.

GTE Cerritos Project

Six Cerritos, CA elementary schoolteachers used video-on-demand and switched, full-motion videoconferencing in their classrooms as part of their everyday activities. The project was a test (now disbanded) of the kinds of voice, data and video educational services an advanced telecommunications network (2,480 mile fiber optics with a prototype broadband switch) would make possible and it helped GTE identify how telecommunications might provide immediate and long-range benefits to education. With video-on-demand, teachers could bring the power of video to the classroom when they could most effectively use it. It fostered teachers' and students' creativity because they used on demand videoconferencing to share ideas and projects with other classrooms and schools on the system. A teacher who was an expert in a certain subject could broadcast lessons to the other classrooms. Teachers accessed services "on demand" by using a remote-control device to highlight menus on a standard television set and made their selections from a video library with 12 video-on-demand titles designated by the education system. Titles were pre-arranged, based on the teacher's curriculum planning and requests and loaded into a central library. A short notice videosystem had 200 additional titles that could be viewed within two minutes of the request. The

remote controls fast forward, reverse, and pause. Classrooms were equipped with microphones and a camera for spontaneous videoconferencing. The same technology was used to provide video-on-demand and video phone services to selected Cerritos residents.

Higher Education Channel - HEC, St. Louis, MO

Students at the University of Missouri-St. Louis and St. Louis Community College can view telecourses on HEC which covers the metropolitan area of St. Louis. HEC is funded by a tax on video cassette rentals.

Illinois Central College - ICC

Two campuses of Illinois Central College are linked to each other through fiber-optic cable. A telecommunication distance learning system links the Peoria downtown campus. The system transmits classes, seminars and teleconferences. The linkup for ICC's two campuses allows classes to be taught via a two-way communication system. In other phases of the project, the college will be linked with area high schools and local businesses. Thirteen area high schools have expressed an interest in establishing a hookup with ICC's telecommunication system.

Interactive Distance Education Alliance Network - IDEANET

The Interactive Distance Education Alliance Network (IDEANET), an alliance of four experienced distance learning providers, will provide five channels of daily educational programming beginning in September 1994. Four channels give students at all levels access to a range of live, interactive class selections.

Arts enrichment, foreign languages and the national Young Astronauts program all are available for elementary and middle school students. Middle schools can also select from science, technology and language courses. High school students can study one of four foreign languages (German, Japanese, Russian, or Spanish), take advanced placement courses in physics, calculus and English, learn career and job skills, study environmental and marine science, or prepare for the PSAT or GED exams.

Many of the high school classes feature a toll-free homework hotline so students get help from their teaching assistant from 8 a.m. to 10 p.m. (Pacific). One channel provides staff-in-service training.

IDEANET has a data communications channel with lessons, homework, and tests for many of the courses transmitted, graded and returned electronically. The basic membership fee includes access to Internet, SATLINK On-Line, CNN Newsroom, and Postlink, the daily on-line newspaper published by the St. Louis Post-Dispatch.

IDEANET was formed by the Missouri Schools Boards Association, Northern Arizona University, Oklahoma State University's ASTS system, and Educational Service District 101's Satellite Telecommunication Educational Programming (STEP) with the Pacific Northwest Star Schools Partnership.

Interchange, Ziff Davis

The Interchange service will accommodate Windows and Macintosh users. It features a point-and-click interface that's rich and uncluttered, providing the user with many options without overwhelming. Interchange will concentrate on computer information. The Washington Post has announced it will set up on-line editions via Interchange. Ziff will have an upcoming sports service. Subscribers to any Interchange service will also have access to Interchange Center, which offers e-mail as well as general news, sports, weather, and securities quotes. Interchange will search through all of its data banks to find every article, bulletin board item, shareware file, and discussion group that meets your criteria. You'll also be able to download files in the background while you browse other parts of the service. Interchange lets you create links between all materials because every item on the service is a distinct object that can be represented by an icon. Linking one object to another is as simple as dragging an icon. For example, if you want to send an article to a friend, drag the article's icon into the e-mail message (without downloading the data). The receiver clicks on the icon embedded in your message and goes directly to the linked material. The service is being tested now.

Internet

Although not an education-focused project, the Internet deserves special mention because it is a means for accomplishing most of the current network educational activities. The Internet is a global network of networks. Users connected at local sites are able to "talk" to colleagues worldwide. The Internet uses the high-speed capability of the National Science Foundation's NSFnet, which serves as a backbone to the Internet, to link the various national, regional, and local networks. BITNET, for example, is a national network of colleges, universities, and research sites. Through the Internet, BITNET users can interact with colleagues all over the country and the world. Many of the network educational services

use the Internet. Increasingly, access to the Internet is a key part of plugging into the world of networking. Millions of people currently use the Internet, and the number of nonprofit institutions and commercial services offering Internet connections is growing every day. This means that educators seeking to get involved with networking should aim, whenever possible, for access to the Internet.

Illinois Rivers Project, Southern Illinois University - Edwardsville

This telecommunications-based educational program is used by 87 Illinois schools. Students collect scientific data on various Illinois rivers and then communicate via modem with each other and with a central data base at Southern Illinois University - Edwardsville, the project coordination site.

Indiana Public Schools - IPS

Indiana Bell and the Indianapolis Public Schools (IPS) designed a fiber optic interactive distance learning network. The closed-circuit television system, designed by Indiana Bell, uses more than 600 miles of fiber optics to provide interactive distance learning, videoconferencing, and television programming to more than 90 IPS locations. The system, linked by and end-to-end fiber optic network, enables two-way video communications between IPS sites as well as multipoint video communications. A research project explored the questions of how distance learning can enhance education. A telecommunications class is the vehicle for the project, designed to research the educational and technological possibilities of interactive distance learning. Broadcast-quality, two-way video technology links the instructor with students in Bloomington and Indianapolis.

Iowa Star Schools Project

In 1992, a partnership of Iowa educators formed the "Iowa Distance Education Alliance: Partnerships for Interactive Learning Through Telecommunications in Iowa's Elementary and Secondary Schools." Iowa carries out the instruction over the Iowa Communications Network (ICN) a statewide two-way full motion interactive fiber optic telecommunications network. The goals of the project include coordinating distance education over the ICN; promoting awareness and understanding of the ICN; preparing and supporting teachers as they use distance education; electronically connecting educators to state, national and international telecommunications networks; improving and increasing the opportunities for instruction in mathematics, science, foreign languages, literacy skills, and vocational education; and establishing a program of research and evaluation to document the impact and effectiveness of distance education over the ICN.

Jason Project

The most spectacular of the field trips has been the Jason Project in which Jason, a small robot submarine, traveled over the floor of the Mediterranean transmitting images of the ocean and sunken artifacts to more than a quarter of a million students in the United States and Canada via satellite. Dr. Robert Ballard of the Woods Hole Oceanographic Institute and the expedition crew provided live commentary on the ocean environment being explored and answered students' questions through the satellite link. A special curriculum was developed for the project which covered oceanography, physical sciences, biology, history, geography, and telecommunications technology. When the Project moved to the Galapagos Islands, broadcasts of undersea and land life in these extraordinary islands were transmitted live to American students in schools and museums through satellite and cable networks. Selected teachers had the use of an interactive fax service, which allowed them to receive the project's curriculum materials on demand (NEA, 1991).

K12Net

K12Net is one of many networks for elementary and secondary school students and their teachers. K12Net establishes "echo" forums around major curriculum areas. Educators and students interested in a particular topic can communicate and work cooperatively with other interested individuals throughout the world. For example, Global Village News is a K12Net news service involving students from around the globe. K12Net is open to anyone who has access to a local computer bulletin board. Access to K12Net is through FidoNet, a free general-interest computer network that joins more than 15,000 computer bulletin boards in more than 50 countries.

Kansas Regents Educational Communications Center

The ECC offers full-year high school Spanish language courses via satellite. Course instruction focuses on the four basic skills of speaking, listening, reading, and writing as well as cultural awareness. Learning is enhanced through a variety of instructional formats, including textbooks, audiocassettes, videotapes, lesson plans and computer software. Life broadcasts are on Tuesdays and Thursdays for 45-minutes each. Phone interaction between the television instructor and students occurs regularly during broadcasts. The Spanish via satellite staff offers daily assistance to teachers partners and students via toll-free telephone.

Schools designate a teaching partner who is a certified teacher, but does not need to know Spanish. The teaching partner is highly valued as a key element in the success of the program. In addition to supervising students and assigning grades, the partner is encouraged to attend a training workshop or watch a training videotape; they receive year long support though regular mailing sand telephone contact with the instructor and staff.

Due to the unique nature of the course and the demand upon the teaching partner, ECC limits enrollment to 15 students per class, saying that field research "indicates that class sizes of 15 or less create an environment between the teaching partner and students that nurtures a 'learning team' approach; larger classes tend to create a dichotomized, adversarial environment."

KIDLINKLISTSERV @vm1.nodak.edu (or LISTSERV@NDSUVM1 on BITNET)

KIDLINK is a grassroots organization, which in two years has had 6,200 children from 45 countries participate in a global dialog. The work is organized in 12-month projects with names like KIDS-91, KIDS-92, and KIDS-93. KIDLINK's purpose is the dialog itself. There are no political objectives. All children in all countries between the ages of 10-15 are invited. Participation is free, but the children have to reply to the following four questions before being allowed to join the dialog with the other children: 1) Who am I? 2) What do I want to be when I grow up? 3) How do I want the world to be better when I grow up? 4) What can I do now to make this happen?

KIDSNET

Accessible through the Internet, KIDSNET is an international discussion group for teachers and others interested in networking for children and education. Participants discuss general questions about computer networking, user interfaces, and specific projects that link teachers and students using the Internet. KIDS, an associated list just for children, was set up after children's messages to each other began appearing on KIDSNET.

Kids Network, National, Geographic Society Geography Education Program

NGS Kids Network is an innovative computer and telecommunications-based science curriculum for grades 4 through 6 in which student-scientists investigate new ideas and exchange information with students around the world. The Geography Education Program, or GEP, is the heart of the National Geographic Society's effort to restore geography to the nation's classrooms. Working in tandem with the National Geographic Society Education Foundation, the GEP mobilizes a wide range of Society resources to improve geography instruction. It focuses these resources in five strategic areas: grass-roots organization, teacher education, materials development, public awareness, and outreach to educational decision-makers.

The NGS Kids Network area on America Online is designed to provide information about NGS Kids Network to those who are not already participants. It also provides a place for NGS Kids Network participants to continue to communicate long after their scheduled time together has ended. "WKID Communications" is a place where Kids Network participants can send messages to each other and organize new joint projects. "NGS Kids Network Laboratories" allows participants to exchange project work, can be used by NGS Kids Network participants and others for conferences or informal chats. Content areas include solar energy, acid rain, weather, and environmental concerns.

In the GEP Online area, geography teachers can read articles from Geography Education UPDATE, the nation's leading newsletter for geography teachers; exchange ideas with other teachers on a geography teaching message board; network via the National Geographic Alliances; plan for the coming year with a regularly updated GEP Calendar; exchange lesson plans in a Geography Teaching Lesson Plan Library; and hold conferences or have informal real-time chats with each other in the Explorers Hall conference room.

Lake County Science, Math, and Technology Project, Lake County, IL

Students at more than 30 schools in Lake County, IL, learn how to use and share computer information by participating in simulated space missions. The project involves the entire school community.

Lansing-Jackson, Michigan

A pilot distance learning network linked seven schools and colleges in the Lansing-Jackson, Michigan area, enabling teachers and students in different locations to interact by audio and video as if they were in the same classroom. About 20 activities were conducted on the network each week, including specialized courses that otherwise could not be justified for smaller groups of students at individual schools.

Laredo Junior College, Larado, TX; Incarnate Word College, San Antonio, TX

A solution to the nursing shortage in Laredo, TX was created through a joint effort between San Antonio's Incarnate Word College and Laredo Junior College, thanks partly to a new multimedia education network. Students in Laredo link with students and teachers on the Incarnate Word College campus in San Antonio through an interactive two-way video classroom that will help prepare nurses to receive their bachelor's or master's degrees. A major benefit of the link is the ability to attend classes in Laredo while earning a bachelor's or master's degree in nursing. That capability is crucial to retaining graduating nurses who will remain in Laredo to serve the health care needs of the community with their advanced skills.

New Mexico State University, Las Cruces and Universidad Autonoma de Chihuahua, Mexico

The Universidad Autonoma de Chihuahua in Mexico and New Mexico State University (NMSU) installed uplinks at both sites to link their students, faculty, and researchers. The Mexican government donated a full-time transponder on the Morelos satellite. Ultimately, the network will connect seven branch campuses throughout the state of Chihuahua. Video delivery of courses will begin when construction on the television studio is complete. Applications include data, voice, and fax. Lehigh University Students at Lehigh University have a television lounge called the World View Room. It brings them live broadcasts in 30 languages and they can immerse themselves in a foreign language and culture. They can hear a language as it is spoken in the country and pick up clues about the culture that they wouldn't ordinarily find in a textbook. The programming is provided by SCOLA. Students can view the programming Monday through Friday on a large rear projection television. The lounge also has monitors for shortwave radio broadcasts and racks with a variety of foreign language periodicals.

Massachusetts Corporation for Educational Telecommunications (MCET)

The Learning Community is a project of MCET. In response to Goals 2000, it uses telecommunications technologies to improve science, mathematics, and literacy education for children and adults. It is an integrated education program which connect learning in schools with learning in out-of-school settings for all members of the community; students, teachers, administrators, parents, families, and adult learners.

Three communities are participating in The

Learning Community as demonstration sites; Boston, MA; Hartford, CT; and New York City. Among the organizations involved in the school-community partnerships are public and parochial schools, centers for adult and community education, human service agencies, cable public access stations, local businesses, and technical assistance providers. It provides technical assistance to these teams and their communities through a range of activities, including workshops, summer institutes, electronic mail, meetings via picture telephone, interactive teleconferences, and site visits.

School-community teams work with MCET staff to select innovative programming from the extensive menu of new and existing programs and to develop community-wide education pans that incorporate technologies and provide expanded educational opportunities for all learners.

Learning Community instructional programming is being delivered through the Mass LearnPike, MCET's educational satellite network, as well as through a variety of other innovative technologies. Community sites in Boston, Hartford, and New York City are linked via PictureTel systems for two-way video teleconferencing. To expand the outreach, MCET collaborates with other networks including those in Maine, New York, Missouri, SERC, TEAMS and BCSN. Twenty states are served through satellite, telephone, and computer technology for students of all ages in science, technology, math, language arts, social studies, and staff development.

Mid-Minnesota Telecommunications Consortium - MidTeC

The seven technical colleges in MidTeC, a leased 400-mile system, pool instructional and administrative resources via interactive television on fiber optic links provided by US WEST and analog video technology to interconnect the schools. St. Cloud Technical College delivers 80-100 classes per week. The system also connects with K-12 consortia. The system allows MidTeC members to offer low-enrollment and "off-sequence" courses that would be impractical for any single consortium member. Students attending one school may receive credits in their chosen field from other MidTeC schools. The result? More educational opportunities for students, and moreefficient use of instructor resources. The system is also used for administrative meetings and a variety of community meetings.

Mind Extension University - ME/U

Mind Extension University (ME/U): The Education Network was launched in 1987 by Jones International Ltd. on of the country's largest

telecommunications companies. It delivers fully accredited college and graduate level courses and degree programs via cable television to approximately 26 million households. ME/U also cablecasts live, interactive elementary school classes for credit, many of which are language courses, five days a week to schools across the country.

The Jones Education Networks include Mind Extension University, Mind Extension Institute, a subsidiary that develops video interactive training programs; and Jones Computer Network (JCN).

Mississippi 2000

Broadband fiber optic networks are used in the Mississippi 2000 project. This public/private partnership involves the state of Mississippi, South Central Bell, Northern Telecom, ADC Telecommunications, IBM, and Apple Computer. A public fiber optic network links secondary schools in Clarksdale, Philadelphia, Corinth, and West Point with the Mississippi State University in Starkville, the University for Women in Columbus, and the Mississippi Educational Television Network studio in Jackson.

Montana

Montana's compressed video network links four sites: University of Montana, Montana State University, Eastern Montana College, and the state capital in Helena. A Missoula-Billings connection will be used to continue the successful MBA program which has been delivered to Billings via microwave for the past four years. Support has come from the Montana Department of Administration which allows the use of the state telephone network for the transmission. By utilizing the state network's video dial tone capability to spread out the compressed digital signal, transmission costs are expected to remain affordable for educational users. A multi-site control unit at Helena will permit program origination from any of the four sites.

Montana Educational Telecommunications Network - METNET

METNET involves a number of activities and projects targeted primarily at K-12 education schools and agencies. METNET supports equipment purchases for the compressed video network and for K-12 satellite programming. It has a statewide electronic bulletin board system with 15 sites (higher education and K-12) serving as regional bulletin board nodes.

National Archives

The National Archives Internet Gopher server provides key information relating to the National Archives, including descriptions of facilities nationwide; information on agency holdings; publications and general information leaflets; and some Federal records regulations. The Gopher menu structure is designed to reflect the breadth of the National Archives organization. The Gopher is to serve government agency personnel, educators, librarians, historians, genealogists, researchers, job-seekers, and the public. Future plans call for a bulletin board service for dial-in access to the Gopher menu. The Gopher is an excellent introduction to the holding of the National Archives for first-time users. Researchers can access information on some of the most widely-used collections, including the Nixon Presidential Materials, Ansel Adams photographs from the Still Picture Branch, captured German sound recordings, electronic records, and an index of selected census records. Research questions can be directed by e-mail to inquire@nara.gov. Text-based information can be accessed with a Gopher client by connecting to the address gopher.nara.gov. This information can also be found on the WorldWideWeb (WWW), using a client such as Mosaic, at http://www.nara.gov.

National Distance Learning Center On-Line Database (NDLC)

NDLC operates as a center within the University of Kentucky's Owensboro Community College. It is funded by a series of grants and contracts with the U.S. federal government as a public information service dedicated to reducing the barriers between the users and providers of educational programming and materials. The Center's core activity is the operation of an electronic information clearinghouse. The database is available to the public via computer modem and Internet. There is no charge for searching the database electronically.

Primary and secondary school users may find listings for an entire semester/year courses, audio/visual materials on more than 185 different subjects to supplement in-class lessons, curriculum guides and development aids, teacher and school staff in-service training materials. Adult education facilitators may find listings for literacy courses available via audio/visual media, remedial skills courses and curriculum guides, and career training courses and curriculum guides.

Higher education faculty and staff (undergraduate and graduate) may find listings for introductory level courses in all basic college subjects delivered via audio/visual and satellite media, professional develop-ment teleconferences of topical interest for faculty, staff and administrators, advanced undergraduate and graduate-level courses, degree programs, and seminars in a variety of disciplines (e.g., engineering, all physical science, humanities and social sciences), and continuing education programming for a variety of professions as well as general interest which may be delivered through existing college and university continuing education channels.

National Education Association - NEA

One of the most compelling advantages of telecommunications technology for teachers is the ability to link teachers with one another and with research institutions. The "School Renewal Network" links the NEA's Learning Laboratories and Mastery in Learning Project schools with each other and with educational researchers. The multiconference bulletin board allows participants to exchange information and discuss issues related to school restructuring, technology, and a variety of other topics. Teachers are benefiting from programming provided by educational networks. For example, "Teacher TV", a weekly 30-minute broadcast co-produced by NEA and The Discovery Channel, contains information on school restructuring efforts around the country and ideas for teachers to use in their own classrooms. TI-IN carries over 200 hours of professional development programming per year. Teachers have the opportunity to see broadcasts of well-known figures in education discussing contemporary issues. If the broadcasts occur at an inconvenient time, they can be videotaped and viewed later. Some networks provide specialized training to teachers who receive their broadcasts. MCET is doing a statewide professional development workshop for teachers in Massachusetts on portfolio assessment.

New American Schools, The New American Schools Development Corporation - NASDC

NASDC has selected the proposals from eleven design teams for Phase One Funding. A few of the features cited by NASDC in the winning proposals include new teacher/mentor relationships, moving "classrooms" into the community and holding classes in technology centers and libraries, eliminating the grouping of children by age, integration of health and social services into school sites, and the design of curricula to develop good citizens. The New American Schools Project was designed to "break the mold" of traditional schooling.

National Public Telecommunications Network - NPTN

NPTN is a network of free public access community computer systems similar to National Public Radio or the Public Broadcasting Service on television. Many kinds of services are available on each: the Cleveland Free-Net alone has over 350 distinct information or communications services. Each system, however, is free to the user. There is no cost to register, no cost to use them.

National Technological University - NTU

NTU is one of the recognized leaders in satellite-based distance learning. It was formed in 1984 as a private, non-profit institution, based in Fort Collins, CO. The NTU consortium consists of over 40 U.S. universities with strong engineering programs. NTU's charter is to serve the advanced educational needs of the nation's engineers, scientists, and technical managers. NTU offers ten master's degree programs in engineering plus a variety of non-credit short courses and seminars for engineers and other technical professionals.

The NTU network of students is a large (over 4,000 for credit students and 85,000 continuing education students). NTU was a pioneer in the use of half-transponder satellite television using two Ku-band satellite transponders. It converted to a digital, compressed video network to allow the live transmission of many more simultaneous programs per transponder. The system allows it to broadcast on one satellite at a savings of $1 million per year in transponder fees. A network of almost 500 different faculty has taught NTU academic courses over the years. During the 1990-91 school year, 260 faculty delivered 370 different graduate or undergraduate bridging courses over the satellite network to 4,155 enrollees. Except for the Management of Technology curriculum, all of these courses are regular offerings on the campuses of NTU's member schools and are uplinked from those campuses in real time — 16,894 broadcast hours of credit courses during 1990-91. The network of NTU customers stretches around the world, but is predominantly a North American network, with the vast majority of its 385 sites in 130 organizations located within the footprint of its satellite. Today a new vision is emerging of NTU as a transnational university, a significant force promoting American technological competitiveness in the global economic system through strategic partnering with the transnational corporations which are NTU's principal clients. The goal of this partnering is to achieve "just in time" delivery of information and knowledge by NTU to its industrial partners. Knowing "what" information and knowledge are needed, "when" they

are needed and getting them there on time in the most convenient and accessible form is the objective of NTU strategic partnering (Martin, 1991).

New Wave Series, Applied Business teleCommunications

Applied Business teleCommunications offered the first corporate sponsored for-credit course over the ME/U network during 1991. Called New Wave, the course included six three-hour programs spread over a nine-month period of time. Programs were broadcast on Saturday mornings from studios throughout the U.S. including Kodak, Westcott Communications, Oklahoma State University. The intent was to present content that would help participants create a lifestyle to fit the decade of the 90s. The text for the course was "Megatrends 2000." Guest lecturers included Marjorie Blanchard, Ph.D. (author of "The One Minute Manager Gets Fit"), Stephen R. Covey, Ph.D. (author of "The Seven Habits of Highly Effective People"), Anthony P. Carnevale, Ph.D. (author of "Workplace Basics" and "America and the New Economy"), Sonja and William Connor, M.D. (authors of "The New American Diet") and others. Credit for the course was granted through Oxnard College to the many corporate sites that enrolled their employees as well as the ME/U cable students.

NREN - National Research and Education Network

This new network will build on to existing networks, but will find the greatest backbone from the Internet, the largest worldwide network in existence. NREN has stirred the imaginations of people in government, industry and education to form a network that can eventually be as commonplace as using the telephone system. Congress passed the bill in November, 1991 to set up the High Performance Computing (HPC) program which will involve many government agencies, including the National Science Foundation(NSF), National Aeronautics and Space Administration (NASA), Department of Energy (DOE) and Defense Advanced Research Projects Agency (DARPA). Part of the HPC program's goals is to build NREN. The network is seen as something that reaches beyond the U.S. to serve a different world than we are used where confining knowledge within certain boundaries will not be done. To realize the full benefit of the information age, high-speed networks must be built that tie together millions of computers, providing capabilities that we cannot even imagine. NREN's speed will enable extensive use as it will eventually be able to handle speeds of up to at least 1 gigabyte per second. The Internet is currently spread over 35 countries, with more than 2 million users logging on each day. Internet traffic is increasing at a rate of more than 1,000 percent a year, indicating the need for a high-speed network, but it lacks the speed to share large volumes of information and key research data. NREN will narrow the gap.

NYCENET, NYC Board of Education

NYCENET, the New York City Education Network, run by the Board of Education, sponsors "Electronic Partners" projects for New York teachers and children to connect with classrooms in other cities, states, and countries for curricular projects. NYCENET offers conferences on many topics, databases, curriculum guides, an encyclopedia and even access to lesson plans for CNN Newsroom programs.

Ohio University

A three-year distance learning research and development experiment is being conducted by Ohio Bell and Ameritech in partnership with GTE and Ohio University's College of Education. The program links the University and three elementary schools in Ohio's Appalachian region. In addition to improving instruction at the third and fourth grade level, the project will enable the University to train future teachers in the use of this technology. Similar networks are being built in Findlay, Ohio, and Columbiana County. The Findlay network will connect 15 schools; the Columbiana County network will connect five high schools.

Pacific Bell's Knowledge Network

Pacific Bell's Knowledge Network includes two transport systems to make the best use of new telecommunications technologies called EdLink and University Hub. They use the public switched telephone network which is already in place, with transport hardware located at the central office. Knowledge Network will use new telecommunications technology and ordinary telephone lines to help connect all California schools. Universities and school districts can use digital Centrex and Pacific Bell Local Area Network (PBLAN) to link their phone systems for voice and data transfers. Advanced Digital Network (ADM) gives them the flexibility to reconfigure the shape of their network with data transport speeds of up to 64 kilobits. Two-way, full-motion interactive video is carried on the copper wire already in place (a T-1 line at 1.544 Megabit transport), making video available to even small and remote schools. Voicemail and electronic messaging can help communications between administrators,

teachers, students and parents and allow electronic transfer of documents.

PBS Ready to Earn, Going the Distance

Ready to Earn is aimed at enhancing Americans' career opportunities and strengthening workforce competitiveness. It is a broad programmatic banner that PBS flies over a whole range of adult education services.

Going the Distance, the first of these services allows students to earn an associate of arts degree from their local college through distance learning telecourses. The degree program requires minimal time on campus, providing maximum flexibility for the working adult. Going the Distance and future services launched under the umbrella of Ready to Earn, are targeted to the National Education Goal that "every adult American be literate and possess the knowledge and skills necessary to compete in a global economy and exercise the rights and responsibilities of citizenship" by the year 2000.

The pilot phase of Going the Distance launched in the summer of 1994 with 22 public television stations partnering with 60 colleges to offer the degree program locally. Telecourses are licensed through PBS' Adult Learning Service (ALS), the world's largest distributor of college-level telecourses and the nation's largest provider of video conferences for higher education. Student enrollment in ALS telecourses has grown more than 600 percent, from about 55,000 in 1981 to almost 350,000 in 1993. More than 2.8 million students have enrolled in ALS telecourses to earn college credit.

Going the Distance is supported in part by the Annenberg/CPB Higher Education Project, which funded many of the telecourses. The Project is committed to expanding access to higher education through telecommunications. Telecourses are produced by a variety of educational organizations, public television stations, and independent producers, including Dallas Telecourses, INTELECOM, Coast Telecourses, and WGBH. Future Ready to Earn services could include: additional higher education opportunities, video models and on-line databases of new industry-based skills standards; career and labor market information for working adults as well as school-age children; and multimedia demonstrations of occupations.

PBS Learning Link

PBS Learning Link is a computer-based information and communications network which provides access to educational resources as well as connections to teachers and students across the globe via Internet mail exchange. Serving K-12, it is the first in a family of on-line service in development under PBS ONLINE. Constantly evolving local and national content regularly features topic-specific discussions centered around curriculum issues; lesson plans and on-line classroom projects related to timely events and issues; and a searchable database of public television programs and services.

Content is broken down into curriculum-specific forums covering topics such as the Rainforest, space exploration, current events, the environment and the arts.

PBS MATHLINE

MATHLINE is the first discipline-based educational service offered by PBS. Based on the mathematics standards set by the National Council of Teachers of mathematics (NCTM) and endorsed by the National Education Goals Panel, MATHLINE offers a core national service for mathematics education to which local stations may add services uniquely suited to the schools in their areas. For students, MATHLINE offers standards-based, high-quality instructional programming that will help them become math literate adults who can live and work successfully in a society and workplace increasingly shaped by math, science, and technology. For teachers, MATHLINE offers professional development opportunities via distance learning courses, video conferences, and electronic learning forums. There is an electronic math teacher resource center with e-mail, bulletin boards, discussion forums, and databases of resources for teaching mathematics.

For parents, there will be programs and products to help them understand that all children can and must learn school mathematics, as well as programs illustrating how they can help their children develop their mathematical abilities. For education leaders, public policy makers, and the general public, there are programs to inform them as to why a math literate population is essential to the social and economic well-being of all Americans, and how they can support students and teachers in pursuit of outstanding mathematics achievement for all students.

MATHLINE launched its services in 1994-5 with a year-long professional development program for teachers of mathematics at the 5-8 middle grades. Teachers view videos of classroom teachers modeling standards-based practices and using mathematical tasks that engage students in important and meaningful mathematics. Content specifications for 25 video were developed by a working group of prominent math teachers and teacher educators. Each video is accompanied by printed resource materials.

Prodigy

Prodigy is the on-line service for families and it's one of the first services to offer multimedia. Prodigy is best known for its consumer services. There are on-line travel guides, kids' forums (encompassing everything from games to humor to contests), and other family-oriented services. It provides access to the Atlanta Constitution and, soon, to the Los Angeles Times and other Times-Mirror publications. Prodigy also features a number of nationally known writers and media personalities, who hold sessions online. Prodigy has added Windows support and ways of displaying full-color news photos. You can get sound bites from stories and hear experts talk. Prodigy downloads the sound files in the background while you browse other parts of the service. Once the file is downloaded, it plays via the sound board, and can be saved as a WAV file.

Project Homeroom, Chicago, IL

Project Homeroom, a two-year trial program provides 500 Chicago-area elementary and high school students with computers, printers, modems, and network services for home use to extend the learning experience beyond normal school hours and to help students improve problem-solving and information processing skills. Project Homeroom technology enables students at seven schools to communicate with each other or with data bases to complete homework assignments. Some schools have integrated Project Homeroom into existing curricula, while others have created new interdisciplinary classes that span several periods and are taught by teams of teachers.

Project Enable, Indianapolis, IN

Project Enable provides computers and establishes a data communication network for teachers and students at Lew Wallace Elementary School #107 in Indianapolis. Teachers use the computers to manage their day-to-day administrative duties, and students can check out the portable personal computers through a lending library program. The electronic bulletin board system exposes the students to electronic mail, topic discussions, interactive games, simulations and other applications. Students and parents at the school use the Education Hotline voice mail system to communicate with teachers after hours. Each teacher has his or her own voice mail phone number to distribute to students and parents. The teacher can record messages listing homework assignments and any other pertinent information. Students and parents can call the number to hear the recording and then leave a message.

Reaching Out Around the World

In "Reaching Out Around the World," students at Brown Point School in Tacoma, WA, collaborate in research and writing projects via satellite with students at Collingwood Intermediate School in Invercargill, New Zealand. Among other things, the students have produced a "Declaration for the 21st Century," which was presented to the United Nations Conference on Children in 1990. The students catalogued a number of critical problems, including environmental pollution, AIDS, the depletion of the ozone layer, and drift net fishing. In their declaration, they urged world leaders to address these issues (NEA, 1991).

RETAC

RETAC is a full service instructional-television agency, delivers, via broadcast and videotape, more than 1,300 instructional programs to approximately one million students in eight California counties.

■ RURAL DEVELOPMENT TELECOMMUNICATIONS NETWORK - RDTN

The Rural Development Telecommunications Network opened in December 1992 with six network connections and will have a total of 19 sites in place by December 1993. The compressed video fiber-based network initially will interconnect the University of South Dakota, South Dakota State University, Northern State University, South Dakota School of Mines and Technology, Southeast Area Vocational Technical Institute in Sioux Falls, and the state capitol in Pierre. During the construction phase, educators have been trained in instructional design, interaction techniques and given hands-on sessions in the video classroom in place as SDSU's Brookings campus.

Rutgers University

Rutgers University has organized a research laboratory to study new technologies for academic libraries. The goal of the laboratory is to preserve what is essential while bringing the benefits of technology to scholars and students. The laboratory is a membership organization open to academic libraries and corporations with interests in library technology.

Satellite Education Resources Consortium, Inc. - SERC

SERC is a partnership of educators and public broadcasters that provides quality courses to students and staff development for teachers and administrators. State and local department of education and state and local public broadcasting agencies combine curriculum expertise, technical and production capabilities to deliver live, interactive courses, via satellite. SERC is a non-profit corporation partially funded by the U.S. Department of Education, private agencies and foundations.

SERC has members in 23 states. It offers high school credit courses in math, science, foreign languages, and social studies as well as middle school programs. Staff development offerings and graduate studies programs provide the avenue for continuing education and enrichment for teachers and administrators. SERC also provides interdisciplinary seminars and special programming that address important issues facing educators in the 90s. The primary focus is on economically and geographically disadvantaged schools as well as those schools with high course interest but few anticipated student enrollments. Through SERC, schools in rural communities, remote locations and major cities may now offer courses that would otherwise not be available.

SCOLA

SCOLA is a non-profit organization that re-transmits the news and other programs from 40 countries. It is an ideal tool to help students better understand the language and culture of a country. SCOLA offers four different channels each with a special focus on language and cultural education.

Channel 1 carries the national news from around the world. Channel 2 offers a wide variety of programs including documentaries, entertainment programs for children, films, novellas, soaps, game shows and talk shows. Channel 3 offers courses in math, science, and humanities from around the world in the language of the country of origin. It offers students the opportunity to participate in other educational systems, as well as the opportunity to raise the level of world-wide understanding. Channel 4 offers classes in less commonly taught languages which are becoming more important to global communications. Languages include Arabic, Chinese, Hindi, Indonesian, Swahili, Malaysian, and Korean. Native American languages include Cherokee and Lakota.

Shamu TV, Sea World, Inc.

Shamu TV is a free, live, television program that takes students into the field with scientists, backstage with animal trainers, on rounds with veterinarians, and face-to-face with sharks, killer whales, and other wonders of the sea. Shamu TV is designed to help fill the need for quality science programming by showing real-life role models and sharing accurate scientific data for the kindergarten to high school levels. The program can help instill in students an appreciation for science and a respect for all living things and environments. Programs are available closed captioned.

South Carolina ETV, Rural Star Schools Project

South Carolina ETV has produced a "how-to" program for teaching via distance education. The series with accompanying text will help teachers and schools understand the value of distance learning and guide those interested in building programs of their own. The series includes videos on the studio classroom, technology and interactivity, student participation, lesson planning and facilitator activities.

Star Schools Program, Office of Educational Research and Improvement, United States Department of Education

In December 1987, the Star Schools Program Assistance Act was authorized for five years (1988-92). The act provided federal funds that allowed grantees to acquire, build, design, program, and operate educational programs, requiring only that services be provided to schools eligible for Chapter One aid. The project is administered by the Office of Educational Research and Improvement (OERI), of the U.S. Department of Education. Grants are for two years, and it is assumed that educational services will continue after the grant is completed. The grants were made to eligible telecommunications partnerships and only multi-state applicants won grants in earlier years. A single state grant was added in 1992. The project has recognized that team efforts are required to provide the best services for the students and teachers (Withrow, 1992). The program was originally designed to reach rural areas but has since been expanded to provide for otherwise unserved populations in K-12. There have been several rounds of funding which accomplished a great deal for distance learning and the 500,000 students and 50,000 teachers that it has served to date. Projects offer instructional modules that can be integrated into classroom curricula, video field trips, and enrichment activities. They also offer formally structured semester-long and year-long courses. To deliver services, projects use many distance education technologies — satellite delivery systems, open broadcasts, cable, fiber optics, computer conferencing, digital compression, interactive videodiscs, FAX machines, and the ordinary telephone. Several include activities using the Internet. Services include instructional programming, staff development, and classes for parents. There have also been Star Schools Dissemination Grants awarded to establish the Distance Learning Resource Network (DLRN) which is a clearinghouse for all types of education about distance learning including research, programming, telecommunications plans, training materials, and consultation.

Currently, the Star Schools Program operates in 48 states, the District of Columbia, and Puerto Rico, with at least 20 origination sites and 5,000 participating schools. The principal technological problem faced by the grantees has been the inadequacies of the domestic telecommunications infrastructure in the locations most in need of distance learning. Some of the rural telephone systems have functioned very poorly. In addition, it is difficult for many inner-city schools to have a telephone installed in classrooms (Withrow, 1992). The StarSchools projects are a major step in creating hypermedia interactive-learning resources. In the new vision of education, as articulated by Secretary Richard Riley in Goals 2000, StarSchools can play an essential role in demonstrating how, as the telecommunications infrastructure of America expands, it can deliver high-quality interactive multimedia programming on demand of the learner to the home, school, and classroom.

In the 1994-96 round of funding, major emphasis was put on pre-service and in-service for teachers which, for the first time, recognized an educational system composed of K-12 and schools of education.

Based on the success of the first years of the Star Schools projects, they have been successful and students in the distance education classrooms are achieving as well as or better than students in traditional classrooms (SWRL, 1993).

State Networks

Networks are often planned and implemented on a state level. A recent survey found that approximately 60 percent of the states now operate a statewide computer or telecommunications network. Some states have established their own networks to serve schools and other nonprofit organizations within their specific boundaries. The Texas Education Network (TENET) and Virginia's Public Education Network (PEN) are examples of statewide systems that connect K-12 administrators, teachers, and students for no cost or at a nominal fee. The state networks generally provide some degree of statewide services such as bulletin boards, conferencing, curriculum resources sharing, and administrative data transfer as well as a gateway to the Internet and other networks such as AppleLink, CompuServe, MCIMail, AT&T Mail, FrEdMail, and FidoNet. Check with your state education agencies for information on the network capabilities.

The Classroom Channel - TCC

TCC is a non-commercial educational and

enrichment service providing up to two hours of programming to secondary schools across the nation four days a week. Programs cover every curriculum area with emphasis on math, science, language arts, social studies and teen issues as well as a thematic core of programming that addresses such specific teen needs as AIDS awareness, violence intervention, media literacy, careers and college prep. Ku band equipped schools receive TCC the night before using it in the classroom. Programs can be used at the teacher's discretion within the rights period which is usually one year or more.

TCC is a service of Pacific Mountain Network, a public television regional network owned and operated by PBS' affiliates in the western US. Funding is provided through a grant from Whittle Communications, L.P., as part of the Whittle Educational Network. The installation of individual classroom TV monitors through the Whittle Educational Network ensures that TCC programs can be used by millions of students. TCC transmissions are available on Ku band via Galaxy 4.

TEAMS

TEAMS is a national distance learning consortium with members in urban, suburban and rural school districts. It produces live, interactive direct student instruction in mathematics and science for grades 2-6. TEAMS programs are designed for active viewing and participation. Programs provide participants many opportunities for hands-on activities and discussion, interaction with other students, interaction with the classroom teachers, and interaction with the studio instructor via phone or fax. TEAMS provides staff development opportunities for teachers which cover effective mathematics and science instruction and offer techniques for teaching language minority students as well. Participating teachers benefit from increased knowledge and proficiency in mathematics and science education. TEAMS studio instructors model effective instructional strategies for teachers in a team teaching approach while providing lessons which motivate and challenge students to master new skills. The site teacher is an active participant in the instructional process.

TEAMS distance learning provides successful learning experiences for all students. They learn to value mathematics and science, become confident in their own ability as scientific thinkers and mathematical problem solvers. TEAMS programs have been successfully implemented with students having a wide range of learning abilities, including, gifted, special needs and limited English proficient because the classroom teacher is an important member of the TEAMS instructional team and can freely adapt the programs to the students' needs and abilities. They

are designed to be used by an entire class and are not pull-out programs. The programs encourage active rather than passive learning. During a "Your Time" feature of each program, students engage in activities at their own site. They share theories, construct their own learning, and develop strategies for problem solving. Students can also call in to the studio to report data and observations. TEAMS parent program focus on helping parents support their children's learning of mathematics and science, are broadcast in English and Spanish, and can be used as an integral part of a school district's outreach to their parent constituents.

The Education Coalition - TEC

The Education Coalition is the only organization of its kind in the U.S., which is established solely to promote educational systemic reform through collaboration and the use of multiple technologies. It is the only organization which brings together K-12, post-secondary and broadcast agencies, with libraries, museums and other resource agencies to connect learners of all ages to the National Information Infrastructure. TEC has affiliate and resource agencies in 17 states which represent over 300 school districts, 2,000,000 students and 50,000 teachers. Many are from low income rural or urban environments.

TEC agencies develop and share programs and services for students, parents and staffs via satellite and computer networks to support the National Educational Goals, specifically in relationship to: parent skills to improve literacy and prevent violence, world-class learning opportunities is social studies, geography, literature and the arts, school to work transition, violence prevention and safe schools. The TEC University and College Collaborative (UCC) in cooperation with the K-12 districts, provide pre-service and in-service course work in those areas. Student programs provide teachers with opportunities to see theory translated into the classroom and they to practice new instructional methods for important, meaningful content.

National resource agencies to TEC include the Smithsonian, the Kennedy Center for the Performing Arts, the John F. Kennedy Library, and the Peace Corps. TEC is guided by a National Advisory Board with members representing school reform, technologies, the business sector, and public and private agencies.

Texas Education Network - TENET

An ambitious statewide effort to enable the state's K-12 educators and students to communicate electronically, share resources, and gain access to databases available on the Internet. More than 15,000 educators and administrators currently use the network to collaborate and to tap into various sources of electronic information. TENET has many classroom applications. Students in the Cajun area around Beaumont have exchanged information about their community and economy with students in the west Texas city of El Paso. High school students have used the network to respond to letters to Santa Claus written by elementary students. In some middle schools, students have downloaded weather information and tracked hurricanes. TENET includes traditional information sources such as a daily newspaper and an encyclopedia. Computer equipment also is in short supply in some schools in some parts of Texas. The Texas state legislature has established a fund to help these schools buy equipment. TENET will be available to all 200,000 of the state's K-12 teachers. Participating teachers or schools will pay a small start-up fee and on-line connect charges to 1 of 16 Internet sites across the state.

TERC

TERC is an education research and development organization committed to improving mathematics and science learning and teaching. The strands of their work include creating innovative curriculum, fostering teacher development, conducting research on teaching and learning, and developing technology tools. It is a private, non-profit.

TERC supports a national movement toward increased use and understanding of data through the dissemination of a model of computer-supported, data-based inquiry in grades 4 through 8. The model uses Tabletop, a new-generation data software tool that provides dynamic visual representations for organizing, exploring, and analyzing data. The project is training, providing support, and getting feedback from a community of teacher leaders who use the software in their own classrooms.

For the Lawrence Berkeley Laboratory Hands-On Universe project, TERC is developing curriculum units that support student investigations in astronomy ranging from studies of the solar system to supernovae in distant galaxies. Using telecommunications, the project links high school classrooms to a remote telescope and an image database. Students can request and download their own astronomical images and manipulate and analyze their images using WinVista, an image processing program.

TERC is involved with other projects including LabNet: The High School Science Network, literacy in a science context, The Global Laboratory, the National Geographic Kids Network/Middle Grades, and the Princeton Earth Physics Project (PEPP).

TI-IN Network

The TI-IN Network offers programming for elementary, middle, and high school students. High School courses include anatomy & physiology, Latin, Spanish, German, French, physics, AP physics, marine science, astronomy, Japanese, sociology, psychology, AP psychology, and environmental science. Most high school sites have voice and data interactive keypads so that students can be in direct contact with their teacher to ask questions or make comments, to privately signal comprehension difficulty, or to respond to oral quizzes during class. The keypads allow students to become active participants in their class sessions, enhancing learning through improved interactivity.

Elementary programming includes The World Around Us and Spanish. Middle school programs include Languages Around the World, a cultural and linguistic awareness program. Staff development includes learning styles and strategies, technological awareness and implementation, school administration and management, health/welfare of student, and curriculum specific topics.

TVOntario

TVOntario offers a number of telecourses including teacher education and a wide variety of curriculum-based series for pre-school through postsecondary that can be used as components of distance learning courses.

United States Education Network (USEN)

This network is composed of two services; the Central Educational Telecommunications Network (CETN) and the Black College Satellite Network (BCSN). USEN represents the shared vision, commitment and collaboration of four major urban city school districts; New York City, Philadelphia, Washington, D.C., and Dallas. CETN is a distance learning provider includes some of these cities most respected cultural/scientific and educational institutions. They provide greater access and academic opportunities through the use of advanced telecommunications applications. Technology is used to reach elementary students in grades K-6, limited English proficient (LEP), home-bound and institution bound students, in addition to teachers, administrators and parents. Technologies include compressed video, cable, satellite, computers, audio graphics and other advanced technology.

Each of the targeted LEA's provide live, interactive, interdisciplinary telecourses in the areas of mathematics, science, history, geography, foreign languages and the humanities. Recognizing the crucial role of parents in the teaching/learning environment, programming has been created to involve all parents more fully in their children's education. These programs are designed to help them better understand their role and function as a parent and citizen, and to improve their literacy and employment skills. The live, interactive telecourses for parents are transmitted to the homes via cable curing evening hours and weekends.

Other programs for staff development focus on instructional materials which prepare teachers and administrators on how to use advanced technologies in the classroom, as well as provide exemplary models on methods of teaching curriculum content. Staff development sessions provide hands-on experiences which integrate multiple technologies in the classroom. USEN partners also produce live satellite field trips to museums and cultural institutions in each of the participating cities.

Black College Satellite Network (BCSN) is a comprehensive telecommunications organization utilizing state-of-the-art technologies to transmit programs to and from colleges and universities, public and private schools serving students in Pre-K through grade 12, as well as in-service programs for teachers and administrators serving at all educational levels. The broadcast coverage includes 23 states, Washington, D.C. and the U.S. Virgin Islands.

The network offers and array of foreign language, science, and mathematics telecourses for elementary, middle and high schools students in urban, suburban and rural communities. Pre-service and in-service teacher education programs are offered to faculty and administrators working at all educational levels.

BCSN has installed over 1000 television receive only (TVRO) units at public schools, colleges and universities, churches and other sites nationwide. It has constructed a fixed C Band uplink and a fixed Ku Band uplink. It also has a compressed video network which links educational institutions for the transmission of educational programs, using interactive video, voice and data transmission.

University of Missouri Schools of Nursing, St. Louis and Kansas City, MO

Nursing students at the four campuses of the University of Missouri share classes and instructors in the bachelor's and master's nursing programs. When the master's program at the University of Missouri-St. Louis was authorized, funding was not available to provide additional instructors. However, the campus in Kansas City was funded for master's level instructors. The two campuses cooperated and the instructors at Kansas City taught the graduate nursing students in St. Louis over the University's compressed video network.

University of New Mexico

A new consortium will link institutions that have substantial Hispanic student enrollments to share credit and non-credit courses via satellite. The ten other universities are from Arizona, New Jersey, New Mexico, New York, Puerto Rico, Texas, and Washington.

University of Northern Colorado

The University of Northern Colorado is using compressed video technology (T1) to develop a statewide two-way interactive distance learning network. The network is designed to bring graduate level courses to teachers throughout Colorado. Three sites are in operation. When all fourteen are implemented, no teacher in Colorado will have to travel more than 50 miles to access education graduate programs.

University of Phoenix Online Division

The University of Phoenix Online Division in San Francisco offers graduate and undergraduate degrees in business entirely through computer conferencing. Called "Online" because it relies on computer mediated communications, the courses are uniquely interactive. Students are not isolated from one another and they are able to benefit from each other's wealth of experience and knowledge. Online students participate in class groups composed of eight to twelve working professionals from around the country. Because the Online virtual classroom is open 24 hours a day, students have the flexibility to schedule their learning time around their other commitments — including extensive travel. Because the program is asynchronous, students can log on to the system anywhere and anytime they have access to a computer, modem and telephone line. Faculty members are also scattered throughout the U.S. Before beginning to teach on the system, instructors receive extensive training on the system, in adult education methods, and interaction skills appropriate to the computer conferencing classroom. Faculty and administrative meetings are held on the system as well. A second distance education group utilizes an audio-graphics system.

Western Cooperative for Educational Telecommunications

The Western Cooperative for Educational Telecommunications has direct links to educational and political leaders through its relationship with the Western Interstate Commission for Higher Education (WICHE). It offers an array of projects and member-driven activities to improve the efficiency, impact and quality of educational telecommunications as well as publications.

Whittle Communications' Channel One

Whittle Communications' Channel One news program covers worldwide events and incorporates them into a format that will have an impact on young people. It gives teachers the opportunity to involve their students in those events with the help of the network. Channel One, which will air in 11,861 middle schools and high schools across the country during 1992-93 reaching 8.1 million teenagers, makes world news relevant to the concerns of teenagers. Channel One is beamed out every school day by satellite to its subscribers. Schools pick up the program on the Ku band satellite dish provided by the network. The newscast is then recorded on one of the school VCRs, also furnished by Channel One. It is later broadcast on 19-inch color televisions, part of the network's equipment package, in all participating classrooms. In addition to the 12-minute newscast (10 minutes of news to two minutes of advertising), Channel One produces timely one-hour specials on topical issues.

World 2000

In World 2000, a senior high school in the U.S. links with a class in Moscow. The students use a combination of mail and computer conferencing to share research and writing on global health issues. Ultimately, participants in the project will develop a shared understanding of the world's health in the year 2000 (National Foundation for the Improvement of Education 1990).

WorldClassroom, International Telecommunications Network

WorldClassroom is an international educational telecommunications network that prepares students, K-12, to use real-life data to make real-life decisions about themselves and their environment. Participating countries have included Argentina, Australia, Belgium, Canada, Denmark, France, Germany, Hungary, Iceland, Indonesia, Kenya, Lithuania, Mexico, Russia, Singapore, Taiwan, the Netherlands, U.S. and Zimbabwe. The international community of teachers and student serves as a resource of information, knowledge, and experience for the others. Each member is encouraged to "give" to the network in order to "get" responses from others. Students are encouraged to learn by doing. Their interaction on-line results in learning.

With a personal computer, modem, communications software, and a regular phone line, WorldClassroom engages teachers and students from around the globe in unique learning opportunities. Innovative lessons provided on the network are supported by dynamic data resources, expert assistance and peer support. Addressing many content areas, the lessons emphasize the educational task, rather than the technology used to accomplish the task. Students use real data and receive timely, personal feedback on their projects while communicating with their peers around the world. In the process, they develop a better understanding of each other, as well as gain a global perspective of real world issues, concerns and cultural awareness. Teachers are able to choose the activities they want from a varied listing, and the curriculum is adaptable to individual schedules and teaching styles. Classes work independently or with each other in small clusters of six to eight classes. Clusters are organized with a WorldClassroom moderator coordinating each of the lessons. Course content covers science international issues, a writing exchange and special project. Each section contains detailed overviews, lesson plans and authentic data for the activities in these sections as well as areas for teachers to get involved in planning and discussion for the activities.

University of Wyoming and Wyoming Community Colleges

The State of Wyoming is in the midst of implementing a higher education distance learning system including nine sites which connects the University of Wyoming with Wyoming Community Colleges. The system is using T1 lines with compressed video equipment and plan to extend their system to the state government offices.

■ PROPOSAL WRITING AND SOURCES OF FUNDING IN DISTANCE EDUCATION FUNDING SOURCES

There are a number of foundations, government agencies, corporations, Regional Bell Operating Companies (RBOCs) and cable companies funding projects in distance education. The process is quite competitive as dollars for all types of educational projects are scarce. The most successful grants are those which "encourage collaboration among institutions and within communities" (Krebs, 1991). Krebs suggests including libraries, community halls, homes, and corporate locations.

Krebs points out that the funding research she did revealed that education is being transformed throughout the country at local, regional, and national levels. "From school reform and restructuring efforts to take-home computers for students and parents, from electronic pen pals to interactive course instruction, from voice mail systems for family-school linkages to video conferences for teacher training — the diversity of applications is overwhelming." Hundreds of innovative educational projects have been funded across all educational institutions, agencies and professional associations. She recommends examining the types of projects that are funded to gain a perspective on the imaginative applications of technology's potential and understand the logic behind the historical funding pattern of the institution.

The federal government has undertaken major initiatives to support distance education and funds cover projects throughout the range of education. One emphasis is to involve citizens in addressing their community's educational needs. School reform and restructuring are important issues and can be addressed effectively by distance education. Proposals in science, mathematics, technology, humanities, languages, English as a second language (ESL), Adult basic education (ABE), and health care are subjects that are receiving grants.

Be aware that the competition is keen and that identifying other subjects that need attention may mark your proposal as worthy. Proposals which target or include prison education projects are beginning to receive more attention as we learn about the correlation between more education and recidivism. Be aware that even though an institution may have funded distance education projects, they may not be familiar with terms such as distance education/learning or teleconferencing. Carefully explain these terms as they relate to the proposal. In some cases you may be able to identify an institution with an interest in innovative educational programs where distance education is not specifically mentioned.

She believes that a key area for funding concerns support for teacher training to transform the traditional curriculum and classroom practices into interactive teaching because distance education requires more preparation, more materials production, and more out-of-class time with students. Grants can provide distance education faculty development to strengthen personal teaching styles and to develop effective course materials.

Funding can be granted to extend the network's outreach and to enhance participation through the use of interactive technologies and the integration of multimedia, as well as evaluation of the program's outcomes. A key is to target funding requests to specific funding agencies. For example, request equipment donations from manufacturers, and request foundation and government grants to support specific curricula areas and teacher training and recruitment. Proposals will also be more successful if partnerships are created and the program can be economically self-sustaining. Another useful resource is the annual "Foundation Grants Index."

Proposal Writing: These pitfalls to avoid in proposal writing and ideas to consider in your search for funding were suggested by Constance M. Lawry, Arts and Sciences Extension, Oklahoma State University.

1. You will be writing up to the end of the deadline period.

2. Most agencies try to show what they want in the proposal guidelines. Many are reluctant to talk to grant writers over the phone because they might say something off hand that could lead the writer astray or to avoid the appearance of showing favoritism. However, do not assume that the agency will not speak with you until you call and they refuse to answer your questions.

3. When the proposal guidelines are first announced, call the project director and ask questions. This will accomplish several things; the project director will know your name and the institution's name; the reviewer will get to know you and the institution as well as understand your needs and the proposal you are submitting. It will help you clarify the purpose of the funding and whether your institution's goals match the funding agency's goals. Ask the project director if he/she would review a draft proposal and when it would be most convenient to receive the draft. This favor should be asked well before the project staff becomes involved with the application deadline submissions. Some agencies may perceive reviewing a draft proposal as showing favoritism, but others will review a draft. If you get agreement to read the draft proposal, consider meeting with the project director to deliver the draft proposal in person or meeting several days after the draft has been received and read. While this is an added expense, it maybe more helpful to get the project director's reactions in person rather than through a brief letter or telephone conversation.

4. Criteria for draft proposal review. Ask that it be read to see if it complies with the proposal guidelines, format and other requirements. Ask if it is clear and if there are areas which should be rewritten to provide more information. Ask if it meets the purpose of the funding. Ask for other suggestions that will make the proposal more worthwhile for consideration for the grant.

5. Write grants that are of interest to you. Regardless of how well funded a grant may be, it will

still involve sacrifice of your time, energy and enthusiasm. If you are not interested in doing the work, do not apply for the grant. Instead, find a source of funding for projects that you personally want to do.

6. Look at the objectives of the grant to aid you in determining how your grant proposal will be judged. When the objectives seem to be asking for the same information in two sections, state in the information in the first section and reference that with the page number in the second section so that the reviewer will know that you have supplied the information. If you do not do this, the reviewer may not remember that the information has been supplied and could disqualify your proposal from further consideration. Describe the outcome of the grant in measurable terms. Describe the population that will benefit from the funding.

7. Adhere to the format requirements which may use the words "should" or "must" in describing how the proposal should be formatted. For example, instructions that state you "should not exceed 15 pages" implies that the proposal might not be disqualified for page overage. Instructions that state you "must not exceed 15 pages" implies that an overage would disqualify the proposal. If you have questions, contact the project director.

8. Do not write by committee. Assigning the writing by section to individuals may result in sections that do not relate to each other. Writing styles may change and the proposal may not "hang together." Instead, collect information from all those involved in the proposal, but assign the final writing to one writer who has the authority to make decisions on what is important to the success of the proposal. This writer should understand the total proposal, the proposal writing process and how best to present the information.

9. Proposal image. There was a time when it was suggested that the proposal not be overly fancy in binding, appearance or graphics. With the general acceptance of desk top publishing systems and the reviewers' knowledge that this method of presentation is not expensive, many grants are now prepared with this method. In fact, reviewers may react adversely to proposals prepared on traditional typewriters as they are now so accustomed to desktop publishing documents.

10. Topics for Research. To develop new research in distance education, Lawry suggests that you offer to conduct research on distance education students through vendors of distance education programming. Most vendors are not conducting research and may be willing to provide funding or negotiate a lower rate for programming if research is conducted on students studying through their programming. Lawry suggests that you search for the answers to the questions that befuddle you. Research examples include the following:

a) tracking graduates to see if they tested out of college courses due to their advanced place ment high school distance education classes;

b) tracking characteristics of distance education students to determine if there are discernible traits for those who do well in distance edu cation and those who do not do well;

c) determining if there are differences in learn ing which can be attributed to the way a course is taught or the school environment;

d) determining how and if tapes are being used, track how many students and which students view or review the course on video tapes; do struggling students make more use of tapes to review and understand the con tent; do excellent students make more use of the tapes and succeed because of that;

e) documenting the use of software provided for content study;

f) conducting in-depth interviews with stu dents and teachers to determine factors con tributing to success or failure;

g) conduct in-depth interviews with teachers and administrators to determine factors leading to adoption of distance education;

h) conduct in-depth interviews with non-involved teachers to determine how they feel about distance education

11. Preliminary proposal. Write a brief preliminary proposal and submit it to an organization or company which might be interested in the topic. If they seem interested, write a full proposal. Your preliminary proposal may also elicit other topics in which the organization is more interested. If you can fulfill their immediate research needs they may be more open to funding there search you want to conduct.

12. Conducting classroom research may first involve interesting and training teachers in conducting classroom research and research techniques.

13. Future funding. Any proposal you submit should include your future financial needs to keep the program operating after it is established. Consider funding for maintenance and repair, salaries, additional equipment and other necessities. You should describe how other funds will be obtained if the granting agency will not fund future costs. Attach letters of commitment, if necessary.

14. Anticipate the devil's advocate questions. The proposal should answer the questions: Why should I fund you? What is your track record? Why you are credible?

15. Department of Education Reviewing Panels have been reduced to two people in many cases. This puts even more pressure on you to state your case clearly and make it stand out from the other proposals.

16. Rejection? If your proposal is rejected, call

and find out why it was rejected. Talk to the project director or the reviewers to determine exactly what was wrong. Were there 15 similar proposals, was it not clearly written, or was it too expensive. Why it was rejected may be helpful to you in writing future grants.

Selection Criteria (as outlined in a typical U.S. Department of Education RFP)

The request for proposal (RFP) offers the following advice on writing the program narrative: "While there is no standard outline for a program narrative, applicants are encouraged to prepare the program narrative by addressing the criteria listed. Please note that the narrative portion of the application should not exceed 15 double-spaced, typed, pages. The total application should not exceed 25 pages, including appendices and letters of support.

1. Meeting the purposes of the authorizing statute (30 points)
    i) The objectives of the project
    ii) How the objectives of the project further the purposes of the authorizing statute
2. Extent of need for the project (25 points)
    i) The needs addressed by the project
    ii) How the applicant identified those needs
    iii) How those needs will be met by the project
    iv) The benefits to be gained by meeting those needs
3. Plan of Operation (15 points)
    i) The quality of the design of the project
    ii) The extent to which the plan of management is effective and ensures proper and effi cient administration of the project.
    iii) How well the objectives of the project relate to the purpose of the program
    iv) The quality of the applicants plan to use its resources and personnel to achieve each objective
    v) How the applicant will ensure that project participants who are otherwise eligible to par ticipate are selected without regard to race, color, national origin, gender, age or handi capping condition; and
    vi) for grants under a program that requires the applicant to provide an opportunity for par ticipation of students enrolled in private schools, the quality of the applicants plant to provide that opportunity.
4. Quality of key personnel (7 points)
    i) Quality of key personnel
    A) Qualification of project director
    B) Qualification of other key personnel
    C) Time that each person will commit to project
    D) How the applicant, as part of its nondis criminatory employment practices, will ensure that its personnel are selected without regard to race, color, national origin, gender, age or handicapping condition.
    ii) To determine qualifications above, secretary considers
    A) experience and training in fields related to objectives of the project.
    B) any other qualifications that pertain to the quality of the project.
5. Budget and cost effectiveness (5 points)
    i) The budget is adequate to support the project
    ii) Costs are reasonable in relation to the objec tives of the project
6. Evaluation Plan (15 points)
    i) Extent to which methods are appropriate to the project
    ii) To extent possible, methods are objective and produce data that are quantifiable.
7. Adequacy of resources (3 points)

Adequacy of resources that the applicant plans to devote to the project, including facilities, equipment, and supplies. (Includes 15 extra discretionary points added to a base of 85, which were distributed as follows: 5 to need; 10 to evaluation)

It should be noted that the evaluation plan garners 15 points in the example above. Most grantors of funds insist upon a good evaluation design. Some agencies even provide extensive information on the types of evaluation that will be accepted and include research designs that they consider effective for the types of projects that they fund. In addition, make sure to include an adequate amount of funding to implement the evaluation including instrument development time, site visits, travel expenses, statistical evaluation and writing the final report. The entire design should be included with the proposal so that the readers can see how the evaluation will determine the effectiveness of the program and the students' level of learning outcomes. Most funders also want staff development, parent and community programs evaluated. It is usually wise to specify an outside evaluator to oversee the evaluation. Several of the U.S. Department of Education regional laboratories are staffed with personnel who have extensive experience in evaluation of programs using educational technologies. Far West Laboratory in San Francisco is conducting the TEAMS and Hughes Galaxy Project evaluations. The Southwest Regional Laboratory in Los Angeles is conducting the overall Star Schools Evaluation. Northwest Regional Laboratory is conducting the STEP Star Schools Project evaluation.

Distance Education Legislation, Accreditation and Regulation

America faces many problems and challenges in education. From coast to coast, from local school districts to large universities, educators are being asked to do more with less and budgets are being significantly reduced. There is overcrowding in urban areas, and a lack of access to educational opportunities in both urban and rural areas. Many states have undertaken efforts to plan and coordinate for distance learning and have formed distance learning consortia, but until all the users are aggregated on a national level, they will not have enough market power to attract commercial interest for a telecommunications infrastructure to facilitate distance learning growth. The education sector is also limited by short-term planning because education budgets are formulated primarily at the state and local levels which are done on an annual or biannual basis. Since funding levels are uncertain from year to year, educators and administrators find it difficult to enter long-term agreements which would gain the discounts that they need.

There is an expanding interest in the use of distance learning for courses and enrichment for students, for teacher training and professional development, networks for teaching and learning and for courses for adult learners and continuing education. Dr. Linda Roberts of the U.S. Department of Education believes that we are just at the beginning stages of understanding what we can do with telecommunications. In continuing to contribute to the growth and utilization of the field of distance education, she feels that the vision has to be maintained and nurtured. "You can't assume that the public is on your side," she said. You have to convince them of the difference that it can create.

There are many opportunities for distance education to provide service to the more than two million teachers. "They are the most isolated professionals I know," Dr. Roberts said. The systems should serve teachers and students. As examples, she suggested that teachers should be involved in teaching specialized courses and reaching more students. Other opportunities include developing new ways of teaching, expanding classroom resources, connecting with others, pre-service mentoring and in-service to advance the profession. Roberts observed that one of the obstacles to adopting distance learning technologies is the common but nevertheless erroneous perception that teachers are delighted when distance education is adopted. "I never found that teachers were delighted because of distance education. Usually, the numbers increased." Technology has transformed every sector of our lives, It can transform education as well. It will not replace teachers, it will empower them with better teaching tools.

Boundaries are changing between schools, states, private institutions, higher education and districts. There are now many groups which are creating new relationships and coalitions. The new relationships emphasize the importance of sharing resources and brings up the question of state teacher certification.

Rural and urban populations are different but the needs are the same and it emphasizes the importance of providing equal access for all students. Networks should be used 24 hours a day.

Competing interests and fragmented regulatory authority are detriments to distance learning. Regulations usually lag behind the use of technology and because of this educators should take the initiative and inform those who need to know about telecommunications policy that will enable distance education technologies and their use. Telecommunication policy decisions should address the costs, capacity and services available for distance education. Issues include the development of the nation's telecommunications structure for a mix of technologies, telecommunications policies, providing technology that teachers want that is easy to use and educational satellites among other topics. More and better courseware needs to be developed and copyright issues must be addressed.

In at least six states (Kansas, Missouri, Oklahoma, Texas, Washington and West Virginia), legislation has been developed to specify the kind of telecourse evaluation to be done and the criteria to meet for telecourses and their delivery systems (Kressel, 1986). The fact that states are mandating evaluation procedures underscores the fact that the available media selection model and evaluation instrument are not being used across the board (Lane 1989). Holt (1989) warns against seeking government entitlement programs to fund the production of distance education telecourses and believes that it should remain entrepreneurial to force bankruptcy on the producer of deficient programming. He strongly feels that the administration, faculty and student consumers must judge quality. Holt's message is that both the buyer and seller should beware. Holt demands a partnership between producers and consumers that amounts to unreserved commitment to distance education as most failures occur because student support systems are not in place (Holt, 1989).

Holt predicts that state controlled accreditation will be established for political reasons rather than for the quality control, which he endorses. He predicts that state accreditation is here to stay since more credit programming is being brought in by satellite from other states. Holt warns that for accountability, state education personnel should be used in a guidance role, but distance educators should perform the evaluation.

An example of an accreditation policy that has been acknowledged as being ahead of its time is the 1985 Project ALLTEL, an acronym for Assessing Long Distance Learning Via Telecommunications. It was created to "get ahead of that giant wave" when accreditors and regulators alike found themselves unable to adequately respond to changes in the environment they were charged with regulating (Goldstein, 1991). Over a period of two years, a policy board, three task forces and a committee struggled to deal with the inextricably intertwined issues of the licensure and accreditation of telecommunications-based distance learning. One task force looked at the accreditation issues, a second look at the state authorization concerns, and a third considered the legal issues arising out of attempts to control and regulate this arena. The Project was an effort to bridge development in the use of technology in delivering higher education services with the traditional roles and responsibilities of state agencies, accrediting bodies and institutions. The object was to set in place reasonable policies to ensure the growth, development and quality of the technology.

Goldstein points out that most programs which have been developed to cross borders have been developed with an overabundance of caution and wonders if we have been spared the charlatans because the regulatory environment filtered them out because of the oppressive regulatory regime which was developed and few dared to challenge. If it is the latter, he states, then "what we are accomplishing is the denial of educational opportunity at precisely the time when this nation can least afford such impediments." One of the legal problems to consider is whether a provider of distance learning services via telecommunications can be barred by a state from bringing those services across its borders. Another legal question is whether the delivery of educational services via a particular modality be constrained to protect local, traditional institutions from perceived competition? Another question is at what point a technology-based distance learning service is sufficiently present in a state to give that state the right to control its conduct?

Goldstein also points out that one cannot ignore the issue of competitiveness among institutions reflecting a sense that the market is unlikely to support every school involved in distance education. Competition for students will increase as the children of the baby-boomers graduate. While the number of traditional college age students has been declining, college enrollments have not reflected the magnitude of the decline because of the increase in the number of adult students returning to college.

Reilly and Gulliver (1992) argue that the distance learning experience cannot necessarily be evaluated by the standard measures applied to traditional education such as seat time, amount of face-to-face contact with the instructor and the immediate availability of massive library collection and extensive laboratory facilities. "In fact, since measurement of these inputs has produced little empirical evidence of the effectiveness of conventional classroom learning, using them as the baseline to evaluate distance learning is problematic at best." When coupled with the different pedagogical assumptions implicit in distance education (Granger, 1990) and "it becomes clear that new evaluative criteria are needed for distance education" (Reilly and Gulliver,1992).

A symposium "Emerging Critical Issues in Distance Higher Education" resulted in a number of recommendations ranging from ensuring that quality in all education be measured on outcomes rather than inputs, to developing a set of principles of good practice for distance higher education, to establishing a research agenda that would inform policy development in distance learning (Granger 1991). The recommendation to develop principles of good practice is being implemented by a task group of educators, regulators and accreditors. Another of the symposium's central recommendations reaffirms the basic message of Project ALLTEL: Regulating and accrediting agencies should develop ways of cooperating among states and regions to facilitate approval while ensuring quality in distance education programs that cross state and regional accrediting lines, with a goal of advancing distance education while protecting the "consumer" (Regents College Institute for Distance Learning,1990).

Another result of the meeting was that participating states (Arizona, Colorado, Connecticut, Georgia, Illinois, Minnesota, New York, Pennsylvania, Puerto Rico, Tennessee, Texas, Vermont and Virginia) agreed to use the review of distance learning institution in the institution's home state as part of the approval process in other states where the institution seeks to operate (Reilly & Gulliver, 1992). Other states are being encouraged to sign the agreement in which states would rely not only upon the home state review, but also upon additional reviews that may have been carried out by other states. This agreement does not entail automatic reciprocity of recognition and each state retains the right to make its own decision but does commit states to build upon the work of other states to minimize duplication of review processes. This avoids the chief impediment to the implementation of Project ALLTEL which called for receiving states to accept the decision of the state of origin.

If states agree to use a common information collection instrument for distance education institutions which is being circulated to all states and accrediting associations, this would reduce some of the duplication. Accrediting agencies are also being asked to use more fully states' approval of distance education programs for accreditation purposes. Reilly and Gulliver (1992) assert that for these steps to be effective "states must first accelerate their movement toward acting more like units of a nation and less like sovereign entities in their regulation of interstate distance higher education via telecommunications."

State education agencies are both gatekeepers

and catalysts for distance education. Stringent teacher certification requirements may prevent skilled instructors from teaching electronically in areas experiencing teacher shortages. Varying state curriculum and textbook requirements make it difficult to share teaching between schools that could be linked. State leadership is critical to foster the efficient use of resources to meet educational needs. In the process of developing plans for distance learning, states have the opportunity to forge cooperation between agencies, encourage sharing of costs among users, and build new linkages between schools, higher education and the private sector. Federal and state regulations guiding the development of telecommunications infrastructure and services significantly affect distance education according to an OTA (Office of Technology Assessment) Report Brief (Nov. 1989). The nation's schools represent major markets for applications of technology and should be in a powerful position to influence telecommunications policy. Because of conflicting interests and fragmented telecommunications authority, educational needs may not be fully served. As distance learning expands, education has a growing stake in shaping future telecommunications policies.

According to the OTA, federal funding for distance education has been important but modest. The Star Schools Program, begun in 1988 to develop multi-state, multi-institutional K-12 distance education, has focused attention on distance learning, and spurred planning and development beyond the projects now under way. Programs at the National Telecommunications Information Administration and the Rural Electrification Administration support distance education by funding telecommunications technologies. Other programs provide limited support for curriculum development, special programming, technical assistance and research. Growth of distance learning can be expected to continue for some time with out increased federal involvement. A major commitment to expanding the nation's distance learning infrastructure will require a change in the federal role.

The growing interest in distance learning comes as calls for improving education increase. States, localities, the federal government and the private sector can plan, fund and implement distance education. Four factors (OTA, 1989)that will most affect the future are;

1. Telecommunications policy. This affects costs, capacity and types of services available. Congress must review and shape policies to reflect the nation's educational needs.

2. Research, evaluation and dissemination. With the dramatic proliferation of distance learning projects, many questions regarding effectiveness, methodology and design have been raised. Federally funded research can contribute to the understanding and improvement of distance education.

3. Support for teachers. More emphasis should be placed on how to help prepare new teachers and encourage others to enter the profession. Funding for teacher preparation could support the use of distance learning technologies. Congress could encourage use of technologies to reach teachers who need to upgrade skills in fields such as math and science.

4. Expanded infrastructure. National leadership could expand distance learning to communities without resources and extend the reach of installed systems. Congress could specify expenditures for distance education in current federal programs or make funds available through new programs. National leadership could focus investments toward the future, ensuring that today's distance learning efforts carry our educational system into the 21st century.

A national infrastructure, projected to be fiber optic cable, would provide enhanced broadband services to every business, home and school. A broadband backbone would have the ability to provide voice, data and video services. It holds great promise for education because such a system would give every student and teacher in the nation access to the same opportunities. With a fiber optic network, schools could access any library in the United States or the world. Students could browse through instructional texts, graphics and video on any subject, any school could have guest teachers from anywhere in the world via a two-way interactive audio and visual network.

Other legislation has been introduced for an educational satellite which is a cost-effective way to deliver instructional programming to a great number of schools and students. Satellite transmission provides a way to reach students no matter how remote. In today's satellite market the education sector is fragmented and commercial market practices leave educational institutions without low-cost, dependable and equitable access to services. For the most part, schools, school districts, state education agencies, colleges and universities all operate independently. A dedicated education satellite would ensure instructional programmers that they will be able to obtain affordable satellite transmission time without risk of preemption by commercial users. It would allow educators using the programming to have one dish focused on one satellite off of which they could receive at least 24 channels of instructional programming. In the legislation, the federal government's role is to take the risk from the private sector in order to encourage the development of a dedicated satellite system.

The National Education Telecommunications Organization (NETO) was formed after the EDSAT Institute held seven regional meetings in 1991. Through these meetings they recognized the need to aggregate the education market for distance learning and concluded that an education programming users organization was needed. Its board is committed to the goal of developing an integrated telecommunications system, dedicated to education with the first step of acquiring a dedicated satellite. Some have suggested that the Public Broadcasting System (PBS) could meet the infrastructure needs of the distance learning community but PBS and NETO have very different missions. PBS is in the business of broadcasting programming and acquires satellite time to deliver its own programming. In contrast, NETO's focus is on the distribution of distance learning, much of it live and interactive and will not generate programming. NETO's sole concern is the creation of an infrastructure which will distribute instructional programming created by others at an equitable price to all users. NETO will aggregate the market so that it will be of sufficient size, but the problem of being a short-term user still faces the education sector. Educators cannot enter into the five or ten-year commitments that satellite vendors look for in long-term users. This legislation solves that problem by offering federal loan guarantees to NETO so that they can, in turn, offer the satellite vendors the long-term commitment they need. Our legislation basically guarantees the vendor an anchor tenant. Without that guarantee, it is likely that even an aggregated education market would not be able to secure a long-term lease or purchase arrangement with a satellite vendor.

A dedicated satellite system will bring instructional programming which is not scattered across 12 to 15 satellites into one place in the sky. This co-location will allow educators to receive a variety of instructional programs without having to constantly reorient their satellite dish. Proponents of an educational satellite say that by making the investment in a dedicated system on the front end, distance learning costs will be reduced for educators at the state and local levels. Programmers will benefit because they will be able to market their programming to a wider audience and will be guaranteed reliable satellite time at an affordable rate that will be equal no matter how much is bought. Users will benefit because their investment in equipment to receive instructional programming may be reduced because of the technological advantages of focusing on one point in the sky.

Satellite technology can expand educational opportunity for students in areas with "teacher shortages in important subjects — such as foreign languages, math and science. We should capitalize on technology's potential for supplementing curriculum, without allowing it to, in any way, replace students' one-to-one interaction with teachers.

In addressing the role of technology, we must deal with the question of whether there should be a mix of technology and if so which media should be

used? If only a single technology can be used, what technology would be the most appropriate? Selection tools and guidelines are available which can help to clarify the role of technology. Those which are recommended because of the national experts which participated in their formation appear in a later chapter (Lane 1989, 1991).

∎GOALS 2000: EDUCATE AMERICA ACT

In March 1994, President Clinton signed the Goals 2000: Educate America Act. He said that, "We insist, with Goals 2000, that every student can learn. We insist that it's time to abolish the outdated distinction between academic learning and skill learning. We know now that most academics has practical application, and that, more and more, practical problems require academic knowledge. And I hope to goodness we don't do anything else — we've finally erased that divide so that we can teach our young people to learn in the way that best suits their own capacities and the work they have to do. But I am absolutely convinced that there is not a single, solitary problem in American education that has not been solved by somebody, somewhere. What we have done as a nation is to resist learning from each other, to resist institutionalizing change, to resist, therefore, holding ourselves accountable for results as a nation.

Clinton added that what we the government was trying to do with Goals 2000 is to say, here are the goals, you figure out how to get there, you learn from each other. Come up with aggressive plans, we will help you fund them and go forward, but you are in charge. The federal government can't tell you how to do it, but we can help you get it done. What this Goals 2000 movement, with the School to Work program, with the adult education program, with the retraining program and the re-employment program, what it all seeks to do is to give America a system by which at the grass roots level we can fulfill the promise of Brown v. Board of Education for all our people. "

Secretary Richard Riley observed at the kick off meeting for Goals 2000 that, "some of you will use the new provisions of Goals 2000 to expand what you have started. Some of you will use it to reinvigorate and connect existing reforms. And others of you may use it to launch a comprehensive new effort to improve teaching and learning. That's really what we're about isn't it? But how you use Goals 2000 to encourage learning back home is really your choice. I urge you to think big, to think comprehensively, to recognize that this won't happen in a year or in just a few years. We spent ten years getting to the point where we had the support to pass Goals 2000. A Nation at Risk was ten years ago. We will probably spend another ten years making it work for all of our children."

Three times in the last six years, Congress has attempted to pass education reform legislation and each time it has been unable to resolve its differences. The strong bipartisan support for Goals 2000 demonstrates that we are ready to move from "a nation at risk" to a nation on the move.

The enactment of Goals 2000 is the beginning of a new era in school and education reform — a revolutionary, all-inclusive plan to change every aspect of our education system, while at the same time aligning its individual parts with one another.

It offers an opportunity for those concerned with the state of American education to become involved in the implementation of real change and improvement of our nation's education system, working at the local community and state levels. It will create and improve learning opportunities for everyone from pre-school to those who return to school.

By generating enthusiasm in schools and states throughout this nation, it will create thousands of community-based reform efforts, each working for the betterment of our educational system, and each allowing every school and every student to be the best they can be — to learn to world-class standards.

Goals 2000 will move the nation toward a system that is based on high standards that all students can meet — a system that will provide both equity and excellence for all of the students in this country.

When we fail to hold all students to high standards, the results are low achievement and the tragic experience of children leaving school without ever having been challenged to fulfill their potential.

High standards lets everyone in the education system know what to aim for. It allows every student, every parent and every teacher to share in common expectations of what students should know and be able to accomplish. Students will learn more when more is expected of them, in school and at home. And, aligning teacher education, instructional materials, assessment practices and parental involvement, will create coherence in educational practice.

The American people have said they are ready to move from the old assembly line version of education to a better way of educating their children. They want their children to be part of the new, emerging high-tech, high-knowledge economy of the 21st century.

By transforming the national education goals into a policy for which committed people across our nation can work, President Clinton has helped to ensure that the future of this nation will remain strong and secure and that its citizens will be able to compete and prosper in this new global economic era that is already upon us.

Since early in our history, the public education system of this nation has been a magnet and a model for people throughout the world who yearn to make something better of their lives. It is a beacon of light across the globe, a symbol of our democratic and egalitarian traditions. Unfortunately, in recent years, this standard has slipped; the beacon has dimmed. That is why the Goals 2000 law is so important, as well as the subsequent enactment of additional education reform legislation like the School-to-Work Opportunities Act, and the revolutionary reauthorization of the Elementary and Secondary Act, both of which are designed to dovetail with Goals 2000. Each of these important changes in the law will offer federal assistance in implementing local education reform...help that is designed to assist, but not interfere with the traditional local character of education.

It has been nearly thirty years since this nation has seen the kind of reform in education that Goals 2000 offers. It is up to us to ensure that we maximize the opportunities this law offers us and work to guarantee a challenging education for every student. For the future of our children and our nation, it is the least we can do.

Goals 2000 provides resources to states and communities to develop and implement comprehensive education reforms aimed at helping all students reach challenging academic and occupational skill standards.

The Goals 2000: Educate America Act is not an experiment; it incorporates the lessons of education reform from communities and states in the 1980s.

• Raising standards and making course content more challenging really works. When more is expected of students, they work harder and achieve more. When employees know what skills they need to succeed on the job, they will work to achieve them.

• We must change our expectations of teachers. They cannot teach to new standards using the same old ways. We must overhaul teacher training and make continuing professional development an integral part of their job.

• Accountability is essential. Schools must be given the tools and the flexibility they need to get the job done and then be held accountable for the results they achieve. There must be real rewards for high performance and significant consequences for failure.

• Schools can't do the job alone. Parents, businesses, families, community organizations and public and private agencies that provide health care, counseling, family support and other social services must be part of community-wide efforts to support students.

• All children in America will start school ready to learn.

The objectives for this goal are that...All children will have access to high-quality and developmentally appropriate preschool programs that help prepare children for school. Every parent in the United States will be a child's first teacher and devote time each day to helping such parent's preschool child learn and parents will have access to the training and support parents need. Children will receive the nutrition, physical activity experiences and health care needed to arrive at school with healthy minds and bodies, and to maintain the mental alertness necessary to be prepared to learn, and the number of low-birth weight babies will be significantly reduced through enhanced prenatal health systems.

• The high school graduation rate will increase to at least 90 percent.

The objectives for this goal are that...The Nation must dramatically reduce its school dropout rate, and 75 percent of the students who do drop out will successfully complete a high school degree or its equivalent; and the gap in high school graduation rates between American students from minority backgrounds and their non-minority counterparts will be eliminated.

• All students will leave grades 4, 8 and 12 having demonstrated competency over challenging subject matter including English, mathematics, science, foreign languages, civics and government, economics, the arts, history and geography, and every school in America will ensure that all students learn to use their minds well, so they may be prepared for responsible citizenship, further learning and productive employment in our nation's modern economy.

The objectives for this goal are that...The academic performance of all students at the elementary and secondary level will increase significantly in every quartile, and the distribution of minority students in each quartile will more closely reflect the student population as a whole. The percentage of all students who demonstrate the ability to reason, solve problems, apply knowledge and write and communicate effectively will increase substantially.

All students will be involved in activities that promote and demonstrate good citizenship, good health, community service and personal responsibility. All students will have access to physical education and health education to ensure they are healthy and fit. The percentage of all students who are competent in more than one language will substantially increase. All students will be knowledgeable about the diverse cultural heritage of this Nation and about the world community.

• United States students will be first in the world in mathematics and science achievement.

The objectives for this goal are that...Mathematics and science education, including the metric system of measurement, will be strengthened throughout the system, especially in the early grades. The number of teachers with a substantive background in mathematics and science, including the metric system of measurement, will increase by 50

percent. The number of United States undergraduate and graduate students, especially women and minorities, who complete degrees in mathematics, science and engineering will increase significantly.

• Every adult American will be literate and will possess the knowledge and skills necessary to compete in a global economy and exercise the rights and responsibilities of citizenship.

The objectives for this goal are that...Every major American business will be involved in strengthening the connection between education and work. All workers will have the opportunity to acquire the knowledge and skills, from basic to highly technical, needed to adapt to emerging new technologies, work methods and markets through public and private educational, vocational, technical, workplace or other programs. The number of quality programs, including those at libraries, that are designed to serve more effectively the needs of the growing number of part-time and mid-career students will increase substantially. The proportion of the qualified students, especially minorities, who enter college, who complete at least two years and who complete their degree programs will increase substantially. The proportion of college graduates who demonstrate an advanced ability to think critically, communicate effectively and solve problems will increase substantially. Schools, in implementing comprehensive parent involvement programs, will offer more adult literacy, parent training and life-long learning opportunities to improve the ties between home and school, and enhance parents' work and home lives.

• Every school in the United States will be free of drugs, violence and the unauthorized presence of firearms and alcohol and will offer a disciplined environment conducive to learning.

The objectives for this goal are that...Every school will implement a firm and fair policy on use, possession and distribution of drugs and alcohol. Parents, businesses, governmental and community organizations will work together to ensure the rights of students to study in a safe and secure environment that is free of drugs and crime, and that schools provide a healthy environment and are a safe haven for all children. Every local educational agency will develop and implement a policy to ensure that all schools are free of violence and the unauthorized presence of weapons. Every local educational agency will develop a sequential, comprehensive kindergarten through twelfth grade drug and alcohol prevention education program. Drug and alcohol curriculum should be taught as an integral part of sequential, comprehensive health education.

Community-based teams should be organized to provide students and teachers with needed support. Every school should work to eliminate sexual harassment.

• The nation's teaching force will have access to programs for the continued improvement of their professional skills and the opportunity to acquire the knowledge and skills needed to instruct and prepare all American students for the next century. The objectives for this goal are that...All teachers will have access to pre-service teacher education and continuing professional development activities that will provide such teachers with the knowledge and skills needed to teach to an increasingly diverse student population with a variety of educational, social and health needs. All teachers will have continuing opportunities to acquire additional knowledge and skills needed to teach challenging subject matter and to use emerging new methods, forms of assessment and technologies. States and school districts will create integrated strategies to attract, recruit, prepare, retrain and support the continued professional development of teachers, administrators and other educators, so that there is a highly talented work force of professional educators to teach challenging subject matter.

Partnerships will be established, whenever possible, among local educational agencies, institutions of higher education, parents and local labor, business and professional associations to provide and support programs for the professional development of educators.

• Every school will promote partnerships that will increase parental involvement and participation in promoting the social, emotional and academic growth of children. The objectives for this Goal are that...Every State will develop policies to assist local schools and local educational agencies to establish programs for increasing partnerships that respond to the varying needs of parents and the home, including parents of children who are disadvantaged or bilingual, or parents of children with disabilities. Every school will actively engage parents and families in a partnership which supports the academic work of children at home and shared educational decision making at school.

Parents and families will help to ensure that schools are adequately supported and will hold schools and teachers to high standards of accountability.

Goals 2000 Funding

Only weeks after enactment, the U.S. Education Department invited states to apply for first-year funding under the new Goals 2000: Educate America Act. Legislation made $86.5 million available to states in 1994 to begin developing school improvement plans. An additional $5 million was made available to develop plans to use state-of-the-art technology to enhance teaching and learning. The President has asked for $700 million in his 1995 budget proposal to be administered by the Department of Education and $12 million for the Department of Labor to support the National Skill Standards Board. The administration expects to seek $1 billion in 1996.

• For first-year funding, state educational agencies (SEAs) will be asked to submit an application that will describe the process by which the state will develop a school improvement plan and how the SEA will use the funds received, including how the SEA will make subgrants to local educational agencies (LEAs) and awards for education pre-service programs and professional development.

• In year one, SEAs will use at least 60 percent of the allotted funds to award subgrants to LEAs for the development or implementation of local improvement plans, and to make awards for education pre-service programs and professional development activities.

• In succeeding years, at least 90 percent of each state's funds are to be used to make subgrants for the implementation of the state and local improvement plans and to support educator pre-service and professional development.

• In year one, LEAs will use at least 75 percent of the funds they receive to support individual school improvement initiatives.

After year one, LEAs will pass through at least 85 percent of the funds to schools. The state improvement plans will address, among other things:

• a process for development or adopting content standards and performance standards for what students should know and be able to do in core subjects;

• a process for developing and using valid, non-discriminatory and reliable assessment;

• standards or strategies for ensuring that all

Table 9.1.3

How Goals 2000 Will Change American Education

U.S. Education System Today	U.S Education System Under Goals2000
• Academic standards are often too low	• High standards for all
• Many students aren't expected to learn	• High expectations for all
• Curriculum is often watered down	• Enriched course content for all

students have a fair opportunity to achieve to the established standards;

• strategies for assisting school-aged children who have dropped out of school;

• strategies for coordinating Goals 2000 efforts with other programs such as the new School-to-Work Opportunities Act and the Perkins Vocational and Applied Technology Education Act;

• strategies for involving parents and the community in planning, designing and implementing the plan.

"Goals 2000 creates an unprecedented opportunity for all Americans to contribute their ideas and energy to make schools work for kids," Riley said. "The states will be seeking ways to reach out and promote bottom-up reform."

In developing an improvement plan, states will establish a broad-based reform panel, with members appointed by the governor and the chief state school officer. If a state has already made substantial progress in developing a comprehensive and systemic improvement plan with an existing panel, the secretary may elect to recognize that existing group. The application deadline is June 30, 1995. States are encouraged to submit their applications as soon as possible. A state's award amount is calculated under a formula based on previous allocations under Chapter 1 and Chapter 2 of the Elementary and Secondary Education Act. Any funds not requested by states will be reallocated among states with approved applications.

In the first year that it participates in Goals 2000, a state will be required to pass at least 60 percent of its funds to local educational agencies (LEAs), which will develop or refine local improvement plans and work with higher education or other organizations to improve teacher training and professional development. In succeeding years, at least 90 percent of funds a state receives under the program must pass through to local agencies.

The simple four-page application asks states to describe how improvement plans will be developed, including milestones, products and timelines. States also are asked to provide descriptions of how subgrants will be made to LEAs and how plans to increase the use of technology will be developed. A budget, describing how funds will be spent, and a signed assurance of compliance with applicable federal laws complete the applications.

Goals 2000 is related to other federal education programs in the following ways:

• State participation in all aspects of Goals 2000 is voluntary, and is not a precondition for participation in other Federal programs.

• Goals 2000 is the first step toward making the Federal government a supportive partner in state and local systemic reforms aimed at helping all children reach higher standards.

• Other new and existing education and training programs will fit within the Goals 2000 framework of challenging academic and occupational standards, systemic reform, and flexibility at the state and local levels. The aim is to promote greater coherence among Federal programs and between Federal programs and state and local education reforms.

• For example, the School-to-Work Opportunities Act will support state and local efforts to build a school-to-work transition system that will help youth acquire the knowledge, skills, abilities and labor-market information they need to make a smooth transition from school to career-oriented work and to further education and training. Students in these programs will be expected to meet the same academic standards states establish under Goals 2000 and will earn portable, industry-recognized skill certificates that are benchmarked to high-quality standards such as the skill standards that will be established under Goals 2000.

• Similarly, the Administration's proposed reauthorization of the Elementary and Secondary Education Act of 1965 (ESEA) allows states that have developed standards and assessments under Goals 2000 to use them for the ESEA, thereby providing a single set of standards and assessments for states to use for their reform needs and to meet Federal requirements.

• In the future, the Administration's proposals for the reauthorization of education programs also will fit within the same framework of challenging standards and comprehensive reform.

■ PREPARING STUDENTS FOR THE HIGH-WAGE JOBS OF TOMORROW SKILL STANDARDS: WHAT THEY ARE AND WHY WE NEED THEM

Many Americans are not equipped with the academic and occupational skills that an increasingly complex job market requires. Often, they do not find stable, career-track jobs for five to ten years after leaving high school. The cost to them, to businesses and to the American economy is staggering.

American students, workers, employers and educators must be aware of the knowledge and skills that the workplace of today and of the future will demand of them. The Goals 2000: Educate America Act encourages the development and adoption of a system of skill standards and certification of an individual's attainment of such standards. Skill standards identify the specific knowledge, skill and ability levels needed to perform a given job in a given industry.

With a system of skill standards in place, these groups will benefit:

• Students in education and training programs

will know what skills are needed for high-wage employment and they can earn a credential that is portable and recognizable by employees and demonstrates they have acquired such skills.

• Employers and businesses will have reliable information to assist in evaluating workers' skill levels in making hiring and training decisions. This is especially important for small and medium-sized businesses that cannot afford to develop their own skill assessment systems.

• Training providers and educators will be accountable for the services they provide because there will be a method in place to evaluate whether the participants or students have attained skills that are relevant to the demands of the workplace.

• Unemployed Americans can seek retraining with the confidence that the skills they gain will lead to new employment opportunities.

• Labor organizations can better determine which skills and training are vital to their members' employment security.

Skill Standards and the Goals 2000: Educate America Act

Goals 2000 contains two major components — a system for helping states and localities establish high, voluntary academic standards, and a system to support business, labor, educators and the public in the development of occupational skill standards. The two are inextricably linked. A new generation of workers — those prepared for high-skill, high-wage jobs — will emerge from a restructured American education system that produces workers firmly grounded in core academic subjects and equipped with skills that are in demand in today's labor market.

To further these goals, the legislation establishes a National Skill Standards Board to encourage and assist partnerships in developing and adopting standards that are relevant to industry. The partnerships — including broad-based representation from business, labor and education — would actually develop the standards. The Board's function would be to provide financial and technical assistance in the development of the standards and to endorse standards that meet objective criteria. Standards endorsed by the Board would be linked to the highest international standards and would promote the transition to high-performance work organizations.

Through the development of broadly defined skill standards, the U.S. will be able to set goals for skill achievement, competencies and performance that will help create a lifelong learning system for all Americans and will drive our nation's economic growth into the next century and beyond.

How the New Voluntary National Education Standards Will Improve Education

When we think about how to improve our schools, one of the most important questions is: What do we want our children to know and be able to do?

Not everyone leaves school with the skills and knowledge necessary to succeed. Too many of this nation's schools offer students watered-down curricula, inadequate textbooks and outmoded teaching methods. And we have, until now, often gauged student achievement by the number of courses taken — not actual learning — and by scores on multiple-choice tests that often measure little more than low-level skills.

The results of international assessments in the 1980s and 1990s show that the skills and knowledge of American students do not measure up to their international peers. Other developed countries have something we don't: clearly defined high standards.

American students can learn more if they are challenged — both in school and at home. If students and schools are not held to high standards, they will not work hard enough and achieve as much as they can. If their parents don't show them the importance of learning, they may not have the will to learn.

■ WHAT NATIONAL STANDARDS ARE AND HOW THEY'RE BEING SET

The voluntary national standards will describe what all students should know and be able to do at certain grade levels. The standards will encourage students to use their minds well, to solve problems, to think and to reason.

Voluntary national standards will provide a focus, not a national curriculum; a national consensus, not federal mandates; voluntary adoption, not mandatory use; and dynamic, not static, applications.

Mathematics standards are already in use in many classrooms. Standards in science, history, civics and government, geography, English, economics, foreign languages and the arts are now being developed by teachers and scholars. The input of state and local leaders, parents and citizens is also being sought.

The voluntary national standards are meant to be a resource to be used by parents, teachers and all citizens as one guide to high standards. They can be used by schools, districts and states to guide and revise curricula, assessments, teacher preparation and instruction. All of the elements should be aligned so that everyone and everything involved in education work together to help students learn more.

National standards do not have to be in place before states and communities can begin to develop their own standards. Indeed, some states have already introduced high standards into their classrooms. States and communities can develop their own standards or modify and adopt those developed under national consensus.

Under the Goals 2000: Educate America Act, federal funds will flow to states and communities to help them develop their own rigorous standards and implement their own programs of school reform to help their students achieve the higher standards.

•   Higher expectations for all students. High standards and   enriched course content produce better student performance. All students can learn more than we currently ask of them. When we expect more of students, they work harder and achieve more.

•   New approaches to teaching. Helping students meet challenging standards requires new ways of teaching. Teacher   preparation and professional development programs need to be   overhauled and improved.

•   Making schools accountable We need to give schools the tools and flexibility to do their job, and then hold them responsible for results.

•   Building partnerships. We've learned that schools can't do it alone. Parents, educators, students, business, labor and public, private and nonprofit groups need to be active partners in the reform effort.

•   Supports the development of challenging voluntary academic standards that define what students should know and be able to do and offers states and local communities the support they need to put those higher standards to work in their classrooms.

•   Encourages the development of a new generation of student performance assessments and new methods of gauging student achievement that will be linked to national, state and local standards and which will be valid, reliable and free of discrimination.

•   Supports the creation of voluntary national occupational standards that, with the help of business and labor, will   define the knowledge and skills needed for the complex, high-wage jobs of tomorrow.

•   Supports a "bottom-up," grassroots approach to school reform, with the federal government assisting states and local communities in the development and implementation of their own comprehensive and innovative reform programs.

A New Federal, State, and Local Partnership...

•   Each participating state and community will develop and implement a comprehensive improvement plan that raises standards and helps students achieve them. A broad-based leadership team composed of policy makers, educators, business and civic leaders, parents and others will help create each reform plan.

States may adopt national content and performance standards or they may develop their own.

•   Federal funds will be provided to support state and local improvement efforts. By the second year of funding, 90 percent of the money will flow to local schools and districts to support their reform plans.

•   Supports the establishment of parent information and   resource centers, in order to help provide parents with the   knowledge and skills needed to effectively participate in their child's education.

The New National Partnership for Educational Excellence

The bill encourages a bottom-up approach to reform. States and local communities will develop their own improvement plans, tailored to their special needs. Business and labor will work together to define the knowledge and skills needed to create secure economic futures for employees and employers alike. The federal government will use its resources to assist local reform efforts and help them implement their improvement plans and will support the development of model standards against which states, communities, schools and individuals can measure their progress.

The Federal Role — Setting High Standards

A National Education Standards and Improvement Council (NESIC), comprised of teachers, parents, business groups, civic leaders and others, will be created to:

•   Review the efforts by national organizations of subject-matter experts to develop voluntary national content and performance standards in each subject area, such as math, science, history and geography. These will be clear statements of what students should know and be able to do as they progress through school. The standards will be far more rigorous than what is currently expected of students and will be as challenging as those in other countries.

•   Lead the effort to develop better measures of student progress and performance, measures that really reflect what we expect them to learn. New and promising assessment programs are being developed throughout the country; NESIC will keep track of changes and encourage those that advance the state of the art.

■ THE STATE ROLE — IMPLEMENTING COMPREHENSIVE STRATEGIES FOR REAL IMPROVEMENT

Each state choosing to participate will be asked to develop and implement a comprehensive improvement plan that raises standards and helps all students achieve them. Many states have already begun this work, though few have undertaken anything as ambitious as called for in this legislation. Every state will be challenged to participate and to build on local reforms already under way.

• States will be asked to form a broad-based and representative leadership team, comprised of policy makers, educators, business and civic leaders, parents and others at the grassroots level. Real and lasting change requires new partnerships working together.

• Many states will want to use the national standards as a benchmark for their own efforts. On a voluntary basis, states may submit to NESIC their content and performance standards for certification that they are as rigorous and challenging as national standards.

* In no state can all students meet challenging new standards as the schools currently operate. A fundamental overhaul is required. States will develop comprehensive reform plans and implementation strategies that will affect every aspect of the state's education system — curriculum, technology, teacher training and licensure, parental and community involvement, school management and accountability.

■ THE LOCAL ROLE - PUTTING REFORM INTO ACTION

To make a difference, reform has to occur in every school. Local school districts and individual schools also will develop and implement comprehensive improvement plans, reflecting unique local needs and circumstances, in conjunction with the state's efforts.

For the first year, $105 million in federal funds is available to implement Goals 2000 with additional funds requested in subsequent years. By the second year of funding, states will be required to use at least 90 percent of their funds to support the development and implementation of reform plans in local school districts.

■ CREATING A WORLD-CLASS WORK FORCE

American students, workers, employers and educators must know what knowledge and skills are required in the workplace. The bill encourages the development and voluntary adoption of national skill standards and certification. This effort is a critical step in establishing a lifelong learning system for all Americans, including high school students not planning to attend a four-year college, unemployed and dislocated workers and employed workers who want to upgrade their skills.

The standards will allow us to build an education and training system that ties schools, colleges and other post-secondary institutions, other job training providers and employers together in an effort to create a high-skills, high-wage work force.

■ MEDIA RESEARCH

Research evaluation and dissemination. Further research should not focus on whether the technology is as good as traditional face-to-face courses which has been established. What is needed is better and more compelling research.

Chu and Schramm (1967) stated that the effectiveness of television had been demonstrated in well over 100 experiments and that adults learn a great amount from instructional television. In 1977, after reviewing over 300 studies, Schramm also concluded that there was no significant difference between learning in classroom and television; this was again validated by Johnston (1987). Levine (1987) argues that the conclusion to draw from these studies is that television instruction is equivalent to traditional, classroom instruction in its learning; there are "good and bad television courses as there are good and bad campus-based courses" (p.16). The question is on what basis should one separate good and bad telecourses?

Bates (1974) observes that this type of research proves nothing and has been totally useless. He believes that the weakness in this research has been that the variables of content taught and styles of teaching have not been controlled; as a result, differences cannot be attributed to one medium over another. The main weakness of comparative studies is that they do not help producers or teachers to improve the product since they do not tell what is wrong or what can be done about it (Bates,1974).

We know very little about how to use television and how to support students in their use of television (Bates, 1974). The educational use of technology cannot reach its full potential until research uncovers more about the learning process and how it varies in each individual with different instructional treatment (Costello & Gordon, 1961; Saettler, 1979). For years, investigators have attempted to identify those media best suited to teaching various instructional objectives. The research has not yielded results that permit definitive statements about the superiority of one medium over another in a particular situation (Chu & Schramm, 1967; Schramm, 1977).

The pattern of research results obtained may have come about for a variety of reasons. In many studies, two media are used to present instruction and the relative effectiveness of the two are compared. Often, students learn equally well from either medium (Chu & Schramm, 1967). Kumata (1961) contends that hundreds of studies have attempted to discover an effect which is directly attributable to the delivery method; most conclude that it makes no difference whether television is absent or present.

Wagner and Wishon (1987) state that media research has not been able to provide concrete selection guidance and that research designs have decreased the ability to generalize the findings. Research has focused on the media as a product rather than on component interaction, or processual aspects which lead to learning outcomes. Others, believe that the findings reflect the situation and it does not matter which medium one chooses to teach a particular objective, as any can do the job equally well. Gagne states that "most instructional functions can be performed by most media" (1970, p. 364) but the statement in no way denies that in a given situation one medium may be more useful than others (Schramm, 1977).

There is a need to address issues of access in selecting technology. How should an institution choose between using a high technology "leading edge" media and one which is "low tech." By addressing the issue of access, an institution will be able to make the choice that provides access and maintains "open-ness" for its students.

Media research does not provide a clear direction, as it does not address the question of how media components of a course interact to produce learning outcomes where there are differences in learners, instructional treatments and content.

Mayor and Dirr reflect that we need to realize the demands that all of these changes place on learners to function independently. It is too easy to say, "Here, work on your own and integrate the materials at hand" (1986, p. 101). Mayor and Dirr assert that ways must be found to prepare students for the challenges that the new opportunities provide. To earn faculty support, we must demonstrate how to use the new tools and materials to serve them and their students by improving telecourse quality and making education accessible (Mayor & Dirr, 1986).

Withrow (1992) reports that evaluation from the Star Schools Program indicate that distance learning students enrolled in high school courses function slightly better than students in traditional classes. One reason may be that students who choose such courses are higher risk takers and more self-directed than the average student.

To be effective, distance education teachers report that they must change their style and create new opportunities for interaction (OTA, 1989). While reaching a small number of teachers today, distance learning will greatly affect the teaching force of tomorrow. Distance learning provides a variety of tools for teaching and a means to upgrade teachers' skills and encourage professional development.

Teachers can team teach with colleagues across town or across the country, discuss problems and challenges over electronic networks, observe master teachers in action, participate in professional meetings and courses, develop new skills and earn advanced degrees — all without leaving their home school. Teachers must have training, preparation, and institutional support to successfully teach with distance learning technologies. Their concerns about technology and the quality of instruction must be taken into consideration in planning distance learning efforts. Teacher input not only shapes development, it assures long-term commitment.

■ ADOPTION OF THE INNOVATION OF DISTANCE EDUCATION

Historically and currently, there has been little emphasis by educators on how to plan for, prepare, evaluate and utilize media. In the past, most university staff have been suspicious about technology and seem apathetic toward and unaware of the potential of the more sophisticated devices. Insights, wisdom, perception and precision applied in the process of media selection are an index of professionalism among educators.

Even though the greatest technological revolution of this era is considered to be information, educators seem oblivious to the potential as well as to the impact on their field. Knowles (1983) predicts that by the end of this century most education will be delivered electronically — if educators learn to use media in congruence with adult learning principles. Moore and Shannon's 1982 study of adult educators supports Knowles and reveals the adult educators' media inexperience.

Media technology has soared ahead of utilization. While media experts are using increasingly sophisticated technology, the fact is ignored that the teacher is bewildered by technology and still does not know how to use it.

Chu and Schramm (1967) found that instructional technology required instructors to learn new roles and processes which they tend to resist because they perceive difficulties in using new techniques. Russell (1979) and Coder (1983) found that faculty tend to teach by lecture as they were taught, not as they were taught to teach by using media. Coder (1983) found that due to a lack of courses, faculty were unfamiliar with learning theory, instructional design or media utilization, a fact supported by Doerken (1977) who states that studies indicted that only 17 percent of all teachers had any training in the use of media.

In 1983, Doerken reported that it would take an estimated $400 million to provide training. During the 1984-85 Annenberg Study (Riccobono, 1986) about half of the institutions offered faculty only two to seven hours of training in media but ten to 15 hours of training in the instructional uses of computers. The figures did not report how many faculty members were trained. Bates (1987b) observes that there is a major requirement to train instructors in the selection and use of media. Brey's 1991 study reported that opposition by faculty, administrators and boards of trustees is not an important impediment to starting programs; the cost of starting programs was the most often reported obstacle by community colleges.

Lack of training in the use of educational technology has also been a problem in Japan (Nishimoto, 1969) and in Britain (1966). The lack of training in the effective use of educational technology and the tendency of faculty members to continue to teach by lecture contributed to technology resistance in these countries. Studies in Japan (Nishimoto, 1969) and in Great Britain (Britain, 1966) concluded that the effectiveness of education can be improved by training the teacher to use technology. Tanzman and Dunn (1971) state that media supervisors do not provide the leadership to encourage faculty to use media and that in-service training in effective media use has not resulted in skill transfer because; 1) media techniques vary markedly from the way teachers have been taught; 2) mechanical fear; 3) lack of professional acceptance; 4) lack of funding for media experts; and 5) decision makers' resistance to technology primarily due to a lack of understanding of its use and value.

There is still a pressing need to train them in media selection and utilization. There is a need to help faculty utilize media so that learners are central to the process; mastering the technology will take time and commitment.

Unwin (1969) states that the faculty's function is to organize learning situations and interpret them after students experience them through technology. Faculty who are untrained in media selection do not effectively plan media use or ways to support instructional objectives. Only a few instructors in a thousand are now equipped to do this. Educators have had little practical help with purposeful selection. During the evaluation and adoption process the judgment of the effectiveness of media is too frequently based upon general impressions, isolated praise or criticisms, personal hunches, intuition, imitation, comfort with the media or its availability . It is vital that adopters be aware of the importance of evaluation and develop evaluation skills since sophisticated instructional demands dictate that judicious use replace hit-or-miss selection. Kressel (1986) questions how educators are to know which packages to select for which students or how to select packages adaptable to their teaching style.

Lewis' (1985) study of faculty involved with instructional media identified problems of how to: 1) convey abstract concepts and relationships between abstract concepts and concrete experience; 2) motivate

students and encourage active learning; 3) deal with learning differences; 4) encourage generic skill and ability development; and 5) obtain funds to train faculty. Faculty consider media appropriate to address the most difficult instructional problems, but also value course management, student contact and providing experiential learning (Lewis, 1985). He reports that faculty who used technology more were likely to agree that it can overcome instructional problems and are less bothered by obstacles that frustrate untrained colleagues. Faculty identified lack of training, funds, access to hardware, and lack of descriptive and evaluative software information as obstacles to effective use of technology (Lewis, 1985).

If the use of media and technology is to be increased, educators must learn how to reach educational goals and objectives through the media (Meierhenry, 1981). A 1979 EPIE study found that fewer than five percent of teacher training institutions surveyed offered courses in the selection of teaching materials. Hezel(1987) assumes that the most effective uses of technology will be made by faculty members who understand its potential, and strongly recommends that telecourse adoption should be preceded by educational technology seminars for faculty and education administrators. Henault (1971) recommends that training for adult educators be extensive while others suggest that instruction should include video production.

Methods to train instructors in media selection and use include in-service professional development provided by qualified media center personnel or qualified distance education specialists. Research found that teachers need guidance in interaction and learning style skills (Lane, 1989). In many cases the production team will also need to be trained so that curriculum and instructional design is not foreign to them. Distance education program administrators will also need to be trained so they can create the appropriate support systems for students and faculty (Lane, 1989). Other methods include publishing a newsletter to share information about instructional technology, viewing tapes on effective media use, reading about and observing the media, and visits to production houses. Teague (1981) recommends instructor training in the basic dynamics of learning, student motivation, adequacy of teaching techniques and timing of learning tasks.

Sive (1978) notes that few writers have analyzed what makes a workable selection tool. She observes that existing selection procedures may be among the factors causing the second class status of media as the purchases are made without the benefit of a thoughtful reviewing process (1983). For media to be instructional rather than supplemental aids to instruction, more sophisticated media selection procedures are indicated. Sive (1978) observes that methods to find out about media do not exist such as those for books; most media is not reviewed or rated for its suitability for use for a specific purpose; telecourse reviews receive significantly less space than educational computer programs and only a fraction of non-book media reviews; library catalogs and bibliographic tools such as Books in Print do not exist for media; there is little comparison of new and existing products; and cross-media approaches are unknown where two products on the same subject are compared. The end point of what can be done when television is combined with other media has not yet been reached; this is the beginning of an educational revolution involving many forms of telecommunications (Hewitt, 1982).

No instrument for evaluation will totally ensure that every resource used will be a positive learning experience for every user (Teague, 1981), however, using a model and an evaluating instrument based upon the model will help ensure the selection of resources that will make genuine contributions to student learning.

Two national studies were conduced to develop a model and evaluation criteria for distance education materials and teleconferencing (Lane, 1989, 1990) are discussed elsewhere. The models and evaluation instruments take into consideration how the material will function with all of its elements, the institution's services, the instructor's skills, and the target population. Both instruments serve as a bridge to inform and train distance education staff in aspects of distance education which are relevant to course adoption or program purchase.

The evaluation forms will: 1) ensure that selection committee members evaluate the same items and use the same scale for judgments; 2) guide adopters through the selection phase so that components are evaluated with the goal of student learning clearly focused; and 3) act as a training instrument for adopters who frequently do not have a media background and are not media selection experts (Lane, 1989).

The instrument trains the evaluator in media selection skills as components are evaluated and enables the evaluator to make an informed decision to adopt or reject the material after the instrument is completed. The instruments are valuable aids in achieving consistency in previewing and are usable by instructors, administrators, production team members, teleconference producers and their clients, as well as others in electronically mediated instruction fields. The instruments standardize evaluation and the models sets standards of excellence for distance education, teleconferencing and electronically mediated instruction.

As the cited literature suggests, the message that an evaluation method should be used has been regularly repeated since the inception of the telecourse. With hundreds of telecourses and 350,000 pieces of instructional media available, choosing suitable material is a problem.

■ THE CONVERGENCE OF COMPUTERS AND VIDEO = DIGITAL FUSION

Digital fusion - multimedia* - desktop media - is described as the convergence of computing, television, audio, printing and telecommunications. Each of the components has been significant. Bringing them together results in the whole having greater impact than each individual part and is one of the industry's most significant developments. The convergence of digital technologies and their use will impact the future of teleconferencing, distance education and business (Arnett & Greenberg, 1989). (*Not to be confused with the multimedia shows of the 70s-80s which were slide and film live presentations.)

Multimedia systems are those that are able to control some or all of the tasks associated with creation, development, production and post production via a single easy-to-use universal graphic console. Multimedia promises to make desktop video as significant in the 90s as desktop publishing was in the 80s. Spurred by personal computing, desktop publishing brought to millions of users the ability to create high-quality printed materials at low cost. The result has been a phenomenal rise in the amount of printed matter which is an irony in an age of electronic information.

Impact of Multimedia Computing

The fusion of these technologies will radically change how communications occur in most organizations. The demand for computers, high-capacity data communications and storage will soar as publishing of electronic documents rises. Fiber optic networks and optical storage systems are the interstate highways and warehouses of the information age. Fiber optic can transmit voice, video and data by light wave signals. With a fiber optic infrastructure, video can be exported from a studio or imported to the home. Other technologies are being explored which will also provide the bandwidth required to carry voice, data and video. These include Asynchronous Transfer Mode (ATM) and integrated services digital network(ISDN.

Computer users will have greater control over the information they receive from news, information and entertainment services. Interactive educational and reference systems will offer students new ways to learn and research but at a high market price, unless industry and government work put the technology into schools and libraries.

Audiovisual, graphic production and communications skills will become increasingly valuable in education, business and social circumstances, shifting power structures away from "left brain" (details) achievers toward a middle ground. Nicholas

Negroponte, MIT Media Lab director, says that to regard some students as "learning disabled" may disappear in favor of regarding their schools as "teaching disabled" because they do not have interactive audiovisual educational systems.

More information will be available about institutions and individuals. The ability to defend the right to privacy will continue to erode. Accelerating information overload will drive the demand for digital storage of more kinds and amounts of information, as well as expert systems and related technologies to sort through the variety of sources. The obsolescence of intellectual property rights will continue as an even greater problem. Digital communication networks that treat information as a commodity will rise; individuals and corporations will make information and services available in a free market. Electronic mail is close to becoming a standard mode of communication rather than "snail mail."

Digital Fusion

Digital fusion has given computers the ability to communicate and TV the ability to manage information. Importantly, digital fusion will give control to the non-specialist. Just as desktop publishing opened publishing to ordinary people, desktop video will also open video to ordinary people. This includes productions done by elementary school children and university professors.

The future is being shaped by new hardware and software that apply the personal computer's low-cost information management capabilities to television's uncanny ability to motivate and communicate. TV is not just video images. It is a complete commercial mass communications form, which includes sound, music, computer graphics and animation, as well as highly developed styles of editing and production.

The merger of computer and video technologies spells an economic challenge at the core of the U.S.'s serious trade conflicts with Japan according to Arnett and Greenberger (1989). The predominantly U.S. based personal computer industry will be forced to compete or cooperate with the largely Japanese consumer electronics industry as the two find themselves pushing toward similar goals.

The technological merger feeds upon itself. Driving the demand is the need to deal with the proliferation of information flowing from the same technologies that are being joined: computers, TV, printers and telephones (Arnett & Greenberger, 1989). TV, which makes us laugh and cry, brings feeling to the coldness of data processing. The computer, which helps us manage information, encourages us to stop and think in the midst of — and as an antidote to — television's frantic and frequently captivating stimulation. Using TV was a matter of choosing one

of a few channels. Information overload has made watching TV more complex, with the addition of cable channels, tapes, camcorders, remote controls and programmable VCRs. Yet, TV remains largely superficial and unable to offer information on demand.

By joining TV and computers, we combine the best aspects of each technology. The result is a powerful communications and information system that joins TV's ability to introduce and highlight a subject with the computer's ability to provide in-depth information tailored to needs. The computer changes existing media by helping one find, store, search and re-use many kinds of information. "Interactivity" is the term to describe this ability, or need, to control what is happening. Two-way communications have the highest level of interactivity, whether the communication is with a person or with a machine.

A prime objective of multimedia computing is for television's superficiality to become more appropriate and function so that it would serve as an introduction, or motivator, to the detailed information that the computer makes available through its ability to store, search and retrieve. (Arnett & Greenberger, 1989).

It is important to understand what motivates the way that each form of media presents information. Printed text (newspapers, magazines, books) allow you quickly scan to see which items you want to read. Newspapers use the inverted pyramid where the most important information is at the top of the story or at the front of the issue; readers aren't expected to read everything, but read as much as they like without worrying about missing anything. This encourages readers to stop and think about whether they want to learn more about the subject or go on.

Television's organizational style encourages one to watch everything, without scanning or considering whether the appetite for it has been satiated. It does this through the use of levels of emotional and sensory stimulation. The focus on feeling and drama is not bad; but it is an incomplete perspective. Analysis and action are discouraged by TV's goal to be exclusive unto itself (Arnett & Greenberger, 1989).

Computers tend to present narrowly focused information in the form of spreadsheets, databases or words. It's hard to establish a relationship between different kinds of computer information because the data format of each is likely to be different

■ HYPERMEDIA

Hypermedia, refers to software that can accept information in a variety of formats — text, images, graphics, sounds and video, and to link related things together interactively to create a single multimedia presentation. Multimedia can branch to a motion video presentation within a window on the computer screen, to text, or any combination of information.

Interactive documents represent a fundamental shift in the way computers are used because interactivity demands that the computer become the delivery medium. An interactive document requires the user's active participation or it stops. Like television, a multimedia document can present an overview with sensory and emotional information; like a newspaper, it lets one browse and choose how much information is wanted; like a computer file, it will allow the re-use of parts and add others to create new documents; and it allows and facilitates communication with others.

These capabilities lend themselves to a four-layer structure for multimedia information products. They are: an audiovisual surface, an information navigation system, information content, and creative tools (see chart). For a student, the four layers would correspond to hearing lectures, research, reading and writing. For a shopper, they correspond to watching advertising, shopping, purchasing and using products (Arnett & Greenberger, 1989).

The Audiovisual Surface

Audiovisual introductions, using video or graphics, are effective because they make few assumptions about the viewer's interests and motivations. An audio-visual introduction helps viewers find something in which they are interested, or motivates them to become interested. For example,the first thing that happens on an ABC News videodisc is that Peter Jennings appears and explains what the disc is about and how it works (Arnett & Greenberger, 1989).

After the introduction, the user decides where to go within the multimedia document. This could range from a single database or text file to all of the world's databases accessible by telephone. The challenge here is to determine which to search. The computer's storage and programmability turn indexes into active, two-way links among parts of a document or even between different documents. A reference to "Hamlet" in a book about politics can trigger on demand a link directly to the text of "Hamlet", and back. Powerful hypermedia software can create these kinds of contextual links, but there is a tremendous amount of manual labor involved. Global indexing and linking requires standards that are only beginning to be discussed by the international standards organization.

Content

ABC videodiscs which are viewed with a Macintosh computer and HyperCard, were developed to give teachers a compelling way to present information about recent events ahead of new textbook edi-

tions. They do not replace text; computer-stored text supplements the video. Television's motivational power draws the students into reading. Users play the disc interactively, using computers that provide the ability to access simultaneously the graphic or textual information. ABC produced a videodisc on the California earthquake in two weeks and sold it six weeks after the quake; it has produced discs on the 1988 presidential campaign and the Israeli-Palestinian conflict.

Most exciting is the use of videodiscs by students to create video reports which will not require skill in computers or video. Rich material is available from ABC, Apple's Visual Almanac, and WGBH's interactive NOVA series. For many students, the transformation from being a passive television consumer to being a creator of a viewable production, will be a watershed event. Students effectively become "programmers" in the computer and television meanings of the term and are likely to become more discerning and critical television viewers, just as improved writers become better readers (Arnett & Greenberger, 1989).

The graphic user interface has an importance that is often overlooked in that it provides a kinesthetic component. The mouse adds movement to the computer interface. Using the mouse to put something on the screen, helps you remember it, just as it is easier to remember a phone number if you punch in the numbers with your fingers. The kinesthetic component involves motor-skills to move the mouse and type words, linguistic skills to create the words, visual graphic skills (seeing the graphic element on the screen), and visual linguistic skills to read the words on the screen (Arnett & Greenberger, 1989). Research is showing that the more areas of the brain that are involved in learning, the easier it is to retrieve the information in the future.

Creative Tools

This is the most active layer. There are two kinds of production: one is production that happens entirely inside the computer, such as writing, creating graphics or animation; the second involves capturing real-world events, such as sounds and images, and storing them on computer disks. In this step, the materials that have been gathered may be synthesized, new material may be created and a new production results. To combine source materials software tools are needed. This software is likely to be increasingly "object-oriented," because it allows users to deal with information realistically (Arnett & Greenberger, 1989).

The difference between a computer paint program and a draw program is object orientation. When the user paints with the computer, the only information stored is whether a dot (pixel) is, black, white or a color. To change the pixel, it has to be erased and re-painted. In an object-oriented program, the computer stores the mathematical description of the object (such as a circle or square); changing the position or size is done without erasing the old one.

Apple's HyperCard uses on-screen buttons and text fields that are programmed by the stack's author in an endless number of situations. (A stack is a completed file that runs under HyperCard). Graphics can be imported. HyperCard has a degree of object orientation that lets users copy function buttons from one HyperCard stack to another, yet retain their functions. It allows ordinary users to create customized programs when they would not have considered programming with structured languages and gives them control that previously was beyond their reach. Reusability means that users do not have to reinvent the wheel for each program. A typical stack is like a pile of index cards that can be linked together for quick reference. The top card might show a television; clicking on the fine-tune button takes the user rapidly (with visual and sound effects) to another card that explains fine-tuning. Clicking on a return arrow sends the user back to the main card.

Critics of multimedia argue that the average user will never be able to create quality productions. This argument overlooks present and future computer capabilities, as well as acceptance of lowered quality levels. Given the content of personal communication, the demand for production quality goes down (Arnett& Greenberger, 1989). Example: home video of children; the video may be jumpy, but the viewer is so interested in the content that he or she does not focus on the production values. There will always be a need and desire for higher quality provided by professional applications and hypermedia offers professionals faster, easier and less expensive ways of doing production.

Data Storage

Information is stored on magnetic and optical discs and magnetic tape. Some forms of storage are removable such as floppy disks and CD-ROM disks. Some kinds of removable disk can be written onto, but others, such as CD-ROM cannot be changed, and thus are appropriate for publishing. Removable media are useful for distribution and storage. Non-removable (or fixed- storage systems, such as hard disk drives), are useful for local storage of data.

Multimedia systems will demand high-capacity data storage because of the huge amount of data necessary for high-quality color image, animation, audio or video, and the need to keep it in a digital electronic format to preserve interactivity. Using the most advanced video compression systems, it's possible to store just over an hour of digital video on the highest capacity, widely available kind of disk, CD-ROM. Technology is still short of being able to offer a full-length movie. What is really needed is the ability to store libraries of images and other information. The average computer's disk drive cannot store more than a few minutes of digital video data. Complex multimedia productions will require substantial storage (Arnett & Greenberger, 1989).

■ FOUR-LAYER MODEL OF MULTIMEDIA PRODUCTS AND SYSTEMS

Every multimedia product doesn't include all four layers. For multimedia computing to achieve its greatest potential, products will have to be designed to fit into this framework, to address basic market needs (Arnett & Greenberg, 1989).

I.   Audiovisual Surface - Viewing, Browsing Offers overviews and motivation
    A.   Less Interactive - Viewing
     1)   Examples:
       a)Audiovisual introductions
       b)Still graphic presentations
       c)Text overviews
     2)   Hardware
       a)Videotape players
       b)Video projectors
       c)Video encoders
       d)scan converters
       e)monitors
     3)   Software
       a)Multimedia/video post-production systems
       b)Animation
       c)Digital audio
       d)Modeling
    B.   More Interactive — Browsing
     1)   Examples
       a)Video"samplers,"
       b)Animation
       c)Graphics
     2)   Hardware
       a)Videodisc players
       b)Video digitizers
     3)   Software
       a)Hypermedia
       b)Menuing systems
       c)Synthetic actors and voices
II.   Navigation — Hunting, Grazing Information navigation to find additional materials
    A.   Less interactive — Grazing
     1)   Examples
       a)Hypertext applications
       b)Hypermedia applications
       c)tables of contents

2) Software
   a) Hypermedia
   b) Expert systems
   c) Contextual linkers
   d) Recognition systems (voice, image, etc.)
B.   More interactive — Hunting
   1) Examples
      a) Query by example
      b) Key word search
      c) Index search
   2) Software
      a) Database query generators
      b) Indexers
      c) Authoring Systems
III. Information = Reading, Viewing, Copying, Storing Content — the actual information that is stored in the system
   A.   Examples
      1) Electronic mail messages
      2) On-line libraries of text graphics, video
      3) Non-reference materials
      4) Encyclopedias
      5) Databases
      6) Reference materials of all kinds
   B.   Hardware
      1) Optical disk drives
      2) Magnetic drives
      3) Other storage systems
      4) Paper
      5) Videotape
IV. Creative and Publishing Tools — Using Creative and publishing tools that are used to manipulate existing information or create new material for distribution.
   A.   Production
      1) Everything above that creates new material
   B.   Post-Production
      1) Everything above that helps edit and assemble material
   C.   Duplication and distribution
      1) Examples
         a) Downloading files from on-line services
         b) Copying diskettes
         c) Stamping CD-ROMs
      2) Hardware
         a) Removable magnetic and optical disks
         b) Telecommunications
      3) Software
         a) Electronic mail
         b) Network system

■ COPYRIGHT AND MULTIMEDIA

Intellectual property rights in the multimedia environment are going to be a major problem. Technologies are changing, market practices are still evolving, the size of the market is unknown and the relationship of multimedia markets to traditional markets is undefined or ambiguous. To create the perception of choice in multimedia requires much more material than linear media. IBM has already invested millions of dollars in its Multimedia Developers Program and is committing millions more to establishing standards.

It may be to the developer's advantage to produce all program components rather than try to deal with an evolving industry. Acquiring intellectual property is costly and problematic. Sometimes the developer finds that the content owners do not own the rights to the content they are licensing. Multimedia developers want licensing to give them the ability to buy and use existing intellectual property rather than create it themselves. However, it is not clear what rights they will need. Content owners are equally puzzled by which rights they should sell.

Today multimedia developers use primarily stand alone storage-based publishing devices. When wide-band transmission is easily available publishing via wide area networks may be commonplace. This is already the emerging model in higher education. Per copy licensing does not work for network-based publishing. Per-use licensing does not work well for storage-based publishing even though it looks fair and it has worked well for the on-line database industry. Usage-based pricing normally requires a networked system with considerable technological and marketing overhead. It discourages new users and the experimentation and exploration that is needed to stimulate and build demand. Nobody likes to hear a meter ticking. Education networks are increasingly ubiquitous and there is a great resistance to metered information. The preferred payment method is a fixed cost which can be budgeted.

"Transclusion" is one solution — once you pay to access something in the system you have permanent personal access to it. Transclusion attempts to make network access more like buying a product.

In concurrent-use licensing, a flat fee is charged based on the number of users who can access the service at the same time.

Kaleida, a joint venture between Apple Computer and IBM, is expected to usher in multimedia — video, sound, text, graphics and animation. The new venture will develop, license and make available specifications and technologies to promote the exchange of multimedia information between a variety of computing and consumer electronics devices. It seeks to create a multimedia standard that will cross existing platforms, along with distribution networks for the data. With no current standard in place, it has been risky for producers to create expensive programs.

Once those standards are set and distribution systems are in place, multimedia could become a preferred communications vehicle for education. The authoring tool and language currently in development, called Script X, will allow multimedia producers without computer science backgrounds to write programs. Easy access by multimedia creators to libraries of program content will be a key feature of Kaleida.

The Commerce Department has released recommendations for rewriting copyright law to protect the creators of books, recordings, movies and other forms of information in the digital age. Existing copyright law doesn't make it clear that it is a violation of the copyright owner's rights to distribute a protected work over the Internet. Without this, it is felt that copyright owners will not use the Internet or the NII. Digital equipment which converts text, pictures and sound into a series of zeroes and ones that are read by a computer, can be used to easily reproduce works ranging from sound recordings to database indexes, making copies that are virtually identical to the original.

The changes in the draft copy of the recommendations are fairly minimal, but would afford copyright owners a solid legal basis to pursue violators through civil lawsuits. Although it would be virtually impossible to stop individual bootleg copies of material from being transmitted electronically, the goal is to create a legal deterrent. The fair-use privilege contained in an informal set of guidelines, now provides for the unauthorized use of copyright material in cases involving comment and criticism, news reporting and classroom teaching.

■ DISTANCE EDUCATION APPLICATIONS AND TECHNOLOGIES

The use of telecommunications technologies for distance education will continue to increase as educators grapple with decreasing dollars. It will become even more apparent that the ability to share resources through technology is a viable alternative to building more buildings.

The need to retrain 50 million American workers will be a driving factor in the continued adoption of distance education. Distance education will be used to bring credit and continuing education programming into the workplace and into the home.

As the nation's workers move to telecommuting two or more days a week, it will become accepted practice to telecommute for their education as well.

The impact of the new technologies will be felt in all areas of education and will take distance education to a different kind of level as desktop video with the new desktop video conferencing systems lead students into more involvement with one another. It will help students develop a better sense of the world. As the new technologies stabilize, the expense will

drop and make access to others as well as learning resources very cost effective.

Communication technology in use or on the shelf today can enhance the efficiency and effectiveness of education and the learning process in and out of the classroom. Intelligently applied communications capabilities, including hardware, software and network services, can be integrated to address the needs of the education community. Awareness of how communications technology can benefit education is increasing. Communications technology linked with computers can improve communication between parents and teachers, boost home learning and generate excitement in the classroom.

Equipment may seem out of reach for schools operating with tight budgets. Yet scaled-down applications are affordable and becoming increasingly popular because educators are starting to look for ways to share resources. That is crucial for urban and rural students because they often do not have access to the educational resources they need. By utilizing distance-learning applications, educators can ensure greater equity in the way resources are distributed.

New technologies and applications are developing even as we struggle to gain acceptance for older technologies and applications.

Groupware

One new category of computer software is called groupware. It allows a number of people to work on the same document at the same time. They are also called electronic-meeting systems, group-support systems, computer-supported collaboration,and collaborative technology. In 1986 with a $2 million grant from IBM, the University of Arizona developed the electronic meeting technology. Today IBM has 64 rooms equipped with the technology and the University has a Collaborative Management Room which has been used by more than 30,000 people for demonstration and research (Watkins, 1992). Computer-meeting rooms usually have a workstation for each team member and the facilitator, a large screen video at the front of the room and other audio visual equipment.

The GroupSystems software is a package of two dozen tools for group activity, such as generating and organizing ideas, establishing priorities and voting on options (also called Team Focus by IBM). Team members sit at their workstations which display two windows; one displays the question; one provides space for typing in the solutions. Responses are sent to the facilitator by pressing a function key. Using key words, the facilitator organizes the solutions so that team members can read the solutions written by others. If the project is a ranking process, each member ranks the topics. In a one hour meeting, when everyone participates simultaneously, everyone gets 60 minutes to present their solution.

Jay F. Nunamaker, Jr., developer of the electronic-meeting technology, calls the process "human parallel process" because everyone talks simultaneously. It is anonymous and allows everyone to honestly record their feelings without fear of reprisal or without the meeting being dominated by several people. Studies conducted by the University of Arizona and IBM have shown that electronic meetings take 55 percent less time than traditional meetings (Watkins, 1992). The process is efficient and team members leave the room with a transcript of the meeting in hand. Nunamaker said that in complex global organizations, no one person has all of the information. "Groups have to have a shared vision if you are going to accomplish anything. People have to challenge each other's ideas and build consensus" (Watkins, 1992).

Educational uses of electronic meeting technology is increasing. It allows group brainstorming sessions. One English professor at Gallaudet has used the technology for several years in his freshman composition courses to brainstorm topics for papers. As we move toward an increasingly interactive model, groupware use will increase as well as the innovative use of it. Nunamaker is also working on the Mirror Project which will have teams seated at one-quarter of a round table in each of four locations and make other teams visible on large screen for distance meetings so that they all appear to be seated at the same table.

■ ELECTRONIC MAIL AND DESKTOP VIDEO-CONFERENCING PROMISES TO BE PART OF THE GROUP SOLUTION.

Multimedia

To date, it has been suggested that multimedia will become a market only when the telephone companies provide us with a national fiber optic infrastructure capable of handling the massive bandwidth that each of us will need. On the other hand, we may not have to wait years for multimedia to become a telecommunications reality.

AT&T Paradyne has devised a system that uses standard existing telephone copper wire to deliver interactive multimedia applications into the home. The new system — carrierless, amplitude-phase modulation — can transmit data at a bandwidth suitable for meeting huge data capacity video images demand.

Bell Atlantic is working on a prototype that it calls Project Edison. It uses a digital technology called ADSL - asynchronous digital subscriber loop. Through the Project Edison prototype, Bell Atlantic is creating a technology that will enable multimedia — voice, data and video — to be available over the traditional copper cable (or twisted pair) that is already in place.

The promise of multimedia is to move more information more easily by doing it electronically and to provide more resources to everyone. The enabling technologies are not all in place, but it is becoming clear that the true multimedia platform is more likely to be something different. It will house a microprocessor, but we probably won't think of it so much as a computer as we will think of it as a telecommunications instrument.

Computer Simulations

Computer simulations will continue to increase in use as multimedia improves. Computer simulations give students a hands-on experience in working with the content. It has the advantage of helping students learn the content as well as apply it. Simulations are already used extensively in chemistry and physics computer labs which has been a major change in the last ten years. Computer simulations can act as a bridge between the theoretical world of simplified equations and actual experiments. They characterize the complexity of the real world. Using a simulation, the complexity of any system can be maintained and it is transformed into graphic display. They are a valuable way of learning about many types of subjects and can open a completely new way of looking at things with all of the variables in place.

■ CABLE SYSTEMS

Cable systems have a major advantage over broadcast television; the audience has a larger choice of programs to watch at any given time. This advantage is gained because a single cable can carry more than one-hundred channels. This is achieved by selecting a portion of the possible spectrum available on the cable and dividing it into frequency bands each of which can carry a full video program. In order to receive the desired channel, a converter is used. A cable company can send digitally coded signals to the converter and program them to disable or enable the reception of any channel (Rahimi, 1992).

Instructional Television Fixed Service (ITFS): The ITFS frequencies are located in the microwave region of the spectrum at six megahertz spacing. They work almost exactly the same as broadcast television. In order to receive these frequencies, special inexpensive antennas are required that are three feet across, lightweight and easy to install. The signal requires line-of-sight access between the transmission antenna and the receiving antenna. In flat terrain this could translate to between thirty to fifty miles or more with a reasonably tall transmission tower. The transmission power is much lower than broadcast television and usually varies between ten and one hundred watts. In a typical situation, where ITFS frequencies are assigned to different institutions in a given area, care is taken not to assign adjacent frequencies so as to avoid interference. Interference often occurs when the transmission sites of these frequencies are different and the strength of the signals received at a receive antenna from each transmission site, is different. In such a case, the more powerful signal will interfere with the weaker signal. Broadcasting all signals from the same antenna and at the same power eliminates this problem, allowing the assignment and operation of adjacent frequencies. Even for non-adjacent frequencies, transmission from separate towers will create problems with the positioning of the receive antenna that can only be solved by substantial increases in cost at the receive sites.

Wireless Cable

Wireless cable systems work exactly the same as a cable system, except that the wire is replaced by a portion of the airwaves spectrum. The portion of the spectrum that lends itself to this application is in the microwave region. The frequency range in this region, dedicated to educational and other public uses is the ITFS. A portion of ITFS channels were reallocated by the FCC for commercial use in the early 1980s as Multi-channel Multi-point Distributed Services (MMDS). MMDS-based wireless cable systems for entertainment do exist, but for the most part they have not proven to be financially successful. To build a wireless analog cable system, one needs to acquire a large number of ITFS frequencies, which no single institution can do. FCC rules prohibit the allocation of more than four frequencies to any one institution. Thus, multi-institutional cooperation is necessary both to acquire the frequencies and to operate from the same transmission antenna. In nearly all other respects, creating and operating a wireless system is the same as with a cable system. The one major difference is working with the FCC instead of local governments (Rahimi, 1992).

■ INTERACTIVE NETWORKS

New interactive devices will allow viewers to play along with their favorite game show, sporting event, or murder mystery. Interactive Network of Mountain View, CA, has been testing a product in Northern California. The heart of the system is a $200 portable control unit. GTE InfoTrak has a similar product in development for educational use. While TV shows are being broadcast, Interactive Network employees sit at computers and program the information that is sent to the control unit. For example, during a show when an answer is shown on the TV screen, the Interactive Network technicians send four possible questions over an FM radio signal that is picked up by the handset. The questions are shown on a small liquid crystal display screen and one answer is chosen by pressing a button on the control unit. When the correct answer is given on the show, the Interactive Network technicians immediately send the information over the radio waves to the handset. The handset adds points to the score. When the games are finished, the control unit adds up the points. By connecting the telephone cord that comes with the unit to the phone outlet and calling the score to Interactive Network, a player can compete with other interactive players for prizes....or in our case, for grades. The applications for education are exciting.

The company has been loaning units at Giants and Athletics games in an effort to get people familiar with the technology. To play, you first predict the outcome of the batter's trip to the plate. If you guess an out, you then predict how that will happen. If you guess a fly out but the batter grounds. you'll receive points for being half-right. Throughout the game the control unit displays the latest scores from other baseball games, much like the scoreboard at the ballpark. You can also get information on a particular player's batting statistics — number of hits, strikeouts, etc. — and team statistics at any time during the game.

How it works: Interactive Network producers watch the telecast and enter game calls and statistical information. From the central computer of the Network, game control data is shipped to FM stations and Interactive Network game data is simulcast along with the television broadcast. The control unit uses a telescoping antenna to receive the FM radio signal that carries the information (it may be necessary to use an FM booster). At the conclusion of an event, subscribers connect the handset to their phone cord for a 20-second call which is transmitted over a telephone digital switching network. All participants' scores are collected, results and standings tabulated, and then broadcast back to each subscriber in four minutes. The control unit has a long-lasting rechargeable battery.

Desktop Videoconferencing

At least three desktop videoconferencing products have been introduced which have the ability to project an image of the person who was called in a small window on the computer screen. The software programs allow users to share and work on documents at the same time. Common components include a software program and tiny cameras that set on top of the computer screen. Each product enables the user to see another user with the same equipment. Compression Laboratories Inc. product is called Cameo and will be available on the Macintosh and IBM platforms. The product requires a small codec and video boards for a price approaching $5,000. Northern Telecom has a desktop multimedia teleconferencing product called VISIT. Prices start at $2,900 and include a camera, video circuit board and Windows compatible communications software. It requires 56 kilobit-per-second digital telephone service. IBM has Person to Person (about $2000). The strange thing about all three products is that you have to make a separate telephone call to the person in order to talk. These products are already due for microphone upgrades. These new products will enable leaving video mail messages, answers to students questions and provide another way to increase interaction.

Desktop video will transform the video post production industry, and computers are already having a huge impact on producers of video programming. The recession forced people to take a look at alternatives for producing video. The traditional video production studio with several rooms and $2 million in equipment may be history in five years. Break throughs in digital storage technology and better video compression techniques will continually advance desktop video.

Computer Conferencing

Computer conferencing software is improving. Off-line processing is an important feature for students who do not want to run up big long distance bills. Most institutions provide an 800 number for faculty so reducing faculty on-line time is also important. As a result, read and respond off-line features are important as well as automatic filing features. When the student or faculty member uploads messages, the best programs automatically dial the service, upload the new messages to the appropriate mailboxes and download new messages automatically to the appropriate files on the student's or faculty member's hard disk, then disconnect from the service. The range of prices for these programs can vary by over $50,000. The new products are easy to use because all of the telecommunications settings are preformatted on a disk which the student transfers to his or her hard disk. Older programs required students and faculty to spend a minimum of seven hours learning the system. Training on the new programs take 30 minutes or less.

PDAs — Personal Digital Assistant

Apple's Computer wants you to call its new

Newton product a personal digital assistant — or PDA, rather than a computer. It weights about a pound, uses a special, unattached pen for data entry and runs on standard penlight batteries. The core of the Newton is a powerful microprocessor, an advanced (object oriented) operating system, a liquid crystal display and the ability to process data and electronic mail. While it can't run conventional business software, it can capture handwritten notes and sketches, sort them and organize them using artificial intelligence techniques, then send the information to facsimile machines, electronic mailboxes or other computers using wireless communications. The operating system is constructed so that it can be trained to recognize what the user is trying to accomplish and complete much of it.

Apple is planning a whole family of Newtons, each modified for different uses and with prices starting at under $1,000. Newton's might be tiny notepads, smart refrigerator magnets, electronic maps for cars, engineers' sketch pads and pocket communicators.

Tandy Corp. and Casio of Japan are teaming up with Geoworks, a software company in Berkeley, to form a more conventional and probably cheaper rival for Newton.

Libraries

Perhaps the greatest impact on civilization since the creation of the printing press will come from the ability to digitize information regardless of format. Digitization allows for the transformation of information from one form to another (e.g., print to sound) and for a range of retrieval options (Stahl, 1992). Technological advances have resulted in many changes in the library and is both a cause and cure for the information explosion. Digital technology has enabled us to generate information at faster rates and 90 percent of all information produced since 1979 has been produced in digital format. Digital retrieval systems offer the only hope of managing the information. All libraries will have to adopt new information technologies to survive; the transition rate will vary by library based on resources, need for the most current information and understanding of the paradigm shift in information technologies (Stahl, 1992). The focus has shifted from acquisition, to access through telecommunications. Rather than every library in a system buying a periodical or book, only one will buy it in electronic form and make it available to others on the system. Libraries will be able to build virtual collections and reference works specifically tailored to their users. As the ability to build comprehensive collections declines, the importance of building virtual collections using relative strengths of several libraries will grow.

Great strides have been made by the library community in sharing descriptive cataloging through systems such as OCLC and Research Library Information Network (RLIN). New electronic access software programs make it easier for patrons to do their own reference searches, then download the abstracts or full text articles from CD-ROMS attached to telephone lines. Electronic access means that libraries can remain "open" 24 hours a day. More journals and magazines are becoming available for electronic access because of the move to electronic desktop publishing. Once the publication's content has been digitized, it can be easily accessed if the publisher makes it available, or the material can be scanned into a digitized format. To get copies of visual materials, many libraries will provide fax or mail service for specific pages if necessary. Students studying at a distance from an institution need electronic access. The University of Phoenix Online Division provides an Academic Information Service (AIS) through which students request searches. Results are sent to them through the ALEX computer E-mail system.

Until now, most data could only be moved as text (ASCII) files. The future for moving print electronically is in moving text as images. A page treated as an image preserves the layout. Table 9.1.4 shows the amount of time to transmit a 25 page article with ten color images (960 megabits) over networks of different speeds (Blatecky, 1991).

Table 9.1.4

Network Transmission Speed for a 25-Page Article

Network	Speed	Time Required
Standard today	9600 bps	28 hours
T1	1.5 Mbs	10.7 minutes
T3	45 Mbs	0.36 minutes
New high-speed	1 Gbs	0.02 minutes

While access to materials in electronic format via networks will likely become a viable substitute to ownership of materials, convincing various campus constituencies is not always easy. Both libraries and academic program-accrediting bodies will need to find measures other than volume counts to determine library quality. The degree of access a library provides to needed materials will be much more meaningful than the number of volumes on the shelves (Stahl, 1992).

Research and discussion is underway to determine how charges for access to digitized materials can be handled to ensure that authors and publishers receive appropriate compensation. Software can be adapted to keep track of fees due to authors and publishers. The creation of wide area networks (such as Internet and NREN) are vitally important to libraries

if they are to provide access to the growing array of electronic resources that are available (Stahl, 1992).

Compressed Video — Digital Satellite Transmission

One of the solutions to the high price of satellite time is a new group of products which compresses and digitize NTSC video, transmits to a fraction (usually 3-6 MHz) of a satellite transponder. Depending on a system's configuration 12-18 channels of educational programming could be carried on one satellite transponder. NTU installed the Compression Laboratories Inc. SpectrumSaver system in 1992 and projects a $1 million savings in satellite time the first year. The SpectrumSaver system requires only a small decoder at each site which costs about $2,500. Scientific Atlanta's system is an upgrade to existing equipment which uses a different compression technology.

Cellular Computer Networks

Cellular computer networks will provide new options to educators as they rebuild campus infrastructures to access new telecommunications services. The national networks that are planned will extend the outreach of distance learning and continuing education so that education truly becomes an anytime, any place service. As more adults enroll, they will expect to use the technologies that are available to them every day in their workplace.

IBM Corp. has teamed with at least eight cellular phone companies to create CelluPlaNII, a nationwide wireless communications network. Major computer marketers, regional Bell operating companies and other technology vendors have targeted wireless, mobile data communications as the next hot growth area. For educators it is expected to spawn a new class of consumer product; the personal digital assistant that provides computer and communications capabilities. IBM will provide technology to the phone companies and add wireless communications capabilities to its own products. The market potential for cellular-based data services is huge; the field is expected to attract 2.6 million customers nationwide by 1997. The wireless data market will hit $175 million in 1995, up from $18 million in 1992.

Five factors will drive the growth; increased use of laptop computers; availability of small notebook and palm-top machines; introduction of personal digital assistants for the mass market; reduced prices for wireless modems; and expansion of wireless data services like CelluPlaNII.

Digital Equipment Corp. recently teamed with Bell South Enterprises and RAM Mobile Data to create Mobitex, a packet-data radio network that transmits data including electronic mail.

Hewlett-Packard Co., Sony Corp. of America, Zenith Data Systems Corp., Tandy Corp. and Casio have also shown interest in developing products in this area.

Direct Broadcast Satellite

Another new technology that may soon be used to broadcast educational programs is the digital, direct broadcast satellite. Advanced Communications Corporation (ACC) is in the process of building two high-power direct broadcast satellites, which will offer educational programming and other services. The orbital positions granted to ACC by the FCC permits simultaneous coverage of all 50 states and Puerto Rico. Two transponders on each satellite will be provided to the Foundation for Educational Advancement, which will establish the YES (Your Educational Services) Network to create and distribute educational programming. The satellites will be capable of broadcasting full-motion video, high-definition television, electronic text, computer software and digitally recorded music. A microprocessor in the receiving dish will direct the storage of satellite materials and permit the user to view or download educational programs or computer software at a convenient time. Each transponder will be able to deliver up to 100 channels of information. The foundation plans to provide receive dishes to all schools and libraries around the country.

Fiber Optic Networks

Fiber optic networks permit the flawless transmission of a course to many locations at once as well as permitting the instructor to send different kinds of materials — including multimedia presentations — to receiving sites.

■ RECOMMENDATIONS FOR FURTHER RESEARCH

The review of literature suggested that there is very little research done in distance education. Because the area continues to expand and is perceived as a viable way to offer educational programming to the masses of Americans that must be educated or retrained, it is imperative that more research be conducted in the following areas.

The literature suggests that distance education is in an expansion phase with many new postsecondary institutions, kindergarten through twelfth grade, and business joining the ranks of those which are currently offering courses through distance education technologies. Because many telecourses are available,

adopters must make decisions about the quality of the programming and related components. As a form of media, distance learning materials have an equal need for effective evaluation. Research is needed to determine the ratio of hours required to have equivalency between telecourse hours and traditional classroom hours. Research should determine whether 30-minutes of telecourse programming is equivalent to one hour in a traditional classroom. Traditionally, student classroom contact is set at 45 hours for three hours of credit. The telecourse norm for student contact is 15 video hours.

There is little understanding of adult education principles as they relate to distance education. Research should determine if adult education principles do work in distance education. Specifically the areas of interaction, self-directed learning and the use of learning contracts need to be researched.

Research should be done with telecourse students. About 25 percent of distance education program administrators conduct post-course evaluation with telecourse students. A number of post-course evaluation studies have been conducted with students taking telecourses produced by the Annenberg/CPB Project. However, no other studies of this magnitude have been conducted by other telecourse producers and made available for public use.

There is minimal understanding about how learning styles apply to distance education. Educators perceive that visual and auditory styles can be addressed through telecourses. They do not perceive that interactive, tactile or kinesthetic styles can be addressed through all components. Research in this area should ascertain if all learning styles can be addressed through telecourses.

Some research has shown that student attrition rates are reduced if local instructors write the study guide. Research should ascertain if this is a factor in attrition as well as what type of information a student needs in the study guide to motivate the student to course completion.

Some research has shown that the student's sense of isolation contributes to high attrition. Research should ascertain what the specific factors are that contribute to isolation and how they can be effectively addressed by a distance education program. Some research has shown that telephone meetings with the instructor, letters from the instructor and other contacts can reduce the sense of isolation as well as lower attrition.

Further research is needed to determine whether the textbook must be specifically written to accompany the telecourse or whether other texts are as effective for the student. There is a perception that the text should be written specifically for the telecourse which was not substantiated by this study.

As there is confusion over who should produce

portions of the faculty guide dealing with distance education, self-directed learning, student isolation and distance teaching strategies, further research is needed.

The optimal size of assignments, frequency of assignments and time frames in which assignments should be filed needs clarification. There is some research which suggests that if students turn in assignments from 14 to 40 days from the beginning of the course, they will complete the telecourse. This research needs to be replicated and the acceptable filing dates of the first assignment should be narrowed.

The realm of student motivation to complete distance education courses needs research. There is a perception that only motivated students will complete the course. There is some research which suggests that instructor contact is the motivating factor.

In the area of production, further research needs to be done on effective education strategies where only video is used. Producers need to know when graphics intrude or contribute to instruction, when music is effective in setting the pace or motivating the student to continue to pay attention, whether instructors should be paid actors or instructors, when certain treatments work, if learning from television is different form other types of learning and finally, if a "talking head" is effective and if so what makes him or her effective.

Evaluation of software is critical to ensure that quality materials are purchased which meet course objectives. This media selection model and its evaluating instrument should be an aid in the adoption process and ensure that standards of quality and excellence are considered.

The media selection model evaluating instrument (Lane 1989) contains nine sections so that evaluators using it will be required to apply specific evaluating criteria to the telecourse to determine the suitability of its use in the video instructional program (Teague, 1981). The model and the evaluating instrument consider the combination of media and factors related to the general organization of the instructional program, factors relating to the video programs and factors related to the learner (Bates,1980).

■ EDUCATIONAL POLICY

Recommendations

Education and the use of technology are at a point of evolution. Schools and institutions are being forced by changing demographics and tightening budgets to establish new priorities and re-evaluate how education can serve the needs of rural and metropolitan needs. K-12, higher education and corporate training all have issues and interests which cross. Educational technology has become so sophisticated that it permits improvements in teaching productivity and the quality of education. Further investment in research, equipment and educator training are needed to realize the potential. The primary burden to date has been carried by local and state institutions. The burden of developing the national infrastructure has been carried by the telephone companies. Additional issues include policy and licensing which will improve the cost-effectiveness of distance education programming.

Educational organizations have been dealing with policy recommendations for their members, local/state/national legislators. In July, 1991, USDLA as part of its mission to provide national leadership, convened a National Policy Forum.

The Problem

The early 1980s found the country in a severe recession. We were deeply concerned about our economic future. Many Americans were persuaded that our economic prospects were dim as long as the quality of education continued to decline. A litany of private and public studies and reports documented the decline and offered a vast array of solutions. The early 1990s find the country again slowly recovering from a recession. While there has been much debate, there has been only scattered success in restructuring and improving America's schools. Many now recognize that the decline of our educational system at all levels from K-12 through higher education, is only one of many areas of growing concern in our economic infrastructure. The nation's highways and railroads, water and waste systems, communication networks, education and corporate structures all represent areas requiring attention if we are to meet the challenges and global competition of the 21st century.

The global economy of which we are a part is information driven and operates at a pace in excess of our prior experience. A new understanding of the infrastructure standards necessary to support the kind of information based work force our nation must have is required. The new infrastructure requirement challenge the most basic premises of the American economic system.

Our current infrastructure developed the world's most productive economy. Our success was based on a national communications system unequaled by our competitors, mass production technique that made it possible to employ modestly skilled workers to produce high quality, inexpensive goods in large volumes and a transportation system that was fast and efficient. Today, communications and transportation systems are more competitive worldwide, and workers in other nations are willing to work longer hours and for lower wages than their American peers. We can continue to compete in this manner, but only at existing global wage levels with a corresponding massive decline in our standard of living; or we can revise our view of the market and the role of the worker.

High wage level societies will be those based on the use of highly skilled workers backed by advanced technologies and with ready access to a deep array of knowledge bases. Economic advances will be dependent upon improvement in intellectual rather than manufacturing productivity. In order to compete we must rebuild our economy to match the needs of the information age.

This restructuring is clearly linked to economic success and it depends on a strong education system. Redefining our national resources is not only necessary to prepare Americans for work but, even more importantly, to prepare them as citizens in a self-governing society. We must provide access to shared cultural and intellectual experience to enable citizens to make informed judgments about the complex issues and events that will characterize the 21st century. The cost of not doing so may be more than a decline in our standard of living, it may also cause erosion of our democratic tradition at an unprecedented time in history when the world is moving closer toward the democratic model. We cannot fail in our leadership now.

The critical natural resources of the information age will certainly be education and access to information. While the nation cannot ignore the pressing problems of health care, environmental waste or decaying cities, we must create a national vision that will focus on the long term economic health of the country. Without this we will not have the resources needed to combat the myriad of problems facing us in our increasingly small world. A new national vision must recognize the interdependence of education, information access and economic development. One without the others will not produce success.

The Opportunity

The USDLA represents universities, K-12 schools and corporate training interests in an association of over 1,500 members, and is uniquely suited by purpose, membership and experience to help articulate a new vision for responding to these new information age core infrastructure needs. Thus, when we called upon distance learning leaders from all over the nation in July, 1991, for our National Policy Forum, we were able to gather together some of the most experienced practitioners in the U.S. to help us explore policy issues.

Distance learning facilitates high performance education by encouraging new instructional techniques and by allowing electronic access to information from any location. Educational technology, in class or at a distance, is beginning to have a profound impact on the organization of schools, the way students are taught and coursework they can access. It is not only an educational tool, but also a driving force behind restructuring efforts in member organizations.

Many successful corporations and schools have already reorganized with technology in mind to capitalize on its potential as a problem solving and information leveling device. Many of our members represent national leadership in distance learning and their institutions do things differently to accomplish better results, often at the same or less cost. Recognizing the many demands on our national resources, educational technology can be the key to improving student and teacher performance while maximizing the use of resources. While in no way a replacement for the teacher, distance learning can cost effectively be a factor in reducing the monetary burden of rare and traditionally expensive specialized resources to the classroom. Distance technologies can expand teaching resources to include practicing scientists, business people, government leaders, health care specialists, parents and seniors and that helps to involve students. The restructured school must bring these resources to the classroom and substantially supplement or replace the dated, non-interactive material used today if we are to realize the goals of "American 2000" as set out by the President and Secretary of Education.

That students learn in a variety of ways is an accepted fact. Yet, most instruction today uses group lecture techniques that fit the learning styles of only a few. Educational technology allows teachers to customize learning and to move toward individual and small group collaborative learning. These are the very skills needed for high wage earner societies hoping to compete in a global economy.

Distance learning might better be described as personal learning, for it removes the barriers of space, time and location. Since Socrates, effective education has relied on conversation and debate between students and teacher facilitators. Personal learning technologies facilitate both, through a wide variety of resources. Interactive dialogue can happen via interactive television, conference calling or computer electronics. Delayed dialogue can happen via voice or electronic mail. No one technology, delivery system or mode of dialogue is best suited to meet all student needs.

The ubiquity of all distance learning technologies will ensure that we can reach all individuals regardless of their location, learning style or when they are available to learn.

The Barriers

There are, however, significant barriers to using these powerful new tools. Today's education, communication and information policies and regulations were developed long before the advent of distance learning capability. New technologies, particularly computers and digitally processed and transmitted information, have blurred or eliminated institutional boundaries in the once discrete world of voice, image and video. New policies must be put in place that remove these barriers so the nation can realize the benefit of distance learning.

The Recommendations

To define these new policies, the USDLA recently convened a National Policy Forum. Over ninety leaders in distance learning representing educational and corporate distance learning providers and users, equipment and transport providers and federal and state policymakers convened to debate the changes in education and communications policy needed for the 21st century.

The recommendations that follow represent their concerns and were unanimously approved by the USDLA Board of Directors as representing the interests of our membership. They encompass both education and communications policy, the inexorably linked cornerstones of our new economic infrastructure.

In order to accelerate and fulfill the tremendous potential of distance learning and educational technology, federal, state and local government should:

1. Develop a vision for national infrastructure recognizing the critical importance and interdependence of systemic educational reform and advanced telecommunication services.

2. Bring coherence to educational technology and distance learning funding and focus those resources on educational restructuring projects. All future education initiatives or policy should include distance learning as an option.

3. Develop national demonstration sites for educational technology and distance learning that disseminate research results, educational applications and effective teaching strategies.

4. Provide incentives for teacher training institutions to restructure pre-service and in-service programs recognizing the importance of communication and information technologies.

5. Provide incentives for regional and professional accreditation associations to recognize and encourage appropriate uses of distance learning technologies.

6. Ensure that financial aid programs recognize distance learning as a peer to traditional course delivery.

7. Address educational use via distance learning technologies as an issue for special attention within copyright laws.

8. Provide incentives for states to remove barriers to distance learning around teacher certification, textbook adoption and   accreditation practices.

9. Provide incentives for faculty who maximize

resources and achieve quality instruction through use of appropriate educational technologies.

Recognizing that all forms of advanced telecommunication services are critical to supporting distance learning and educational reform, federal, state and local government should:

1. Facilitate the development of a broadband educational network utilizing the public network with an open system architecture and guarantee equal access and governance responsibilities for all educational constituencies.

2. Provide incentives for telecommunications carriers to develop special pricing for educational applications.

3. Provide incentives for telecommunications carriers to provide dynamically allocated broadband service on a common carrier basis to schools, libraries, and other learning sites.

4. Remove the regulatory and business restrictions on telecommunication carriers for distance learning and educational applications.

5. Maintain "set asides" for educational applications in "RF" frequency allocations.

6. Provide incentives to ensure adequate, cost-effective access to satellite transponders for educational applications.

Restructuring American education, like all systemic change, will require significant, long term commitments of time, energy and resources from teachers, parents, students, administrators, business and government leaders. Change must and can occur in virtually all aspects of our current educational system simultaneously. Each recommendation will affect all of the learning constituents in some way. Acted on individually, their implementation will affect each constituency differently and there agenda will be undermined. If acted on as a whole, all of the learning community will benefit and the consensus needed to sustain the complex change process required for restructuring can be achieved.

■ NATIONAL EDUCATION ASSOCIATION

NEA Special Committee on Telecommunications

Clearly we are on the verge of significant growth in educational telecommunications. The participants in the wave of change are the federal government, state and local governments, the cable companies, the phone companies, computer companies, manufacturers of telecommunications equipment, school districts and institutions of higher education, to name some of the most visible. Many of us will soon find ourselves in the position of participating in and evaluating telecommunications-based projects.

Issues and Conclusions

In spite of its potential, the advent of telecommunication as an educational tool poses an array of issues that confront practitioners, their organizations and the education community. The Special Committee on Telecommunications identified issues in several areas and reached conclusions in the form of policy positions and/or recommendations on those issues.

General Principles

Role of Education Employees. Education employees play critical roles with respect to the introduction and use of telecommunications technology in schools (the term schools is intended to include institutions of higher education as well as K-12 schools). They are planners, problem solvers, designers, coordinators, technicians, researchers and evaluators. Most important, they are among the primary users of every existing telecommunications system. Their perspectives, insights, support and commitment to the use of the technology are vital for the successful implementation of telecommunications technology.

Conclusion: Education employees should be represented on committees and in groups making decisions with respect to telecommunications. The local education association should be an active participant in the decision making process. Education employees are essential in the success of any telecommunications project.

Curriculum Enhancement. A growing number of schools in the United States and overseas use telecommunications technology in instruction. The uses to which telecommunications technology have been put have led to the creation of new learning environments with new demands on teachers and students, the development of alternative teaching and learning strategies, and the growth of new communities of learners (geographically separated but electronically connected). The technology is both a new tool for teachers and a means by which scarce educational resources can be shared.

Conclusion: Telecommunications technology is an effective tool to enhance the curriculum and support the restructuring of schools.

Choice of Telecommunications: There is a wide variety of possible uses for telecommunications technology. There is also a wide variety of delivery systems. The particular use to which the equipment is put and the type of delivery system chosen should be driven by the specific educational needs of the school, the resources to which it has access, and its financial capacity. Many systems are hybrids of several delivery systems. The best system is one that meets a school's or school system's needs, is affordable and allows room for change and growth.

Conclusion: No one best model exists for the use of telecommunications technology. Schools must choose the system that is most appropriate for them.

Equity Issues

Access to Enrichment Opportunities: It is often not possible for geographically isolated or small districts to afford the array of educational courses and programs available to larger or wealthier districts. Using telecommunications technologies, school districts have been able to provide courses and enrichment programs to students who would otherwise be denied access. In recognition of the problems of many smaller, geographically remote districts, the federal government's Star Schools program has explicitly focused on the provision of courses in mathematics, science, and foreign languages to small or remote schools through satellite and other distance learning technologies.

Even within districts, educational opportunities can vary considerably. In Prince George's County, MD, the school district operates an interactive television network that brings advanced placement courses to six schools with predominantly minority populations. Minority enrollment in advanced placement courses on the interactive television network is 86 percent of total advanced placement enrollment, compared to 38 percent minority enrollment in advanced placement courses system wide. According to the system's director, many of the students taking classes over the interactive television network would not have access to the courses through other means.

Conclusion: Telecommunications technology has the capacity to reduce educational inequities within and among schools and school districts.

Funding Equity: Telecommunications technology can connect schools and classrooms with a dizzying array of on-line databases, educational courses and programs, professional development opportunities, and potential partners for collaborative educational projects. The inequity that exists when one school has newer or better textbooks than another is magnified many times in the case of telecommunications haves and have-nots.

Conclusion: The NEA should encourage the development of public and private funding to allow schools to purchase, maintain and upgrade telecommunications systems and connections.

Process Issues

Employee Involvement: Employee involvement in planning ensures that, first, the distance education effort is educationally appropriate for the needs of the school and school district. Second, employee involvement contributes to the determination of realistic staffing practices relative to the distance education effort. A distance education effort which places onerous demands upon its staff risks certain failure. Third, employee involvement leads to the development of effective instructional strategies for distance education. Finally, employee involvement in a project's planning phase is more likely to generate the buy-in of the staff as a whole. Larry Cuban in "Teachers and Machines" points out that teachers act as gatekeepers with respect to what comes into the classroom and how available learning resources are used (Cubana,1986). Without the buy-in of the staff, distance learning efforts are not likely to take hold within schools. In support of the need for a staff to buy into distance education, Dr. Linda Robers (formerly of the Office of Technology Assessment's) told the committee that one of the characteristics of a successful program is a cooperative effort among all interested parties (Roberts 1991).

Employee involvement in evaluation of the system is also important. While the educational benefits of distance education are perhaps the most critical aspect of the evaluation effort, problems resulting from staffing levels, logistical problems and specific design issues that staff would be aware of might be over looked if they were excluded from the evaluative effort.

Because a number of the issues that are likely to be generated in connection with a distance education effort concern contractual provisions, at least some of the employee representatives involved in the planning and evaluation processes should be appointed by the local association.

Conclusion: Education employers, including representatives of the local association, must be involved in all aspects of telecommunications projects.

Assignment of Staff: A number of the individuals who made presentations to the committee identified the characteristics of good teachers in a telecommunications environment. Beyond good teaching skills in conventional classrooms, teachers in environments rich in telecommunications technology must be flexible and willing to experiment. In the case of interactive television courses, teachers need to have a strong presence on camera.

In addition, participation should be voluntary. Telecommunications projects may make additional demands on teachers and on the staff involved in the effort. If staff members are not interested and willing participants, the extra demands will not be met.

Conclusion: Participants in projects involving telecommunications technologies should be recruited on the basis of skills identified as necessary for success as well as seniority. Participation should be voluntary.

Professional Issues

Licensure of Distance Learning Teachers. According to Noreen Huante, TI-IN's Manager of Program Services, 18 states (Arkansas, Florida, Idaho,

Illinois, Indiana, Kansas, Louisiana, Maryland, Minnesota, New Hampshire, North Dakota, Oklahoma, Oregon, Pennsylvania, South Carolina, Tennessee, Utah, and Vermont) require that the distance learning teachers be licensed/certified only in the state in which the programming originates. Florida, Oklahoma, and Oregon approve the curriculum, not the teachers (personal communication, October 1991).

Many state licensing agencies have no specific policies for distance learning because they do not know what it is or have not considered it as a special problem. Therefore distance learning teachers must follow the same path for licensure that regular classroom teachers follow. This has created a number of interesting situations.

California, Washington, Nevada and Colorado require distance educators to have fingerprint checks. California and Washington require criminal history checks in addition to the fingerprint checks. Distance education networks must pay the extra fees for the fingerprint and criminal history checks above the fees required for the state license. Some states require that distance educators take special courses in, for example, exceptional learning or human relations. New York requires a child abuse course. In Arizona, Wyoming and Nevada, all teachers (including distance educators) must take an examination on the state constitution, which is administered by the states themselves (and almost always administered only within the states themselves).

TI-IN is the only national network to seek to license its teachers in the states to which it broadcasts its courses. Other networks have handled the licensure problem in other ways. The Midlands Consortium (A Star Schools grantee, which has now ceased to exist as a functioning unit), whose courses were broadcast from Oklahoma State University, had nonlicensed instructors teaching courses and classroom teachers, who were licensed (but not necessarily in the subject being broadcast), serving as facilitators. The Satellite Education Resources Consortium (SERC) which broadcasts in 23 states, has persuaded the chief state school officers in those states to recognize the credentials of the educators that the network uses. The officers, in turn, have secured permission from the state boards of education to accept the credentials of these teachers (Welch 1991).

To reduce the need for emergency certificates and other expedient solutions, states or groups of states might consider the adoption of specific policies for licensing distance education. One of the problems here is that state licensing agencies have had only limited reason in the past to talk with one another about licensing requirements. Distance education highlights the insularity in which licensing requirements have historically been determined. As distance education continues to grow, there will be mounting pressure on states to adopt criteria that will make licenses more portable; however, the educational requirements for licensure should not be compromised.

In addition, because distance education requirements typically involve new competencies not now considered in granting teaching licenses, the committee believes that the NEA should encourage the licensing agencies of states involved in distance education and experts in the field to establish competency standards in the field. These standards could then be shared with school districts to be used as a basis for the selection of distance educators.

Conclusion: States should be encouraged to develop specific policies for licensing teachers involved in distance education.

■ TRAINING FOR TELETEACHERS AND FACILITATORS

Whatever kind of distance education effort is attempted (the delivery of whole courses at a distance, computer conferencing or videoconferencing with other classes around the country or world), teleteachers will have to know how to operate the equipment they are using as well as what instructional strategies may be necessary in the new classroom environments. Stimulating group research and coordinating the exchange of information among a number of remote sites using a computer bulletin board will require new skills for a number of teachers. In delivering whole courses via satellite, cable or compressed video, teachers need to know how to generate a high degree of interactivity among receiving sites (among students and with the teacher), how to stimulate peer interaction at each of the sites, how to plan and organize a course delivered at a distance, and how to make the most effective use of the camera to communicate information. According to the seven case studies cited in "Linking for Learning" (OTA, 1989), almost two-thirds of all teachers involved in the various projects had received no training prior to teaching in their respective distance education systems.

Facilitators involved in the delivery of whole courses must also be trained in the use of the equipment procedures for distributing materials, and protocols for working with teachers. It has been stressed in the literature that a good relationship between the teacher and facilitators has beneficial effects for the class (Gilcher and Johnstone, 1988).

Conclusion: Compensated training should be provided for teachers and facilitators in the use of telecommunications equipment, the development of effective materials and appropriate instructional strategies.

Training for Prospective Teachers

The record of schools of education in training prospective teachers in the use of technology in general and telecommunications technology in particular is not encouraging. According to "Linking for Learning," only 37 percent of teacher training institutions surveyed offered instruction in the use of interactive television for instruction (OTA 1989). Only 26 percent offered similar instruction in the use of audio technologies, and fewer than 20 percent of all institutions surveyed required this instruction of its students.

Another area in which teacher preparation seems to be lacking is in the integration of telecommunications and computer technologies. Both in the delivery of whole courses and in enrichment efforts, learning can be considerably enhanced by combining telecommunications technologies with materials designed to be used with in-class personal computers. A restructuring of teacher education in this area is critical.

Conclusion: Prospective teachers should receive training in telecommunications technology and the instructional strategies to be employed in its use.

■ RESTRUCTURING ISSUES

Telecommunications and International Opportunities. While many of the most visible telecommunications projects simply expand the courses available to students, others like AT&T's Learning Network or National Geographic's Kids Network actually reconfigure relationships in the traditional classroom. Both the Learning Network and the Kids Network serve to link classrooms around the country and the world in order to facilitate joint research and writing projects. Through the use of telecommunications technologies, a single classroom can literally become part of a global learning community. These technologies make it easy for teachers to organize classroom instruction around multifaceted group projects. Teachers serve as coordinators and resources for students collecting information to be shared with other classes on the network. In fact, teachers using telecommunications technology in the classroom experience substantially the same kinds of new roles that the NEA Special Committee on Technology identified in its 1989 report: collaborator, mentor/mentee, planner, researcher and seeker of new ideas (NEA 1989).

These technologies may facilitate restructuring but do not guarantee that it will occur. Restructuring requires a conscious and committed effort on the part of classroom teachers. In the committee's investigations into telecommunication technology and restructuring, it became clear that it is not necessarily the technical sophistication of the links among classrooms that is critical to change, but the way the links are used. The computer conferencing technology used in the Learning Network or the Kids Network is not as sophisticated as the fiber optic technology used in other networks. Nevertheless, often fiber optic networks simply allow a class to be taught in the traditional way in several locations at once, whereas less sophisticated technologies such as computer conferencing may allow a class to be taught differently. For restructuring to occur, teachers have to be made aware of the possibilities that are available to them. This awareness can be further reinforced by paid training and time to experiment with the new tools.

Conclusion: Telecommunication technology should be used to enhance the roles and instructional opportunities of teachers.

Student Learning: When teachers are placed in new roles with respect to their students, the students also have new opportunities open to them in the instructional process. Through the use of joint research and writing projects using multisite links, students have the opportunity to become active learners, collaborating with other students in remote sites to produce something that no one site could produce alone. Telecommunications can promote student roles such as researcher, coordinator and collaborator.

Conclusion: Telecommunications technology should be used to support the development of critical thinking and collaboration skills as well as to expand opportunities for students.

Overview:

In the images we hold of American education, none is as prominent as the self-contained classroom. The classroom is an island on which a teacher, a group of students, standardized textbooks and other limited resources determine the educational process. From time to time, the teachers and students make forays into the world outside to do research or take field trips to relevant sites. On the whole, though, the educational process is focused inward on the resources that exist within the classroom and the activities that occur there.

The use of telecommunications technology in classrooms literally inverts the typical locus of educational activities. Classrooms face the world outside rather than the world inside. Instead of islands, classrooms have become links in communications highways transmitting data, video and voice to thousands of other sites. Teachers and students have easy access to vast databases and participate in joint activities that involve classes in other states and countries by traveling on these highways.

The committee overwhelmingly agreed that education employees are central to the success of any telecommunications project and must be involved in every state — planning, design, curricular objectives, equipment selection, staffing, implementation, evaluation and future policy direction — if the project is to have any educational value.

The committee found that distance learning networks, which can broaden the horizons of both teachers and students through course offerings and educational enhancement programs, have the capacity to reduce inequities and to be effective tools in school restructuring efforts. To further develop the ties between the school and the community that are basic to successful restructuring as well as to permit access to the growing world of electronic bulletin boards and on-line databases, the committee strongly recommended telephone service for every classroom.

Noting the tendency of school systems to fund the acquisition of equipment and neglect the ongoing provision of training and maintenance, the committee emphasized the need for both technical assistance and compensated training at all levels coupled with sufficient preparation time for the development of effective learning strategies.

Although the committee found no evidence to date that education employees have been replaced by telecommunications technology, it recommended that a policy of no direct or in direct reduction of positions, hours or compensation be a part of any telecommunications project. Agreeing that there is no one best model for the use of telecommunications technology, the committee recommended that schools choose the system that is most appropriate for them and that participation on the part of education employees be voluntary and based on skills as well as seniority.

The committee recommended that states be encouraged to develop specific policies for licensing teachers involved in distance education and that standards and policies regarding facilitators in telecommunications projects be developed by the district and the local association.

The committee concluded that telecommunications providers should have the opportunity to develop, produce and distribute products and services relevant to public education; and that NEA and its state affiliates have a responsibility to inform their members concerning current telecommunications regulations and the impact of proposed regulation changes on the accessibility and affordability of telecommunications technologies. Further the NEA and its affiliates should work with appropriate agencies and government bodies to secure low-cost access to telecommunications services for schools and classrooms.

Policy Recommendation

The Special Committee on Telecommunications recommends

1. That the NEA adopt the following policy positions:

•   Education employees are essential to the success of any telecommunications project.

•   Telecommunications technology is an effective tool to enhance the curriculum and support the restructuring of schools.

•   No one best model exists for the use of telecommunications technology. Schools must choose the system that is most appropriate for them.

•   Telecommunications technology has the capacity to reduce educational inequities within and among schools and school districts.

•   Education employees, including representatives of the local association, must be involved in all aspects of telecommunications projects.

•   Maintenance, technical support, training, evaluation and staffing, as well as equipment purchases must be fully funded.

*   Participants in projects involving telecommunications technologies should be recruited on the basis of skills identified as necessary for success as well as seniority. Participation should be voluntary.

•   Standards and policies regarding facilitators in telecommunications projects should be developed by the district and local association.

•   Compensated training should e provided for teachers and facilitators in the use of telecommunications equipment, the development of effective materials, and appropriate instructional strategies.

•   Prospective teachers should receive training in telecommunications technology and the instructional strategies to be employed in its use.

•   Telecommunications technology can be an effective vehicle for professional and staff development.

•   The local association has three fundamental roles with respect to telecommunications technologies; to support efforts to improve the quality of instruction in local schools, to enhance the working conditions of its members and to protect their rights• N o reduction of positions, hours, or compensation should occur as a direct or indirect result of any telecommunications system.

•   Individuals who teach classes over interactive telecommunication networks should be given sufficient time to prepare for their classes.

•   Additional preparation time should be granted to teachers using telecommunications technology to enrich their regular programs. Class size and load should be educationally sound and determined by agreement between the district and the local association.

•   Education employees should receive compensated training provided by the district. If training occurs outside the school year, staff members should be compensated at their normal hourly rate for time spent in training.

•   The discipline of students at remote sites should be the responsibility of the district at the remote site.

•   Teleteachers should be held harmless from any and all actions, suits, claims, or other forms of liability that arise from their involvement in a telecommunications network. Employers of teleteachers should be obligated to provide a legal defense for them in the event that they are named in a negligence action.

•   The evaluation of teleteachers should be conducted openly and meet the requirements of the local collective bargaining agreement or evaluation policy.

•   Telecommunications technology should be used to enhance the roles and instructional opportunity of teachers.

•   Telecommunications technology should be used to support the development of critical thinking an collaboration skills as well as to expand opportunities for students.

•   Every classroom should be equipped with a telephone.

•   Every classroom should have access to the resources necessary to make full use of telecommunications.

•   Telecommunications providers should have the opportunity to develop, produce, and distribute products and services relevant to public education.

Action Recommendations

The Special Committee on Telecommunications recommends

1. That the NEA encourage the development of public and private funding to allow schools to purchase, maintain and upgrade telecommunications systems and connections.

2. That states be encouraged to develop specific policies for licensing teachers involved in distance education.

3. That NEA state affiliates monitor changes in state telecommunications regulations that will impact accessibility and affordability of advanced telecommunications technologies to public schools.

4. That existing NEA publications be used to inform NEA members regarding the impact of proposed amendments to current telecommunications regulations.

5. That the NEA support the efforts of the Consortium for School Networking to promote K-12 inclusion in the development of a nationwide fiber optic network backbone.

6. That the NEA and its affiliates encourage state legislatures and state public utility commissions to grant school districts reduced rates for telephone service.

7. That the NEA and its affiliates work with appropriate agencies and organizations to ensure that

telecommunications providers and networks offer free or low-cost access to schools and classrooms.

8. That the NEA and its affiliates begin to explore the ways in which schools and classrooms are prevented from receiving the benefits of telecommunications technologies and develop appropriate strategies for dealing with these problems.

■ COUNCIL OF CHIEF STATE SCHOOL OFFICERS POLICY STATEMENT (1991)

Improving Student Performance Through Learning Technologies

Potential: Learning technologies have an enormous capacity to support and advance restructuring of teaching and learning. Our nation must use technology's potential to improve elementary and secondary education and to provide all learners with the knowledge, skills and experience they need to be responsible and caring family members, productive workers and informed global citizens.

"Learning technologies" encompass a wide range of equipment and applications that directly or indirectly affect student performance. Learning technologies range from ordinary telephones, which connect parents with teachers to complex networks of satellites, cable and fiber optics, which deliver interactive, multimedia learning opportunities. Technologies are tools. Their power as learning instruments is not inherent; their effectiveness is derived from the teachers and students who use them. This effectiveness is measured by whether they improve student performance and help students reach their full potential.

Technologies offer information in a variety of formats — text, video and audio — so students can use the medium most effective for their learning. General or standard transmission of information through the technologies enables teachers to focus their energies on coaching students with their individual growth. Teachers can give special attention to certain individuals without neglecting the progress of others who are successfully guiding their own learning. Technologies enable students working individually or in small groups to take advantage of vast sources of information and work with complex connections among varied disciplines. Technologies stimulate students as active learners controlling the pace and direction of content, questions and responses.

Learning technologies can provide students and teachers equitable access to learning no matter the geographic location or fixed resources of the school. Telecommunications provide students and teachers with the information resources of distant libraries, museums and universities. Telecommunications offer courses, degree programs and career development. Learning technologies expand the opportunities of teachers, students and parents to connect learning activities in school with those in homes, community centers and other institutions. They provide access to colleagues and specialists around the world and connect student work to the problems and real work of other students and adults.

Learning technologies are the tools for productive, high-performance workers in the 21st century. In the "Information Age," the work force must be prepared to manage substantial amounts of information, analyze complicated situations for decision making, and react rapidly in a well-informed manner. Equitable availability of learning technologies is essential to prepare students to be adults with access to productive employment and community and political power. To keep up with the tools of the future workplace and the technologies of the home, all students must have access to them and master their use.

Technologies are productive tools for teachers and administrators to automate record-keeping, student information and data for accountability. They help provide convenient and timely access to essential information on student outcomes thereby helping teachers tailor instructional programs to meet specific student needs.

State Action

Most states, districts and schools have successfully used some technologies to develop effective, exciting and innovative learning environments. To stimulate systemic change and move beyond isolated model programs toward widespread integration of technology into learning, we must commit our efforts to these activities; planning at the state and local levels; funding; ensuring equitable access to technology; human resource training and support; expanding telecommunications networks; developing technology-based assessment tools; and establishing national leadership for learning technologies. To realize the potential for learning technologies, states must take action, both individually and together, as stated in the following "recommendations for Implementation."

States are at different stages in the development and use of learning technologies. Some have made bold moves or are ready to make a quantum leap in their actions. Many have completed steps such as those recommended below. Where bold actions have been taken they are applauded as examples for other states to emulate. The comprehensive order of this papers is in no way intended to slow progress of any state to back-track or adjust the previous actions to the systemic approach suggested here. Quite the contrary, the intention is to encourage the leaders who have accomplished certain steps to maintain their leadership toward complete implementation.

The recommendations that follow provide guidance for a comprehensive approach to incorporating technologies into the center of teaching and learning. These are generic proposals, intended for all states but not detailed to apply to any specific state. Each state must develop its own application, informed by this comprehensive design and cognizant that each of the components must be included in some form to ensure a complete and effective state strategy.

Recommendations for Implementation

1) Develop a State Plan for Use of Technology in Education.

States should establish a clear, long-term, strategic plan for learning technologies. The plan should provide a vision of technology's role in education services, propose effective uses of funds, ensure equitable access to technology, and maximize connections among technologies.

Provide a state vision of technology's role in education. States must communicate a clear and persuasive vision of technology's role in education to ensure that all key persons — the governor, legislators, state education agency staff, higher education authorities, school board members, administrators, teachers, parents and students — work toward a common goal for technology use.

Include certain key components in state and local plans. State and local plans for implementing the use of learning technologies should include an identification of needs; clearly defined goals and objectives; an evaluation of each selected technology's capabilities and cost-effectiveness; a description of the governance structure and systems operation; a delineation of current and future funding sources; a strategy for teacher, administrative and support staff training; strategy and schedule for implementing the plan; procedures for assisting local education agencies in the development of local technology plans; an evaluation plan; and a mechanism for modifying the plan itself. Planning is an ongoing process. Plans should be continually re-evaluated based on program outcomes, analysis of program effectiveness, new research and technology development.

Outline the responsibilities at the state, district and building levels to ensure that the technology plan is successfully developed and implemented. Each state should determine the planning process that best fits its needs; there is no single planning process for all states. With a trend towards site-based decision making, district and schools are increasingly responsible for planning and implementing technology programs to meet their specific needs. At the same time, the economies of scale derived from aggregate purchases and the use of telecommunications networks for large-scale delivery drive planning to higher levels within and among states.

While specific responsibilities vary by state, educators from different levels of each state's education system should participate in planning to achieve full integration of technology into education and to ensure clarity of responsibility and action at each level. Technology plans at each level should be developed by teams that include financial and policy decision makers, teacher and administrator representatives; post-secondary and higher education representatives; technical experts; individuals with experience in curriculum development, instructional management and assessment; and other major stakeholders in education.

Ensure that plans for other programs within state education agencies incorporate the use of appropriate technology. State education agency plans for the state and federal programs should incorporate the use of appropriate technology to ensure that technology is effectively integrated into each state education service and across the services.

Share the state vision and plan with other state agencies.

2) Ensure that the State, Districts and Schools Have Sufficient Funding to Initiate and Sustain On-Going Use of Technology as Articulated in the State Plan

Develop a bold new plan to provide steady funding for learning technologies. Technology is an integral part of education; consequently, the federal, state and local governments are responsible for providing funds to initiate and sustain the use of educational technology. Funding should cover all costs associated with the technology and the necessary support for continuing effective use, such as training, maintenance and upgrades. Avenues to decrease the cost of technology by aggregating purchases across the nation, state or regions should be developed to the full extent. To supplement federal, state, and local funds, alternative funding options, such as business-state partnerships and foundation grants, should also be pursued.

Initiate state development of learning technologies. States are in a unique position to stimulate and initiate the development of learning technology products. States should use this opportunity to undertake cost-effective funding options, such as business-state partnerships and foundation grants, should also be pursued.

Include expenditures for technology as part of capital outlay. Investments in educational technology should be considered capital expenditures, which may be depreciated over the life of the product.

3) Ensure that Students and School Personnel Have Equitable Access to Technologies for Their Learning, Teaching and Management Needs

Federal and state policies should ensure access to learning technologies. Many current federal and state policies were developed prior to the introduction of new technology into education. Such policies may now limit access to technology and, therefore, it is imperative to review them for currency and equitable access. The following issues are especially important for policy review and update.

Cost of access: To ensure that students and school personnel have affordable access to technology and information networks, it is necessary for technology providers to establish rates and other policies specifically for educational purposes. What is affordable for education may not be what is affordable for profit-making corporations. For example, state public utility commissions and the Federal Communications Commission should establish special telephone rates for education. The rates must be low enough to enable students and school personnel to take advantage of the voice, video and data services transmitted over the telecommunications systems. And the rates must also be sufficient to ensure continued investment in development of future applications for the education market. In addition, telecommunications costs should be equitable regardless of the factor of geographic location.

Information access: Intellectual property and copyright laws must be revised to increase student and school personnel access to information and provide them the flexibility to use the information for instructional purposes. These laws must also ensure that the owners and originators receive adequate recognition and financial reward. In addition, these laws and other policies should encourage development of electronically accessible information sources.

School facilities design: School facility design requirements, whether for new schools or for building rehabilitation, must support the use of learning technologies. Electrical outlets and voice, video and data lines are critical components of the modern school. School facilities must also support new instructional strategies that use technology (including individual or small-group learning, and varied workstations).

Use of federal and state funds: Federal and state policies should authorize purchase of learning technologies with funds currently earmarked for textbooks, instructional materials and learning resources.

Provide access to learning technologies both in and outside the school building just as access to textbooks is provided both in and outside school. To compensate for unequal technology resources in the home and among schools, extra effort must be made by states, districts and schools to provide all students access to learning technologies both in and outside school buildings. Schools should establish programs to loan equipment to students and school personnel for home use. Schools, libraries and other information sources should make their resources accessible during extended or non-school hours.

4) Ensure that Educators Have the Staff Support, Training, Time Authority, Incentive and Resources Necessary to Use Technology Effectively

Encourage local districts and schools to develop "technology teams." To effectively integrate technology into the classroom, teachers need to work closely with strong support teams that include principals, library media specialists, technicians and other support staff. Technology teams should include individuals with decision-making authority and expertise in technology, curriculum design, instructional design and student assessment. Technology teams should provide teachers with technical support to keep equipment operating; inform them about emerging technologies and programs; suggests ways to renew the curriculum through technology; and assist in assessing the outcomes of the learning technologies.

Provide professional development activities to facilitate full integration of technology into education. States must provide rigorous, continuous training to ensure that all educators develop the skills necessary to use technology in their work. Ongoing professional development activities should be offered cooperatively by states, local districts and vendors to provide training along with technology purchases and upgrades.

As learning technologies become more powerful and complex, teachers must increase their capacities to use technology. Teachers must learn how to operate available equipment and applications; evaluate the potential of instructional applications; integrate the technology into the curriculum; use technology for administrative and assessment purposes; and develop a willingness to experiment with technology. They must receive training to develop the group management, decision-making and coaching skills necessary to help students use technology effectively.

State and local education agency staff must be provided training that helps them understand technology's potential as an instructional, administrative, and assessment tool. They must also be encouraged to experiment with technology-based programs.

State education agency staff must join with higher education authorities to ensure that licensure requirements encourage professionals to use technology effectively in the learning environment.

Provide the time, authority, incentives and resources necessary to use learning technologies. The integration of learning technologies at the center of teaching and learning requires substantial changes from the practice of the traditional classroom. Many of the changes pertain to the role of teachers — their use of time, incentives, relationships between colleagues and the resources available to them. Examples of necessary changes follow:

Educators must have convenient access to a wide range of technologies in their schools, classrooms and homes. These include the technologies of the contemporary workplace of other professionals as well as specialized learning technologies. The more opportunity educators have to become comfortable with and competent with technology, the more likely they are to use it in teaching.

Many elements of the school day must be reviewed. Use of learning technologies may require substantially different class schedules, class lengths and class sizes. Such changes cannot be made in isolation but must be part of decisions that authorize different arrangements for cooperation and logistics.

Evaluation criteria and processes for teachers must ensure that they are fairly judged in the effective use of technology and are encouraged to use it. Current criteria and processes may effectively penalize teachers who use technology. For example, if the criteria is to require a teacher to deliver instruction, the teacher who coaches the students to use technology for "delivery" may be penalized.

5) Encourage the Development and Expansion of Telecommunications Networks

State, inter-state, national and international telecommunications networks are critical for providing students and educators equitable access to resources outside the school and establishing connections between the school and the home, the community and other outside resources.

Plan, fund and build telecommunications networks. Governors and state legislators, the President and Congress are encouraged to provide support for the coordination and expansion of current telecommunications networks and to develop new statewide, inter-state, national, and international telecommunications networks to serve education.

Advocate national standards to increase connections among and use of voice, data and wide-band video networks. Telecomputing networks should operate as national, non-proprietary standard telephone networks do. A telephone user can communicate with another user regardless of which telephone companies provide the service. A routing system is needed to communicate across telecomputing networks. National standards and policies for telecommunications are needed to ensure that the networks service education.

To expand distance learning and ensure that it meets acceptable standards, multi-state cooperative agreements are necessary for teacher qualifications and course specifications. Varying state requirements for certification and course approval currently require teachers of distance learning to meet multiple state certification and course approval requirements. In some cases, teachers are required to take physical exams and demonstrate knowledge of the state's history and government, even though they are not teaching those subjects. Multi-state cooperative agree-

ments are needed to promote high standards for teachers in a manner that facilitates the expansion of distance learning.

Multi-state agreements on standards for courses offered by distance learning are also needed to ensure effective expansion of learning opportunities.

6) Support the Use of Technology in Student Assessment to measure and Report Accumulated Complex Accomplishments and New Student Outcomes

Learning technologies are valuable tools for strengthening the teaching and learning of critical thinking and problem-solving skills and for measuring these capacities. Technology-based assessments help educators monitor student performance by allowing for: clear statement of multiple student outcomes; measurement of complex indicators of student learning; collection of data; management of information in such forms as portfolios; and the analysis processing and timely reporting of testing.

7) Develop National Leadership for Learning Technologies

The federal government should establish leadership in learning technologies. The federal government should institute processes to develop a coordinated vision for the effective use of technology in education. This vision should be based on the Office of Technology Assessment's reports, "Power on!" and "Linking for Learning." Federal leadership is essential to the nation's efforts in research and development; to provide direction in the development of the national telecommunications infrastructure; and to ensure that all federal education programs incorporate the use of technologies as summarized below.

The federal government should provide increased investment in research and development of learning technologies. To realize the full potential of learning technologies, systematic research must be conducted on how students learn, the capabilities of current and emerging technologies, and the effect of the technologies on student outcomes and the learning environment.

A national research agenda related to technology in education must be developed collaboratively by federal, state and local education agencies with the federal government playing the primary role in providing increased and consistent funding for research and development of learning technologies and instructional strategies.

The federal government should take leadership in the establishment of an infrastructure to support learning technologies. The use of learning technologies across the nation requires the federal government's leadership in establishing an infrastructure that includes fiber optic cable and other carriers to transmit all signals throughout the nation. This infrastructure must have the capacity to handle all signals including telephone calls, data transmission, fax, graphics, animation, compressed television, full-motion television, and high-definition television.

The federal government should ensure the transfer of technologies from federal agencies to state and local education agencies. The Department of Education should lead an effort to identify and disseminate learning technologies developed and used by the Departments of Defense, Energy, and Commerce and other federal agencies. The federal investment in learning technologies in such agencies is far more extensive than in the Department of Education. State and local educational systems need access to these technologies through a coordinated dissemination program. CCSSO and other national education organizations should increase advocacy for education's technology needs at the national level. Federal policies and actions on learning technologies are critical to the availability of such technology at the state and local levels.

CCSSO and other national education organizations must increase their efforts of advocacy to ensure that federal telecommunications and technology decisions support improvement of teaching and learning. Strong appeals need to be made to the President, Congress, the federal courts, the Federal Communications Commission and the Departments of Education, Commerce and Agriculture. Specifically, CCSSO should continue to take positions on learning technologies authorization, appropriations and legislation that effect the national information infrastructure and education's access as it has on the recent legislation and court rulings concerning the Bell Operating Companies' right to manufacture telecommunications equipment and provide information services. As Congress debates future actions concerning such issues as cable interconnectivity, spectrum allocation debates and intellectual property rights, CCSSO should represent educational concerns.

■ CONCLUSION

The potential for technological advances in support of teaching and learning seems limitless. Each new generation of computers, each advance in multimedia applications and each gain in telecommunications delivery opens more opportunities. Information Age realities seem close to the reach of some students, but the gap between current opportunity and actual use of technology in most schools is enormous.

We hope this paper captures a vision of the opportunities that learning technologies might provide for all. This vision will keep changing as invention follows new paths of technological creation. The vision will help us only if the states and our nation take the steps recommended here to bring the next generation's tools to the hands and minds of our students. The Council of Chief State School Officers is committed to bringing the vision and recommendations here to reality for all American students.

The Instructional Telecommunications Consortium (ITC), an affiliate of the American Association of Community Colleges (AACC), was established in 1977, and is a national leader in advancing the instructional telecommunications movement. ITC represents over 400 educational institutions from the U.S. and Canada. ITC held a Symposium on Telecommunication and The Adult Learner in 1991 (Brock, 1991). The goal of the Symposium was to bring the experience and intellect of the participants to bear on defining the major needs of technology-based distance education and to recommend appropriate future actions. The meeting planners recognized that the participants could not cover the entire field of postsecondary learning via telecommunication in one day. While the use of many kinds of technologies, both on and off campus, are burgeoning, the symposium participants focused the discussion on preproduced television courses. This decision was based on four major assumptions:

1. Colleges and universities must continue to reach out to adult learners where they are. Colleges must continue delivering education to adults in their homes and workplaces in addition to attracting adult students to campus classrooms. Television courses, among all the technology-based courses, have the best chance of reaching into every home now.

2. Lifelong learning is no longer just a pleasant, perhaps somewhat shopworn slogan. Lifelong learning must be the new reality. "Keeping America Working," a well-known AACJC project, was successful because it was predicated on keeping American learning. The education and training requirements of the workplace and professional office will continue their steep climb and rapid change. Equally important is the accelerated expansion of the knowledge base needed in both the neighborhood and global arenas. A first-class life requires continuing a first-class education. Postsecondary education must become even more sensitive to adult needs and ever more innovative in tailoring education to the realities of life.

3. The successes of preproduced television courses in meeting adult learning needs are well-documented, and they must continue. Hundreds of studies support this assumption, as does the experience of most of the symposium participants.

4. Television is the most widely used technology for college courses. Television courses enroll more adult students through more colleges than any other technology-based course. A new research study by Ron Brey, scheduled for publication in the summer

of 1991, indicates that television will increasingly be the technology choice for colleges.

The initial presentations and the following group discussion converged on five major ideas:
- Think strategically
- Identify learning needs
- Redefine a television course
- Raise the level of awareness
- Secure new sources of funding

American Association of Community Colleges (AACC)

In its publication "Building Communities: A Vision for a New Century" (1988)American Association of Community Colleges, AACC states that "In community colleges, technology is an important tool for teaching and for learning. Television extends the classroom electronically far beyond the campus. At colleges from coast to coast, students can, through on-line terminals, gain access to library resources and the word processor extends the creative power of faculty and students.

As we look to the twenty-first century, the challenge of technology in support of teaching will grow even more intense. Technology, if well used, can democratize the learning environment. Through technology, all students, regardless of their backgrounds, can travel to the moon or travel to the bottom of the ocean. Through technology, every college can provide a nine million volume library for its students. With programmed learning, students can learn at different rates.

The effective use of technology increases retention. Through technology, college officials can keep better track of who their students are, what they need and how they are progressing. The system can flag a student who is absent or who is performing unsatisfactorily. Within the classroom, in learning labs, and in tutoring sessions, new interactive computers make drill and practice much easier for the student. By increasing feedback, they expand faculty ability to improve learning. Appropriately used, technology can increase the quality of human interaction. It can for example, handle more of the routine and thus open up time for discussion.

Finally, technology should surely encourage innovation. Electronic teaching may provide effective exchanges of information, ideas and experiences. New technologies promise to enrich the study of literature, science, mathematics, and the arts through words, pictures and auditory messages. But television, calculators, word processors and computers cannot make value judgments. They cannot teach students wisdom. That is the mission of the faculty, and the classroom must be a place where the switches are sometimes turned off. To achieve this goal, the sup-

port of technology must be linked to college objectives.

The goal should be to use technology as a means, not an end. And the challenge for the community college will be to build a partnership between traditional and nontraditional education, letting each do what it can do best. If technology is not made evenly available to all students — if some colleges leap ahead while others lag behind — the gap between the haves and have nots in education will increase.

Thus, we conclude that each community college should make clear the assumptions on which its use of technology is based, and there should be a plan that precedes the hardware and is regularly updated. Most important, faculty must be involved both in establishing priorities and developing the mechanisms to support these priorities.

- We recommend that every community college develop a campus-wide plan for the use of technology, one in which educational and administrative applications can be integrated.

- We also propose incentive programs for faculty who wish to adapt educational technology to classroom needs.

- Further, we recommend that a clearinghouse be established at the American Association of Community and Junior Colleges to identify educational software of special value to the community college.

- The community college — through technology — should continue to extend the campus, providing instruction to the work place and to schools, and scheduling regional teleconferences for community forums in continuing education.

- Finally, we recommend that new uses of technology be explored. Specifically, community colleges should lead the way in creating electronic networks for learning, satellite classrooms, and conferences that connect colleges from coast to coast, creating a national community of educators who transcend regionalism on consequential issues."

OTA Report

The United States Congress' Office of Technology Assessment (OTA) released a comprehensive study of distance education. The report, "Linking for Learning: A New Course for Education." (1990) was requested by the Senate Committee on Labor and Human Resources and endorsed by the House Committee on Education and Labor. OTA has made a significant contribution to the literature regarding distance education.

The report focuses on distance learning in elementary and secondary education. It analyzes various technological options, examines current developments and identifies how federal, state and local policies could encourage more efficient and effective use of distance learning.

Chapter headings include: Distance Education in

Today's Classroom, Technology Links: Choices for Distance Learning Systems; The Teacher Link: New Opportunities for the Profession; States: Catalysts for Change; and Federal Activities in Distance Education. The appendices include a state-by-state profile of distance education activities; sample costs of transmission systems, a glossary, and contractor reports.

To summarize the document, federal activities in distance education should focus on the following:

1. Federal government funds have accelerated the growth of distance education in this country, through direct purchasing power as well as the leveraging power of the federal dollar. The Star Schools Program (Department of Education) and the Public Telecommunications Facilities Program (Department of Commerce) are the primary federal programs directly affecting distance education in elementary and secondary schools.

2. Other federal agencies have interests in distance learning through their responsibilities for technology development, training and education. Yet, no agency-wide strategy or interagency coordination is now in place. The fragmentation of telecommunications regulation and policy making may inhibit development of a coherent plan for educational telecommunications. Since the education community is diverse and speaks with many voices, it may be difficult to have its concerns articulated over the din of other stakeholders more fluent in these issues. On the other hand, the volatility of the telecommunications policy making environment may work to the advantage of education interests. Because the nation's schools represent a major market for new technology applications, the education community could create a powerful position from which to influence telecommunications policy.

3. Federal agencies will have increased opportunities to accomplish agency missions via distance delivery in the near future. The largest providers together can reach a great number of American schools and communities today, and that number will increase in the next few years. Agencies may find distance delivery an attractive way to reach national audiences for a variety of missions including education.

4. Federal telecommunications regulations are central to distance education, because they affect costs, availability and types of services. In light of the rapid growth of distance learning, it is time to review and shape federal telecommunications policies to ensure a more effective and flexible use of technology for education.

■ REFERENCE LIST

AACJC (1988). "Building Communities: A Vision for a New Century," pp. 27-28. Washington, DC. American Association of Colleges and Junior Colleges.

American Association of School Librarians (1976). Policies and procedures for selection of instructional materials, Chicago.

Anderson, Ronald H. (1976). Selecting and developing media for instruction. New York, Van Nostrand.

Armstrong, Jenny R. (1973). A sourcebook for the evaluation of instructional materials and media. Special Education Instructional Materials Center, University of Wisconsin, Madison, WI. ED 107 050.

Armstrong, M., D. Toebe and Watson, M. (1985). Strengthening the instructional role in self-directed learning activities. Journal of continuing education in nursing 16(3): 75-84.

Arnett, Nick and Greenberger, Martin (1989 work in progress). "Multimedia Computing."

Annenberg/CPB Project (1985). Research on Student Uses of the Annenberg/CPB Telecourses, Washington, DC., Annenberg/CPB Project.

Barron, D. D. (1987) Use and Barriers to Use. "Jour. Ed. Lib. Inf. Sci. 27, p. 4.

Bates, Anthony, (1974). Obstacles to the effective use of communication media in a learning system. Keynote address to the International APLET Conference, Liverpool University. Paper No. 27.

Bates, Anthony, (1975a, July). Designing multi-media courses for individualised study: the Open University model and its relevance to conventional universities. Speech at the Northern Universities Working Party for Cooperation in Educational Technology at Grey College, University of Durham, July 7, 1975. IET papers on broadcasting; Paper No. 49. Open University, England.

Bates, Anthony, (1975b, November). The British Open University: Decision-Oriented Research in Broadcasting. Speech to the National Association of Educational Broadcasters Convention, Washington, DC, November 17, 1975. Milton Keynes, Great Britain, Open University. IET papers on broadcasting: Paper No. 53.

Bates, Anthony, (1980). Towards a better theoretical framework for studying learning from educational television. Instructional Science, 9, pp. 393-415,

Bates, Anthony, (1982). Roles and characteristics of television and some implications for distance learning. Distance Education, 3, 1, pp. 29-50.

Bates, Anthony, (1987a, May). The Open University: Television, learning and distance education. Text of inaugural lecture, Open University, May 29, 1987.

Bates, Anthony, (1987b, September). Teaching, media choice and cost-effectivenss of alternative delivery systems. Speech to the European Centre for the Development of Vocational Education, Berlin, September 3-4, 1987. Milton Keynes, Great Britain, Open University. IET Paper No. 264.

Beaudoin, Michael, (1985, April 24). Chronicle of Higher Education.

Bergeson, John (1976). Media in instruction and management manual. Central Michigan University, Mt. Pleasant, MI, ED 126-916.

Berkman, D. (1976, May). Instructional television: The medium whose future has passed?" Educational Technology, pp. 34-43.

Bernard, Edward G. Evaluating media resources for urban schools. In Hitchens, Howard, Ed., (1974). Selecting Media for Learning: Readings from "Audiovisual Instruction," Washington, DC. Association for Educational Communications and Technology. Reprinted from Audiovisual Instruction, September 1971.

Blatecky, Alan, (July 1991). Presentation before the University of North Carolina University Library Advisory Council.

Blythe, N. and Sweet, C. (1979, April). The thrill of victory: A commercial TV format you can use. Audiovisual instruction, p. 22.

Borg, Walter R. and Gall, Meredith Damien. (1983). Educational research, 4th Edition, New York, Longman. PP 413-425.

Boucher, Brian G., Gottlieb, Merrill J. and Morganlander, Martin L. (1973). Handbook and catalog for instructional media selection. Englewood Cliffs, NJ, Educational Technology Publications.

Bowsher, Jack E. (1989). Education America: Lessons learned in the nation's corporations. New York, John Wiley.

Branson, R. K., Rayner, G. T., Cox, J. L., Furman, J. P., King, F. J., and Hannum, W. H. (1975). Interservice procedures for instructional systems development (5 vols.) TRADOC (Pam 350-30). Ft. Monroe, VA: U.S. Army Training and Doctrine Command, August 1975.

Bretz, R. (1971). The selection of appropriate communication media for instruction: A guide for designers of Air Force technical training programs. Santa Monica, CA: Rand.

Brey, Ronald and Grigsby, Charles (1984). Telecourse student survey 1984. Austin, TX: The Research Group.

Brey, Ronald (1988, October). Telecourse utilization survey: First annual report: 1986-87 academic year. Austin, TX. Annenberg/CPB Project and the Instructional Telecommunications Consortium.

Brey, Ronald (1991). U. S. Postsecondary Distance Learning Programs in the 1990s. Washington, DC. Instructional Telecommunications Consortium.

Briggs, L. J., and Wager, W. W. (1981). Handbook of procedures for the design of instruction (2nd ed.) Englewood Cliffs, NJ: Educational Technology Publications.

Brinberg, David and Louise H. Kidder, (eds.) (1982, June). Forms of validity in research, San Francisco, Jossey-Bass.

Brown, James W., Norbert, Kenneth, and Srygley, Sara K. (1972). Administering educational media: Instructional technology and library services, 2nd ed. New York, McGraw-Hill.

Brown, James W. (1977). AV instruction, 5th ed. New York, McGraw-Hill.

Brock, Dee (1991). Symposium on Telecommunications and the Adult Learner, Washington, DC, Instructional Telecommunications Consortium.

Bruder, I. (April, 1989) Dist. Learning "Electronic Learning" Carnegie Commission, (1979). Public trust: The report of the Carnegie Commission on public broadcasting. New York, Bantam Books. pp. 255-256.

Carpenter, Ray. (1972). Form for evaluating the instructional effectiveness of films or television programs. In Quality in instructional television, Wilbur Schramm (Ed.) Honolulu, East-West Center Book, University Press of Hawaii, pp. 205-210.

Carpenter, P. (1973, May). Cable television: A guide for education planners, R-1144 NSF, Santa Monica: Rand Center for Learning and Telecommunications, (1984). Telecourse Inventory, Washington, DC.

Chu, G. C. and Schramm, Wilbur (1967). Learning from television: What the research says. Washington, DC National Association of Educational Broadcasters.

Clark, F. E. and Angert, J. F. (1981). Teacher commitment to instructional design: The problem of media selection and use. Educational Technology, 1981, 21(5), 9-15.

Clark J. and Clark, Margaret (1983). A statistics primer for managers New York, Free Press, pp. 26-28.

Cohen, V. (1983, January). Criteria for the evaluation of microcomputer courseware. Educational Technology, 23(1), pp. 9-14.

Corporation for Public Broadcasting (1980). Telecourses: Reflections '80 Executive Summary. Washington DC. Corporation for Public Broadcasting. p. 5.

Crow, Mary Lynn (1977). Teaching on television. Arlington: The University of Texas, p. 8.

Curtis, Cally (1989, April). Dull is a four-letter word. Training Media Association Resource Supplement to Training, pp. 9-13.

DeNike, Lee and Stroether, Seldon (1976). Media prescription and utilization as determined by educational cognitive style. Line and Color publishers, Athens OH.

Diamond, Robert (1961, December). Single Room Television, Audiovisual Instruction, 6:526-27, p. 194.

Diamond, Robert M. (Ed.) (1964). A guide to instructional television, New York, McGraw-Hill.

Dirr, Peter and Katz, Joan (1981). Higher education utilization study phase I: Final report. Washington, DC.: Corporation for Public Broadcasting. Dirr Peter J. (1986 May 24) Changing higher education through telecommunication, presentation for The World Congress on Education and Technology, pp. 1-2

DiSilvestro, F. R. and Makowitz, H. J. (1982). Contracts and completion rates in correspondence study. Journal of educational research 75(4):218-21.

Doerken, M. (1983). Classroom combat: Teaching and television Englewood Cliffs, NJ. Educational Technology.

Duchastel, P. (1983). Toward the ideal study guide: An exploration of the functions and components of study guides. British journal of educational technology. 14(3):216-37.

Eash, Maurice J. (1972, December). Evaluating Instructional Materials. Audiovisual instruction, p. 37. In Hitchens, Howard, (ed.) (1974). Selecting media for learning: Readings from "Audiovisual instruction," Washington, DC. Association for Educational Communications and Technology.

Educational Products Information Exchange (1973). Improving materials selection procedures: A basic "how to" handbook. EPIE Report No. 54. New York.

Ehrmann, Stephen C.; Lemke, Randall A; and Warner, Amy Conrad (1992). "Using Technology to Open the College: The 'New Pathways' Program," in Metropolitan Universities, Vol 3 No 1, pp. 51-61.

ELRA Group, Inc. (1986, August). Executive summary: The adoption and utilization of Annenberg/CPB Project Telecourses, Washington, D.C. Annenberg/CPB Project.

Erickson, C. (1968). Administering instructional media programs. New York, Macmillan.

Erickson, C. (1972). Fundamentals of teaching with audiovisual technology, 2nd ed. New York, Macmillan.

Eurich, Nell P. (1985). Corporate classrooms: The learning business. Lawrenceville, NJ, Carnegie Foundation for the Advancement of Teaching and Princeton University Press.

Evans, R. I. (1982) Resistance to Innovation, "Inform. Tech. Jossey Bass, NY.

Farnes, Nicholas (1975, May). Student-centred (sic.) learning. Teaching at a distance. Milton Keynes, Great Britain, The Open University/Technical Filmsetters Europe Limited., No. 3, pp. 1-6.

Finkel, A. (1982). Designing interesting courses. In Learning at a distance - A world perspective, eds. J. Daniel, M. Stround, and R. Thompson. Edmonton, Canada: Athabasca University.

Flinck, R. (1979). The research project on two-way communication in distance education; An overview. EHSC Workshop paper. Malmo: Liber-Hermods.

Forrer, Stephen E. (1986). The Annenberg/CPB project; An Interview with Robben Fleming, National forum: The Phi Kappa Phi journal, Summer ,Volume LXVI, No 3. pp. 2-3.

Frankel, Martin M. and Gerald, Debra R. (1982). Projections of education statistics to 1990-91 Volume I — Analytical report. Washington, D.C.: National Center for Education Statistics.

Gagne, R. M. (1970). The Conditions of Learning (2nd ed.) New York: Holt, Rinehart, and Winston, p. 364.

Gagne, R. M. and Briggs, L. J. (1979). Principles of instructional design (2nd ed.) New York: Holt, Rinehart, and Winston.

Galagan, Patricia A. (1989, January). IBM gets its arms around education, Training and development journal, pp. 35-41.

Gallagher, M. (1977). Broadcasting and the Open University student. Milton Keynes, England: The Open University (mimeo).

Gardner, H. (June 1988) "Newslink."

Glatter, R. and Wedell, E. G. (1971). Study by correspondence. An enquiry into correspondence study for examinations for degrees and other advanced qualifications. London: Longman.

Goldstein, Michael B. (191). Keynoe address "Distance Learning and Accreditation." COPA Spring 1991 Professional Development Program. Washington, DC. COPA Professional Development Series pp. 7-17.

Gomez, Louis M. (Summer, 1992). "Erasing barriers to learning through telecommunications, Review," pp. 4-6, U S West, Portland.

Granger, D. (1990) "Open Universities: Closing the distances to learning." Change Vol 22 No. 4, pp. 44-50.

Granger, D. (1991). Symposium report: "Regulation and accreditation in distance education," The American Journal of Distance Education, Vol 5 No 1, pp. 77-79.

Gropper, G. (1976). A behavioral perspective on media selection. AV Communication Review, 24, 157-186.

Grossman, David M. (1987). Hidden perils: instructional media and higher education. In Occasional Paper, National University Continuing Education Association.

Grossman, Lawrence K. (1982, April 30). Coming together — Public television and higher education. Speech before the National Telecourse Conference 1982: "Managing Technology for Adult Learners." Dallas: pp. 1-13.

Gubser, Lyn, (1985, February/March). Is technology education's last hope? TechTrends.

Gueulette, David G., (1980). Television: The hidden curriculum of lifelong learning. Lifelong learning: The adult years, Vol. 3 (no. 5), pp. 4-7 and 35.

Gueulette, David G. (1988, January). A better way to use television in our classes. TechTrends. 33/1, pp. 27-29.

Gueulette, David G. (ed.) (1986). Using technology in adult education. Washington, D.C., Scott, Foresman/AAACE Adult Educator Booklet.

Haney, John B. and Ullmer, Eldon J. (1975). Educational Communications and Technology. Dubuque, Iowa; William C. Brown Co. p. 29.

Harman, Alvin J. (1975, July). Collecting and analyzing expert group judgment data. P-5467, Santa Monica: Rand.

Havighorst, Robert J. (1960). Developmental Tasks and Education. New York; Longman, Green.

Heidt, E. U. (1978). Instructional media and the individual learner: A classification and systems appraisal. London, Kogan Page.

Henault, Dorothy, (1971). The media; Powerful catalyst for community change. Mass Media and Adult Education. John A. Niemi, Editor. Englewood Cliffs, New Jersey; Educational Technology. pp. 105-124.

Hewitt, Louise Matthews, (1980). An administrator's guide to telecourses. Fountain Valley, CA, Coast Community College District. pp. 6-7.

Hezel, Richard T. (1987, November). Statewide planning for telecommunications in education; Executive summary: Washington, D.C., Annenberg/CPB Project.

Hewitt, Louise Matthews, (ed.), (1982). A telecourse sourcebook for the 80s. Fountain Valley, CA, Coast Community College District.

Holmberg, Borje, (1980). Aspects of distance education. Comparative education 16(2): 107-19.

Holmberg, Borje (1981). Status and trends of distance education. New York: Nichols.

Holt, Smith (1989, April). Speech at Learning by Satellite IV Conference, Tulsa, OK. San Ramon, CA. Applied Business teleCommunications.

Holt, Smith (1992). "Barriers to Quality Distance Education," in Metropolitan Universities, Vol 3, No 1, pp. 43-50.

Honey, Peter and Mumford, A. (1982). Learning Styles Questionnaire, The manual of learning styles. Berkshire: Peter Honey.

Jones, Brynmor (1965). University Grants Committee, Department of Education and Science, Scottish Education Department: Audiovisual aids in higher scientific education. London: H.M.S.O., p. 8.

Jones, Glenn (1991). "Make All America A School." Englewood, CO. Jones 21st Century, Inc.

Johnston, Jerome, (1987). Electronic learning: From audiotape to videodisc. Englewood Cliffs, NJ.: Lawrence Erlbaum Associates.

Kalton, Graham (1983). Introduction to survey sampling, Beverly Hills, Sage, p. 69.

Keegan, Desmond J. (1982). From New Delhi to Vancouver: Trends in distance education. In Learning at a distance — a world perspective, pp. 40-43. J. Daniel, M. A. Stroud and J. R. Thompson (Eds.) Edmonton, Canada: Athabasca University/International Council for Correspondence Education.

Keegan, Desmond J. (1983). On defining distance education. In Distance education — international perspectives, pp. 6-33. David Sewert, Desmond Keegan and Borje Holmberg (Eds.) New York: St. Martin's Press.

Kemp, Jerrold E. (1975). Planning and producing audiovisual materials. New York, T. Y. Crowell, p. 47.

Kemp, J. E. (1971, December). Which Medium? Audiovisual Instruction, 32-6, p. 36.Kemp, J. E. (1980). Planning and producing audiovisual materials (4th ed.) New York; Harper and Row.

Klitgaard, Robert E. (1973, March). Models of educational innovation and implications for research, P-4977, Santa Monica: Rand.

Knowles, Malcolm (1975). Self-directed learning: A guide for learners and teachers. New York, Cambridge.

Knowles, Malcolm (1983). How the media can make it or bust it in education. Media and Adult Learning, vol. 5, no. 2 Spring. In Gueulette, David G. ed. (1986). Using technology in adult education. Glenview, IL. American Association for Adult and Continuing Education, Scott, Foresman/AAACE Adult Educator Series. pp. 4-5.

Komoski, Kenneth (1977). Evaluating nonprint media. Today's Education 66:96-97 March-April. Krebs, Arlene (1991). "The USDLA Funding Source Book for Distance Learning and Educational Technology," San Ramon, CA. USDLA.

Kressel, Marilyn (1986). Higher education and telecommunications. National Forum: The Phi Kappa Phi Journal Summer, Volume LXVI Number 3. pp. 4-6.

Kumata, Hideya (1961, October 8-18). An inventory of instructional television research. Ann Arbor, MI: Educational Television and Radio Center. A report presented at the International Seminar on Instructional Television, at Purdue University, Lafayette, IN.

Ladd, Barbara (1989, April). Why self-study video training makes sense. Training Media Association supplement to Training, pp. 19-22.

Lane, Carla (1988). "Student Attrition in Distance Education Programs," Unpublished manuscript.

Lane, Carla (1989a). "A Selection Model and Pre-Adoption Evaluation Instrument for Video Programs," American Journal of Distance Education, Vol. 3 No 3, 1989, pp. 46-57.

Lane, Carla (1989b). "A Media Selection Model and Pre-Adoption Evaluation Instrument for Distance Education Media." Ann Arbor, UMI.

Lane, Carla (1990). "The Use of Audio Interaction and Self-Directed Learning Contracts in a Telecourse Offered by Satellite: Foundations of Adult Basic Education. Proceedings, Midwest Research-to-Practice Conference in Adult, Continuing & Community Education.

Lane, Carla (September, 1990). "Research Establishes National Teleconferencing Standards." San Ramon, CA. "Ed," pp. 10-12.

Lane, Carla (1992). "Model Program: Interaction Through a Mix of Media," Proceedings, Global Trends in Distance Education, University of Maine at Augusta, September 24-26.

Lesser, Gerald S., Lundgren, Rolf, and Carpenter, Ray. (1972). In Quality in instructional television, Wilbur Schramm (Ed.) Honolulu, East-West Center Book, University Press of Hawaii. pp. 213-217.

Levine, Toby Kleban, (1987). Teaching telecourses: Opportunities and options, a faculty handbook. Washington, D.C. Annenberg/CPB Project/PBS Adult Learning Service.

Lewis, Raymond J. (1983). Meeting learners' needs through telecommunications: A directory and guide to programs. Washington, D.C.: American Association for Higher Education.

Long, Thomas J., Convey, John J. and Chwalek, Adele R. (1986). Completing dissertations in the behavioral sciences and education, San Francisco, Jossey-Bass, pp. 94-95.

Lundgren, Rolf (1972). What is a good instructional program. In Quality in instructional television, Wilbur Schramm (Ed.), East-West Center Book, University Press of Hawaii.

Lynton, Ernest A. (1992). From the Editor's Desk, Metropolitan Universities, Vol 3 No 1, p. 2.

Martin, Thomas J. Jr. (1991). "NTU Annual Report 1990-1991." Fort Collins, CO., NTU, p. 1.

Martino, J. P. (1972). Technological forecasting for decision making, New York, American Elsevier, p. 27.

Matthews, E. W. (1972). Characteristics and academic preparation of directors of library-learning resource centers in selected community junior colleges. Carbondale, IL: Southern Illinois University. ERIC Document Reproduction Service No. ED 110 127.

Matsui, J. (1981). Adult learning needs, life interests and media use: Some implications for TVOntario. Toronto: TVOntario (mimeo).

Mayor, Mara and Dirr, Peter J. (1986). "Telelearning" in Higher Education. National forum: The Phi Kappa Phi journal, Summer 1986 Volume LXVI Number 3. pp. 7-10.

McCutcheon, John W. and Swartz, James (1987, September). Planning for Cablecast Telecourses, T.H.E. journal, pp. 98-102.

Meierhenry, W. C. (1981, Fall). Adult education and media and technology. Media and adult learning,

Vol. 4, no. 1. In Gueulette, David G. ed. (1986). Using technology in adult education. Glenview, IL. American Association for Adult and Continuing Education, Scott, Foresman/AAACE Adult Educator Series. pp. 2-3.

Menmuir, K., (1982). Educational technology by distance learning. Media in education and development 14(4):9-11.

Merrill, M. David, and Goodman, Irwin (1972). Selecting instructional strategies and media: A place to begin. Provo, UT, Division of Instructional Services, Brigham Young University.

Moore, Richard L. and Michael C. Shannon. (1982, February). Meeting needs for continuing education through advances in technology; Lifelong learning, vol. 5, no. 6, pp. 4-6, 35. In Gueulette, David G. (ed.) (1986). Using technology in adult education. Glenview, IL. American Association for Adult and Continuing Education, Scott, Foresman/AAACE Adult Educator Series. pp. 17-21.

Munshi, Kiki Skagen (1980). Telecourses: Reflections '80. Washington, D.C.: Corporation for Public Broadcasting.

Myers, Sheldon (1972). A study of the educational technologies of computer-assisted instruction, instructional television, and classroom films, based on tour sites. EPIE Report No. 435. National Committee for Citizens in Education, (1974). Fits and misfits; What you should know about your child's learning materials, Columbia, MD.

National Education Association (1976). Instructional materials; Selection for purchase. Rev. ed, Washington D.C. ED 130-380.

Niemi, John (1971). The labyrinth of the media: Helping the adult educator find his way. Mass Media and Adult Education. Englewood Cliffs, New Jersey: Educational Technology Publications, Inc., pp. 35-47.

Nishimoto, M. (1969). The development of educational broadcasting in Japan. Tokyo: Charles E. Tuttle.

Nolan, Ernest I., (1984, April). Planning for telecommunication in the liberal arts college, T.H.E. journal, pp. 82-85.

Norman R. F., (1967, July). Assets and liabilities in group problem solving: The need for an integrative function. Psychological Review, Vol. 74, No. 4, pp. 239-249.

Northcott, Paul and Holt, Dale (1986, February). Professional development programmes for accountants through distance education: An Australian study in programmed learning and educational technology, Journal of the Association of Educational and Training Technology, Vol 23, Number 1.

Office of Technology Assessment (1989). "Linking for Learning: A New Course for Education." Washington, DC, U.S. Government Printing Office.

Pascarella, E. T., and Chapman, D. W., (1983). A multi-institutional, path analytic validation of Tinto's model of college withdrawal. American educational research journal: 20:87-102.

Parlett, M. and Woodley, A. (1983). Student drop-out. Teaching at a Distance, 24.Pease, P., and Tinsley, P. (1986) Eval. Diff. and Adoption. "Teleconf. Elec. Comm, pp. 399-404.

Perrin, D. G. (1977). Synopsis of television in education. In J. Ackerman and L. Lipsitz (Eds.), Instructional television: Status and directions. Englewood Cliffs, NJ. Educational Technology. pp. 7-13.

Perry, Walter, (1977). The Open University, San Francisco: Jossey-Bass.

Pfeiffer, J., and Sabers, D., (1970). Attrition and achievement in correspondence study. National Home Study Council News, February supplement. Washington, D.C.: National Home Study Council.

Portway, Patrick, (1989, April 1). Speech at Learning by Satellite IV Conference, Tulsa, OK. San Ramon, CA. Applied Business teleCommunications.

Powell, J. T. (1983). A practical program to use media for staff development. Media and Methods, pp. 12.

Powell, J. T. (1982a, September). Faculty development through use of media: Part I. General planning precepts. Media and Methods, pp. 18.

Powell, J. T. (1982b). Faculty development through use of media: Part II. A general plan in five phases. Media and Methods, pp. 36-38.

Purdy, Leslie N. (1980). The history of television and radio in continuing education. New directions for continuing education: Providing continuing education by media and technology. No. 5. San Francisco: Jossey-Bass.

Quinn, Pamela K. and Adams, Sandy (1984). A guide to Dallas telecourses, Dallas, Dallas Community College District, p 4-7.

Rahimi, M. W. (1992). "Cable Technology to Enhance Service to a Metropolitan Area" in Metropolitan Universities, Vol 3 No 1, pp. 28-34.

Regents College Institute for Distance Learning (1990). Proceedings of the Regents College Institute for Distance Learning Invitational Symposium, Thornwood, NY.

Reider, William L. (1985). VCRs silently take over the classroom, TechTrends, Nov/Dec. pp. 27-29.

Reilly, Kevin and Gulliver, Kate (1992). "Interstate Authorization of Distance Higher Education via Telecommunications: The Developing National Consensus in Policy and Practice," American Journal of Distance Education, Vol 6 No. 2, pp. 3-15.

Reiser, R. A. (1981). A learning-based model for media selection: Development (Research Product 81- 25b). Alexandria, VA: Army Research Institute.

Distance Education, Vol 6 No. 2, pp. 3-15.

Reiser, Robert A. and Gagne, Robert M. (1983). Selecting Media for Instruction, Englewood Cliffs, NJ, Educational Technology Publications.

Rekkedal, T., (1982). The drop-out problem and what to do about it. In Learning at a distance — A world perspective, eds. J. S. Daniel, M. A. Stroud, and J. R. Thompson, Edmonton, Canada: Athabasca University.

Riccobono, John A. (1986). Instructional technology in higher education; A national study of the educational uses of telecommunications technology in American colleges and universities; Executive summary; Washington, D.C. The Corporation for Public Broadcasting, The Annenberg/CPB Project and The Center for Statistics, U.S. Department of Education.

Romiszowski, A. J. (1974). The selection and use of instructional media. London: Kogan Page.

Sackman, H. (1974, April). Delphi assessment: Expert opinion, forecasting, and group process. R-1283-PR, Santa Monica: Rand.

Saettler, P. (1979). An assessment of the current status of educational technology. Syracuse, NY: Syracuse University. (ERIC document Reproduction Service No. 18-30).

Salomon, Gavriel (1983, September). Using television as a unique teaching resource for OU courses, England, Open University, IET Papers on Broadcasting No. 225.

Schoch, L. A. (1983). Author's guide to independent study. Bloomington, Indiana: Trustees of Indiana University.

Schramm, Wilbur, (1967, January). Instructional television promise and opportunity, Monograph Service, 4, pp. 1-20.

Schramm, Wilbur (ed.) (1972). What the research says. In Quality in instructional television. Honolulu: University Press of Hawaii.

Schramm, Wilbur (1977). Big media, little media, tools and technologies for instruction. Beverly Hills: Sage Publications.

Seidman, (1986) Survey of Teacher Utilization, "Ed Tech," p.19.

Sewart, D. (1981). Distance teaching: a contradiction in terms? Teaching at a distance 19:8-18.

Sewert, David (1982). Individualizing support services. In Learning at a distance — a world perspective, pp. 27-9. J. Daniel, M. Stroud, and J. Thompson, (Eds.) Edmonton, Canada: Athabasca University/International Council for Correspondence Education.

Sive, Mary Robinson (1978). Selecting instructional media: A guide to audiovisual and other instructional media lists. Littleton, CO. Libraries Unlimited, Inc.

Sive, Mary Robinson. (1983). Selecting instructional media: A guide to audiovisual and other instructional media lists, 2nd edition. Littleton, CO. Libraries Unlimited, Inc.

Sleeman, P. J., Cobun, T. C., and Rockwell, D. M. (1979). In Instructional media and technology . New York: Longman.

Smith, M. H. (Ed.) (1961). Using television in the classroom. New York: McGraw-Hill.

Sonquist, John A., and Dunkelberg, William C. (1977). Survey and opinion research: Procedures for processing and analysis. Englewood Cliffs, Prentice-Hall. p. 7.

Stahl, Wilson M. (1992). "The Future Impact of High-Performance Networks: Library Collections, Facilities and Services," Metropolitan Universities, Vol 3 No 1, pp. 66-74.

Stephens College, (1962). A Stephens Challenge, Information Brochure, Columbia, MO, Stephens College. pp. 20.

Stephens, D. (1979). Motivating students in correspondence courses. Continuum 43(3): 27-38.

Stoffel, Judith A. (1987). Meeting the needs of distance students: Feedback, support, and promptness, Lifelong Learning: An omnibus of practice and research, Vol. 11, No. 3.

Svenning, L. L., and Ruchinskas, J. E. (1986) Decision Factors, "Teleconf. Elec. Comm. pp. 258-266.

Tanzman, Jack and Dunn, Kenneth. (1971). Using instructional media effectively. West Nyack, NY Parker.

Teague, Fred A. (1981). Evaluating Learning resources for adult. Media and Adult Learning, vol. 4, no. 1, Fall. pp. 27-33. In Gueulette, David G. ed. (1986). Using technology in adult education. Glenview, IL. American Association for Adult and Continuing Education, Scott, Foresman/AAACE Adult Educator Series.

Thompson, J. J. (1969). Instructional communication. New York, Van Nostrand.

Thompson, Loran T. (1973, June). A pilot application of Delphi techniques to the drug field: Some experimental findings,   R-1124, Santa Monica: Rand.

Tosti, D. T. and Ball, J. R. (1969). A behavioral approach to instructional design and media selection. AV communication review, 17, 5-25.

Tough, Allen. (1979). The adult's learning projects. Toronto: Ontario Institute for Studies in Education.

Turner, Philip M., (1985). A school library media specialist's role, Littleton, Colorado, Libraries Unlimited.

U.S. Department of Education (1987, February). Office of Educational Research and Improvement. Center for Education Statistics. Enrollment in colleges and universities, Fall 1985. Bulletin: OERI, 5-6.

Unwin, D. (1969). Media and methods: Instructional technology in higher education. London, McGraw-Hill. pp. 136-142.

Vehige, B. (1989, April). Speech at Learning by Satellite IV Conference, Tulsa, OK. San Ramon, CA. Applied Business teleCommunications.

Wagner, Ellen D. and Wishon, Phillip M. (1987). In International journal of instructional media, Vol. 14, No. 4.

Ward, Terry A. (1986, December). Statview converts raw data into useful information. MacUser, p. 91.

Watkins, Beverly T. (September 16, 1992). "Universities Try Electronic Format to Make Meetings More Productive," The Chronicle of Higher Education, p. A22.

Weingartner, C. (1974). Schools and the future. In T. Hippie (Ed.) The future of education: 1975-2000. pp. 182-206. Pacific Palisades, CA: Goodyear.

Whisler, J. S. (1987). Dist. Learn. Tech. Mid Cont. Reg. Lab. Report.

Withrow, Frank (1992). "Distance Learning: Star Schools" in Metroplitan Universities, Vol 3 No1, pp. 61-65.

Wong, A., and Wong, S. C. P. (1978-79). The relationship between assignment completion and the attrition and achievement in correspondence courses. Journal of educational research 72:165-68.

Zigerell, James J. (1986). A guide to telecourses and their uses, Coast Community College District: Fountain Valley, CA., p. 35.

Chapter 10

CURRICULUM DESIGN

by Carla Lane, Ed.D.

Editor's Note: It is not the purpose of this chapter to cover traditional curriculum and instructional design considerations. The information in this chapter covers the content as it relates to teleconferencing and distance learning.

During the last several years research has suggested that there is more that can be done to make curriculum design and instructional design more effective. The ensuing discussion has taken place on a number of planes.

1. Mimetic learning (where the student mimics the instructor) needs to become transformational (where a change is made in the student).

2. Education must become learner centered rather than teacher centered.

3. Students learn differently and have preferred learning styles. Teaching to these learning styles results in a better learning environment for the student.

4. The instructor must compensate when there is a mismatch between his or her teaching style and the student's learning style.

5. There is a need to write better behavioral objectives or learning outcomes.

6. Students do not know how to learn because they do not understand their learning style. Their learning style should be evaluated and they should receive education in how they can choose learning resources that work best with their learning style. They should also be encouraged to broaden their preferred style so that they can learn from many types of inputs and will not be limited in their learning ability. Learning how to learn was identified as the most important skill in an American Society of Training and Development (ASTD) and Department of Labor (DOL) national study (Carnevale et al, 1990).

7. Instructors do not know how to learn, do not understand their learning style, and therefore do not understand the concept of learning styles or how their preferred teaching style meshes with student styles.

8. Teachers have not taught to their students' individual learning styles but have relied on the traditional role model of teaching — lectures, reading assignments and worksheets.

9. Experiential learning is a valid method of teaching and learning, Real problems should be used.

10. Instructors are not competent in media selection, media use or media development because they have not received formal training in media.

11. There is a theory that a certain type of media should be used to teach a certain type of content. There is very little work being done in what type of instruction or instructional media works best with a particular learning style.

12. We do not have a firm understanding of how a mix of media affects learning.

13. It has become accepted that the use of technology should be content driven. We are just beginning to understand that the use of technology should also be driven by the student's learning style.

14. It has become accepted that live face-to-face interaction is necessary in distance learning because it seems to more closely resemble the traditional classroom model. Because distance learning is easier to sell to faculty if it has an interactive component, real-time interactivity through a two-way audio system has become the acceptable minimum for distance learning systems. However, several research studies have shown that asynchronous interaction is equally effective and that students will learn and that teachers can teach even if they cant see one another in real time.

15. The development of self-directed learners as a component of a distance learning program has largely been ignored.

The resolution of all of these questions is not yet at hand. As courseware is created and the use of media increases, it is important to be aware of these questions and apply the knowledge that we have to date.

■ BEHAVIORAL OBJECTIVES

To write behavioral objectives for a class, one must know and precisely describe the kind of performance learners should be capable of doing at the end of the training/instructional period (ASTD, 1985). Objectives provide the framework for the course and may help in the selection of the media and methods that are used in the course, as well as how testing will be constructed. Behavioral objectives are also called learning, training or performance objectives. They are an integral part of the instructional design, but are equally important for the instructor and student working through the materials. Exact statements will help learners know how much they have accomplished and how much more they are required to learn.

Objectives are classified according to learning domain. Objectives are classified under the cognitive (knowledge) domain in which the learner might critique, analyze or evaluate something; psychomotor (skills) domain in which the learner might operate equipment or coordinate; and affective (attitudes) domain in which learners might show, demonstrate or exhibit a change of attitude or outlook through specific behaviors.

Objectives in the cognitive domain include all task performances and behaviors that use knowledge of certain information. This domain may involve knowledge of terminology, specific facts, conventions, trends, classifications, methodology, principles, generalizations and theories. Use objectives in the cognitive domain to: develop classroom instruction; organize instructional content on the basis of increasing difficulty of subject matter; and describe intellectual aspects of learning such as knowledge, information, thinking, naming, solving, analyzing, evaluating and synthesizing.

Objectives in the psychomotor domain focus on skills. Performance requires adept use of objects, tools supplies, machinery or equipment. Statement of psychomotor performances include constructing models or operating a computer. Use objectives in the psychomotor domain to: focus on actual skill performance, focus on the finished product; and specify accuracy within limits, level of excellence and speed.

Objectives in the affective domain require demonstrations of attitudes, feelings and emotions. They enable instructors to identify aspects of instruction that can help learners on a personal or social level. An example might be to increase collaborative team skills. These objectives involve paying attention to people and events, responding to them through participation, expressing values by showing either support or opposition and acting according to those values. Use objectives in the affective domain to: demonstrate listening, perceiving, tolerating and being sensitive to someone or something; show a willingness to cooperate, follow along, reply, answer, approve and obey; select, decide, identify and arrange values in order of importance as they relate to specific situations; and translate feelings and attitudes into observable behaviors.

Course objectives are primary objectives and are usually set out in broad statements that describe the general outcome of an entire instructional endeavor. Enabling objectives support the broad course statements but are more detailed and focus on short-term, specific results of lessons, units or modules of the larger instructional endeavor. Enabling objectives are seldom tested or evaluated at the end of the instructional period because their purpose is to enable learn-

ers to do what is necessary to complete those tasks required by the course objectives.

To write clear behavioral objectives, the instructional designer must be able to distinguish between the current and expected ability of the learner and then state that in the objective.

Selecting the most suitable action verb to describe the behavior being taught to learners is critical. Helpful guidelines for choosing appropriate verbs are found in Bloom's (1977) six intellectual levels of the cognitive domain. The lowest level is knowledge and the five increasingly intellectual ones are comprehension, application, analysis, synthesis and evaluation. Action verbs describing the mental abilities in these categories are:

Knowledge — arrange, define, list, memorize, name, organize, relate and recall.

Comprehension — classify, describe, explain, identify, indicate, locate, report, restate, review, select, sort and translate.

Application — apply, choose, demonstrate, illustrate, interpret, operate, prepare, sketch, solve and use.

Analysis — analyze, appraise, categorize, compare, contrast, diagram, differentiate, distinguish, examine, inventory, questio and test.

Synthesis — arrange, assemble, construct, create, design, formulate, organize, plan, prepare, set up and synthesize.

Evaluation — appraise, assess, choose, compare, defend, estimate, evaluate, judge, rate, select and value.

Avoid words that can be misinterpreted or are too vague such as understand, grasp, believe, internalize and appreciate.

Behavioral objectives are used to:

1. describe competent performances that should be the result of training or instruction

2. select and design materials, content or methods

3. direct and organize the learner's course of study

4. Implement the instructional process

5. stimulate review of the value and usefulness of the instructional process

6. communicate instructional intent

7. design tests and procedures for performance appraisal and

8. provide a framework to evaluate the success of the instructional process.

The following steps outline the process of setting and formulating objectives:

1. Analyze the source of objectives; needs and task analyses.

2. Make a list of the major objectives. What should the learner be able to do by the end of the instructional period?

3. Distribute the list to the instructional design team and request their comments on clarity and sequencing. Are the statements clear and are they in the proper order?

4. Establish criteria and draft action plans for objectives.

5. Draft objectives. Use action verbs describing what must be learned. Follow the verb with a description of what is being treated. Keep the statement short and simple and where possible, state objectives in measurable terms. Analyze the first draft of objectives by focusing on the performance part of the statement before the condition and criterion parts. Decide how the team will ascertain whether or not an objective has been satisfied.

6. The design team should discuss differing opinions and reach consensus.

7. Delete or reword objectives for which criteria and action plans cannot be developed.

8. Where necessary, break objectives down into components. They should be logically connected.

9. List and describe any conditions which are out of control of the design team that may influence the accomplishment of the objectives.

10. Add conditions by answering questions based on what equipment will be available, time limits, learning resources that are available. To define achievement levels, answer such questions as how comprehensive, how accurate, how well, the conditions are met.

Putting objectives into a logical sequence will provide learners and instructors with orderly guidelines to accomplish activities. Put enabling objectives in a sequence that will be most helpful as a logical and orderly procedures to accomplish the major task of the course. Use common sense in sequencing by following the usual sequence patterns of simple to complex or known to unknown. Arrange the order of the objectives to provide learners with basic skills early in the course.

■ LEARNING STYLES

Harvard University psychiatrist Howard Gardner (1983) developed the theory of multiple intelligences. In simplified form, this theory asserts that seven distinct thought processes exist in separate sections of the brain and can be assessed independently. This theory has emerged from recent cognitive research and "documents the extent to which students possess different kinds of minds and therefore learn, remember, perform and understand in different ways," according to Gardner (1991).

"There is ample evidence that some people take a primarily linguistic approach to learning, while others favor a spatial or a quantitative tack. By the same token, some students perform best when asked to manipulate symbols of various sorts, while others are better able to display their understanding through a hands-on demonstration or through interactions with other individuals." According to this theory, "we are all able to know the world through language, logical-mathematical analysis, spatial representation, musical thinking, the use of the body to solve problems or to make things, an understanding of other individuals and an understanding of ourselves. Where individuals differ is in the strength of these intelligences ," (Gardner, 1991).

Gardner said that these differences "challenge an educational system that assumes that everyone can learn the same materials in the same way and that a uniform, universal measure suffices to test student learning. Indeed, as currently constituted, our educational system is heavily biased toward linguistic modes of instruction and assessment and, to a somewhat lesser degree, toward logical-quantitative modes as well." Gardner argues that "a contrasting set of assumptions is more likely to be educationally effective. Students learn in ways that are identifiably distinctive. The broad spectrum of students — and perhaps the society as a whole — would be better served if disciplines could be presented in a numbers of ways and learning could be assessed through a variety of means."

Visual-Spatial - thinking in terms of physical space, as do architects and sailors. They are very aware of their environments. They like to draw, do jigsaw puzzles, read maps, daydream. They can be taught through drawings, verbal and physical imagery. Tools to use with them include models, graphics, charts, photographs, drawings, 3-D modeling, video, videoconferencing, television, multimedia, texts with pictures/charts/graphs.

Bodily-kinesthetic - using the body effectively, like a dancer or a surgeon. They have a keen sense of body awareness. They like movement, making things, touching. They communicate well through body language. They can be taught through physical activity, hands-on learning, acting out, role playing. Tools to use with them include equipment, real objects.

Musical - showing sensitivity to rhythm and sound. They love music, but they are also sensitive to sounds in their environments. They may study better with music in the background. They can be taught by turning lessons into lyrics, speaking rhythmically and tapping out time. Tools to use with them include musical instruments, music, radio, stereo, CD-ROM and multimedia.

Interpersonal - understanding and interacting with others. These students learn through interaction. They have lots of friends, empathy for others and street smarts. They can be taught through group activities, seminars and dialogues. Tools to use with them include the telephone, audio conferencing, time and attention from the instructor, video conferencing, writing, computer conferencing and E-mail.

Intrapersonal - understanding one's own interests and goals. These learners tend to shy away from others. They're in tune with their inner feelings; they have wisdom, intuition and motivation, as well as a strong will, confidence and opinions. They can be taught through independent study and introspection. Tools to use with them include books, creative materials, diaries, privacy and time. They are the most independent of the learners.

Linguistic - using words effectively. These learners have highly developed auditory skills and often think in words. They like reading, playing word games, making up poetry or stores. They can be taught by encouraging them to say and see words and read books together. Tools to use with them include computers, games, multimedia, books, tape recorder and lecture.

Logical - mathematical, reasoning and calculating. These learners think conceptually, abstractly and are able to see and explore patterns and relationships. They like to experiment, solve puzzles and ask cosmic questions. They can be taught through logic games, investigations and mysteries. They need to learn and form concepts before they can deal with details.

At first, it seems to be an impossibility to teach to all learning styles. However, as we move into using a mix of media or multimedia, it becomes easier. As we understand the different learning styles, it also becomes apparent why multimedia appeals to learners. It satisfies the many types of learning preferences that one person may embody or that a class embodies.

■ EXISTING MEDIA SELECTION MODELS

Knowles (1983) states that two models of media selection have been followed; the pedagogical model of learning and the entertainment model of media use. As a result, the media have not been used effectively as resources for learning and there is less than optimal learning. He suggests following the andragogical model of learning and the educational model of media use.

The features are interaction; task centeredness organized around the acquisition of the knowledge that is applicable to performing life tasks; individualization which takes into account learner differences in backgrounds, readiness to learn, motivation to learn, learning styles, developmental stages and learning pace; and self directedness, as adults have a need to take responsibility for their lives so that media which involve learners in making decisions about what they are going to learn, how they are going to learn it,

when they are going to learn it and how they are going to verify that they have learned it will be more effective than those in which all these decisions are made for the learners.

Reiser and Gagne state that in order to make instruction minimally effective, selection of media has become a "burning" question (1983, p. 3). They conclude that for a given instructional task various media will differ in instructional effectiveness (1983). Emphasis on effectiveness and cost effectiveness, as well as accountability of instructional programs, necessitates that media selection be considered a critical issue (Reiser & Gagne, 1983; Bowsher, 1989). Reiser and Gagne (1983) assert that much instruction is not planned to ensure effectiveness.

Reiser and Gagne (1983) point out that there is no generally accepted media selection model even though much has been written about instructional media. Sive (1983) points out that few media books list "selection" in the index. Schramm (1977) observes that no procedure can be applied automatically in every instructional situation and guidelines should consider local needs, situations and resources. Clark and Angert (1981) conclude that available models reflect a preoccupation with technical considerations such as the convenience and portability and lack substantial instructional design considerations.

Reiser and Gagne (1982) reviewed nine media selection models which attempt to answer how educators should go about selecting media. Reiser and Gagne (1983) concluded that information concerning the usefulness of existing models was limited due to the rarity of finding detailed information about situations in which selection models were employed and that there was limited empirical evidence about the relative merits of media selection models. Their conclusion was that choosing a media selection model is not simple.

The literature clearly states that it is vital that media selection criteria cover educational objectives, instructional design, student study guide, computer software, video production, content, textbook, faculty guide and cost. There is agreement in the literature that media should be evaluated; however there is little agreement on what constitutes good media evaluation The models are more useful to designers than to telecourse adopters who are not selecting media for production but are faced with a pre-produced package of media to be adopted or rejected (Holt, 1989). The model and the evaluation instrument should require the evaluator to consider all phases of the telecourse, including student needs.

■ THE ADULT EDUCATION MODEL

Meierhenry (1981) observed that adult educators must have a clearer understanding of how media and technology contribute to achieving educational objectives and how the teacher's responsibility is to integrate the human and nonhuman resources. Gueulette (1986) describes this collaboration between adult educators and technologies as an "imperative mission," (p. vi). Farnes (1975) pointed out that the Open University, a distance education institution, was operating under an authoritarian system so that the only responsibility the student had was to select courses. Farnes states that the extrinsically motivated system lead to high attrition, withdrawal of personal commitment and other forms of "pathological behaviour" (p. 3).

Farnes observed that the course development teams experienced exciting and immensely demanding learning tasks as they acquired and organized knowledge. He states that "if it is in the course that there are genuine learning experiences, should we not allow the student to participate in these learning experiences by delegating more of the job to him (p. 3)?" Farnes concluded, "It is a tragedy that as soon as normally responsible adults come into contact with education they expect to be told what to do and what to learn. Worse still, we as teachers play along with this and find it much easier to meet these expectations than to create the conditions in which students will take responsibility for their own learning" (p. 3).

Potvin (Clennell, 1975) found it helpful in planning adult distance education programs to create a climate conducive to adult learning. He maintains that learning improved when adults lost their dependence, which was created by the traditional educational methods. Potvin attributed the increased learning to the student's experiences being used as a learning resource, intrinsic motivation and knowledge sought being related to immediate problems. As they learned to assess their learning needs they became increasingly self-directing.

According to Knowles (1983) the media have not been used effectively for adult education because they have been seen as one-way transmissions of teacher-controlled instruction which does not result in optimal learning; they are based upon the pedagogical model of education and the entertainment model of media use. Knowles recommends the andragogical model of learning and the educational model of media. Andragogy is defined as "the art and science of helping adults learn" (Knowles, 1975, p. 19). Knowles makes the distinction among the andragogical and pedagogical models of teaching based upon sets of assumptions about learners which teachers make. The teacher who makes one set of assumptions will teach pedagogically whether he or she is teaching children or adults, whereas the teacher who makes the other set of assumptions will teach andragogically whether the learners are children or adults (Knowles, 1975).

The pedagogical model revolves around teacher-directed learning where the learner is seen as having a dependent personality, the learner's experience is built on rather than used, readiness to learn varies with levels of maturation, orientation to learning is subject-centered and motivation is gained through extrinsic rewards and punishments controlled by the teacher (Knowles, 1975).

The andragogical model revolves around the learner who is seen as becoming increasingly more self-directed. The learner's experience is considered to be a rich resource for learning, readiness to learn is developed from life tasks and problems, the orientation to learning is task or problem-centered and the learner's motivation is intrinsic and driven by curiosity (Knowles, 1975).

Following the andragogical model, the teacher sets an informal climate which is supportive, collaborative, consensual and mutually respectful. Planning is conducted by participative decision-making, needs are diagnosed through mutual assessment, goals are set by mutual negotiation, learning plans are carried out by learning projects executed by learning contracts which are sequenced in terms of learner readiness. Learning activities are conducted through inquiry projects, independent study and experiential techniques. The evaluation of the learning which has occurred is done through mutual assessment of the evidence which is prepared by the learner (Knowles, 1975). Knowles sees the model as being a process design rather than a content plan (1985) so that there is no attempt to cover particular content areas; instead the student samples content in relevant problem situations (1984). He explains that it is useless to have a stockpile of content information without having a process or method by which to handle it (1984). The key features of the model are interaction, task-centeredness, individualization and self-directedness.

Construction is a similar theory dominating K-12 education. Students construct their own learning from reality based problems, work in collaborative groups and the teacher facilitates the learning process.

Interaction

Knowles states that learning is most effective when learners engage interactively in the inquiry process. Interaction can be introduced between the learner and the program using interactive videodisc and CD-ROM, computers, computer conferencing, audio conferencing, video conferencing and interactive reading materials. Knowles attributes the failure of learning machines in the late sixties to the lack of learner involvement. He feels that branching programs, such as hypermedia or multimedia, are being adopted because of learner involvement. Interaction can also be introduced between the group and the program by telephone where the learners discuss content by phone with an instructor who is in a studio. Knowles observes that this is being done successfully now in teleconferencing and in computer-assisted learning with superior results in terms of learner involvement and learning outcomes.

Task-centeredness

Knowles observes that adults are usually motivated to learn in order to perform tasks associated with their lives. He states that they seldom learn something for its own sake, or to accumulate academic credits. For media programs to be effective with adult learners, Knowles states that they must be organized around the acquisition of the knowledge, skills, understandings, attitudes and values that are applicable to performing the life tasks with which adults are concerned. Knowles (1975) states that one of the most significant findings from research (Tough, 1979) about adult learning is that when adults go about learning something naturally, rather than being taught, they are highly self-directing. He finds evidence is accumulating to support that what adults learn on their own initiative, they learn more deeply and permanently than what they learn by being taught. The same in K-12 education.

Individualization

Knowles believes that the individual differences among adults and especially among adults of different ages, are great. To accommodate these differences, media programs need to provide a wide range of learning options which can take into account differences in backgrounds, readiness to learn, motivation, learning styles, developmental stages and learning pace. Knowles believes that if these factors are missing, a structured media program based upon the standardized curriculum of traditional education will not attract adult learners.

Self-directedness

Knowles states that adults have a deep psychological need to be responsible for their lives and develop the self-concept of being responsible; this leads to a need to be seen and treated by others as being capable of making their own decisions. In media programs a way of doing this is through actively involving and making the learner responsible for decisions about what they are going to learn, how they are going to learn it, when they are going to learn it and how they are going to verify that they have learned it. This is

the basis for the self-directed learning contract in which adult learners reconcile imposed requirements from institutions with their need to be self-directing (Knowles, 1975). Through learning contracts, the mutual responsibilities of the learner, the teacher and the institution are made visible (Knowles, 1975).

Knowles has also described the benefits that androgogical methods can bring to teaching in kindergarten through twelfth grade. This is because androgogical methods are learner-centered rather than teacher centered.

Dwyer (1984) observed that present methods, principles and guidelines for organizing content may not adequately use the possibilities suggested by andragogical theory. If based upon adult learning theory, Dwyer feels that media would be very differently organized than it is now so that the tightly integrated, cohesive, consecutive and fast-moving instructional sequence may be less effective for adult learners than the discrete unit with intermittent presentation. Dwyer suggests that adult learners will profit most if instructional sequences can provide a range of responses. Dwyer notes that perception is now depicted as constituting boundaries of interactions between adult learners and the media from which they select, ignore, or reject cognitive inputs. The interaction depends on student need, activity in progress, personality characteristics and goals. Dwyer concludes that this has profound implications for the design, development and implementation of meaningful instructional experiences which help the adult learner meet educational goals (1984).

■ MEDIA SELECTION MODELS

A number of media selection models have been developed to help educators evaluate and select media. How useful these models are is questionable as it is rare to find information about where the models were used (Wagner & Wishon, 1987). Reiser and Gagne (1983) concluded that there is limited empirical evidence about the merits of models; their final conclusion was that choosing a model is not simple. They suggest that an approach to media selection is to identify model features, decide which features are important and select a model containing them.

Bates (1974) argues that it has been a mistake to consider media merely as a service to subject disciplines. He states that media use should evolve from a partnership where the academic and media specialist have equal status and responsibility to organize and produce media. Bates feels that a great mistake has been made in casting media into competitive models so that the media are misused. Rather, he suggests that media should be used in conjunction with one another which allows different things to be done through a whole new range of teaching objectives and methods which to some extent will allow the media to determine what is to be taught and how it is to be taught (Bates, 1975). This may seem to make media take precedence over subject matter, but he argues that if there were no books, teachers would teach differently through dialogue; since books and dialogue exist, both are used. He continues that if media is considered to be primarily a means by which information is distributed, then books and television have a functional equivalence, even if their characteristics are different. If courses are designed from the beginning with media in mind, one is still free to reject their use.

The methods available for teaching will inevitably influence what is taught; form and content are interactive (Bates, 1974). Bates (1987b) contends that there is a lack of sound theory of media selection based on pedagogic criteria partly because of differences among educators about the best way to teach and partly because media selection has not until recently been a major problem facing educators. Consequently, Bates (1987b) observes, most instructors have not bothered to use media to a significant extent; those that have used media have acted purely on intuition and were influenced considerably by what is conveniently available. They will also choose media that reflect their own learning style preference. Bates (1974) concludes that at Great Britain's Open University, it is impossible for an academic to teach without making use of media because the system ensures that courses are designed from the beginning with the potential of media in mind by a course team which is composed of the academic and media specialist who have equal status and responsibility.

Two studies compared selection techniques. Braby (1973) compared the usefulness of ten media selection techniques. Models judged superior were Briggs (1970), the Training Analysis and Evaluation Group (1972) and intuitive techniques. Romiszowski (1974) found that using a selection technique which he developed helped users made better choices than did an intuitive approach (Reiser & Gagne, 1983).

Reiser and Gagne (1983) reviewed nine media selection models (Anderson 1976; Branson, Rayner, Cox, Furman, King & Hannum, 1975; Bretz, 1971; Briggs & Wager, 1981; Gagne & Briggs, 1979; Gropper, 1976; Kemp, 1980; Romiszowski, 1974; Tosti & Ball, 1969) to yield the comparisons described here. The models presented media features in flowcharts, matrices, or work sheets. An essential difference among these formats is the procedure for decision-making each demands. Flowcharts lead to a progressive narrowing of media choices. Questions about media selection are posed in a particular order and as each is answered, the number of candidate media is reduced. The matrix display includes all of the selection criteria so that one tallies the criteria met and their relative importance. Work sheets present a tabular array of media characteristics against desired criteria and require that media selection be deferred until all criteria have been considered. A number of media classification categories have been devised including audio, print, still visual, motion visual and real objects. Media categories are usually connected with the idea that one medium can best present a task having a similar classification. Most classifications depend upon characteristics of the display such as visual, motion, or auditory. Tosti and Ball propose classifying types of interactions and Gropper categorizes by feedback capabilities.

Visuals

Visual media help students acquire concrete concepts, such as object identification, spatial relationship, or motor skills where words alone are inefficient.

Printed Words

All models required decisions on the use of printed words. There is disagreement about audio's superiority to print for affective objectives; several do not recommend verbal sound if it is not part of the task to be learned.

Sound

A distinction is drawn between verbal sound and non-verbal sound such as music. Sound media are necessary to present a stimulus for recall or sound recognition. Audio narration is recommended for poor readers.

Motion

Models force decisions among still, limited movement and full movement visuals. Motion is used to depict human performance so that learners can copy the movement. Several assert that motion may be unnecessary and provides decision aid questions based upon objectives.

Color

Decisions on color display are required if an object's color is relevant to what is being learned.

Realia

Realia are tangible, real objects which are not models and are useful to teach motor and cognitive skills involving unfamiliar objects. Realia are appropriate for use with individuals or groups and may be situation based. Realia may be beneficially used to present information realistically but it may be equally important that the presentation corresponds with the way learner's represent information internally.

Instructional Setting

Design should cover whether the materials are to be used in a home or instructional setting and consider the size of the learning. A rationale for these decisions is not included except for print instruction delivered in an individualized mode which allows the learner to set the learning pace. The ability to provide corrective feedback for individual learners is important but any medium can provide corrective feedback by stating the correct answer to allow comparison of the two answers.

Learner Characteristics

Most models consider learner characteristics as media may be differentially effective for different learners. Although research has had limited success in identifying the media most suitable for types of learners several models are based on this method.

Reading ability should be considered. Pictures facilitate learning for poor readers who benefit more from speaking than from writing because they understand spoken words; self-directed good readers can control the pace; and print allows easier review.

Older or more experienced learners may have developed learning strategies that enable them to manage instruction and may need fewer external aids. Dale's (1969) cone of experience tool helps identify suitable media by age group. The cone lists 12 media categories and experience in an ordered hierarchy. For cognitive objectives, it is efficient to use abstract media with older learners and concrete media and experiences with younger learners. For attitude formation objectives abstract media should be used for younger learners and concrete media and experiences used for older learners.

Categories of Learning Outcomes

Categories ranged from three to eleven and most include some or all of Gagne's (1977) learning categories; intellectual skills, verbal information, motor skills, attitudes and cognitive strategies. Several models suggest a procedure which categorizes learning outcomes, plans instructional events to teach objectives, identifies the type of stimuli to present events and media capable of presenting the stimuli.

Events of Instruction

The external events which support internal learning processes are called events of instruction. The events of instruction are planned before selecting the media to present it. Two models use charts to indicate the degree to which a medium is appropriate to present instructional events.

Informing the learners of the objectives provides them with an indication of learning expectations to maintain their task orientation. Visual media which can portray motion are best to show psychomotor or cognitive domain expectations by showing the skill as a model against which students can measure their performance.

Many models discuss eliciting performance where the student practices the task which sets the stage for reinforcement. Several models indicate that the elicited performance should be categorized by type; overt, covert, motor, verbal, constructed and select. Media should be selected which is best able to elicit these responses and the response frequency. One model advocates a behavioral approach so that media is chosen to elicit responses for practice.

To provide feedback about the correctness of the student's response, an interactive medium might be chosen, but any medium can provide feedback. Learner characteristics such as error proneness and anxiety should influence media selection.

Testing which traditionally is accomplished through print, may be handled by media. Media are better able to assess learners' visual skills than are print media and can be used to assess learner performance in realistic situations.

Summary of Reiser & Gagne (1983) Model Comparison

Media selection can be affected by a model's physical form of display; matrices or work sheets defer media choices until all criteria are examined. Flowcharts progressively narrow media choices and are easier to use if the user has minimal media selection experience. Decisions about media are influenced by the selection factors included in a model. In reviewed models, selection criteria focus on the medium's physical attributes such as the ability to present sounds or motion while others focus on learner characteristics, instructional setting and the learning task.

Proper identification of the media attributes is dependent upon consideration of learner characteristics, instructional setting and the learning task.

The choice of learning theory as a basis for rational model derivation means that other groundings have been rejected. While it is evident that several other characteristics of media cannot be ignored, they do not appear to have been successful as bases for the generation of positive media selection procedures. This includes a variety of categories pertaining to media attributes as mode of sensory stimulation (Romiszowski, 1974), physical nature of stimulation (Bretz, 1971), type of learning experience (Dale, 1969), function with respect to the learner (Tosti & Ball, 1969), or some combination of these (Anderson, 1976; Kemp, 1980).

The main problem is that ready-made classifications claim to be reliable instruments appropriate to all instructional situations and applicable without modifications. In reality each instructional situation contains a set of factors which may determine the media. Attempts to classify media should not aim at the development of a media taxonomy as a final, generally applicable multi-dimensional decision matrix. What is required is a detailed description of each medium, which uses more specific ratings than "applicable - partly applicable - not applicable" so that the user can develop a decision model tailored to an instructional problem. Because of the number of instructional situations, there are a multitude of factors which determine the media

Reiser and Gagne's model (1983) is a flowchart containing six panels, each representing an instructional situation which cover student competence, delivery method and instructor or self- instruction with readers or non-readers. The procedure used to arrive at a fewer media is to answer questions on course objectives; the objective's domain of learning outcome, instructional setting, student reading competence, cost, availability and convenience.

Bates (1980) suggests that the efforts to demonstrate that there are linkages between type of media and objectives intended to teach are fruitless. He states that the impact of television must be seen as depending on a combination of media and factors related to the general organization of the instructional program, factors relating to the video programs and factors related to the learner. The positive effect of the variety of media has been confirmed in a number of studies. Bates (1982) showed the clear advantage of using radio and television to supplement readings versus readings alone. Schramm (1977) reviewed several studies of multi-media programs in higher education environments; his general conclusion was that students who work with a combination of media do significantly better than others.

While there is agreement in the literature that media selection is important, the models illustrate the range in opinion on how it should be selected. The models are more useful to designers than to telecourse adopters who are not selecting media for production but are faced with a pre-produced package of media to be adopted or rejected.

■ CURRENT MEDIA SELECTION METHODS

It is not clear that any formal evaluation using appropriate media selection methods is being used (Kressel, 1986, Lane 1989a, 1989b). Distance education professionals could not recommend and are not using a telecourse evaluation procedure because an evaluation instrument does not exist (Lane 1989) . A related problem is that a critical analysis of what is effective when delivered by technology is unavailable. Bernard (1974) notes that evaluation problems and results are massive and complex; traumatic experiences indicate that these are not abating. Evaluation problems which have been overlooked have led to the misuse and overuse of inconsequential telecourses and other instructional media. As a result, the decision to use media is not well grounded and may lead to minimal student learning because the media is ineffective in its instructional design, inappropriate for learners, or does not fulfill course objectives. It is probable that the result is a great deal of ineffective instruction. Since it is likely that more video and other forms of electronically mediated instruction will be used, it must be used judiciously and correctly.

■ EVALUATION INSTRUMENTS AND MODELS FOR DISTANCE EDUCATION MATERIALS

Criteria for the Evaluation of Distance Learning Materials

The purpose of evaluation is to find out the extent to which the goals or objectives of an educational activity are being achieved. Reiser and Gagne state that selection of media is a "burning" question in order to make instruction optimally effective (1983, p. 3) and they observe that much instruction is not planned to be optimally effective.

Existing media selection models variously emphasize physical features or human senses. Clark and Angert (1981) reviewed media selection models and concluded that they are preoccupied with technical considerations such as convenience and portability and are weak on instructional design considerations. Schramm (1977) points out that no procedure can be applied to all situations and guidelines should consider local needs, situations and resources. Bates (1980) states that the primary concern is how the media interact. A literature review produced the following concerns.

Instrument Terminology

When the term understanding or appreciation is used, it should delineate the specific nature by student behaviors.

What values are highlighted? The program is in keeping with the principles that guide the user institution. Materials represent artistic, historic and literary qualities.

How are educational objectives selected? The student is central to the learning experience; evaluation should be done within the total context of student learning; educational needs are defined so that they can be met for the educational system and individual programs; expected changes in student behavior, attitudes or interest are defined; curricular objectives are stated; media contributes to specific instruction goal achievement; the extent to which stated objectives are achieved; objectives are stated by cognitive, affective and psychomotor domain; objectives are measurable and can measure success or failure; lesson objectives give adequate direction for student study; and whether students can correctly identify educational objectives.

Characteristics of students should be known including their initial competence in the topic. Material should be suitable for learners with an appropriate level of content complexity and vocabulary which accommodates ability differentials.

Compare the similarity of the campus class with the electronically mediated class; objectives, course experiences and content should be equivalent. Supplements or experiences can be developed or adapted to make the courses similar; the course should be adaptable to many teaching situations, populations and methods; and the course should be of interest to students as a required, elective, or interdisciplinary course. A report, such as a program producer's field evaluation of student learning, should be available to provide learner verification data on the product's effectiveness. The method for evaluation and assessment which has been validated should be described and the evaluation should be directly related to the course objectives.

The delivery method should be considered: loaned tapes are available when needed, facilitate repetition, search and mastery, analysis, relating and reflection, are easier to integrate. Broadcast programs are shown perhaps once at a fixed time, do not facilitate repetition, search, mastery, analysis, relating or reflection and are more difficult to integrate. The control characteristics of cassettes should be exploited,

of segment use, clear stopping points, use of activities, indexing, close integration with other media (text, etc.) and concentration on audio-visual aspects so that the video cassette is to the broadcast what the book is to the lecture.

Instructional Design

Consider the schedule of learning set up for the student so that students are not overloaded. Consider the time required to complete the course; the number of lessons; appropriate segment length stated instructional objectives; it is fully planned and has an appropriate level of abstraction; uses visual, audio and tactile components; directs student activity toward specified learning outcomes by frequent overt and covert responses; the familiar is used as a bridge to the unfamiliar; and a range of direct and indirect methods is used. The material should be broken into manageable chunks; the first two lessons are shorter; lesson size is easily managed, not too long or difficult to discourage students; lessons are self-paced to allow student planning; and the production pacing maintains interest. Telecourse components should be examined for high quality; components should make learning experiences occur; accomplish individual objectives for which they were created; utility of each component part; provide realia (real objects); effectively use graphics; components should be easy to use; useful; well packaged; transportable; available; have an appropriate quantity; should include concepts of appropriate difficulty; relate ideas and link discussion.

Components should be examined for relevance of reading rates, speed vs. critical reading; readability; use of unexplained technical terms; overall coherence and consistency; argumentative and indices of fallacious reasoning; does not make assumptions, draw conclusions in error, or masquerade examples as definition or opinions as fact; clarity; well phrased instructions and questions; have complete, adequate and useful proofs; show a balance of active and passive assignments; should contain self assessment questions and activities to make the student think and evaluate progress; and have appropriate role, position and function of summaries. Material should appeal to the students' interests, achievement and background; and provide a stimulus to creativity. Components should correlate well with one another so that they are integrated.

Self instruction should be encouraged through strategies which motivate student learning, hold student attention and stimulate students. Students should be provided with help to develop basic learning skills such as fast and selective reading, essay writing, development of objectivity and knowing how to learn from television and radio. In the early stages of students experience with self- instruction, there should be a progression from a structured situation to a situation where students are able to organize material in their own learning package including more responsibility for deciding which areas to study, how to organize the study; and how to present it.

The programs should move from highly didactic to open ended; the structured learning should not limit the students learning so that students should do creative thinking. The presentation should avoid using many facts so that students find contexts and causal connections to create the students' ability to critically analyze what they see and hear and help them find their own way to knowledge. Emotional experiences should be provided. Student work should be based upon andragogical (adult education) principles.

Media can be used for learner interaction and feedback by providing for student drill and using techniques to motivate students to work and study; by actively involving learners through writing, talking, manipulating, competing, cooperating, critical viewing or activities on tape or in print components, or in some way respond to the teaching material to considerably increase learning effectiveness. Feedback should be immediate and timely to induce lesson submission; the assignment turn around time should be no longer than five days to increase student completion rates. Feedback should provide the correct response and a commentary on the incorrect response. The presentation sequence and rate should be learner controlled with branching to alternative units after incorrect answers. The instructional strategy should vary as a result of both current and past learner behavior and portions should repeat at the learner's volition. Students should have activities such as answering questions.

All student learning styles should be addressed as individuals may be primarily visual, auditory, tactile, conceptual, or quantitative in various combinations to focus on human learning and ensure learning for all students. Audio components should be provided for auditory students; visual components for visual students; and realia, models and other objects provided for tactile students.

Strategies should match student cognitive styles, previous experience and presentation factors. Cross-modal reinforcement should occur frequently where the same message is given through two modalities - words and pictures. Strategies should meet adult viewing styles which are open learners (about 33 percent who are interested in the world and learning, slightly older, more highly educated, who see television as one source of information), uninterested learners (50 percent of viewers who are not interested in learning, watch television for entertainment and have a low level of formal education) and instrumental learners (15 percent of population who are interested in learning as a means to a better job, young, upwardly mobile, blue collar or office workers, mid range in formal education, but do not consider television as a knowledge source).

Assignments should be specific to course content and may be created by students through the use of self-directed learning contracts. Assignments should help students become self-directed and adapt to local needs by utilizing faculty expertise through syllabus development and suggesting successful assignments for distance learners. Students should not be overloaded with more material than can be handled. Facilities should be available for laboratories. The first assignment should be due early, within 14 days, or within 40 days. There should be a great number of small assignments due rather than one project or several large projects, or one major assignment due each month. Computer marked assignments should be used.

Content

Content should be examined for appropriate scope of content; accuracy; authenticity; typicality; in good taste; reflective of research in learning; utilizes innovations in instruction; authoritativeness of materials; clarity; and illustrative of the interplay of process and growth of content. The same thing should be said more than once in different ways to replicate the central points. The course should be interesting and stimulating and provocative; lessons should be exciting to positively influence completion; and the video should have a long shelf life.

Differing viewpoints should be provided; controversial issues should be handled fairly without evidence of bias. The pluralistic society of multiple ethnic, racial, religious, social, geographic and sexual characteristics should be represented. The material should be relevant to today and the copyright should be recent and not older than two years. The material should be important and interesting to the learners.

Textbook

The textbook should be recommended by the producer; be acceptable; be as attractive as other textbooks to hold attention; be high quality, well presented and lavishly illustrated; be up to date; available on time; have further editions planned; have a clear role in course design; be widely used and the author's credentials should be appropriate and recognized. The textbook should encourage students to learn. The textbook should correlate well with other components and should match video revisions. If the text must be augmented a second text will have to be found or written if one is not recommended by the producer.

If a reading anthology is recommended, it can be used to tailor the course to a particular focus by eliminating reading assignments.

Faculty Guide

The electronically mediated course should have a faculty guide to act as a mentor for the new instructor; provide in-depth discussion about instructional design; discuss content embodied in the components and how they relate to one another; present detailed teaching strategies and evaluation strategies; contain background information on course development, developers, consultants and advisors along with their credentials; and course goals. The guide should contain a course outline by lesson; weekly student activities for each week of the academic terms; test bank or suggested tests; alternative course structure; recommend varied uses of course materials; list required or suggested materials and sources; bibliography and sample promotion material. It should contain segments to guide students in learning from electronically mediated instruction, viewing holistically, finding patterns, developing analytical skills and other explanation about the course. If the faculty guide does not exist, local staff should have the experience to supply the necessary faculty support.

Human resources to support the course should be considered including whether the local instructor is competent and whether the course matches the instructor's teaching style. The instructor should write the course syllabus, assign additional readings, make assignments and grade them, hold an opening structured seminar, hold face to face meetings with individual students, call class meetings, maintain contact with students by mail, phone and meetings to add content for students' consideration; maintain student interest through study groups to provide support and raise completion rates. The instructor should be interested in and encouraging to the students. Technical facilities should be considered including, library access, physical circumstances and other logistical considerations.

A test bank should provide questions which are suitable for correspondence or proctored testing and based on the content. Viewing video programs should be linked to student assessment. Test keys should include a listing of where answers are found in the content. The test bank should have many types of short answer questions which can be graded by computer and suggest short essay questions. Test validity should be described. Students should be allowed to choose and provide evidence of learning.

Student Study Guide

The study guide should be recommended by the course producer and be acceptable to the institution. The guide should be an important component of the course which ties all course elements together to help the student complete the course. It should be written by content specialists as the course was developed and contain lesson-by lesson guides to meet course objectives, list additional readings, optional activities and be easily augmented by faculty by adding sections or deleting section depending on curriculum. Research shows that student completion rates increases by 10 percent if the study guide is written by the instructor. The guide should teach students how to use the course by explaining the function of the video and other electronically mediated content and give students guidance in what to look for and how to approach the program. It should train the student to look at video events holistically, to use analytic processes, what to focus on and how to discern patterns and self directed learning strategies. The guide should contain segments on objectives, components, lesson outlines, video outlines, glossary, key concepts, references, exercises and self-tests with explanations.

Pre-broadcast notes should be brief, but should clearly state the purpose of the program and what students are supposed to do before during and after seeing or hearing the broadcast or tape. Audio cassettes are not lectures but are tightly integrated with print to talk students through diagrams, illustrations, statistics or provide discussion material for analysis.

Computer Software

Recommended software should be suitable; easily available with appropriate site and home licensing at a suitable cost. Software is appropriate to content and used to present and test rule based procedures, areas of abstract knowledge where there are clearly correct answers so that educational objectives are achieved. Computers can be loaned to students. Logistics, including computer access to provide software to students should be suggested and the software should be available in many versions for many types of computers.

Computer Conferencing Software

For courses which include the ability for students and faculty to use computer conferencing by dialing into a central computer via modem, the choice of the computer conferencing software is important. E-mail systems may suffice, but it is usually better to have software specifically designed for teaching over the computer. Programs should be easy to operate by inexperienced students and faculty. It should not take as long to master the telecommunications program as it does to master the course content. Students may find that having to train on how to use a telecommunications program is an obstacle that they don't want to tackle. This is particularly true for short courses or continuing education courses. New programs are entering the market regularly.

Points to consider include: User friendly - pull down menus. Telecom-munications program can be downloaded by the student or mailed to the student. The computer program should support 2,000 or more students and faculty and the many messages they will generate. Ability to segment classes and allow entry only into authorized mailboxes. Ability to segment assignment mailboxes so that the flow of assignments and interaction can be easily followed. Ability to support private mailboxes for students and faculty, bulletin boards, faculty forums, students forums, registration, filing grades, delivering contracts to faculty and access by other administrators and interested lurkers. High points should be given to programs which dial the computer, upload new messages to appropriate mailboxes, download new messages and file them in existing files on the student/faculty members hard disk. Filing new messages is time consuming and this should have a high priority. The program should encourage the students to become interactive because a up to 40 percent of the students grade may be based upon this.

Video

Video programs should use the full presentational power of video; words, still and moving pictures, events occurring in real time, slow or accelerated motion, animation and text. Production should be high quality as this correlates with lower attrition and higher grades particularly for borderline students. The technical quality should be acceptable or excellent, balanced and satisfying, meet professional standards or meet national broadcasting production standards; this is essential because of its motivational impact on students as the pleasure of watching the programs breaks the students' inertia of beginning to study. The video format should not differ too much from what is considered to be a good general commercial television program with an expensive appearance to compete with commercial television. Programs should be one-hour or can be shown as one-hour to meet normal programming times.

The number of programs should be high as more programs correlates with lower attrition. Tapes should be available for student loan as this has considerable advantage over a pre-scheduled distribution by cable. The video should not rely heavily on the lecture/talking head format or show students in a video class unless it is a teaching method class; the instructor should talk to the viewers for interaction.

Chemical experiments should be performed in an industrial laboratory to show the experiment's industrial application to demonstrate experiments or experimental situations where equipment or phenomena to be observed are large, expensive, inaccessible or difficult to observe without special equipment. The video should use the medium's unique possibilities to give students content that they would otherwise not get or see. The plot should not be wild or slapstick. The use of video material should be influenced by relevance more so than dramatic quality. Video is not used for dense, abstract ideas, comprehension of detailed arguments and facts; it is used to deal with abstract ideas through the use of concrete examples, stimulates sophisticated level of thinking which leaves interpretation and analysis open to the student. Programs should have structure, organization, sequential progression, be well paced to provide variety and a content development rate which holds attention and facilitates learning so that they are more swift than real life but not frenetic. Video should be used to increase the students' sense of belonging.

The video should demonstrate human interaction and time-space relationships to illustrate principles involving two, three or n-dimensional space; to act as a bridge between the concrete operational and formal, more abstract stages of learning; words (audio and written), dramatizations and music should generate attitudes and interest; use case illustrations, dramatization and supplantation (formulas, scope, rotation, animation, etc.) to advance content; complete coordination and integration between audio and video should exist; video should present unique material not found in the classroom; video should present well known content in unique forms; video takes society to the student to form links between class and life; video should use many open ended methods to encourage student inquiry; to change student attitudes towards a particular subject area by presenting material in a novel manner or from an unfamiliar viewpoint; and allow students to look into something otherwise inaccessible. For student comprehension and instruction on how to approach television, video sequences should show the whole sequence, then repeat it with each sentence presented as a separate entity which is explained and elaborated upon; in later programs the elaboration should be decreased to give the student more independence. The video should encourage students to interpret, analyze and problem solve by facilitating the students' ability to apply knowledge, evaluate evidence or arguments, analyze new situations, bring insights to portrayed situations and suggest solutions. The camera work should be considered for appropriate and imaginative use of video which advances the content. Video should visualize the abstract to provide contrived images that present in visual form the concepts and relationships for which students cannot conjure images on their own. The screen should be used to its full potential with camera angles (single and two shots, point of view, over the shoulder, close-ups, wide shots and camera focus changes) and techniques (zooms, pans, swish pans, cuts) to attract attention through pictures, sound bites, demonstrations, diagrams and graphics. The video should show the world to create authenticity and effectively use color and motion. Effects should provide pace change and the material should dictate the use of effects such as wipes, freeze frames, flips, computer graphics, split screens; effects should not be used because the technology is available. Styles of clothing etc. should not detract. Clarity should be maintained by smooth bridges between segments and programs. Clear demarcations between discontinuous segments should be apparent in settings, presenters, etc. The use of sound should be considered so that sound, music and sound effects emphasize content. Sound should be imaginative, advance content, add variety and pace and not use a continuous music bed. Pictures are provided with clear verbal narratives for clarity.

The video instructor is important to the telecourse; is on camera; is competent; conveys interest in the content; transmits enthusiasm; and personality and appearance add to the effectiveness. The instructor does not lecture or preach; so that concepts are difficult to grasp and understand but simplifies the message by using understandable language, humor to motivate, make content palatable and act as change of pace; humor is situational, not slapstick. A diversity of experts, talent and characters provide variety and good acting with believable dialogue.

Costs

Costs should be considered as they relate to funding. Costs should be considered as to their appropriateness for a given media system and the proportion of money and resources to be devoted to various aspects of a media system; capital costs and recurrent expenditures, equipment obsolescence, staff, space and overhead, cost and delivery should also be considered. The cost effectiveness of the program to other programs on the same subject should be compared by projecting student per head costs and relationship to shelf life and student per head program costs to purchase and deliver. Media costs versus face to face instruction should be considered, broadcast costs versus loaned tape costs and other economies of scale where more students will make the media more cost effective.

■ MEDIA SELECTION MODELS AND EVALUATION INSTRUMENTS FOR ELECTRONICALLY MEDIATED MATERIALS

Major Findings

Statements which contained adult education principles, methods or teaching strategies reached consensus except for the use of learning contracts (rejected at 72.1 percent). Based upon the number of respondents ranking themselves with high knowledge (35) as well as the low group knowledge (35.7 percent) and importance scores (33.9 percent), there is little knowledge about learning contracts or how they can be used with telecourses.

One statement questioned whether strategies should be evaluated for their ability to reach all student learning styles and was accepted (80 percent). The comments and the lower group score (48 percent) suggest that learning styles and how they may be applied to media are not well understood. Respondents perceived that visual and auditory styles can be addressed, but do not perceive that tactile, kinesthetic or interactive styles can be addressed through all components.

All statements regarding shelf life were rejected (69.8 to 74.6 percent). Producers indicated that a two to three year shelf life was normal; users wanted a much longer shelf life for cost effectiveness which indicates that producers should minimize aspects which directly affect shelf life. A possible solution is to package content with the shortest life in one video program or in print supplements which could be updated at minimal expense to producers or users.

There were two statements on video technical quality which received consensus. One stated that the video technical quality should meet professional broadcast quality standards (82.9 percent). The second stated that video technical quality should meet professional broadcast quality standards appropriate to the delivery method (cable, ITFS, broadcast, learning center, etc.) (90.9 percent). There has been a long-standing discussion about video technical quality and learning differences, if any, between educational programming produced with low or high production values. Opinion polarized into two areas. One group believes that production values are acceptable if they fit the delivery method. The second group believes that production values must meet broadcast standards because students are constantly exposed to it and will judge educational programming by their experience with commercial television. The implication for producers is that they will need to produce programming which meets broadcast quality standards in order to meet both needs. If production is less than broadcast quality, they will lose part of the market.

A statement that a total of 15 hours of video programming is ideal was rejected (47.1 percent). Comments documented that a new regulatory factor has emerged where some accrediting agencies have set student contact hours at 45 hours, the same amount required for traditional classes. This has made the 15-hour telecourse unacceptable to institutions operating under these regulations as additional projects cannot be substituted for contact hours. The 15-hour telecourse carries one credit hour; respondents felt that students would not register for this. This has implications for program design and funding and is a warning that the 45-hour contact regulation may be broadly applied. Respondents' concerns are that it increases tape costs, replacement courses are difficult to find and broadcast time becomes limited. Solutions may be lecture audio tapes or classes conducted via audio bridge. If the 15-hour telecourse is to be defended, the defense will have to be based on its quality, excellence and research showing that equivalent educational objectives can be achieved in 15 hours by a multi-media program as that obtained from 45 contact hours.

A statement regarding experts being nationally recognized was rejected (44.3 percent). This has implications for producers who use experts.

Statements were rejected about profit projection (45.0 percent) and comparison of cost effectiveness of similar courses (70.4 percent). Comments indicated that it is difficult to project income. Shrinking resources would seem to mandate that distance educators develop this management ability.

A statement about the inclusion of marketing concepts and materials was rejected (64.6 percent).

Conclusions

The media selection model and its evaluating instrument require evaluators to apply specific criteria to the telecourse to determine the suitability of its use in the program. The model and the evaluating instrument consider the combination of media and factors related to the course's general organization, video programs and learners. It can be used with training or post secondary materials to evaluate more than one medium and is short enough to be of practical use.

■ NATIONAL TELECONFERENCING STANDARDS

During the years that teleconferencing has emerged as a viable educational method, efforts have been made to set teleconferencing standards. The scope of past efforts was not large enough to encompass a representative sample of the professional teleconferencing population. Because the field was still emerging and experimenting with methods, a consensus could not be gained about what elements were inherent in a teleconference for quality standards to be set. As the successful model of a quality teleconference emerged through trial and error, more producers entered the field and some produced bad teleconferences. Few refunds were made and teleconferencing professionals acted like consumers and avoided them. The problem still exists because new organizations are entering the field as receive sites. Because a few bad teleconferences can ruin teleconferencing for a campus or for a corporation, the need for quality standards has increased.

Teleconferencing professionals undertook this study to provide the field with quality standards that teleconferences should meet in order to minimally satisfy the market.

This research (Lane, 1990) ultimately led to the Program Standards adopted by the National University Teleconferencing Network (NUTN) which won the Teleconference Magazine Most Significant Advance in Distance Learning Overall for 1991.

Method: The study took place over a period of one year during which a survey was developed which consisted of about 200 preliminary statements. It was sent to 250 NUTN members. A Delphi two-round method was used as it allowed for group process and expert input. Approximately 50 statements did not meet the first round 50 percent retention figure. The second round retention figure was 84 percent which further reduced the number of statements to 81. The interquartile deviation showed that consensus had been reached and that further rounds would not be productive.

The population (n=100) consisted of NUTN teleconferencing professionals (receive site coordinators and administrators) and originators.

Conclusions

Based upon the results of the research, a teleconference provider would have to meet at least 85 percent of the criteria in each section to meet these quality standards.

■ HOW TO USE THIS STUDY

Receive Sites

In selecting teleconferences, use this form to review the teleconference before signing the contract. If it does not meet these minimal quality standards, contact the producers to determine the needs it will meet.

Originators

These are the minimal standards that receive sites say they need and which producers say that they can meet. In preparing your next teleconference, the form should be useful in helping you create a teleconference that meets current quality standards for the industry.

A. Initial Program Offering Announcement
(Requires seven points to comply with standards)

1. Identifies originator's name, facility, location, C-and/or Ku-band delivery.

2. Timely, cutting-edge content which is not widely available in literature or offered extensively via seminars.

3. Developed with receive sites or other groups (via surveys, etc.) to steer development, states the need for program and how the need was determined.

4. Narrowly focused and clearly defines target audience. Provides moderator's and presenter's credentials (content expertise and media experience).

5. Describes format in detail (live, tape, panel, etc.), content level and knowledge prerequisites for participants.

6. Provides thorough wraparound recommendations so that facilitators without content expertise can easily adapt the materials without major research and planning.

7. Contract includes: broadcast date/times by time zone; fees for member, non-member, per head/site/multi-site; transmission methods; copyright release for taping and printing with use restrictions and fees; and cancellation policy and fees. Includes a summary of rights in the offering brochure, contract and facilitator's guide.

8. Will provide a consortium with a sample tape of video work for screening prior to final endorsement.

B. Production
(Requires eight points to comply with standards)

9. Production format is content driven and developed in conjunction with instructional designers.

10. Production professionals with appropriate training and experience in their areas of responsibility will be utilized in all key roles.

11. Production will have high production values.

12. Designed for TV and not just cameras pointed at "talking heads" or "old boy" discussions.

13. Graphics designed for TV (no overheads, flip charts, transparencies unless carefully designed for TV).

14. Will use short taped segments to enhance content; produced with professional broadcast equipment.

15. Will adhere to published agenda times.

16. Breaks will be 15 minutes long and scheduled at least once every 90 minutes.

17. Program's pace and variety will maintain high interest.

C. Educational Objectives
(Requires eight points to comply with standards)

18. Program design is content driven and developed by an instructional designer or design team.

19. Will state outcomes, goals, objectives, content level, informational or instructional program (college degree credit or CEUs which provide specific learning experiences and skill transfer).

20. Instructional programs will provide exercises, small group work and ways to apply content during the program and wraparound.

21. Instructional objectives will state the specific learning outcomes which participants will be able to achieve after viewing the teleconference and participating in the wraparound. (Ex. Participants will learn how to ____ and apply ____. They will learn the skills of ____, ____.)

22. Content will be clearly and narrowly focused with a definite purpose aimed at specific target groups so that participants have a clear sense of a learning experience.

23. Instructional programs should use adult education teaching methods which provide ways to interact, apply knowledge immediately, work in small groups and openly participate in the learning experience (for program and wraparound). 24.Will use a variety of presentation techniques to reach the participants' varied learning styles (visual, auditory, experiential, tactile).

25. Activities and/or presentation methods will help participants apply and/or integrate the new knowledge with existing information.

26. If the video is longer than two hours, consider a program series, a more narrow focus or use longer breaks for site content exercises to help participants assimilate and apply the information.

D. Presenters
(Requires four points to comply with standards)

27. Presenters will be selected for their unique and recognized content expertise and presentation skills; describe organization's background which they represent.

28. Additional consideration will be given to any unique credibility that the presenter has.

29. Presenters will demonstrate professional presentation skills whether or not they are nationally recognized in their area of content expertise.

30. Moderator will have content expertise and professional broadcast experience, control the program (stop long-winded answers and interruptions, keep to the scheduled time and agenda) and help participants accept a less experienced content expert.

E. Participant's Handout

(Requires ten points to comply with standards)

31. Participant's handout will be provided.

32. Will provide each registered site with one camera-ready copy of excellent quality on white paper four weeks before the program. Send by Federal Express to Canadian sites.

33. Will key the handout to the program agenda and print in that sequence. Participant's handout will include (35-42):

34. Agenda (times listed by time zones for segments, call-ins, breaks), conceptual synopsis, presenters' brief biographies, segment names with content synopsis, format (panel, interview, lecture, tape, etc.).

35. Form to explain call-in process and write questions.

36. Major graphics used in video with explanations of how they are used in the program.

37. Graphics appropriate for photocopy (line drawings - not half tones).

38. Sufficiently detailed handout material for the content.

39. Ways to apply information during the program (exercises), wraparound and in self-study after the broadcast.

40. Presenters' bibliographies and additional bibliography.

41. Quality photocopies of copyrighted materials (newspaper/magazine clippings, pages from books, reports, etc.) and convey duplication rights (if any).

42. Order form for materials to be offered (books, tapes, software, reprints, etc.) with price and payment method.

F. Interactive Segments

(Requires 13 points to comply with standards)

43. Sites will not get telephone busy signals.

44. Will use telephone bridge (one line per 10 sites) or a minimum of four call-in lines with ringdown circuits (incoming calls automatically transfer to the next line when the main line is busy).

45. Will utilize alternative methods to receive questions – FAX, computer, etc.

46. Will describe equipment used for call-ins, how calls will be handled, Q&A process and how to avoid audio feedback.

47. Will announce upcoming Q&A and time allotted.

48. Will use operators who are familiar with content to screen and/or take questions.

49. Will allow sites to call in questions continuously to content experts who screen and write them down.

50. Will screen calls to ensure that previously asked questions are not asked again or that questions which trivialize the content will not be aired.

51. Will answer all questions – phone in and those sent in after the conference (computer, FAX, audio conference or send a written answer.)

52. All panelists will not respond to every question.

53. If sufficient calls suggest that a segment is unclear, clarification will be provided on air before continuing.

54. Will use calls/questions which strongly agree or disagree with presenters or which ask for amplification.

55. Interaction will help program flow and not lose pace through bad or duplicate questions and feedback from inexperienced callers.

56. Interaction will be conducted with content experts who can be effective in an interactive dialogue.

57. Will provide a method to contact the presenter for clarification after the program.

Figure 10.1.1

Scoring Form

Teleconference Standards

Teleconference Standards	Possible Pts.	Required Pts.	Total Pts.	Standard Met
A. Initial Program Offering Announcemnts	8	7	—	Yes No
B. Production	9	8	—	Yes No
C. Educational Objectives	9	8	—	Yes No
D. Presenters	4	4	—	Yes No
E. Participant's Handout	12	10	—	Yes No
F. Interactive Segments	15	13	—	Yes No
G. Facilitator's Guide	3	3	—	Yes No
H. Wraparound	5	4	—	Yes No
I. Technical	3	3	—	Yes No
J. Marketing	12	10	—	Yes No
Totals	80	70	—	Yes No

G. Facilitator's Guide

(Requires three points to comply with standards)

58. A facilitator's guide will be sent to registered sites 12 weeks prior to the teleconference.

59. Will not assume content knowledge by facilitator.

60. Will provide clear and specific information on how to market this teleconference.

H. Wraparound

(Requires four points to comply with standards)

61. Will provide approaches for wraparound in initial program offering. Wraparound materials will include (62-65):

62. Ways to help facilitator localize information.

63. Suggested appropriate experience and expertise to aid in selecting local wraparound panelists or presenters.

64. Suggested activities which include small group exercises.

65. Wraparound discussion questions.

I. Technical

(Requires three points to comply with standards)

66. Production – audio, video, phone interfaces, tape/studio segments, transmission, etc. – should meet or exceed National Association of Broadcasters' engineering standards.

67. Will display trouble messages (when possible) and an estimate of how long it will take to correct the problem.

68. Will not schedule program during known solar outage periods or, if scheduled, will state in initial program offering and provide flexible site activities in case of disruption.

J. Marketing

(Requires ten points to comply with standards)

69. Marketing information will be received by registered sites 12 weeks before the program.

70. Will provide clear definition of the target and secondary audience (age, minimum education level, job experience, prerequisite courses, skills or required understanding).

71. Will state benefits that will accrue to the participants.

72. Will provide professional marketing materials with space for the site's local information. Marketing materials will include (73-80):

73. Presenter's credentials (experience, associations, professional organizations, etc.).

74. Target group marketing letters asking for participation.

75. Newspaper advertisements (camera ready).

76. Brochures with copy (objectives, benefits, brief presenters' biographies, etc.) for #10 envelope or self-mailers (camera ready).

77. Print marketing materials will be camera-ready.

78. Suggestions (with contact names/addresses) for groups or companies to underwrite the program (Chamber of Commerce, SBA, etc.).

79. Specific target audiences and how to locate them; names of professional organizations, associations and demographic data about the target participants; contact names/addresses for mailing lists.

80. During breaks, will run lists of upcoming teleconferences emphasizing the types of conferences in which the same target audience might be interested.

■ INSTRUCTIONAL DESIGN FOR TELECONFERENCING

Teleconferencing means meeting through a telecommunications medium. It is a generic term for linking people between two or more locations by electronics. There are at least six types of teleconferencing: audio, audiographic, computer, video, business television (BTV) and distance education. The methods used for each differ in the technology, but there are common factors that contribute to the shared definition of teleconferencing; each uses a telecommunications channel, they link individuals or groups at multiple locations, they are interactive and provide two-way communications, they are dynamic and require users' active participation.

Interactive Technologies

The new systems have varying degrees of interactivity - the capability to talk back to the user. They are enabling and satellites, computers, teletext, viewdata, cassettes, cable and videodiscs and all fit the same emerging pattern. They provide opportunities for individuals to step out of the mass audiences and take an active role in the process by which information is transmitted.

The new technologies are de-massified so that a special message can be exchanged with each individual in a large audience such as face-to-face communication. In this respect they are the opposite of mass media and represent a shift in control from the facilitator to the learner.

Many are asynchronous and have the capability to send or receive a message at a time convenient for individuals without being in communication at the same time. This overcomes time as a variable affecting communication . Satellite delivery reduces travel and tape can be shown at anytime or anyplace.

As more educational interactive technologies emerge, the value of being an independent learner will increase. Research shows that learning from new technologies is as effective as traditional methods. With large enrollments, there is a greater cost-effectiveness and the advantage that all get the same information.

Audio Teleconference: Voice-only communication; sometimes called conference calling. It interactively links people in remote locations via ordinary telephone lines. Systems include telephone conference calls and audio bridges that tie all lines together.

Audiographics Teleconference: Uses narrowband telecommunications channels to transmit visual information such as graphics, alpha-numerics, documents and video pictures as an adjunct to voice communication. Other terms are audio plus, desk-top computer conferencing and enhanced audio. Devices include electronic tablets and boards, freeze-frame video terminals, integrated graphics systems (as part of personal computers), Fax, remote-access microfiche and slide projectors, optical graphic scanners and voice/data terminals.

Computer Teleconference: Usually uses telephone lines to connect two or more participants through computers. Anything that can be done on a computer can be sent over the lines. It can be synchronous or asynchronous. A common example is electronic mail (E-Mail).

Video Teleconference: Combines audio and video to provide voice communications and video images. Can be one-way video/two-way audio or two-way video/two-way audio. It can display anything that can be captured by a TV camera. The major advantage is its capability to display moving images. The most common application is to show pictures of people which creates a social presence that resembles face-to-face meetings and enables participants to see the facial expressions and physical demeanor of participants at remote sites. There are three basic types of video teleconferencing systems: freeze frame, compressed and full-motion video.

Business Television - BTV: One-way videoconferencing often with two-way audio interaction. BTV is an increasingly popular method of information delivery for corporations and institutions. Company updates, news, training, meetings and other events can be broadcast live to any number of locations.

Distance Education: Delivers teleclasses (live or pre-taped) to students in their home or office. Distance education is being utilized by K-12 and higher education. As the cost of delivering quality education increases, institutions find that limited resources prevent them from building facilities, hiring faculty or expanding curricula. They are using distance education to maximize resources and are combining their assets with others to produce programming. This has led to institutions offering courses via satellite regionally or nationally or via local cable and utilizeing other technologies such as qudio and computer.

■ WHY USE A TELECONFERENCE?

To Move Data - Not People: Electronic delivery of data to decision makers is more efficient than physically moving data or people. Conferencing provides the medium to move the data and the forum for analysis. Valuable, time-sensitive data can be disseminated efficiently.

Save Time: You don't have to wait for people to travel and gather. The message is presented from one source and is received in many places simultaneously and instantly. Travel time is reduced and everyone is productive because they do not have jet lag and are not distracted by resort amenities. This results in improved communications and improved meeting efficiency. The bottom line is that teleconferencing adds a competitive edge that face-to-face meetings cannot add.

Lower Costs: Costs must be compared to the productivity savings that can be realized by keeping employees out of airplanes and in their offices, speeding up product development cycles, improving performance through frequent meetings to provide timely information and against the sheer cost of travel. Without travel, meals and lodging, costs are sharply reduced.

Accessible: Anyone can address the audience from any origination site in the world.

Larger Audiences: Because of costs, only a small percentage of a group could attend a meeting, now, all of them can attend. The larger the audience, the lower the cost per person.

Adaptable: Conferencing can be used for business, associations, hospitals and universities to discuss, inform, train, educate or present. Flexible: With a remote receive or transmit truck, a transmit or receive site can be easily situated.

Security: Encryption prevents outside viewers.

Unity: A teleconference provides a shared sense of identity. People feel more a part of the group..more often. Individuals or groups at multiple locations can be linked frequently.

Timely: For time-critical information, sites can be linked quickly. An audio or point-to-point teleconference can be convened in three minutes.

Interactive: Dynamic; requires the user's active participation. Because we are accustomed to working by telephone or letter, we don't consider it impersonal. Once we become accustomed to teleconferencing, it is not impersonal.

■ THE TELECONFERENCE MODEL FORMAT

The following model agenda works well in developing the proper balance between local and national segments and in putting the audience at ease for the program.

Part I: Registration: Have participants arrive 30 minutes before the satellite broadcast. Ask local experts to arrive at registration in time to informally visit with the audience. The satellite portion will start at the scheduled time and late arrivals disrupt others.

Part II: Test Signal: Five-thirty minutes. During the test signal, the technicians ensure that everything is in order to receive the broadcast audio and video signals.

Part III: Local Panel Introduction: Occurs during the test signal. The moderator introduces the local experts, lists the credentials of the national experts and goes over the agenda. Explain how the phone-in question and answer segments will be handled, give information about breaks, refreshments, rest rooms and other pertinent information. Pass out materials. A handout outlining the program elements and giving the names and titles of the wrap-around panel members is helpful for all learning styles of all participants.

Part IV: Experts By Satellite: Usually 90-120 minutes. Satellite broadcast of the national experts. This segment may include pre-produced video which is shot and edited before the live broadcast.

Part V: Audio Interaction: About 30 minutes. The audience telephones in questions for the national experts at the national origination sites.

Part VI: Local Wrap-Around: About 60-90 minutes. This time is used for the local wrap-around panelists to provide their insights on the local aspects of the content, to react positively or negatively to the national panelists and to localize the content so that it is immediately usable for participants. The local wrap-around should be a different experience than that provided in the national segment.

Local Question and Answer Period; after the experts' prepared comments are completed, allow 15-20 minutes for questions.

Part VII: Close: Five minutes. The close of the program should be very clear. Don't suggest that those who must leave may do so while the questions continue. This results in an untidy finish which does not provide the participants with a sense of closure. When the time arrives, announce that only two more questions will be taken. Stop after two questions and deliver a short summary of the day's events. Thank participants for coming and ask them to complete their evaluation form and hand it to a coordinator as they leave.

Planning a Teleconference

The elements of a successful teleconference:

1) Participants focus attention on the informa-

tion being delivered;

2) facilities arranged so that participants can see and hear the information;

3) environment is devoid of distractions;

4) national and local segments are planned so that neither overpowers the other; and

5) participants are comfortable with the technology and their interaction with it.

Buying a Teleconference

How can one make an informed video teleconference purchase? Evaluate the program using the Teleconference Standards and scoring form. Using it should aid you in your selection.

Contract: Register early to receive the teleconference so that the producers will know if the teleconference has enough receive sites to broadcast the program. All contracts are not the same.

Cancellation: Usually, there is a clause stating a date by which the producer must notify receive sites that the teleconference will be canceled. It is wise not to undertake financial commitments (brochure printing, deposits, etc.) until that date has passed. There is no penalty if the producer cancels; if the receive site cancels, there may be a penalty as high as $100.

Copyright: Standard policy is to comply with copyright. Saving fees is not an acceptable reason to violate copyright. Without permission, it is usually not legal to view a program unless one has purchased the rights to view it; tape a program and view it at a later time; sell seats to view a tape at a later time; make copies of the tape to give or sell to others; use portions of the tape in other programs; show portions of the tape in classes; or to make any other use of the program or a tape. All copyrights should be in writing from the producer.

Staffing a Teleconference

Personnel recommended to develop and implement a teleconference via satellite and audio include a coordinator, site facilitator, content specialists and technicians.

Coordinator: The coordinator should have expertise in program planning, marketing and facility coordination. Select a teleconference that will be of value to the target audience and view it as a component of a whole program which includes a local wraparound where participants learn how to apply the information in their job or local area. The local wraparound is equally as important as the satellite component. Establish how the local wrap-around will be staffed and conducted.

The coordinator works with people at the local,

regional and national level. Locally, he or she will work with the viewing audience, site facilitator, facility staff and technicians. Regionally or nationally, the coordinator works with the teleconference producer, network staff and technical coordinator. The coordinator is the organizational and communications link between the producer and the participants.

During the teleconference, the coordinator's duties include seating the participants and briefing them on the schedule, procedures and facilities and coordinating the telephone interaction sessions.

The coordinator's commitment of time, attention to detail and a positive, service oriented attitude will be rewarded with a video teleconference where participants feel that their time/money were well spent. The coordinator's job will have been done successfully if the audience concentrates on the message rather than on how the message is being delivered by satellite. Other duties include:

> Technical Operations (through site facilitator and technician) Publicity, Advertising and Promotion
>
> Facility Arrangements (through site facilitator)
>
> Budgeting and Accounting
>
> Write, photocopy and distribute handouts Be present during video teleconference
>
> Write, distribute and collect evaluations
>
> Troubleshoot problems and answer questions
>
> Moderate teleconference or hire a moderator.
>
> Facilitate local wraparound activities
>
> Arrangements for experts (travel, hotel, fee)
>
> All staffing
>
> Registration and registrars
>
> Video teleconference follow-up

Site Facilitator: Works with coordinator and should have access to all of the materials concerning the teleconference. He or she guides the coordinator in using the facilities and should have expertise in program planning, technologies, troubleshooting and catering. The site facilitator is responsible for assuring the flawless operation of the receive site and should be present during the teleconference.

Site Technicians: Basic knowledge of TV transmission, audiovisual equipment and other minimal technical operations.

Originating Institutions: The producer has the responsibility of advance planning to prepare and distribute the program announcement, promotion packet and instructional materials to the receive sites. The program announcement must address content, marketing, target audience and fees with enough flexibility to adapt the program to all receive sites. The originator usually lists a contact or the producer, who can verify information on the content so that you can plan the wraparound.

Setting Up a Teleconferencing Room

Technical: The technician should have the following information: satellite and transponder; test signal time; start and end times; originator's trouble and question telephone numbers; weather bureau number; room plans; number of registrants; all equipment and its placement; taping instructions and tape; and lighting requirements.

Environmental Engineering: Environmental engineering uses equipment and space to make telecommunications programs more effective for the user by focusing participants' attention on the program and eliminating distractions. The viewing area should make participants feel that they are at a live event and not sitting in front of a television screen. Arrange seating in conference or classroom style. If only one large screen television or projector is used, do not have a center aisle as optimal viewing is directly in front of the screen. Put aisles on the side.

TV Monitors: Placed on stands above the heads of the audience so that all can see clearly. As a rule of thumb, allow at least one inch of diagonal screen for each viewer. Monitors smaller than 25" should not be used. Video projectors are preferable, but are susceptible to room light flooding. Seats should be at less than a 45 degree angle to the screen.

Lighting: Set lights behind projectors to prevent the screen from becoming saturated with light yet bright enough for participants to take notes.

Speakers: Program sound should come from the front of the room. Avoid amplifying the sound through room systems with loudspeakers on all sides or restrict them to the front of the room.

Test: Program materials will include numbers to call for technical problems. Meet with the technician to set problem procedures.

The purpose of the 30-minute test signal is to test the video and audio downlink in the conference room to ensure that technical problems are resolved before the teleconference begins. If the signal from the scheduled transponder is bad, but a good signal is received from another transponder, call the trouble number and report the problem.

Once the best signal is received, the technician should adjust the picture under the lighting conditions, check sound levels (occupied rooms require more volume) and test the telephone. Move a telephone next to the receiver for technical problems.

Broadcast Interruption: Under rare circumstances the broadcast may be lost. Heavy thunderstorms with lightning or snow and ice storms may contribute to the problem. The technician should attempt to correct the problem or call the originator.

The moderator should go to the front of the room and assume control of the meeting immediately. Announce that "we seem to losing the satellite signal but will stay with it as long as we can understand the audio and see some of the video." If the outage appears to be lengthy turn off the monitors and 1) announce a refreshment break or 2) introduce local experts and begin the local wraparound. The technician should monitor the reception on a monitor elsewhere. If the satellite feed returns, he or she should tell you so that you can rejoin the broadcast without cutting off a speaker. Common courtesy demands that the person who has the floor be allowed to finish a statement.

To prepare for a broadcast interruption, have panelists on site throughout the meeting. If you must track them down elsewhere, you will lose the ability to resume control. Go over this set of circumstances with panelists and outline expectations. In short, plan for the worst scenario and go over the contingency plans with everyone involved.

Telephone Equipment: Place the telephone order early to ensure that it is available. During calls, there will be a slight delay between the time the question is asked and the time it is heard over the TV. The delay occurs because the signal must reach the satellite and return to earth.

Telephone Interaction: Teleconferences use a variety of interaction methods. Some producers use questions throughout, while others limit questions to the end of the program. Encourage participants to avoid lengthy explanations and personal identification and to ask questions in less than 25 words. Use a question form to help them structure the question. Some participants are uncomfortable asking questions on national TV. The coordinator may ask the question. Usually, only one question per site is taken. If participants ask their own questions, be sure to tell them how the calls will be handled. To avoid audio feedback from the room sound system, place the phone in the hall.

The call may be placed on hold while other callers questions are answered; don't hang up. A busy signal will be heard if all of the lines are full; wait a few moments and place the call again.

Planning the Local Wraparound

Confirm the moderator immediately and include him or her in all planning and meetings.

Local expert selection: identify people with whom you could consult about the content and its local implications.

Determine what objectives to accomplish with the program. Will these objectives compliment the national segment? What format would best help meet the objectives? A local speaker(s) followed by discussion? A panel of local experts? Small group discussion or activity? A combination? How much time is need-

ed for the local program? Are there local materials that should be handed out to participants in addition to that provided by the teleconference producer such as lists of local resources related to the topic.

The key to the success of the teleconference is the participation of local on-site experts. Without this component, participants may leave feeling that they just paid $75 to watch TV. While they probably could not have seen the program without your efforts, they will learn far more if you put together a well-developed wraparound segment that will enable them to utilize the information immediately.

If the teleconference is educational and an exchange of knowledge, skills or attitudes is desired, a local wraparound program will greatly enhance your chances of success. As in any adult education program the teleconference needs to address concepts that are of interest and meaningful to the participants. It is also more effective if it provides participants the opportunity to discuss and apply the concepts they have heard presented in the teleconference. A wraparound which includes local resources and discussion, can greatly enhance the programs effectiveness. Experienced teleconference producers believe that a well designed local wrap-around is the most important part of the program. Your role in planning and facilitating the local program is extremely important.

The producer should provide suggestions for the wraparound. Consider activities to be done immediately before the broadcast such as a series of discussion questions, work sheets, questionnaires or exercises or pre-taping demonstrations or examples of the content to show the local importance.

Some content may create adversarial positions among experts. Meet with experts before the teleconference and go over how adversarial positions are to be handled. Most audiences are not comfortable with shouting matches that confuse the issue rather than presenting objective credible opinions from experts which create a decision making environment. The moderator should be aware of adversarial relationships and be instructed to control these areas to keep the program on track.

Encourage them to call the producer so that in their preparations they will not duplicate information being covered by the national experts and so that they can develop local solutions which synthesize and apply the national content.

Local sources of experts include executives and faculty. Using local people as on-site experts has two advantages;

1. It creates a source for answers to questions regarding what is available locally and points that are unique to the area and

2. It adds "instant credibility" to the promotion by being able to advertise a faculty member organization or company representative as local presenters.

When asking local experts to participate, be definitive about what they are to do. Be certain that they are a part of the planning process. Hold a meeting with all local experts so that they can discuss the topic and decide who will be responsible for which portion of the local content. Plan the contents of the local handout, assign responsibilities and set deadlines to receive material. This will avoid duplication on the program and save time for the experts.

Program producers will frequently send updates before the teleconference. If appropriate, send these to the local experts so that they will be current with all program aspects.

Solicit questions from panelists to see if they have any uncertainties about how the teleconference will be handled or about their role in the local wraparound segment.

Stay in touch with the panelists to ensure that their work is proceeding as scheduled and that handout material will be on time.

Hold a brief rehearsal with the local experts prior to the start of the satellite program.

Food Services: Planning appetizing and nutritious refreshments at a reasonable cost can be a challenge. Rely on catering to provide menu consultation, delivery and refill services which will stay within the budget.

If the teleconference falls during the lunch hour, you may wish to plan a meal before the telecast, This can be offered as an option so that food does not inflate the ticket cost. A prime consideration is whether the number of people can be served in the time allowed.

Provide coffee, soft drinks and snacks throughout the day, but take specific breaks for refreshments. Many people prefer to have simple refreshments available throughout the day. Some participants become uncomfortable if they must sit too long and appreciate the excuse of a trip to the refreshment bar to stretch their legs. Order extra meals for possible walk-in registrants. Having refreshments available can be helpful if there is an interruption in the broadcast. A five minute refreshment break will provide time to re-group and correct the problem or to gather the local experts to begin the wraparound.

Make arrangements with food service to quietly clean the refreshment area and restock it throughout the teleconference. If time is short, consider having a meal served in one room and having the teleconference in a second room.

In setting up room plans, leave aisles so that people can get to refreshments without walking in front of the monitors or disturbing others.

Registration Procedures: You probably have a standard procedure for program registration. For teleconferences, set up registration tables outside the viewing room. Often, participants arrive at the last minute since they assume the meeting will not start on time. Teleconferences start promptly due to the cost of satellite time. Have one registrar stay on the desk after the start of the program to handle late arrivals.

Contingency Plans: Anticipate the unexpected. It is wise to have a written contingency plan for each of the following possibilities and to share it with the technical crew, other coordinators and the site facilitator.

1. Absence of a local expert. By using a panel of experts, one dropout will leave the local portion manageable. If one is using a regional expert from another city, have him or her arrive the day before. Never book the last flight into town.

2. Failure of support services. Check available options for food service and satellite equipment. 3. Schedule changes. Establishing rapport with the local experts will allow flexibility if the national schedule changes.

4. Walk-in participants. By having an overflow space and extra handouts, an unexpectedly large audience can be accommodated. During the wraparound, the overflow can be seated in the main room.

5. Technical malfunctions. By having local and national technical trouble numbers available, information about malfunctions and the probable repair time will be readily available.

6. Back-up staff: Anyone can become ill. Share all plans and information with the entire staff. Appoint a backup for everyone. YOU should have a backup too!

7. Weather: Satellite reception is affected by weather - particularly thunderstorms or ice/snow buildup on the dish. Listen to the forecast or call the weather bureau.

8. Telephone: Have a back-up telephone in the conference room in case the main phone breaks.

Satellite Communications

Satellite communication has been used for years. Long distance telephone calls, national and international televised sporting events and cable TV movie channels all operate via satellites. Traditionally, these have been analog signals. Digital products have been successfully introduced into the market and require significantly less bandwidth (about three megaHertz) for satellite transmission which has reduced the cost of satellite transmission.

Geostationary Orbit: The Clarke Satellite Belt: Sir Arthur C. Clarke, British physicist and science fiction writer, was the inventor of satellite communication. In his 1954 paper "Wireless World", he explained the geostationary orbit, 22,300 miles above the equator and said that three satellites based in this orbit could provide world-wide communications.

Today, many domestic satellites use the Clarke Belt. Arrayed in an east-west belt over the equator, they appear fixed in space to earth stations.

Satellite Footprint: Because of the height of the geostationary orbit, communications satellites have direct line-of-sight to almost half the earth. The major advantage of satellite communications is the "footprint" of coverage it provides. A signal sent via satellite can be transmitted simultaneously to every city in the US. Any number of downlinks around the country can be aimed at a particular satellite and receive the same program. This is called point to multipoint communication.

Transponders: Video, audio or data signals can be transmitted to a satellite transponder by an uplink. There may be 24 or more transponders per satellite; each can amplify and relay a signal back to earth which is picked up at an earth station.

C Band & Ku Band: Currently domestic communications satellites operate on two frequency ranges designated C- and Ku-band. Each requires specific electronic equipment. C-band is less expensive; it operates at 4 kHz. Ku-band operates at 12 kHz. Some teleconferences are broadcast on both bands.

Receivers: Receivers convert satellite signals into channels that can be viewed (one channel at a time) on a TV monitor. A receiver must be designed to tune-in the format, bandwidth and audio sub-carrier that will be viewed. Some programs are broadcast in a scrambled code (encryption) and decoded at the receive site. Digital transmission require a digital receiver.

Basic Receivers: Lowest cost; limited channel tuning capability, so they may have manual channel selection switches. Usually combined with fixed antennas to form an inexpensive system package.

Multi-Format Receivers: Most versatile; have adjustments for every broadcast format and can receive any satellite video program in six or more bandwidth selections and two agile audio subcarrier switches; usually used with motorized systems. Some digital receivers will receive only digital transmissions.

Fixed Position System: Low cost systems limited to reception from one satellite and one band.

Motorized System: Receives programs on different satellites by moving the dish which is connected to a position adjustment device in the control room.

Automated Systems: Controlled by a microprocessor to allow instant movement to different satellites whose positions are stored in memory.

Video Teleconferncing Transmission

A standard analog video signal with a bandwith of approximately 6.5 MHz must be didtized to be transmitted over digital circuits. The signal is applied to an analog to digital conversiona circuit. The output signal is a digitized version of the analog input signal of 90 Mbs which is expensive to transmit. Codecs (COder/DECoder) compress the signal allowing it to be transmitted over standard circuits.

Standard transmission rates for video teleconferencing are usually multiples of 64 Kbs up to the T1 rate of 1.54 Mbs. Some codecs enable the user to select the speed to match that of the circuit used for video teleconferencing.

T1 circuits are used to connect corporate PBXs to the telephone company's central office facilities. Each T1 circuit can accommodate up to 24 voice channels at a lower cost than 24 voice circuits. A 56 KB or 64 KBS codec operates in the range of one voice channel. A standard video signal digitized at 90 Mbs is comprised of about 1400 voice channels.

The compressed video signal quality decreases as the transmission speed decreases. The cost of transmission decreases as speed decreases. Freeze Frame Video: A freeze frame system uses telephone channels to transmit video information. Because of the narrow bandwidth, the image takes a number of seconds to reach the receive site where it appears on the TV monitor as a still picture. The advantages are lower costs and flexibility in linking multiple sites. Slow scan systems are similar to freeze frame and the terms are often used synonymously.

Freeze frame technologies include a range of features; analog digital, monochrome or color pictures, resolutions, transmission speeds and extra memory. Newer models provide multiple send times to select the resolution and transmission time through digital circuits and compression coding. Some units transmit the video information in a digital format over a data circuit. This reduces the transmission time to about 9 seconds to a 56 kilobit link. Because of the faster transmission rates, many new freeze frame applications use data circuits.

With compression techniques, one model reduces the information about 50 percent. Used on a digital circuit, the color picture is sent in about 12 seconds at 56 kilobits. The unit can also be used on a 1.5 megabit T1 data channel, where the transmission time is 0.3 seconds for a color image. A T1 channel is also used for compressed motion video.

Compressed Video: Compression is performed by a picture processor or codec which makes it possible to transmit a moving video picture on a data channel of 56 kbps to 45 mbps per second. The newest 1.5 megabit codecs also improve picture quality.

Compressed video (near motion) and full-motion video basically differ in two ways: 1) Compressed video uses compression techniques to reduce channel bandwidth requirements; 2) Compressed video images may not appear to be as natural or have as high a quality as full-motion images; there may be blurring or loss of background resolution. Its main advantage is the significant reduction in the amount of bandwidth needed to transmit a moving image which decreases transmission costs.

The compressed video system uses a telephone data circuit – currently a T1 carrier or 1.5 or 3 megabits – to transmit video, voice and data. It reduces the full video information (NTSC Standard-color video) by using a compression technique to eliminate redundant information which reduces the original video signal of 100 million bits to 1.5 or 3 million bits. Compression significantly reduces the cost of transmission. Standard TV signals are broadcast in analog signal format requiring a significant amount of bandwidth, the digital equivalent of 80 Mbps or more. This corresponds to a full satellite transponder or 1820 voice phone lines. This translates into high cost for signal transmission.

Digital video signals are broken down into thousands of individual elements called pixels. Between frames, many elements are the same. A codec takes advantage of this duplication by sending complete information on the first pixel and a brief code to repeat the values for the next pixel. This reduces the information sent and the bandwidth required.

Interframe coding for conditional replenishment compares the changes between two frames and transmits only the changes. Motion compensation predicts changes between frames and transmits only the difference.

The compression algorithm is housed in software allowing the codec to be upgraded without changing major components. An international standard has been set called CCITT Px64 so that future rates will be operable in multiples of 64.

Full-Motion Video: A full-motion video system uses wideband channels – 4 to 6 megahertz for color analog – to send video, voice and data. Because of the large channel capacity, it transmits a picture with the full motion and resolution of broadcast TV. Its disadvantage is cost.

Fiber Optic Systems: The transmission of voice, video and data by light wave signals inside a thin, transparent glass fiber cable, is providing more choices for telecommunications users and is rapidly bringing day-to-day digital communication to the home and office. Fiber-optic systems are one of the latest advances in communications.

Fiber cable has three major components; the center core contains glass or plastic light guide fibers; cladding surrounds the core to reflect the signal back; and a coating encases and protects the fiber against breakage. One pair of fibers can carry up to 10,000 telephone calls simultaneously. A fiber-optic system offers the advantages of clarity of transmission, speed, accuracy, security and volume. The system is smaller

and construction, installation and maintenance costs are much less.

Teleconferencing Planning Calendar

Week 16
1. Make decision to receive the teleconference.
2. Get approvals.
3. Review, sign and return contract originator.
4. Contact potential co-sponsors.
5. Identify and contact key people who can recommend local experts.
6. Preliminary budget.

Week 15
1. Coordinator, facility and room assigned. Others notified of date.
2. Reserve equipment.
3. Arrange for satellite reception.
4. Arrange for experts for wraparound. Make speaking assignments.
5. Do research for handouts to cover local content.
6. Arrange for room moderator.

Week 14
1. Obtain printing estimates for promotional materials.
2. Contact local faculty and other content experts.
3. Initial contact with co-sponsors.
4. Develop budget.
5. Set local registration fee structure to cover costs.
6. Develop mailing lists for the teleconference.
7. Get lists from co-sponsors.

Week 13
1. Prepare brochure for printer.
2. Have artist prepare graphics and dummy layout.
3. Have photos shot if necessary.
4. Call producer for literature or new content.
5. Prepare press releases and other advertising.
6. Set pre- and on-site registration procedures.

Week 12
1. Contact key groups to participate/sponsor.
2. Print labels printed for brochure.
3. Arrange for mail room to handle mailing.
4. Get final approvals on brochure copy.
5. Set type; complete camera ready brochure art.

Week 11
1. Send camera ready brochure art to printer.
2. Distribute initial press releases and PSAs with

number to call to receive brochures.

3. Distribute press releases to co-sponsors and other interested parties who need information before the brochure is ready.

4. Schedule and prepare copy for advertising.

Weeks 10 & 9

1. Follow up and complete above.

Week 8

1. Review facilities and satellite arrangements.
2. Finalize on-site personnel arrangements.
3. Complete labels for brochure mailing.

Week 7

1. Mail brochure.
2. Check with local experts, moderator to see that their portion is progressing.
3. Complete handout materials for local content.
4. Get national content handout materials for duplication without rush fees.
5. Select binder to hold all materials.
6. Send advertising copy to newspaper, radio and TV.

Week 6

1. Follow up and complete above.
2. Process reservations. Keep reservations list on computer so that it can be updated and printed.
3. Arrange promotion interviews on local talk sshows.

Week 5

1. Do a second mailing to key people.
2. Do second press release with updated information
3. Request that the teleconference announcement be read at organization meetings.

Week 4

1. Follow up and complete above.
2. Prepare evaluation form.

Week 3

1. Follow up and complete above.
2. Send out press releases with updated information.
3. Arrange promotion interviews on local media.

Week 2

1. Arrange with printing to duplicate handouts on short notice so that only one run is required.
2. Prepare handout binders.
3. Print name tags for registrants.
4. Confirm arrangements for facility, satellite reception, food, printing, etc.

Week 1

1. Final check with local experts.
2. Make arrangements for second viewing room if walk-in registrants may cause an overflow.
3. Review procedures for on-site activities with staff.
4. Prepare final advance registration list.
5. Have handouts printed and put into binders.

Day Before the Videoconference

1. Reserve for last minute details and problems.
2. Prepare late registration name tags.
3. Pick up direction signs.
4. Confirm number of materials, meals/refreshments, seats and facility arrangements
5. Organize registration/program materials.
6. Confirm local experts and go over new information.

Day of the Videoconference

1. Put up directional signs.
2. Set up registration/program materials.
3. Check equipment and phone.
4. Check facility and room.
5. Conduct test to receive satellite transmission.
6. Confirm number of meals including walk-in registrants.
7. Hold teleconference.
8. Collect evaluations.

After the Teleconference

1. Take down signs.
2. Record and send in feedback to the producer.
3. Remove or secure equipment.
4. Make copies of items to be sent to producer.
5. Invoices for reimbursements sent to producer.
6. Catalog video cassettes with at least two copies of program materials.
7. Have video copies reproduced if necessary.
8. Evaluate evaluations.
9. Send copies of evaluations to those needing them.
10. Update video teleconference statistics sheet.
11. Send thank you letters to experts.
12. Follow up on payment of fees to experts.
13. Follow up on payment for equipment/service.

■ AUDIOCONFERENCING

Approach audioconferencing differently than a face-to-face meeting; there is a big difference in presentation. Use these points to capture the participants' attention in an audioconference.

1. Know the equipment: Know what to do to be heard on the system and what it takes to mute it. Locate the on-off switch and the microphone. Work

with the equipment the day before the audioconference so that its operation is easy for you.

2. Participation: Involve the audience early to make them comfortable with the medium. Get participants talking within the first few minutes or they will act as if they are listening to a radio show. Ask participants to identify themselves by name and location.

3. Create Pace: Alternate short presentations with discussion, visuals or a work sheet. Keep segments shorter than ten minutes.

4. Use Visuals: Throughout the conference, use plenty of visuals such as slides, overheads or video tapes (mail to the receive site before the meeting). Use more visuals late in the program to provide focus and relieve boredom.

5. Handouts are important for visual learners. Use an agenda with paragraphs explaining discussion points.

6. Speaker Variety: Combine male and female teams as the change is pleasant to the ear. Use people with accents who are immediately identifiable. Plan the presentation order to vary the voices.

7. Plan the wrap-up. The worst thing one can ask in a 50-site teleconference is, "are there any questions?" Instead, ask if there are "Any questions in Dallas?" Everyone in Dallas will look at each other and silently nominate someone to ask a question.

■ THE NATURE OF INTERACTIVITY

One of the unique characteristics of many new media is their greater interactivity compared to the conventional mass media. This greater interactivity in mediated human communication provides an appropriate setting for development and testing theories of involvement.

"Involvement" is defined and operationalized in many ways. Broadly it refers to the degree to which an individual actively participates in an information-exchange process. Involvement may be psychological through perceptual and cognitive processes that are either involuntary, such as seeing visual images from an interactive videodisc or voluntary, such as interpreting the meaning of these images to decide which frame to select next. Involvement also can be social — that is, being involved with other individuals by interacting with them, perhaps through a communication medium. These two levels (psychological and social) may interact, as in computer conferencing where the relationships among group members may set norms that influence what kinds of messages are read and what kinds are ignored.

How is interactivity in the new media — due to the form of messages or characteristics of the media — associated with psychological or social involvement? This question is not new: McLuhan (1964) categorized media as s "hot or cool" depending on their level of ambiguity or the degree of information processing that they required from the user. Krugman (1965) suggested that the extent to which an individual made connections (a conscious bridging of the medium's message with one's experiences or personal references) is greatly influenced by characteristics of the medium. Both McLuhan and Krugman felt that television, for example, is a "cool" medium because it does not generally require much involvement by the viewer. This low involvement is one explanation for the susceptibility of television audiences to entertaining, repetitive messages about low-salience products and issues. (Williams, et al., 1988, p. 169).

Another issue is whether interaction is necessary for learning to take place. A secondary issue to that question is whether interaction must take place in real time or whether it is equally effective asynchronously. Guaranteeing sufficient interaction is a key concern of distance educators who use techniques such as E-mail, telephone office hours and peer interaction to provide an opportunity for dialogue to learners. One study (Stone, 1991) and others show student determination and course difficulty, not degree of interaction, to be the strongest predictor of distance learners. Stone's study surveyed over 8,400 graduate-level engineering students enrolled at eight universities belonging to the National Technological University (NTU) and its parent organization, the Association for Media-based Continuing Education for Engineers (AMCEE). The results of the study found that engineering students learning by satellite performed at least as well academically as demographically similar students taking the same courses by conventional classroom methods. Over 300 previous studies have reached the same conclusion. However, the research also showed that students, especially older ones, who watched videotapes of the satellite broadcasts performed better than those who watched the broadcasts live. This suggests that for the older, working adult, flexibility in the time of learning may be more important for learner performance than the ability to talk back directly to live instructors.

While interaction has become significant, it may not be the most significant contributing factor to all students learning. Learning styles are varied. What works well for one student may not be effective for all students. In the past when distance learning was just beginning, early developers attempted to replicate the classroom environment as closely as possible. As the research continues to grow in favor of learning styles, self-directed learning and, asynchronous interaction, we may find that there are many types of content that don't require two way audio and video for learning to take place. The following study (Lane, 1990) is an example of a model of interaction that can be used in synchronous or asynchronous communications and is suitable for audio, video, computer and others distance learning technologies.

■ THE USE OF AUDIO INTERACTION IN A TELECOURSE OFFERED BY SATELLITE

A telecourse, "Foundations of Adult Basic Education" was offered via satellite across the United States by the Video Instructional Program (VIP) at the University of Missouri, St. Louis. It aired for two hours on each of eight consecutive Saturday mornings from Jan. 13 - March 3, 1990 The 15 hours of video tape were up-linked (C-band) by the Educational Satellite Network (ESN), the Star Schools unit of the Missouri School Boards Association. To provide interaction with the instructor, four hour-long audio conferences were used. The course was geared toward adult basic education teachers, as well as corporate and business trainers. It carried three hours of undergraduate or graduate credit. The course was approved for Missouri State ABE/GED Teacher Certification. Missouri students used federal 353 funding to pay most of their tuition. Most students completed the course on May 15, 1990.

Funding was the remainder of a grant to downlink teleconferences. Time on Saturday mornings was used because at $140 an hour (for 15 hours) it was the lowest cost time period. Satellite was not used for the interaction, although it was contemplated and ruled out because of the additional cost. What appeared to be a disadvantage actually worked out well as most dishes are not in use on Saturday mornings and students had few conflicting activities. Normal tuition fees were charged; no broadcast or out-of-state fees were charged. As students registered for the class, they were assigned a permanent group, audio conference call time and conference bridge telephone number and time to place calls to the audio conference switchboard. Students incurred the long distance charges which averaged about $40. The first audio conference was scheduled after students watched programs on the first day. The operator taped the audio teleconference and students could order the tape by telephone for $6.00. The fee covered only the cost of the tape, duplicating and shipping (48 hours after the order was received). Tapes were not available for rental. Conference times were scheduled back to back (eastern time) for Group A: 12:15 - 1:15 p.m.; Group B : 1:30 - 2:30 p.m.; and Group C :2:45 - 3:45 p.m. The switchboard accommodated 20 participants at once.

The telecourse focused on how to teach basic reading, word recognition, comprehension, writing and intermediate math skills. It addressed problems that are characteristic of adult learners, including their needs, interests and learning styles. The video was produced by Maryland Public TV and focused on teaching practices in adult basic education. Students learned how to select and use classroom materials, learning contracts and other methods to individualize instruction. Teaching in literacy programs is not normally taught at the higher education level. There were pockets of six and eight literacy teachers who work in ABE/GED programs throughout Missouri who took the course together as well as smaller groups in other states. Three texts were required for the course, including a 200-page study guide written by the author and John Henschke, Ed.D., associate professor, Educational Studies Department, School of Education, UM-St. Louis, who was the national instructor. Students wrote and conducted their assignments through self-directed learning contracts, a method which had been used successfully by the author during the two years she taught the telecourse locally. Henschke routinely uses learning contracts in all classes. Students used the contracts to create projects that they needed for their literacy programs which added the benefit of hands-on application There were 39 students. VIP assisted students in locating receive sites – sometimes suggesting the local satellite dish dealer or restaurant when educational facilities were not available. All but one facility donated the time. Most students provided videotape to the facility to tape the feed and picked up the tape during business hours, viewing the tape when it was convenient for them. When several students recorded HBO by mistake, other students mailed their tapes to them.

Research Question

Does group audio teleconferencing provide a useful method of interaction with the instructor for a pre-produced telecourse?

Most information delivered by satellite is live and interaction is conducted by the student calling the instructor with one question; conversations, when allowed, are very short. Satellite time can range from $300 to $1,000 an hour. Because of the high cost, it is necessary to find other ways to provide satisfactory interaction with the instructor. An audio teleconference may be one solution as it can be conducted when rates are low. If students must miss a conference, they can listen to a recording.

Review of Literature

The heavy emphasis of past communication scholars on investigating the effects of the newest medium has come at a cost in terms of what was not studied. The telephone, has been almost totally ignored by communications scholars (Williams, et al., 1988, p. 24). Past study of the telephone was shortchanged because the telephone was widely diffused among U.S. households before communication research escalated around 1950. The study of the telephones diffusion and impact was made more difficult after the point at which almost all U.S. households had phones.

Bales has defined interaction as the behavior of one person influencing the behavior of another in a face-to-face situation. Interaction analysis in its broadest sense is a method of describing and interpreting human interaction as it occurs in a specific group setting (Bales, 1950 in Emmert, 1970, p. 373). Interactivity is a widely used term, but it is an under defined concept. As a way of thinking about communication, it has high face validity, but only narrowly based explication, little consensus on meaning and only recently emerging empirical verification of actual role (Hawkins, 1988, p. 110). The most helpful definition for interactivity would be one predicated on the issue of responsiveness. The distinction called for is between interactive, quasi-interactive (reactive) and non-interactive communication sequences. Quasi- and fully-interactive sequences differ clearly from non-interactive communication in requiring that sender and receiver roles be interchangeable with each subsequent message. The complete absence of interaction is marked by incoherent conversation (Hawkins, 1988, p. 110).

The users of interaction analysis techniques have identified three dimensions: the affective, cognitive and multidimensional. The affective systems generally examine such teacher behaviors as positive/negative reaction to students, praise, criticism, encouragement, acceptance and support. The cognitive systems focus on the level of abstraction of a statement, logical processes and the type of logical or linguistic function a particular behavior seems to serve in the classroom. Multidimensional systems attempt to identify factors from both dimensions, affective and cognitive (Emmert, 1970, p. 374.)

Data collection may be handled by taping the interaction, transcribing it and coding from the typed transaction. (Emmert, 1970, p. 374.) A coding or sampling unit may represent a unit of time, a thought or verbal unit, a content area or a sequence of two or more behaviors (Emmert, 1970, p. 374.)

Affective systems are probably the earliest and most widely used types of interaction analysis systems as developed in 1945 by H. H. Anderson and H. Brewer. They used a continuum from integrative to dominative to analyze or classify the behavior of both teachers and pupils. (Emmert, 1970, p. 374.) Using this framework, the Withall (1949) Climate Index was developed and has since been used in various research studies. Withall's system classifies teacher verbal behavior into seven categories: 1) Supportive statements intended to reassure and commend the pupil; 2) Accepting and clarifying statements intended to convey to the pupil the feeling that he was understood and to help him elucidate his ideas and feelings; 3) Problem structuring statements or questions objectively proffering information or raising questions about the problem in an attempt to help the learner solve his problems; 4) Neutral statements - polite formalities, administrative comments, verbatim repetition of something that has already been said - with no inferable intent; 5) Directive or authoritative statements intended to make a pupil follow a recommended course of action; 6) Reproving or deprecating remarks intended to deter the pupil from continued indulgence in present unacceptable behavior; and 7) Self-supporting remarks intended to sustain or justify the teacher's position or course of action (Emmert, 1970, p. 376).

B. O. Smith (1962) pioneered in the area of cognitive dimension with his attempts to develop a method for the logical analysis of the strategies teachers use in the classroom. Another of the early methods used to analyze teaching behavior was developed by Gallagher and Aschner (1963) using Guildford's (1956) framework for looking at the human intellect. This system uses four basic cognitive categories – cognitive memory, convergent thinking, divergent thinking and evaluation – to analyze both the types of questions the teacher asks and the students' replies (Emmert, 1970, p. 376.) The relationship between the types of teacher questions and the types of pupil responses becomes a legitimate research question (Emmert, 1970, p. 377).

With the exception of Withall's system and Joyce and Harootunian's system, all the affective systems use time as a basic sampling unit (Joyce and Harootunian use a shift in content and Withall uses either words or phrases as the basic unit). In addition to the time sampling unit, the Flanders Interaction Analysis and related systems use shifts from one category to another. The basic sampling unit in the cognitive system is always a content or cognitive unit. In the system of Bellack et al, the number of written lines in a one-category sequence is the unit. In the Gallagher-Aschner system as well, the category change indicates the length of the unit In itself, then unit length is the length of time that teacher or student behavior can be classified into a single category without the intervention of a second category. Smith uses a content change to indicate the length of the basic unit. The multidimensional systems of Honigman and Amidon both reflect the same sampling procedure used in the Flanders Interaction Analysis system (Emmert, 1970, p. 378.) For this study, the unit used was one line of type from the transcript.

The adult learner has the potential and desire to increase self-directiveness in cooperation with other learners and trainers. This means that in the learning situation the adult: 1) accepts and wants responsibility; 2) orients toward the future; 3) values initiative; 4) opens up to opportunities when they are presented; 5) solves problems; and 6) is creative. Other adult learning concepts include that the instructor; 1) set a climate for learning; 2) establish a structure for mutual planning; 3) diagnose learning needs; 4) formulate directions for learning; 5) design a pattern of learning experiences; 6) manage the execution of the learning experiences; and 7) evaluate results and re-diagnose learning needs (Knowles, 1970). These were added as coding factors for the study.

Basic Interaction Analysis Categories

For this study, the 10-category Interaction Analysis system (E. J. Amidon and Flanders, 1967) was used. It is a direct outgrowth and refinement of Flanders' original system. (Emmert, 1970, p. 378).

Two guidelines for setting up categories are used by the observer recording interaction. First, enough categories should be available to the observer to describe any occurrence. Second, the categories should be mutually exclusive so that the observer cannot describe an occurrence with more than one category. All verbal behavior is classified into one of three basic divisions: Teacher talk; student talk; and silence, confusion or miscellaneous occurrences. Teacher talk is further classified as indirect and direct. (Emmert, 1970, p. 381). Subcategories were added (E. J. Amidon et al., 1968, in Emmert, 1970, p. 398), which indicates that categories may be added to the existing system. These additional categories show how the original ten categories may be expanded, making it possible for the observer to tailor the observational technique to the particular dimension of concern in human interaction (Emmert, 1970, p. 399). Occurrences are determined by calculating the percentage of time used for all categories (Emmert, 1970).

Validity: The construct validity of Interaction Analysis is based on a conception of group climate or classroom social-emotional climate. This construct is characterized by the feelings and attitudes that the members of the group have toward one another, the subject they are studying, their teacher and the work conducted. Methods used to identify outside criteria for measuring the climate have included supervisory ratings, behavior and reactions of the learners and in some cases objective observer ratings of climate. The early studies of Withall and Flanders indicate that, in general, where learners have more positive attitudes towards classes, teachers tend to be more indirect

than teachers of classes in which learners have more negative attitudes (Emmert, 1970). More recent notions of the validity of Interaction Analysis have had to do with the relationship between teacher influence patterns and achievement of students. If interaction analysis shows that certain teacher behavior patterns do or do not encourage achievement, then it can be used to predict achievement. Interaction Analysis' validity as a prediction tool is much more significant than its validity as a tool simply to describe climate, pupil attitudes toward the class or pupil perceptions of the teacher.

The results of teacher education research using Interaction Analysis have added a further dimension to the question of validity. Results of a number of studies (Hawkins, 1988) clearly indicate that teachers who learn Interaction Analysis are more likely to be accepting, supportive and less critical than teachers who are not taught Interaction Analysis. It is a tool that can be used to identify teaching patterns of teachers and to discriminate between groups of teachers who have received different kinds of training.

Reliability: Traditionally, reliability in Interaction Analysis has been thought of primarily as inter-observer agreement for the same classroom observation or consistency within the same observer across two observations. The problem with the second index of reliability is that it is difficult to assume that a teacher who is observed by the same observer over periods of time will produce the same behavior. The non-obtrusive method was used so that the tapes were recorded and later transcribed. Students knew that the tape was being made but no announcement was made.

Methodology: Tapes were made of each audio teleconference. Students were assigned to one of three sections to test the factor of group size; sections contained, 19, 14 and 6 students. The tapes were transcribed, coded and analyzed for the structure provided by the instructor, types of questions asked and the instructor's responses, discussion, group interaction and other factors. Responses were coded.

Findings: The transcripts were coded as follows, using the unit of one line of type in the transcript.

Teacher Talk: Indirect Influence

1: Accepts, Clarifies, Student Feelings: 93
Used in a number of ways when the teacher communicates acceptance of feelings expressed by learners. First and probably most basic, the teacher simply uses a word or phrase which identifies the feeling of the pupil without criticizing the pupil for having the feeling. Or he relates the pupils feeling to other peoples feelings. Perhaps he tries to relate the feeling to the supposed cause of the feeling: "I guess

we're feeling kind of blue today; I've often felt that way myself when I was disappointed"; "The class seems very excited about our trip to the _____." (Emmert, 1970, p. 381.)

One kind of behavior which is often misclassified into this category is reassurance. Actually, a reassuring statement is rejection of feeling. When the teacher says, "Don't be upset about your test, it's not all that hard and I know you're very bright," he is, in fact, rejecting or ignoring the pupil's anxiety and concern about the test. (Emmert, 1970, p. 382.)

2: Praises or Encourages: 0

Statements which evaluate a student's ideas as right, good or appropriate, for instance, "You are right, that answer is a good answer," or "I like that answerare not praise or encouragement". Encouragement means only statements that actually function to encourage the learner to continue talking. The teacher's "Uh huh," "Okay," "Yes," "Um hmm," "Right," and "All right" during a learner's hesitations are classified 2 only if they do not inhibit learner talk. Also included are statements which cause laughter, jokes not said at the expense of the learner.

3: Accepts or Uses Ideas of Student: statements rephrasing learner's idea

This includes statements repeating, rephrasing, summarizing or restating a learner's idea to communicate that the teacher has heard the learner's statement; they do not communicate that the idea is right or wrong.

3a Acknowledges Student Ideas	13
3c Clarifies Student Ideas	72
3d Diagnoses learning needs	0
3D. Designs pattern of learning experiences	0
3f. Formulates directions for learning	32
3E Evaluates results: re-diagnose needs	0
3s Summarizes Student Ideas	0

4: Asks Questions: to gain information, knowledge or opinion (not rhetorical questions.)

Includes questions about procedure or about content designed to elicit an answer from a student. Only questions which are legitimately designed to gain information, knowledge or opinion from students are classified in this category; they may be broad in scope or very narrow. This does not include rhetorical questions and questions which communicate sarcasm or criticism.

4f Asks Factual Questions	56
4c Asks Convergent Questions	107
4d Asks Divergent Questions	62
4e Asks Evaluative Questions	2
4s Asks for Sharing of experiences	437=5%

Teacher Talk: Direct Influence

5: Gives lecture facts, information, opinions, ideas and orientation (includes rhetorical questions).

"Lecture" signifies facts, information, opinions, ideas and orientation presented to introduce material to the class, review material or focus attention on an important topic. Usually given in extended time periods. It may be given in response to a student question or presented to clarify a question the teacher has previously asked or is about to ask. Rhetorical questions are also included in this category. This is the most frequently used category.

5f Factual Lecturing	1,008=12%
5M Motivational Lecturing	0
5O Orientation Lecturing	229
5P Personal Lecturing	109
5R Gives or asks for Resources	86

6: Gives directions: physical action on the part of the learner.

Used when the observer can predict an observable behavior on the part of a learner or the class as a result of the teacher's instruction. "Observable behavior" is usually some physical action on the part of the learner or a specific response which the teacher has demanded. 643= 7%

7: Criticizes/Justifies Authority: defends position 0

Criticism is a statement designed to change a learner's behavior from unacceptable to acceptable. In effect, the teacher is saying, "I don't like what you are doing, so something else." This also includes statements in self-defense or justification of the teacher's behavior or authority often difficult to detect when the teacher appears to be explaining the reasons for a lesson. Loosely, when the teacher is explaining the reasons why he should be telling the learners what to do, why he is the one who makes the decisions or why he is the one who should be listened to, he is justifying his authority. These also include statements of extreme self-reference in which the teacher asks a learner or the class to do something as a favor to him.

Student Talk

8: Student Talk: predictable response where teacher initiates talk.

Used when the teacher directly initiates the contact or solicits the student's statement and the response by the learner is a predictable response, that is, statements of fact asked for in a question or limited choice responses which give the learner's feeling or opinion. Example: "Columbus, in 1492" to the question "Who discovered America and when?" Another is the response to the teacher's question "Do you think we are doing the right thing in Vietnam?" The answer "Yes" is an "8".

8f	Factual Response	491
8c	Convergent Response	164

9: Student talk - Unpredictable: student initiated. In general, if the learner raises his /her hand, is acknowledged and makes a statement or asks a question, he/she has not been prompted by the teacher to talk. The appropriate category is "9." Also when the learner responds at some length to a very broad question asking for opinion or divergent thinking, the category is "9."

Distinguishing between the two categories of learner talk is often very difficult. The criterion seems to be whether or not the observer can predict the general kind of answer that a student will give in response to a question. If the answer is not predictable, then classify as a "9." If it is predictable, then the statement would be classified as an 8. In general, the kind of question asked gives a clue as to whether the learner statement is an "8" or a "9." A broad question will give a clue that a "9" is likely to follow: a narrow question will give a clue that the learner response is likely to be an "8."

9d	Divergent Response	38
9e	Evaluative Response	253
9i	Initiated Comment	288
9s	Student share experiences and/or solutions	2,141 =24%
9t	Student talking to student	175
9Q	Student questioning another student	73

Miscellaneous:

10: Silence or confusion

This category includes everything not included in the other categories: periods of confusion in communication when it is difficult to determine who is talking, periods when a number of people are talking at once, periods when there is no talking at all and miscellaneous occurrences such as laughter, music, bells ringing.

10s	Silence	120
10c	Confusion	4)
10E	Equip. induced silence/confusion	1,808)
10N	Name/city ID preface	209) = 24%
10Q	Equipment induced Question "Still connected?"	147)
	Total Units (one line)	8,860

Total Interactions

Group A	40
Group B	35
Group C	19
	95

Summary of Findings

The 24 percent downtime caused by the clipping of the audio bridge was a major deterrent. The media was not seamless and transparent, but very apparent to the users. It impeded the process of communication, however, as the students became accustomed to the equipment clipping, the percentage of clipping episodes decreased.

Students spoke 41 percent of the time and of that, shared their experiences 24 percent of the time. The teacher spoke for 22 percent of the time and of that, spent 5 percent of the time asking for shared experiences. This five percent asking for sharing generated 24 percent of the interaction by students.

For all three groups, there were a total of 95 interactions where one statement generated the next statement so that interaction was perceived to be occurring according to the earlier definition. Group A accounted for 40 separate interactions through the four hour periods; Group B had 35 interactions; and Group C had only 19 interactions. Note that the Group C which had the smallest number of students generated the longest periods of interactions.

Recommendations: This research begins to frame the questions regarding the use of audio teleconferences rather than answer all the questions. Students agreed that the audio interaction was productive as opposed to having no interaction at all. Since the students and teachers were located throughout the United States, the audio interaction was the least costly and the only alternative for group interaction.

Future purchasers of audio conferencing or audio bridge systems should note the heavy percentage of time spent that was unproductive due to the peculiarities of the equipment. Users of older audio bridges might consider the feasibility of replacing a heavily used bridge with equipment that does not clip.

The results should assist distance educators in structuring telecourses crossing borders to include the use of audio conferences to provide interaction with the instructor. The literature shows very little research in the use of audio teleconferences to provide interaction.

We should be asking the pragmatic question: What about interactivity and acceptance of innovations in communication arrangements, interactivity and novelty value (does it fade?), interactivity and utility and interactivity and tenacity of use? What is the role of interactivity in the diffusion of media, maintenance of allegiance to channels and the intervening impact of interactivity on use via attention? (Hawkins, 1988, p. 130)

These questions have rarely been asked or researched. Interactive arrangements help overcome barriers to adoption of the use of telecommunications

■ ELECTRONIC CAMPUS

The electronic campus is wired so that video, audio and data are available in all classrooms, offices and dormitories. The electronic campus can also connect branch campuses, unofficial student housing and faculty residences. The key to this connectivity is an individualized system designed for the campus which provides the necessary highway to enable the institution to provide all the services that students, faculty and administrators require. Added services include voice mail, electronic mail, campus security systems (video surveillance, fire alarm and protection, campus emergency phone systems), local and wide area networks (LANs and WANs), computer or phone registration, video teleconferencing, audio conferencing, computer conferencing, distance learning, library database and resource access.

Electronic Dormitories

Many institutions are experiencing space problems, but the funding to build new classroom buildings is not available. When wiring is expanded to the dormitory, each student room becomes a learning center. Live or taped lectures can be delivered to the dormitory. It makes sense to install satellite receive dishes so that the dormitories can receive traditional cable programming as well as a variety of programming either produced or adapted to college curricula. Educational programming in the areas of foreign languages, business administration, medicine, journalism, music, drama and much more is available through satellite technologies. A charge back system to the students for cable programming will pay for the expense of wiring each dorm room to support personal computers and access to campus library resources, personal telephone and cable television. The cable distribution system can be used to transmit announcements, pre-recorded lectures, guest speakers and live sports events campus wide.

When the dormitory is used as a learning unit, students purchase their own computers, provide their own televisions and pay for their telephone line use. These are expenses that are no longer borne by the institution in the form of rapidly expanding computer labs or new classrooms or learning centers in the library. Students can access the library database electronic card file and access a growing number of interconnected libraries throughout the world for abstracts and full text. This not only cuts down on the clerical help in the library, but opens up worlds of information to the student that the on-campus library cant afford to support.

Through computer conferencing programs, students and instructors can interact through messages and assignments. A choice of software services that best fits the curricular demands of the institution can be selected. A student tool kit provides facilities for word processing, database manipulation, spreadsheets, telecommunications, interactive library services and distance learning programs. Voice mail for students and faculty provides another communication mode.

Electronic Classrooms

The electronic classroom is a technology and service package that allows educational institutions to expand the boundaries of classrooms far beyond the traditional bricks and mortar which have encompassed them in the past. They allow educators to access and utilize a wide variety of available electronic resources to strengthen the impact of the curriculum being taught. The resources can range from a video tape located at the school's library to a live multi-point connection to broadcast an expert lecturer to various distant campuses.

The successful integration of video, audio and computer transmission techniques, coupled with educational applications, form the backbone of the electronic classroom. During the design phase of an electronic classroom, many factors should be taken into account: specific classroom application(s); the make-up of the class; size of the classroom; transmission requirements; level of the curriculum; video, audio and computer applications; information expansion; budget and vendor financing; and future needs - upgrades and enhancements.

Educational Video Access

Educational video access is designed to give educators the ability to access remotely stored multimedia material (VCR, CD-ROM, etc.) and display it in a fashion to complement the lecture. This allows for a myriad of information sources to be at the educator's or the student's fingertips. It also includes the ability to connect to resources through audio and computer conferencing abilities.

Computer Access to Resources: As campus libraries move to computer based electronic card catalogs, students and faculty can access the catalog from offices, dormitories or their homes. This provides 24 hour a day access. As the use of CD-ROM technologies increase in libraries, these systems can also be accessed through campus computer networks or by a computer and modem from remote sites anywhere in the world. Students need access to library electronic card files and full text in order to do research and write papers.

Connection to Remote Campuses: This system is designed to allow the educator to connect to a remote campus to deliver video for the purpose of teaching and interacting or to receive video for the purpose of learning and interacting. It also encompasses audio and computer conferencing technologies.

Computer Conferencing: Computer conferencing allows faculty and students to conduct components or entire courses asynchronously. With a computer and modem hookups, some schools have discovered that they can reach students who never would enroll in a traditional class because of unusual working hours, extensive travel schedules, disabilities or other obstacles. Entire degree programs including masters and doctoral programs are being offered. Several institutions have national and international student bodies and some colleges with local computer conferencing classes have realized that they have a new market of students available to them. One college in Los Angeles is applying for funding to their local environmental office as a means of not only providing education, but as a significant way to reduce auto traffic and reduce emissions. Institutions that are pressed for space can provide new classes to students in their dormitory rooms or in their homes or offices without building a new building or maintaining a parking lot.

Conferencing and telecommunication programs have been developed and several have been in use long enough to have track records of dependability. IBM and Macintosh platforms are supported. The software takes care of log on, uploading and downloading files, log-off and filing.

For campuses that are already interconnected with wide band transmission lines, desktop video conferencing is a reality. Videoconferencing groupware products provide a motion picture of the participants and access to documents on which they can work. Prices range from $2500 to $5000 per computer station now, but prices will drop soon.

As the telephone companies provide wide band lines and video dial tone, institutions using computer conferencing now will be in the forefront of the move to multimedia. Instead of text, students will have a full array of learning resources available to them that will include video, audio, text, videoconferencing, video messages, collaborative group work and a multitude of resources that havent been created yet. Courseware will be developed for multimedia and a true electronic university will be launched that will meet the needs of the American work force as it competes in a global marketplace. Just as the idea of just-in-time training has become a reality in the workplace, educators to will be able to provide education when it is needed and where it is needed at the convenience of the instructor and the student.

Evaluation: For programs to continue to succeed, improve and grow, student and faculty response to the technology must be analyzed. Measuring satisfaction, outcomes, retention and overall indications of student learning is imperative. Faculty satisfaction and support are critical to the support of the program. The visibility of the organization, internally and externally, as well as the industry of distance learning is contingent upon the continued reporting of the success and outcomes of the program. Recommendations for continued evaluation include: faculty end-of-course surveys on the new technology; student end-of-course evaluation surveys on the new technology; evaluation of new courses and instructors; systematically planned expansion to targeted geographic areas and information services; inter-institution cooperation; and annual meetings to conduct an evaluation and to provide recommendations for continued improvement and workshops for instructors to achieve standards of excellence.

Summary: In a global society, it is no longer possible for colleges and universities to limit their mission goals to local boundaries and still remain up-to-date on the latest educational information and ideas that are needed to respond to changing conditions. Many of the more progressive schools are moving toward teaching techniques, technology and information transfer to effectively deliver curriculum, share scarce resources and offer a broader educational experience to students. The access to information and ideas is boundless. To maintain a competitive standing in the global marketplace, American business demands that new entrants in the work place be proficient in the use of electronic information and video resources for communications and information access. With video, audio and computer technologies, colleges and universities can lead in the innovative use of technology which enhance academic programs.

Learners in K-12, higher education and business are growing up and working in a rapidly expanding world of electronic information access. From the telephone, radio, cable television and personal computers, students can receive, send and interact through video, voice and data using various communications resources.

■ REFERENCES

Amidon, Edmund J. (1970), Interaction Analysis, in "Methods of Research in Communication," Eds, Emmert, Philip and Brooks, William D, Houghton Mifflin, Boston.

Amidon, E. J. and Flanders, (1970) "10-category Interaction Analysis system" P. 378.

Advancing Communication Science: Merging Mass and Interpersonal Processes, "Sage Annual Review of Communication Research:" (1988) Editors Robert P. Hawkins, John M. Wiemann and Suzanne Pingree.

Armstrong, Jenny R. (1973). "A Sourcebook for the Evaluation of Instructional Materials and Media." Special Education Instructional Materials Center, University of Wisconsin, Madison, WI. ED 107 050.

ASTD (1985). "Write Better Behavioral Objectives," American Society of Training and Development, Alexandria, VA.

Baird, L. S., Beatty, R.W., & Schneier, C. E. (Eds.) (1982) "The Performance Appraisal Sourcebook." Amherst, MA.

Bates, Anthony (1974). "Obstacles to the Effective Use of Communication Media in a Learning System." Keynote address to the International APLET Conference, Liverpool University. Paper No. 27.

Bates, Anthony (1980). Towards a better theoretical framework for studying learning from educational television. "Instructional Science," 9, pp. 393-415.

Bates, Anthony (1986, December). "Creating a Technologically Innovative Climate: The British Open University Experience." IET Paper #251.

Bates, Anthony (1987, September). "Teaching, Media Choice and Cost-Effectiveness of Alternative Delivery Systems." Speech to the European Centre for the Development of Vocational Education, Berlin, September 3-4, 1987. Milton Keynes, Great Britain, Open University. IET Paper No. 264.

Bergeson, John (1976). "Media in Instruction and Management Manual." Central Michigan University, Mt. Pleasant, MI, ED 126-916.

Billings, Diane (1988). "A Conceptual Model of Correspondence Course Completion," American Journal of Distance Education, Vol. 2, #2, pp. 23 - 35

Bloom, Benjamin S., et al. (1977). "A Taxonomy of Educational Objectives. Handbook I: The Cognitive Domain. New York, Longman.

Brey, Ronald (1988, October). "Telecourse Utilization Survey: First Annual Report: 1986-87 Academic Year." Austin, TX. Annenberg/CPB Project and the Instructional Telecommunications Consortium.

Brown, Bernice & Helmer, Olaf (1964, September). "Improving Reliability of Estimates Obtained from a Consensus of Experts," P-2986, Santa Monica, CA: Rand.

Brown, Bernice (1968, September). "Delphi Process: A methodology Used for the Elicitation of Opinions of Experts," P-3925. Santa Monica, CA: Rand.

Carnevale, Anthony; Gainer, Leila J.; and Meltzer, Ann S. (1990). "Workplace Basics," San Francisco, Jossey-Bass, pp. 37-65.

Corporation for Public Broadcasting (1980). "Telecourses: Reflections '80 Executive Summary." Washington DC., Corporation for Public Broadcasting, p. 5.

Dalkey, Norman C. (1969a, June). "The Delphi Method: An Experimental Study of Group Opinion," RM 5888-PR. Santa Monica. Rand.

Dalkey, Norman C., Brown, Bernice, & Cochran, S. (1969b, November). 'The Delphi Method, III: Use of Self- ratings to Improve Group Estimates," RM-6115-PR, Santa Monica: Rand.

Dalkey, Norman C. & Brown, Bernice (1971, May). "Comparison of Group Judgment Techniques with Short-range Predictions and Almanac Questions," R-678-ARPA, Santa Monica: Rand.

Dirr, Peter J. (1986, May 24). "Changing Higher Education Through Telecommunication," presentation for The World Congress on Education and Technology, pp. 1-2.

Dunkin, M. J. & Biddle, B. J. (1974). "The Study of Teaching." New York: Holt Reinhart and Winston.

Educational Products Information Exchange (1973). "Improving Materials Selection Procedures: A Basic How To Handbook." EPIE Report No. 54. New York.

ELRA Group, Inc. (1986, August). "Executive summary: The Adoption and Utilization of Annenberg/CPB Project Telecourses," Washington, DC. Annenberg/CPB Project.

Gagne, R. M. (1977). Analysis of objectives. In L. J. Briggs (Ed.) "Instructinal design: Principles and applications. Englewood Cliffs, N.J., Educational Technologies Publications, Inc. pp. 115-145.

Gardner, Howard (1985). "Frames of Mind," New York, Basic Books. Gardner, Howard (1991). "The Unschooled Mind," New York, Basic Books, pp. 11-12.

Gueulette, David G. (1988, January). A better way to use television in our classes. "TechTrends." 33/1, pp. 27-29. Harasim, Linda (1987). "Computer-mediated cooperation in education: Group Learning Networks." Proceedings of the Second Guelph Symposium on Computer Conferencing, June 1-4, 1987.

Heidt, E. U. (1978). "Instructional Media and the Individual Learner: A Classification and Systems Appraisal." London, Kogan Page.

Helmer, Olaf. (1966, December). "The Use of the Delphi Technique in Problems of Educational Innovations," P-3499, Santa Monica: Rand.

Hezel, Richard T. (1987, November). "Statewide Planning for Telecommunications in Education; Executive Summary "Washington, DC, Annenberg/CPB Project.

Hewitt, Louise Matthews, (1980). "An Administrator's Guide to Telecourses." Fountain Valley, CA, Coast Community College District. pp. 6-7.

Hewitt, Louise Matthews, (ed.), (1982). "A Telecourse Sourcebook for the 80s." Fountain Valley, CA, Coast Community College District.

Holt, Smith (1989, April). Speech at Learning by Satellite IV Conference, Tulsa, OK. San Ramon, CA. Applied Business teleCommunications. "Interpersonal Processes," Sage Annual Review of Communication Research: Editors Robert P. Hawkins, John M. Wiemann and Suzanne Pingree. Page 110.

Kalton, Graham (1983) "Introduction to Survey Sampling," Beverly Hills, Sage, p. 69.

Kemp, J. E. (1985). "The Instructional Design Process." New York, Harper and Row.

Knowles, Malcolm. (1970) "The Modern Practice of Adult Education," New York, New York. Associated Press.

Knowles, Malcolm (1983). How the media can make it or bust it in education. "Media and Adult Learning," vol. 5, no. 2 Spring. In Gueulette, David G. ed. (1986) "Using technology in adult education." Glenview, IL. American Association for Adult and Continuing Education, Scott, Foresman/AAACE Adult Educator Series. pp. 4-5.

Komoski, Kenneth (1977). Evaluating nonprint media. "Today's Education" 66:96-97 March-April.

Kressel, Marilyn (1986). Higher education and telecommunications. "National Forum: The Phi Kappa Phi Journal Summer," Volume LXVI Number 3. pp. 4-6.

Lane, Carla (1988). "Student Attrition in Distance Education Programs," Unpublished manuscript.

Lane, Carla (1989a). "A Selection Model and Pre-Adoption Evaluation Instrument for Video Programs," American Journal of Distance Education, Vol. 3 No 3, 1989, pp. 46-57.

Lane, Carla (1989b). "A Media Selection Model and Pre-Adoption Evaluation Instrument for Distance Education Media." Ann Arbor, UMI.

Lane, Carla (1990). "The Use of Audio Interaction and Self-Directed Learning Contracts in a Telecourse Offered by Satellite: Foundations of Adult Basic Education. Proceedings, Midwest Research-to-Practice Conference in Adult, Continuing & Community Education.

Lane, Carla (September, 1990). "Research Establishes National Teleconferencing Standards." San Ramon, CA. "Ed," pp. 10-12.

Lane, Carla (1992). "Model Program: Interaction Through a Mix of Media," Proceedings, Global Trends in Distance Education, University of Maine at Augusta, September 24-26.

Lewis, Raymond J. (1983). "Meeting Learners' Needs Through Telecommunications: A Directory and Guide to Programs." Washington, D.C.: American Association for Higher Education. Mager, R. F. (1975). "Preparing Instructional Objectives (second edition), "Belmont CA, Pitman Learning, Inc.

Mayor, Mara and Dirr, Peter J. (1986). "Telelearning in Higher Education." "National forum: The Phi Kappa Phi Journal," Summer, Volume LXVI No. 3. pp. 7-10.

McDonald, F. J. & Ellias, P. (1976). "Beginning teacher evaluation study: Phase II Final Report," Volume 1, Chapter 10. Princeton, NJ: Educational Testing Service.

Meierhenry, W. C. (1981, Fall) Adult education and media and technology. "Media and adult learning," Vol. 4, no. 1. In Gueulette, David G. ed. (1986) "Using Technology in Adult Education." Glenview, IL. American Association for Adult and Continuing Education, Scott, Foresman/AAACE Adult Educator Series. pp. 2-3

"Methods of Research in Communication." (1970) Eds, Emmert, Philip and Brooks, William D, Houghton Mifflin, Boston).

National Education Association (1976). "Instructional Materials; Selection for Purchase." Rev. ed, Washington DC, ED 130-380.

Niemi, John (1971). The labyrinth of the media: Helping the adult educator find his way. "Mass Media and Adult Education." Englewood Cliffs, New Jersey: Educational Technology Publications, Inc., pp. 35-47.

Portway, Patrick, (1989, April 1). Speech at Learning by Satellite IV Conference, Tulsa, OK. San

Ramon, CA. Applied Business teleCommunications.

Reiser, Robert A. and Gagne, Robert M. (1983) "Selecting Media for Instruction," Englewood Cliffs, NJ, Educational Technology Publications.

Sive, Mary Robinson. (1978 and 1983). "Selecting Instructional Media: A Guide to Audiovisual and Other Instructional Media Lists," 1st and 2nd editions, Littleton, CO. Libraries Unlimited, Inc.

Stone, Harvey, (1991). "GTE Spacenet: Furthering education through telecommunications." San Ramon, CA, "Teleconference," Vol. 10, No. 6, p. 19.

Tanzman, Jack and Dunn, Kenneth. (1971). "Using Instructional Media Effectively." West Nyack, NY Parker.

Teague, Fred A. (1981) Evaluating Learning resources for adult. "Media and Adult Learning," vol. 4, no. 1, Fall. pp. 27-33. In Gueulette, David G. ed. (1986) "Using Technology in Adult Education." Glenview, IL. American Association for Adult and Continuing Education, Scott, Foresman/AAACE Adult Educator Series.

Williams, Frederick; Rice, Ronald E.; and Rogers, Everett M. (1988) "Research Methods and the New Media," Macmillan, New York.

Winkelmans, T. (1988). "Educational computer conferencing: An application of analysis methodologies to a structured small group activity." Unpublished master's thesis, University of Toronto.

Zigerell, James J. (1986). "A Guide to Telecourses and Their Uses," Coast Community College District: Fountain Valley, CA., p. 35.

Michael Brown is president and CEO of Cognitive Training Associates, which was recently recognized by the "Dallas 100" as one of the fastest growing companies in the Dallas area. Brown had over 12 years experience in sales, marketing, training, and executive management in diversified leading edge companies prior to founding CTA. His executive career was with firms such as Johnson & Johnson and American Hydron. He has served as project monitor/advisor for research projects focusing on job design, employee performance capacity, and organization performance ability indicators.

A keynote speaker and author of several specialty learning programs, Michael has accepted the appointment as Adjunct Facility for a Fortune 100 company. He was recently appointed to the Oxford's Who's Who and the Elite Registry of Extraordinary Professionals. Brown currently is a board member of the United States Distance Learning Association and is president of Project Bluebonnet, a state-wide, grass roots coalition that integrates technology into the classroom and business environments.

As president of Bluebonnet, he founded and directed the first statewide distance learning conference in Texas during the summer of 1992. He is also currently developing distance learning laboratories connecting Dallas and the State Capitol in Austin which will allow "hands on" demonstrations. Brown also serves as the Chairman of the Texas Federal Rural Development Council Telecommunications Task Force.

MULTIMEDIA

■ HISTORICAL PERSPECTIVE

Multimedia has been around for many years. Although recently individuals in the computing professions claim to have invented multimedia, the term has been in common use for over two decades. In the early 70s, the term multimedia was used to describe high-impact visual presentation systems, consisting of some combination of slide projectors, tape recorder, motion picture projector, video screen or projector, and a control system. By the end of the decade leaders in the field were discovering that\at it was cheaper and easier to adapt a general purpose microcomputer as a control device rather than custom engineer a control system to synchronize the various media elements.

Early computer-based devices included slide projector control systems, videotape control systems, and videodisc control systems. Of the available microcomputers in 1980, the Apple II+ had the greatest flexibility to accept interface cards designed by others. Although several manufacturers developed integrated devices, such as level II video-disc players incorporating a CPU and RAM into the player, such devices remained niche products.

Throughout the 80s, multimedia has developed as a continuous stream of variations on a theme — specifically the computer-based control of analog media systems. Several systems have been developed for controlling videodisc players from an external computer system. Numerous systems are available for overlaying the analog video output from a video source on a computer screen. Other systems were developed for controlling the audio output of CD-ROM players. (Although the CD-ROM is fundamentally a digital technology, many applications continue to make use of the audio data stream after the conversion to an analog form with the CD-ROM player.)

As we move into the 90s, an important transition is taking place in the field of multimedia. Rather than using the computer as a control device to synchronize external analog sources, the computer is being used as both the control device and the source of the information; information being stored in a completely digital format. This unified strategy has the potential, in the long term, to decrease the cost of multimedia systems with the resulting increase in such system's applicability.

The transitions currently taking place are only now showing a move toward maturity. At present, the growth of the field is significantly impeded by the lack of standards. In the audio domain, different individuals working with different computing platforms have developed proprietary standards for the formatting of digital audio files. In any field where there is a lack of defacto standards, the capitalistic nature of our economic system drives companies to develop proprietary ways of accomplishing a function faster, better, or at lower cost. As a result, there are numerous pieces of hardware on the market for digitizing and playing back audio materials on a variety of computing platforms.

The motion video domain even more of a challenge than is audio. The vast amount of data required to completely represent 30-frame-per-second full motion video easily exceeds the computational capacity of most general-purpose personal computer systems. Again, competitive pressures have given rise to a number of different methods for digitizing video. There are software-only systems such as Apple's Quicktime video system and IBM's PhotoMotion. There are hardware-assisted systems such as IBM/Intel's DVI (Digital Video Interactive) and N-Cube's video compression chipset. Each of these systems saves information in a unique format that is foreign to each other.

Because of the huge computational requirements of full motion video, it is likely that various proprietary systems for creating video with different sets of compromises (resolution, color depth, frame rate) will continue to dominate the computer scene for sometime. The MPEG (Motion Pictures Experts Group) specification is presently in draft form. The final MPEG specification will provide a recognized standard to unify the field. IBM and Intel have stated that they will continue to enhance their DVI technology to support MPEG, but do so in such a way that current DVI data files will continue to play back on future systems.

Historically, the Apple Macintosh and Commodore Amiga have had the advantage of providing for a richer set of data types defined within the operating system. As a result, adding resources such as sound to the Macintosh has happened in a largely well-defined manner. However, with the Macintosh platform accounting for only slightly more than 10 percent of all personal computers, and the Amiga not a significant mainstream contender, significant progress remains to be made.

The first steps toward maturity in the field can be found in the Multimedia Extensions to Windows and the Multimedia Extensions to OS/2. Both IBM and Microsoft have, through these extensions, defined basic audio services within their strategic operating systems. The impact will be similar to what happened with graphics cards with the introduction of Windows. Once upon a time each video card manufacturer wrote device drivers for their video card and worked to persuade software developers to support their video devices in applications software. With the arrival of Windows, the video card manufacturer need only write to the Windows interface and all Windows programs can access the features of the video card.

Although specialized graphics card drivers are still produced for specific high-powered applications software, such as AutoCAD, the majority of high-resolution video cards rely on Window's support for establishing a large software base. During the next year we are likely to see a parallel development in digital audio. Software vendors will see the increased markets will rapidly open up through support of the standard audio interfaces in Windows and OS/2, and the manufacturers of the audio cards need only produce high quality drivers for two operating systems to access a large library of applications programs.

The second indication that the field is maturing is the Kaleida Corporation created by Apple and IBM. The mission of Kaleida is to develop technologies that allow the owners of creative material to author or master their program material once and have it deliverable across platforms. Cross platform goals certainly include Macintosh System 7, OS/2, and future Open Systems operating systems developed by Apple and IBM through Taligent to run on their RISC-based systems of the future.

Although the Multimedia Marketing Council originally specified a 10mhz 80286 computer system as the baseline system for multimedia, no one seriously looks at the 286 computing platform as a viable multimedia platform. Today's multimedia systems typically require as minimums an Intel 80386 chip in the IBM-compatible world and a 68030 chip in the Macintosh world. Regardless of the claims of various manufacturers to the contrary, serious multimedia work requires 8 mb of RAM as a minimum, regardless of computing platform.

Digital media require large storage capabilities, which translate into either large hard disks, optical media (i.e. CD-ROM) or high-speed network connections.

It is the computer network that spells the death of analog multimedia materials such as the videodisc and audio CD-ROM. The inherently analog nature of the video signal cannot be directly transmitted across digital networks. The future of the multimedia world is in 100 percent digital systems. There are still significant issues to be addressed in the area of networking. For example, although Ethernet is nominally a ten mbps technology and DVI requires only 155 kbps of data transfer, putting DVI on Ethernet is not a trivial task. When our present standard network technologies, i.e. Ethernet and Token Ring, were developed, the maintenance of a time-critical continuous data stream was not a design criteria. Consider, for example, loading a document into a word processor. No one would notice a 1/10th second delay in the middle of leading a document. However, interrupt the audio and video signal in a motion video segment for the same 1/10th second and everyone notices. Significant work remains to be done in extending the data types in our network operating systems and packet construction to optimize them for the time-critical delivery of continuous data streams. The major players in multimedia and networking recognize the need for this work to take place and have engineers hard at work to devise solutions to the problems.

Even with the minor tuning that will be required to reliably transmit full motion digital video across Ethernet and Token Ring networks, simple math quickly shows that it doesn't take many 155 kbps data streams to saturate either an Ethernet or a Token Ring network. As a result, multimedia is likely to quickly drive a move to technologies such as 100 mbps FDDI (Fiber Distributed Data Interface) and beyond.

As the technologies to capture, manipulate, and disseminate information become more ubiquitous, the opportunities for abuse of the intellectual property of others increases. The lossless nature (freedom from generation loss through duplication) of digital nature of multimedia materials creates additional opportunities for neglect of responsibilities with respect to intellectual property. Because of the wide variety of source material likely to be incorporated in a single multimedia presentation, it is likely that the rights to redistribution of the work will become confused. It is likely that, given the complexity of intellectual property laws and copyright permissions, there will be widespread problems with accidental infringement. A need will soon emerge for standard methods for obtaining copyright clearance for multimedia materials.

The present non-system only serves to inhibit the growth of what promises to be a field with explosive growth potential.

■ EDUCATION

Today, there is a significant feeling among many professionals in the education community that teachers will develop their own multimedia materials. Teachers do not write their own textbooks, produce their own films and video materials, or write computer assisted instruction programs. Not that there aren't significant isolated exceptions to that rule — but in general, teachers are hard pressed to find enough time in the day are hard pressed to find enough time in the day to prepare quality class presentations with commercially prepared materials. Although creative teach-

ers will certainly will certainly make use of multimedia technologies to enhance their classroom presentations, production of multimedia-based classroom materials is not likely to become a major activity for classroom teachers in our lifetime. There is, and will be, however, a significant market for high-impact, quality multimedia materials prepared by commercial publishers and targeted toward the K-12 education market.

■ CORPORATE TRAINING

In the field of industrial training, multimedia has tremendous potential. Here the organizational model typically includes individuals with expertise in instructional design and in production who have the creation of training materials as part of their job description. Digital multimedia are inherently more easily modified to respond to changes in manufacturing practices, operating procedures, or management philosophy than their analog counterparts. As a result, corporate trainers will embrace digital multimedia to the extent that it allows them to be more responsive to the requirements placed on them by management.

Digital multimedia creates a significant market opportunity for the creative community. Not only is there potential for the mastering of existing analog resources into digitally accessible multimedia materials, — but this new delivery system provides opportunities for packaging information in creative ways that have not been previously explored.

A classical theory of the adoption of technology describes the implementation of new ideas in terms of an S-curve. The theory maintains that each new technology starts out slowly with a small group of early-adopters. The rate of adoption increases in a linear fashion with a slow, but steady growth rate for some period of time. At some point, the curve stops its linear growth and enters a period of exponential expansion. At some saturation point in the future, the curve again flattens out to a linear growth profile. We are right now rounding the corner from the first linear-growth phase into exponential expansion. For those who choose to ride, plan on becoming dizzy. For those who choose to stand aside and watch, plan on being left far behind.

■ INSTRUCTION THROUGH MULTIMEDIA: AN APPLICATION

The Texas Learning Technology Group (TLTG) develops multimedia based instruction for the education market. TLTG's Division Director, Ms. Paula

Hardy, leads a consortium of school districts whose mission is

"to revolutionize eduction by providing leadership in the deveolpment and use of innovative technology-based instructional-products and supporting services."

To that end, TLTG has produced two complete curricula that are a combination of level III interactive videodisc, print materials, and laboratory activities.

The initial effort, TLTG Physical Science, provides 160 hours of instruction, and consists of 15 videodisc sides, 85 megabytes of program software, a teacher resource guide, and a student guide. The intent of TLTG Physical Science is to provide a comprehensive course that

Increases students' in-depth understanding of physical science,
illustrates the relevance of science to daily life, and
prepares students for further study in science.

The second complete curricula, TLTG Chemistry I, responds to the need for students to have the skills and knowledge to provide scientific leadership on global environmental issues. The product includes seven videodisc sides, 86 megabytes of program software, and teacher and student resource guides. Program goals are to

improve student's conceptual understanding of chemical principles
increase students' awareness and appreciation of chemical processes, and
develop critical thinking and mathematical skills.

TLTG is a unique combination of a qualified development staff with expertise in the content area, education, planning, design, video production, authoring, graphics, and evaluation and a group of school districts with expertise in classroom instruction and student learning. Added to this team are content and instructional reviewers from universities and industry.

The development process is a fascinating study in teamwork. A design team and a production team are necessary to develop the curriculum. The design team includes instructional designers, teachers, content experts, and a scriptwriter. The production team includes programmers, graphic artists, and video production personnel.

The design team develops videodisc-based lessons, computer-based activities (e.g., simulations, labs, and practices), and print-based teacher and student manuals. Design work involves researching the content, writing objectives, analyzing and organizing the content using learning principles, generating ideas for effective presentation, developing student activities, scripting the instruction for video presentation, and storyboarding computer-based activities.

Storyboards are a frame-by-frame depiction of the scripted presentation, and contain video, audio, graphic, animation, and authoring information. Using the TLTG electronic storyboarding program a TLTG graphic artist imports the script, and then creates images representing the final product. There can be anywhere from 500-1000 storyboards per script. Review of the storyboards is performed by the designers, content experts authoring and graphics staff, and the video production company. Responsibility for accuracy of the storyboard falls to the design team.

An external video production company duplicates the storyboard information onto videotape. The company hires the talent, determines locations for shooting, and shoots and edits the video footage. Design and authoring team members review a rough cut and final cut of this footage. During this phase, the TLTG graphic artists create two- and three-dimensional animated graphics for use in video production.

The TLTG graphics team provides two other functions as well. They develop the computer-based graphics and animations for the programmers, and they use desktop publishing software to produce the print materials that accompany the finished product. Production hardware for graphics includes a Silicon Graphics 3-D modeling system, numerous Macintosh, Amiga, and IBM computers, and video recording capabilities.

TLTG programming staff uses a special computer language to tie together the graphics, audio and video with the logic required to make the product work. This process is called "authoring", and provides a way for the computer to control the videodisc, intersperse graphics and text in the program, capture input, and add interactive logic. Utilizing programming tools such as Tencore, IWPS, and C language, many routines are developed and standards are established to make the work efficient and presentation methods consistent.

Programming tasks include programming computer-based animations, scoring, record keeping, and displaying appropriate content and feedback screens. For computer-based practices and simulations, problem sets are randomized, so each time through the student encounters different problems. The program must allow for touch, mouse and keyboard input, which includes significant digits and scientific notation for chemistry.

Each module takes approximately 39 weeks to design and produce. When storyboarding begins on module one, design begins on another module. At the end of storyboarding the first module, video production begins, and storyboarding begins on the second module. Each phase of production overlaps this way, so that three to four modules are in development at once.

Multimedia offers sound conceptual instruction through a combination of video, graphics, audio, and text. Combined with engaging interactivity and impactful sequences, multimedia products result in increased student interest and achievement! More than 90 School Districts are using TLTG's expertise in the development and deployment of Instructional Multimedia programs.

Credits: Dr. Paul L. Schlieve, University of North Texas, Department of Computer Education & Cognitive Systems

Ms. Paula Hardy, Division Director, Texas Learning Technology Group

Chapter 11

INTERACTION IN ONE-WAY VIDEO (BUSINESS TV)

by Patrick S. Portway

■ RETURN AUDIO

The primary means of interactivity in business TV systems is by telephone from the receiving sites. The difficulty arises when you combine satellite video and a terrestrial telephone, each with drastically different transmission delays.

If program audio coming over the satellite is allowed to return over the phone line to the original site of the program, a delayed echo will reverberate through the transmissions.

There are several simple ways to avoid this problem by suppressing the inbound satellite audio. One is a push-to-talk microphone which cuts off program audio when someone asks a question. Electronic systems are available that suppress or cancel the program audio in the return channel and thus, achieve the same result automatically without the necessity of training participants to press the button in order to talk. (See Chapter 2 for a thorough discussion on how these systems operate.)

■ INTERACTION IN BUSINESS TV AND DISTANCE LEARNING BROADCASTS

The most difficult part of achieving interaction is getting started. Very similar to the situation in a large live audience, getting someone to ask the first question can be very difficult.

If you think of interaction as a siphon, you can visualize the need to get interaction flowing. Once a few questions are handled courteously, the participants will join in and interaction will flow naturally.

Some Suggestions

1. Plant the first question. Ask somebody in advance to help. This is no different than if you were speaking in front of a large live audience and asked a friend to ask a questions
2. Use a live studio audience and get interaction started locally.
3. Ask a question of someone at a distant site. Call on them by name as you would in a classroom.
4. Use exercises. A project to be completed at the distant site can actively involve your distant audience.
5. Response Terminals or touchtone phones can be used to poll distant sites or to administer pop quizzes (See Segment 2 of this chapter for details on response terminals.)
6. Compressed video or a videophone could be used to return a video image of the questioner from the distant site. That return video image can then be windowed into the studio video and rebroadcast out over the networks.

Without interaction, a live broadcast might just as well be on videotape. Interaction is what creates the excitement and live dynamic to videoconferencing. Plan interaction throughout your program to keep the audience actively involved. Don't wait for a question and answer session at the end of the program.

Forms of Interaction in Business T.V.

1. Audio
2. Studio audience or Panel
3. Exercises
4. Response Terminals
5. Return Video i.e.
   VSAT Analog Video out
   56 kbps Bideo Back

Forms of Interaction in Business T.V.

1. Audio

2. Studio Audience or Panel

3. Exercises

4. Response Terminals

5. Return Video

      i.e.VSAT Analog Video out 56Kbps

      Video Back

Figure 11.1.1

Lisa B. Perkins received her Marketing/Advertising degree from San Jose State University in 1989. She is currently with Hewlett Packard. She was with One Touch Systems, Inc. for four years and has authored two articles on response systems and their ability to enhance the educational process in a distance learning environment. She has participated on a number of industry panels and has also worked closely with industry pioneers such as Hewlett Packard Company and Tandem Computers. She is also a member of the United States Distance Learning Association (USDLA) and on the advisory committee for TeleCon.

RESPONSE SYSTEMS

Response systems are making inroads into corporate and educational life. Though only a handful of companies make use of them today, the demand is growing as the emphasis toward participant and employee involvement increases. Response systems encourage students to participate and help establish bi-directional communication between instructors and students.

Response systems help remove the geographic barriers between instructors and students and help to create a "virtual classroom" which is not bound by the physical parameters of a traditional single site classroom. Through this virtual classroom, response systems aid instructors by automating the assessment, evaluation and reporting of thousands of students.

■ WHAT IS A RESPONSE SYSTEM

A response system is an interactive network that allows participants to respond and interact electronically with a presenter who may be either in the same room or in a studio thousands of miles away. Interaction is accomplished through a small hand-held or desktop device called a **keypad** that enables each student to establish a one-on-one relationship with the presenter, much like they are able to do in a traditional classroom. The presenter, in turn, immediately has a graphical representation of incoming results, which are accumulated at the center control site (called the **host**). The system achieves a sense of immediacy that makes the instructional process more valuable and effective.

Specific features of the keypad differ among various manufacturers, however, the basic capabilities are the same. Usually there is an LCD display, which may show a variety of information such as the participant's name, messages from the instructor or quiz questions. Calculator-like keys allow participants to enter data and respond to multiple choice questions in real time. Keypads are typically hard-wired in a daisy chain configuration, although some are wireless and use infrared communications.

■ WHAT ARE RESPONSE SYSTEMS USED FOR?

Applications for response systems are virtually unlimited, ranging from taking surveys and voting, to teaching and training and to brainstorming activities and meetings. Users vary from syndicated television programs to Fortune 500 companies.

Response systems assisted in a survey conducted by Bryant Gumbel during a two-night broadcast on NBC called "R.A.C.E." (Racial Attitudes Consciousness Exam), which gauged the racial attitudes of more than 300 people across four cities in the United States. Through responses received simultaneously from New York, Los Angeles, Milwaukee and Jackson, Miss., Gumbel was able to find out how people from various ethnic backgrounds in diverse geographic locations and from specific age groups responded to certain questions, all in a matter of a few seconds.

In a well-known application of the popular vote, a response system aids Bob Sagget, host of ABC's "America's Funniest Home Video" in selecting the evening's winners. Audience participants select their favorite video via keypad and immediately Sagget knows the $10,000, $3,000 and $2,000 winners. "Love Connection" is another example of how response systems have been used in voting on network TV programs.

In education, response systems bring courses to the workplace and to rural communities. East

Exhibit 11.2.1 Example of a response terminal keypad

Exhibit 11.2.2 Host Interface

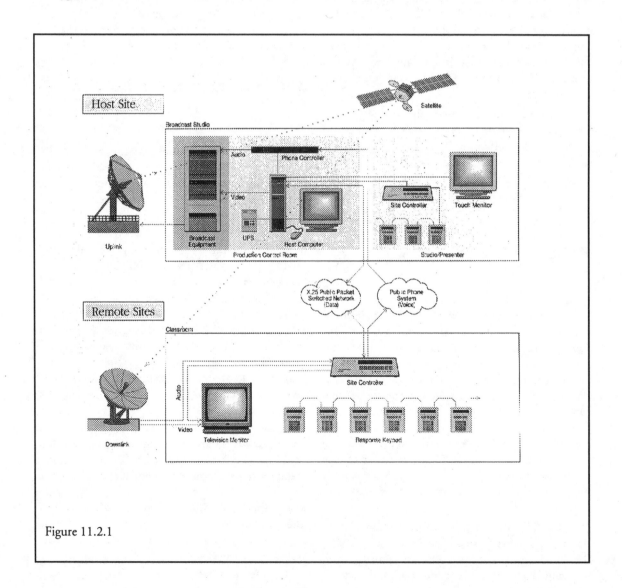

Figure 11.2.1

Lansing-based Michigan Information Technology Network (MITN) uses a response system to bring college and MBA courses to corporate employees. Through the response system, individuals can take engineering, computer science and math classes without leaving their place of employment. Another educational institution uses a response system to offer foreign languages to facilities in rural communities that lack the resources to offer such courses to their enrollees.

Hewlett-Packard Company uses a response system in it's distance learning studio to train its employees. HP has recognized the need to provide effective training in a timely manner to its employees. Using the response system, HP is able to train employees faster and more economically. As a result, engineers and the sales force are familiar with the intricacies of a new product well before its roll-out.

■ HOW DO REPSONSE SYSTEMS WORK

Response systems work in conjunction with a variety of delivery modes, such as satellite, fiber or cable. The delivery, however, is transparent to the integration of a response system, and various manufacturers integrate their systems differently. Most require a phone line and modem into the host for each remote site in the network. One manufacturer simply requires a local connection to an existing X.25 public network. The configuration of the network varies, depending on the manufacturer. To illustrate the basic workings and requirements of a response system, the description below is based on a multi-site distance learning configuration with audio and video being transmitted via satellite.

■ REMOTE SITES

The equipment at each remote site usually consists of a keypad per student and an interface to the host. Each site also requires one or two standard phone lines to establish a connection to the host. Data from student responses are transmitted in real time to the instructor via a modem. The modem may be a separate device, or it may be integrated into the controller.

Integrated audio capabilities enable students to ask questions as they would in a traditional classroom. Audio is typically handled using a service or 800 number. Thus, a standard telephone handset is installed at each remote site for the participants. One manufacturer has integrated audio into individual keypads. In this case, instead of a handset, the phone line is brought directly into the controller to enable

the students to ask questions from their desks with the touch of a button.

■ HOST FACILITY

The host configuration differs dramatically among the various manufacturers. The main differences are the handling of both data and audio lines and the instructor's interface to the remote sites.

Some response systems require that a modem and phone line for each remote site come directly into the host facility to establish data communication. Therefore, in a 30-site network, 30 lines and 30 modems would be necessary. Another option used is an X.25 public data network (PDN), which allows organizations to take advantage of low cost communication. The host connection into the PDN enables a single line running at 9600-56K baud rates to handle all data communication from the remote sites, eliminating the need for direct data dial-up lines for each individual site.

Audio is typically handled using standard telephone handsets. If students have questions they go to the phone and call the host site. At the studio, incoming calls are managed with the support of an operator or phone service. Once a call has been screened, it is passed on to the instructor either verbally or on a typed message. With built-in audio, audio lines interact with a telecommunications control device that automatically manages incoming calls from remote sites to the instructor's console, allowing the instructor to interact with each participant on an individual basis. This "phone controller" eliminates the need for switchboards and telephone operators.

The instructor's interface console is a computer-based system, either touch or mouse driven, that links the instructor to the audience. Normally located at the host site, the console allows the instructor to interact with participants on an individual basis. When identification numbers are used, the instructor has the name and location of each participant. If no ID's are used, then anonymity is maintained. As students respond to a question through their keypads, the instructor immediately has a bar graph of incoming results. If the instructor chooses to share these results with the audience, a video display card can broadcast the results to the entire audience. Depending on the particular response system, the instructor also has a variety of other statistics at hand, such as the number of people in the audience, the number of remote sites in the network and the percentage of people having difficulty with the material being presented.

■ How Response Systems Contribute

No matter how response systems are used or how they work, their ability to transmit and receive real-time data make them valuable. By providing the capability to assess the comprehension levels of audience members and adjust delivery immediately, they make distance learning more effective and they enable the production of more dynamic programs. By giving students the ability to interact with the presenter, they increase attentiveness and promote discussions, benefiting both participants in the field and the home office. Connected by the response system, people no longer feel isolated by the technology.

Besides creating a comfortable feeling, response systems improve performance as a result of closer student monitoring and better instructor preparation. They have proven to be valuable tools for teaching any discipline and their presence has eliminated the instructor's fear of broadcasting to an oblivious audience. Response systems have truly enhanced the quality of distance learning by bringing warmth to a once sterile and passive environment.

■ Benefits for Participants in the Field

Response systems establish a partnership between instructors and students, creating a stronger learning environment by allowing students to actively participate in distance learning sessions. Students no longer have to sit passively relegated to the role of mere observers. They can discreetly communicate their confusion through either a sliding scale or a designated key on the keypad. They can ask or answer questions from their desk through the keypad's built-in microphone.

The dynamic nature of response systems enables students to get immediate feedback on their individual levels of comprehension. Because instructors get quiz results in just a matter of seconds and have the capability to share those results with the students, participants can easily assess their own progress on a particular subject. They no longer have to wait for the instructor or class facilitator to get back in touch with them to answer a question or get direction.

By reducing isolation and putting people in touch with one another, response systems bring a camaraderie to distance learning that was previously missing. Students no longer feel detached or isolated from their co-workers. Those who before were too timid to verbally express their feelings and ideas can anonymously answer questions and respond to ideas without the fear of being judged. For individuals who are more inclined to voice their feelings and attitudes, response systems provide that capability with a touch of a button.

■ Benefits for the Corporation and Instructor

Response systems not only enhance the learning process for participants; they also allow corporations and instructors to offer the best possible product to their customers while reducing the costs traditionally associated with training.

Response systems give corporations a means to evaluate and validate the courses they offer, as well as the effectiveness of the instructors. Before the use of response terminals, corporations were limited in how they could assess the success of a course or broadcast. Hand-tallied questionnaires made the evaluation process time consuming and tedious. Typically, responses could not be tabulated and analyzed before the next broadcast, so weaknesses reoccurred.

Now, with the assistance of a response system, evaluations can be automated and the results tabulated in real time. This immediate feedback enables course designers to better define support material and exams. Developers quickly discover poorly defined questions or exams and can correct problems before the next broadcast. Additions or corrections to support documentation can also be easily implemented. These types of revisions enable corporations to offer the best possible programs and allow instructors to be more effective.

Because of the greater participant involvement and bi-directional communication that response systems offer, instructors and presenters no longer have to wonder if they are broadcasting to an empty room. They can direct questions to the audience and receive feedback instantaneously. The dynamic exchange of ideas wakes people up, increases information retention and comprehension, and thus reduces the need for retraining.

Are Response Systems Expensive?

The cost of integrating a response system depends on a number of variables, both tangible and intangible, that affect the bottom line. Costs vary depending on an individual organization's requirements. In analyzing the total cost, the organization needs to consider not only the initial equipment purchase, but other factors such courseware changes and on-line charges.

The average price of a keypad is about $200. The prices of other components like controllers and hosts vary widely among manufacturers.

There are also a number of other factors that should be taken into consideration when deciding whether or not a response system is affordable. For example, travel is one of the major expenses confronting corporations. The cost of being away from the office, including compensatory time off or overtime pay for extra hours, can have a big impact on the bottom line. Travel expenses can be considerably reduced when suitable courses are taught on the distance learning network. Integrating a response system into the network makes more courses suitable for distance learning, thereby reducing training costs even further.

Factors Affecting Cost

- Travel Expenses
- Current Curriculum
- Audience Size
- Peripheral Equipment
- Training Goals

Additional factors that need to be reviewed include the size of the audience the company wants to reach and the current curriculum's appropriateness for this medium. If the current courseware is not acceptable, then the amount of time and resources required to develop or modify a variety of new courses must be considered.

The cost of miscellaneous peripheral equipment and on-line charges also needs to be considered, as does the cost of installing a host connection to an X.25 PDN and the associated on-line charges. If the system requires a separate phone line and modem at the host for each connected remote site, costs of this equipment, phone installation and long distance charges also must be estimated.

Finally, the organization must consider it training goals. If instructors, trainers and users do not support the technology or believe in its capabilities, then the transition from a passive environment to an interactive environment is bound to be strenuous and expensive.

By looking closely at all of the variables, organizations can avoid unnecessary expenditures as they move into the dynamic arena of interactive learning through response systems.

Chapter 12

Patricia Heffernan-Cabrera, Ph.D. is the Executive Producer-Director of ETN, the educational telecommunications network, a K-12 satellite network owned and operated by the Los Angeles County Office of Education.

Cabrera holds a Ph.D. in communications and education. She has had more than 40 years in elementary and secondary schools, in educational television and in higher education. She has spent more than 16 of those years in administrative posts. Cabrera is a specialist in information technologies, media communications design and resources development. She is now the executive director for ETN, a FCC-licensed, Ku-band satellite network owned and operated by the Los Angeles County Office of Education. ETN programs a teachers channel, a parent channel, an adult learning channel, a classroom channel, a leadership channel...She is responsible for managing all aspects of ETN television production and transmission nationwide.

Celia C. Ayala, Ed. D. is the Director of the Curriculum Programs and Instructional Technologies Division of the Los Angeles County Office of Education. ETN, the educational telecommunications network and TEAMS, Telecommunications Education for Advances in Mathematics and Science, the Federally-funded Star Schools project, are both located in her division.

FROM THERE TO THERE WITH K-12 DISTANCE LEARNING

■ THE WHY...

In the beginning, distance learning for K-12 was conceived as a response to meet the needs of an estimated one-third of the country's school age children who were being denied an adequate education either because of school size or geographic isolation. It soon became apparent that inner-city schools, while not geographically isolated, could not attract sufficient numbers of qualified teachers in specialized areas either. Both rural and urban schools face the impact of economics and teacher shortage while at the same time meeting the challenge of providing children with access to and success in a more vigorous curriculum. They both need a more efficient and effective distribution of learning resources. When well-planned, distance learning is a major solution.

■ THE WHERE...

Across the nation, distance learning activity is either being planned or implemented. The activity, state-by-state, varies from beginning planning to coordinated statewide plans, from involvement of K-12 schools and districts to involvement of state government agencies as well as higher education.Some are reaching beyond their borders to share and to respond to national directives.

"Linking for Learning," (1989) in compiling data from a number of resources to describe state-by-state profiles, noted the information was neither final nor complete; rather represented a first attempt to present the range of distance language activity nationwide. The following is a brief restatement of existing distance learning activities presented by time zones. (See Table 12.1.1)

■ THE WHAT...

The convergence of video and computer technologies as well as the advances in interactive capability have made it possible to increase the variety and to broaden the range of instructional programs for students, to distribute information to educators, parents, students and interested others and to facilitate delivery of inservice education for teachers and administrators.

Table 12.1.1

■ PACIFIC

Alaska:

- The Alaskan Teleconferencing Network and University of Alaska Computer Network operated by the University of Alaska for local schools
- The Rural Alaska Television Network (RATNET) operated by the State Department, both instructional television - ITV - series and live and interactive.

Washington:

- The Satellite Telecommunications Education Programming (STEP), operated by Education Service District 101
- Receiving programming from TI-IN, ASTS
- State legislation for statewide educational telecommunications network.

Oregon:

- TERC participant through Northwest Regional Laboratory
- Receiving TI-IN programming
- Legislation for Ed-Net, statewide telecommunications network (satellite delivery).

California:

- ETN, Educational Telecommunications Network, operated by the Los Angeles County Office of Education, transmitting staff development, parent education and TEAMS programming via satellite
- TEAMS, Telecommunications Education for Advances in Mathematics and Science, is a Star Schools project serving eight states and their districts
- California Technology Project funded by technology dollars for use by schools
- KLCS Channel 58, a PBS station, operated by LAUSD
- TERC participant through Pepperdine University
- Receiving TI-IN, TEAMS, ASTS programming, all Star Schools projects
- Multiple ITFS and cable institutional networks operated by county offices and districts.

Nevada:

- Receiving TI-IN programming.

Hawaii:

- Legislation authorizing distance learning
- TELEclass Project linking students in Hawaii and those in Japan and Australia for voice and video exchanges.

■ MOUNTAIN

Idaho:

- The Idaho Rural Education Delivery System (IREDS), consisting of live video, two-way audio interaction, operated by the SEA.

Arizona:

- Arizona Educational Telecommunications Cooperative (AETC), operated by SDE, et al
- Arizona School Services through Education Technology (ASSET)
- TERC participant through Arizona State
- Receiving TEAMS, TI-IN, ASTS and ETN programming.

Utah:

- EDNET, the state's microwave system, providing two-way audio/video instruction to high school and college students and teachers
- The Intermountain Community Learning and Instructional Services project using audiographics to deliver instruction
- The Carbon County School District Distance Learning Project, a fully interactive two-way cable/microwave system with data transmission capabilities

- Audiographics use is widespread
- Receiving TEAMS and ETN programming.

Colorado:

- TERC participant, Denver with Biological Science Curriculum Study
- Receiving TI-IN, ASTS, TERC programming
- Participating in the Intermountain Community Learning and Instructional Services project
- An interactive television network using local cable interconnecting high schools with district media centers
- Multiple audiographics used for placement courses.

Wyoming:

- A statewide computer network (audiographics) operated by the Governor's Telecommunications Division
- Participating in the Intermountain Community Learning and Instructional Services project (audiographics)
- Receiving TI-IN programming.

Montana:

- Big Sky Telegraph Network (computer network) providing both K-12 courses and teacher training
- Edunet, private non-profit company, produces courses and provides computer networking
- Participating in the Intermountain Community Learning and Instructional Services project (audiographics)
- Receiving TI-IN programming.

New Mexico:

- A statewide data communications network connecting some districts and several institutions of higher education
- A fiber optic network for voice, data and video transfer
- Three public television stations maintain a microwave network
- Receiving TI-IN programming
- Legislation for the New Mexico ITV Network.

■ CENTRAL

North Dakota:

- The Red River Valley Telecommunications Consortium provides (satellite) programming from ASTS/TI-IN
- Member of SERC Star Schools project
- Decision about technology system using audiographics for rural areas.

South Dakota:

- Technology in Education project, a statewide telecommunications consortium, delivers foreign language and other instruction
- Participating in Pennsylvania Teleteaching Project
- Receiving TI-IN programming.

Minnesota:

- Minnesota Educational Computing Consortium (MECC) is one of ten TERC Star Schools teacher   training centers
- Receive TI-IN programming
- Northwestern Minnesota Fiber Optic project
- Sherburne-Wright Educational Cooperative (cable, microwave and satellite), a member of Classroom Earth
- KIDS, Knowledge Interactive Distribution Systems, uses microwave, computers, tele conferencing and ITFS
- Des Moines River Tele-Media project linking districts via fiber optics
- East Central Minnesota Educational Cable Cooperative (ECMECC) using two-way cable

television and microwave
- TERC participant
- Many other sites using fiber.

Wisconsin:

- The Wisconsin Rural Reading Improvement Project, delivered via television, radio, ITFS and the telephone system, operated by SEA, public radio and TV and educational services units
- The Wisconsin Educational Communications Board (WECB) operates a statewide educational television broadcast network to provide programming to K-12, vocational-technical schools and universities
- Member of SERC
- Receiving TI-IN and ASTS programming.

Iowa:

- The Fiber-Optic Communication Instruction System, an interactive television network in Des Moines
- Member of SERC
- Legislation for a statewide telecommunications network.

Nebraska:

- Legislation for developing a statewide telecommunications network
- Member of SERC; also producing programming for SERC.

Kansas:

- University of Kansas and Kansas State University produce programming for the Midlands Consortium Star Schools project
- Receiving programming from ASTS and Midlands Consortium.

Missouri:

- Missouri Education Satellite Network (MESN), a satellite delivery network, receiving programming from the Midlands Consortium ASTS, KSU and STEP
- Member of SERC
- Legislation authorizing a tax on videotape rentals for five years to fund distance learning projects in the state.

Illinois:

- K-12 electronic network links SEA with 18 regional educational centers
- SEA, a partner in the TI-IN Star Schools Project
- Receiving ASTS programming.

Tennessee:

- Upper East Tennessee Educational Cooperative found using microwave, cable television and satellite
- Receiving ASTS and TI-IN programming.

Arkansas:

- AETN, Arkansas Educational Television Network, oversees educational television programming
- Legislation establishing guidelines for satellite-delivered distance learning education and use of educational technology
- Participant in SERC
- Receiving ASTS and TI-IN programming.

Oklahoma:

- ASTS, Oklahoma State University, deliver high school and staff development courses to 250 schools in 20 states
- OSU operates the Midlands Consortium Star Schools project
- The Panhandle Shar-Ed Video Network, a digital fiber optic system telephone/school cooperative.

Texas:

- Legislation required and funded the development of the 1988-2000 Long Range Plan for Technology
- The InterAct Instructional Television Network in Houston uses ITFS/microwave and telephone to deliver courses and programming
- TI-IN, in cooperation with ESC Region 20 (San Antonio), provides live satellite-delivered courses to 700 sites in 32 states. Also operates the TI-IN Star Network, Star Schools project.
- A member of SERC Receives ASTS programming.

Louisiana:

- The Louisiana Educational Satellite Network (LESN) broadcasts interactive video programming
- A member of SERC and a Star Schools participant
- Fiber optics interconnect some parochial and public schools.

Mississippi:

- The Mississippi Authority for Education Television operates a statewide network for video
- Member of SERC and the Midlands Consortium
- Receiving ASTS and TI-IN programming
- Electronic mail participation among districts and SEA.

Alabama:

- Alabama Educational Television (AETV) distributes K-12 supplementary programming
- Member of SERC and the Midlands Consortium

■ EASTERN

Michigan:

- Michigan Statewide Telecommunications Access to Resources (M*STAR) provides instructional television programming to all schools
- Providing Academics Cost Effectively (PACE) an interactive television project using cable and microwave
- Many local projects use cable television and microwave facilities to deliver interactive television courses
- The Archdiocese of Detroit operates a four-channel ITFS network
- TERC participant through University of Michigan
- Member of TEAMS and SERC
- Receiving ETN programming.

Ohio:

- The Ohio Education Computer Network links K-12 school districts for administrative purposes
- Member of SERC
- Receiving ASTS programming.

Pennsylvania:

- The Pennsylvania Teleteaching Network, an audiographics network, operated by Mansfield University, links multiple states
- Penn-Link, a statewide K-12 computer network
- PSU, Penn State University, uses public television cable microwave and satellites to deliver programming
- Receiving ASTS programming
- Member of SERC.

New York:

- Technology Network Ties, a statewide K-12 computing network

- Bureau of Cooperative Educational Services (BOCES) uses audiographics, microwave and computers in more than 100 local projects
- Receiving AST and TI-IN programming
- TERC participant through City College of New York
- Member of SERC.

Vermont:

- The Northeast Kingdom Rural Telecommunications Cooperative provides live interactive audio/video instruction via satellite
- Vermont Interactive Television, using fiber optics, offers interactive audio/video services to education and business
- Receiving programming from ASTS.

Maine:

- Receiving ASTS and TEAMS programming.

New Hampshire:

- Keene Junior High School Project uses microwave to link with five local high schools and a vocational center
- Manchester School District Instructional Television Network uses fiber optics to link three schools.

Massachusetts:

- Massachusetts Corporation for Educational Telecommunications (MCET), founded by legislation
- The Cambridge Teleteaching Group uses audiographics to deliver courses
- Kids Interactive Telecommunications Project by Satellite, a satellite based system, allows two-way video and computer conferencing internationally
- South Berkshire Educational Collaborative uses a two-way interactive cable television system
- TERC headquartered in Cambridge
- Boston School District, a member of TEAMS
- Receiving ETN programming.

Connecticut:

- Legislation established the Joint Committee on Educational Technology and created the Telecommunications Incentive Grant Program
- SEA administers the Telecommunications Incentive Grant Program to seed local distance learning projects
- Links to Learning, a joint project between SEA and SNET, uses voice, computer or compressed video in a variety of applications
- Area Cooperative Educational Services in Hamden uses a fiber optic system
- Talcott Mountain Science Center transmits via satellite to over 300 schools in 35 states.

New Jersey:

- OTIS, Office of Telecommunications and Information Systems, coordinates statewide resources.
- The Educational Technology Network, a statewide computer network, for administrative and instructional purposes
- Member of SERC
- Receiving ASTS programming.

Delaware:

- Project Direct, a statewide K-12 electronic network used primarily for administration.

Maryland:

- Anne Arundel County and Prince George County use local cable television systems for two-way audio and video
- SEA offers K-12 instructional television and staff development programming over public television.

Virginia:

- Fairfax County produces over 100 hours of live interactive broadcast series and seminars for SERC
- TERC participant through University of Virginia
- Live courses are transmitted through a combination of cable, microwave and ITFS from Varina High School.

West Virginia:

- Receiving programming from TI-IN, SERC, ASTS
- Two computer networks, WVMEN and WVNET, serve K-12 and higher education
- Member of SERC.

Indiana:

- Indiana Higher Education Telecommunications Services (IHETS), a statewide fiber optic backbone (leased from Intellenet) delivers three channels of one way video, two way audio interactive courses
- Receiving ASTS.

Kentucky:

- The Kentucky General Assembly voted $114 million to construct the KET Star Channels System, a statewide educational network to serve all 1300 elementary and secondary schools
- Member and course producer for SERC.

North Carolina:

- Legislation created the Agency for Public Telecommunications
- The Distance Learning by Satellite system delivers instruction and inservice programming to 100 counties
- Downeast Instructional Telecommunications Network, uses audiographics
- Member of the TI-IN and SERC Star Schools projects.

South Carolina:

- South Carolina Educational Television (SCETV) uses broadcast television and an extensive ITFS network to transmit programming to public schools; also teleconferencing via satellite and microwave
- The SERC project is based in South Carolina, delivers instruction via satellite in 22 states.

Georgia:

- The Georgia Public Telecommunications Commission in cooperation with SEA, operates an educational television network for K-12 and higher education via microwave
- Member of SERC.

Florida:

- SEA operates Florida Information Resources Network, a data communications network used for computer conferencing, database access and support of computer based courses in K-12 and higher education
- Extensive ITFS facilities
- Member of SERC
- State Satellite Network, operated by SEA, links 28 sites.

Students in rural areas can enroll in advanced placement courses in mathematics, science, foreign languages and more from Oklahoma State University. Students in 22 states take Spanish, physics, computer science, basic skills, psychology, sociology, art history, German, French, English as a second language from TI-IN — complete courses via satellite because economics or teacher shortages make distance learning viable.

Not all distance learning for students is in the form of complete courses. The greatest growth has accrued in partial courses and enrichment materials allowing innovation in distance learning. Some projects enrich existing curriculum in math and science. NASA officials estimate more than 20,000 teachers have participated in programs linking teachers in their schools with NASA astronauts and scientists. More than a quarter of a million students in grades 4-12 explored the floor of the Mediterranean Sea with Jason, a remote controlled robot vessel. The Jason project reverses the traditional classroom setting of distance learning. In Jason, media is not a compromise, or a necessary evil, or a second choice, rather a provocative vision of the future of distance learning-electronic transportation to a new site for learning offering previously out-of- reach experiences.

Seminars, college-level courses, certification classes and workshops are offered to teachers and administrators on distance learning systems. Parents are learning about schooling issues as well as how they can support their children in their education. Policy makers are learning about change, reform, restructuring and other issues important to those who lead K-12 schools.

■ THE HOW

To reach outside people, information, resources and peers, students, teachers and education managers are using telecommunications technologies — databases, homework hotlines, learning circles, students in other countries, curriculum materials exchanges, electronic mail, computer conferencing etc. Most recently, various cable operators have made their presence known — and available — in the classroom in the form of current events programs. .

When considering how distance learning technology is being used, there are two "how's" to be considered: the pedagogy and the technology.

The Pedagogy

The primary use of distance learning has been to replicate the experience of face-to-face instruction —

the teacher teaches in the present (live) and the teacher and students interact. Allowing for interaction or participation is what differentiates this application of education technology from earlier attempts such as ITV. This interaction is considered a critical incident in recreating traditional instruction. It is usually accomplished via telephone, i.e., one-way video, two-way audio. Two-way video which has usually been found in small systems is the closest imitation of the traditional classrooms. In large classes where there is limited time for questions, "office hours" are used to maintain the classroom connection. More and more computer software has been developed to foster interactivity. Simulations and cooperative learning have stretched both student and teacher activities.

Teachers and learners in separate locations whether in classroom, home, office, studio or learning center are using multiple communications technologies — satellite, microwave or fiber optic cable, television (broadcast, cable, or ITFS (Instructional Television Fixed Services), video cassette or direct phone line, audio cassette, printed material — text, guides, computer modem, floppy disk and compressed video. Many are connected into systems that allow interactivity whether via telephone to support one-way video, two-way video or graphics, two-way computer hook-up or two-way audio.

The Technologies

The basic technologies of distance learning are audio, video, computers and supplemental technologies such as the VCR, CD-ROM, compact audio discs and hypermedia programs.

Audio distance learning is possible through the telephone system and audio bridges or broadcast radio. Voice mail often supplements other distance learning systems. The availability of high bandwidth transmission paths over satellite, and fiber optic links and the increased ability to compress video signals has positioned video technologies in the forefront of distance learning.

In "Linking for Learning," the Office of Technology Assessment (OTA) summarized the transmission technologies.

● Terrestrial broadcast is one-way broadcast of audio, video and data. It reaches most homes and schools because no special receiving equipment is required; however, there is limited reception due to geography, limited channels and air time. Production costs are high. It is increasingly used for data/text transmission.

● Fiber optic is two-way video, audio and data. It delivers a high quality signal, is easily expandable and is very fast. However, installation costs are very high with right-of-way acquisition required. Costs are declining allowing for increased fiber deployment.

● Microwave is two-way, point-to-point audio, video and data. Transmission time is inexpensive; however, must be FCC-licensed, tower space location is difficult to acquire, crowded frequencies, costly to expand channels and "line of sight" is required.

● ITFS (Instructional Television Fixed Service) is one-way broadcast or point-to-point audio, video, and data. Video delivery is inexpensive.. Compression or digitalization may triple channel capacity. repeaters increase coverage and can be used to rebroadcast satellite-delivered programs.

● PSTN (Public Switched Telephone Network) is two-way voice, limited data and video. It offers wide coverage with low initial cost. Others handle repairs and upgrades. However, quality is spotty, data and video transmission is limited and cost is distance-sensitive. There is expanding fiber installation and increasing intelligence in the network.

● Satellite is one-way broadcast of video, voice and data with the possibility of audio/data return. It offers wide coverage, close identity to the "TV" image and cost is not sensitive to distance. However, uplinks are expensive downlinks are not cheap; transmission costs are high uplinks must be FCC-licensed and carriage "on the ground" is essential. Compression will increase channels making transmission costs more reasonable. There is increasing use of data and increased interactive capabilities.

● Audio graphics is two-way computer conferencing with audio interaction. It is low cost, with easy exchange of graphics and uses PSTN. However, visual interaction is limited to graphics and still video. More powerful computers, better software and peripherals increase capabilities.

● Cable Television System is one-way broadcast or two-way, point-to-point video, audio and data. It is widely available with low delivery costs. however, it offers educators limited access,there is limited capacity, difficult to interconnect, and not designed for interactivity. Fiber increases capacity, offers more addressability and two-way capability.

■ THE WHO...

Distance learning projects that have been successful have found the following elements to be critical to their success:

1. A coherent, well-planned program addressing significant educational needs
2. Programs designed to meet the real needs of the learning audience
3. Common agreement on project goals, scope and evaluation
4. A transparent governance structure that ensures leadership, participation and decision-making procedures
5. Tele-teachers strong in curriculum, instructional skills and "presence"
6. Thorough training of users of the new delivery system
7. Materials that compliment the technology and augment any deficiencies
8. Computers used in a special role to stimulate interaction
9. A setting complimentary to the technology, i.e., good acoustics, modular furniture, adequate light and relaxing ambience

Following are but a few profiles of successful projects. They are chosen because they represent a mix of clientele, service areas, curriculum, pedagogy and technologies.

Panhandle Shar-ed Video Institutional Network

Located in the eastern end of the Panhandle, Beaver County includes four school districts that serve students spread over long distances. The Beaver School District has 519 K-12 students coming from a 426-square-mile area; Forgan has 191 students across 397 square-miles Balko has 159 students across 305 square-miles; and Turpin has 420 students within a 303-square-mile area. The educational problems faced by Beaver County schools are typical of small, isolated school districts. It is almost impossible to offer advanced courses and specialized subjects on a regular basis.

Four years ago, Beaver County superintendents and school board members learned about distance learning projects in Wisconsin and Minnesota. They visited projects and hired consultants to analyze site and technical requirements, examine alternatives and plan the system. The four districts, with support from the State Director of Rural Education, agreed to seek external funding for a two-way full-motion interactive television system for sharing instruction among their schools. They established the Panhandle Shar-Ed Video Network Cooperative, a partnership between the four school districts and Panhandle Telecommunications Systems, Inc., a subsidiary of Panhandle Telephone Cooperative Inc. (PTCI). PTCI is a co-op owned by 4,200 individuals in the three-county Panhandle region. According to PTCI, the network is an important factor in keeping the schools open and in assuring the economic viability of the region.

To build their distance learning system, the four superintendents sought outside funding. These funds ($340,000) covered startup costs: installation of four-strand and eight-strand fiber optic cable between the four schools, the telecommunications hookups and studio classroom equipment (cameras, television monitors, microphones, VCRs and facsimile machines). They also covered costs of a five-year lease of the fiber optic lines that are owned and operated by PTCI.

Shar-Ed Network offers a way to expand resources for the districts yet maintain local control of curriculum, assure high-quality instruction and keep the local school/community identity intact. Also, they are expanding the network to link the other eight school districts in the Panhandle and to connect Panhandle State University. With the addition of the university, college credit courses and other educational services will be offered for students, teacher and members of the community.

The North Carolina Distance Learning by Satellite Program

In January 1988, the North Carolina State Department of Public Instruction (SDPI) and the TI-IN Network of San Antonio, Texas, entered a contractual agreement to form a statewide satellite network to provide high school instruction and staff development throughout the state. This statewide investment and partnership with a private corporation was motivated by the realization that North Carolina has many small, rural high schools which because of low enrollment and remote locations, cannot offer all the courses mandated by the Basic Education Plan.

In 1987, the North Carolina General Assembly passed the Learning by Satellite bill. The bill appropriated just under $2 million for fiscal year 1987-88 to purchase satellite receiver equipment and hardware for 153 sites. The sites encompassed SDPI chosen sites, each of the 52 smallest high schools in the State, and 100 additional sites chosen by district superintendents for staff development programming. Under the North Carolina plan, a district coordinator oversees satellite programming for all receive sites within each participating district; at each site, one person serves as manager (usually the principal or assistant principal). Each school site also has a classroom facilitator and an equipment manager.

The North Carolina distance learning efforts are expected to expand to more schools and to offer a wider array of programming in the future. This support is grounded in the bill that authorized the program: "It is the intent of the General Assembly that the Distance Learning by Satellite program shall be an ongoing component of the public school system and that operational funds for the program shall be included in future continuation budgets."

TEAMS/ETN, STAR Schools Satellite Project

TEAMS/ETN is a Star Schools project funded by the U.S. Department of Education. It is a service of the Los Angeles County Office of Education and is delivered from the ETN, Educational Telecommunications Network facilities in Downey, California. TEAMS is a partnership reaching out to provide motivational distance learning opportunities to elementary students, teachers and parents throughout the U.S.

The major focus of TEAMS/ETN is direct student instruction for grades four through six in mathematics and science, reinforcing and enhancing subjects already taught in elementary school. This is compatible with self-contained classrooms, and also satisfies the frequently expressed need for elementary teachers to improve their skills and knowledge in mathematics and science education.

The programs follow national directions in mathematics and science. Instruction is characterized by: active learning; development of mathematics and science concepts, problem solving and critical thinking skills; development of oral and written language proficiency in the content area; motivation of students to continue in mathematics and science.

During the 45-minute student programs broadcast twice a week, the studio instructors model effective instructional strategies for teachers in a team-teaching approach. The lessons motivate and challenge students to master new skills while providing a unique staff development opportunity for participating teachers. Specific staff development programs for TEAMS mathematics and science teachers prepare them to effectively use distance learning as part of their ongoing instructional program.

Parent training complements distance learning opportunities for students and involves parents as partners in their children's education. Programs for parents are broadcast in Spanish as well as English.

TEAMS provides live, direct instruction via ETN, Educational Telecommunications Network on Ku-band satellite to students, teachers and parents across the United States. The original partners in the project included the large urban school districts of Boston Public Schools, Detroit Public Schools, District of Columbia Public Schools, Los Angeles Unified School District and select districts within Los Angeles County. Within a year, these original partners were joined by schools in other counties in California. Arizona, Utah, Missouri and Toronto, Ontario also participate in TEAMS.

Mississippi 2000

Mississippi 2000 is a public/private partnership formed to propel education in Mississippi into the 21st century. It is the first distance learning network to use a switched public broadband network with digital fiber as its transmission medium. The network demonstrates how barriers to superior education, such as distance or limited resources, can be overcome by using technology. The network is used by high schools, and for in-service, enrichment and data networking.

Significant federal resources, as well as some state and local resources, are flowing into Mississippi for distance learning. Three of the four projects funded under the Federal Star Schools Program serve Mississippi. Two universities, the state education agency and the state educational television network are partners in the Star Schools consortia. A total of 6,162 schools in the State were served by the Star Schools grantees (Midlands Consortium, TI-IN United Star Network, and Satellite Educational Resources Consortium — SERC) in 1989-91. This interest in Mississippi is due, in part, to the requirements of the Star Schools legislation that at least 50 percent of the funds serve Chapter 1-eligible school districts and the educationally underserved. All of Mississippii's 152 school districts are Chapter 1-eligible. Mississippi's experience with Star Schools may demonstrate interactive distance learning's capacity to offer important educational opportunities to students from resource-poor homes and communities.

Another factor in the Mississippi picture is the aggressive educational reform effort under way in the State. Major components of that effort include full-day, statewide kindergarten, teacher aids for K-3 classes and new procedures for school accreditation, teacher certification, staff development and teacher evaluation.

Bergen County and New Jersey Bell

Eventually, public schools are going to take a major technological leap forward into the world of broadband, a technology that holds vast possibilities for aiding education. Whether broadband comes sooner or later to schools depends in part on how soon it becomes a medium for mass-market consumer services (whose volume will drive the technology's price to affordable rates). Some public school systems, in fact, are ready for this technology today. Bergen County, an affluent section of northern New Jersey, offers a case in point. The county is in the process of implementing an optical-fiber-based network to link its public high schools.

Bergen County adopted it not for the sake of having a "showcase" network but in order to address immediate practical problems. County school officials, concerned about maintaining their high academic standings, wanted to add new programs and preserve special curricula even though declining enrollments were making such courses unjustifiable from a cost standpoint. County educators wished to maximize student access to master teachers; at the same time, they sought to reduce the costs and administrative burdens involved in transporting students from school to school for special programs. The county also wanted to take advantage of opportunities to work cooperatively with local colleges and other institutions for mutual benefit. Not least, Bergen County wanted to address issues of equity, i.e., each student gaining access to the same quality of programming. A fully interactive fiber-based network addresses all of these needs, and much more.

Bergen County signed a 10-year, $4.7 million contract with New Jersey Bell, a unit of Philadelphia-based Bell Atlantic, to construct and maintain the network and provide switching services. Now in the first year of a five-year implementation period, the network already connects 14 Bergen County high schools, 11 of which use the system every day. By 1995, the network will connect 44 high schools, two or more colleges and possibly a number of other public institutions.

New Jersey Bell designed the network using a hub-and-spoke concept. At the hub is the company's Hackensack central office, equipped with a video switch. Each of the central offices serves four high schools by means of custom-installed optical-fiber loops. Every school gets four video channels: one for transmission and three for reception. Each school's interactive television (ITV) classroom is equipped with two cameras, four monitors plus a video cassette recorder, fax machine and teleconferencing sound system. Students scattered throughout the school system can participate in a class and see, hear and exchange documents with each other. Even with only 14 of its 44 high schools connected, Bergen County has the largest education-dedicated switched interactive video network in the U.S.

How economical is the network? Serving a student population of approximately 100,000, the system works out to far less than $50 per student over the course of the 10-year contract.

The Telelearning Project

The Telelearning Project, administered by the Delaware-Chenango Board of Cooperative Educational Services (BOCES), is one of the pioneer distance learning projects in New York State. Begun in 1985, this audiographic network links 10 of the 18 school districts throughout this rural area. Many of the 4 schools in the region are small and have had difficulty providing a full curriculum; many of the students have had little exposure to the world outside of the small villages that make up their remote communities. Sharing instruction via audiographics became an attractive means to expand high school credit offerings and to enhance educational opportunities for students.

Two schools have used audiographics to teach homebound students. At one school, an eighth grade student was given audiographics equipment to use at home while recuperating from back surgery. The computer, graphics tablet and speaker phone made it possible for him to take his regular classes from his bed, to participate in class discussions, not fall behind in his work and keep up with his friends during this difficult time.

The project has also made possible the increasingly popular "electronic field trip." An electronic field trip is a telephone conference call from one of the schools in the project to an outside authority or classroom. Over 50 of these field trips are conducted by participating schools in the BOCES region each year. The BOCES administrator explains that rural students are so isolated and have so little, if any, cross-cultural contact. The electronic field trip is a very simple and inexpensive way to give students that contact.

STEP, Member Supported Satellite Network

STEP, Satellite Telecommunications Educational Programming Network, was created to meet a relatively localized but common problem. School superintendents in Washington's Educational Service District (ESD) 101 service area needed help in delivering night school credit courses in subjects where they were unable to provide certified teaching personnel. They went to their ESD, which traditionally acts as a liaison between local schools districts and the State education department and assists local schools in meeting instructional and administrative needs.

In considering distance education as a way to meet the need for courses, a number of technologies were considered. Satellite was chosen because ESD administrators wanted the capability to broadcast to the entire State of Washington, and because, fortuitously, an uplink capacity was available through Eastern Washington University (EWU) in Cheney, 20 miles south of Spokane. STEP operates as a public, nonprofit cooperative. Costs are distributed among users. The operational budget is financed from the subscribing districts through installation, annual subscription and tuition fees. While the direc-

tor's salary is paid by ESD 101, all other costs (teachers' salaries, production costs and other support costs) are financed through member fees. ESD 101 makes available to new subscribers an equipment package that includes a downlink satellite dish and associated television classroom and telephone equipment and maintenance. Districts can also make arrangements to obtain their own equipment. In either case, first year capital equipment items required typically range between $5,000 and $6,000. New members pay a $4,750 initial membership fee, which is renewed at $3,000 per year. Charges for high school credit courses are based on a per student, per course, per year fee.

Enrichment courses are also offered, and reach down into elementary and middle schools. These programs are broadcast on Fridays, when regular STEP classroom instruction is not being broadcast. Student enrichment programming is optional, assessed at $350 per program or $1,000 for a total package of 10 programs.

Staff development is an important part of STEP programming. Course credit is available through EWU and Whitworth College, or teachers may take the courses for "clock hours" credit for advancement. Efforts are under way to make it possible for teachers to complete a master's degree program via distance learning.

TERC Star Schools Project

The TERC, (Technical Education Research Center) Star Schools Project uses a computer and a commercial teleconferencing network to connect students studying science or mathematics. TERC is a collaborative of Boston Museum of Science, the Northwest Regional Labs, Minnesota Educational, Computing Consortium, City College of New York, Biological Sciences Curriculum Study, and five universities: Arizona State, Michigan, Pepperdine, Tufts and Virginia.

The curriculum approach is to engage students in data collection and problem solving, exchanging observations and date with other classrooms around the country. The telecommunications network, basically an electronic mail network, allows students to share results, write reports and ask questions of leading scientists who are serving on the project as models and collaborators.

The schools are responsible for providing the computer, modem and telephone line for the project, although some subsidies are available. Teacher preparation, software, scientific experiment supplies and telecommunications are paid for by TERC and the teacher training center partners. The overall hardware costs (one computer per two classes) and telecommunications costs (computer network hookup at off-peak hours) are low.

In the TERC program, the classroom teacher remains the subject expert. Whereas in other projects, the teleteacher provides most of the instruction, supplemented by the attendant classroom teacher. Therefore, inservice teacher training is considered an especially critical component to the TERC model.

The Jason Project

The Jason Project is a multi-technologies, multi-instructional partnership among many sectors. Woods Hole Oceanographic Institution, a private nonprofit marine research facility, is the coordinating organization for the project, and also is responsible for the 7-year development of the ARGO/JASON vehicles (funded by the Office of Naval Research). Electronic Data Systems provided the communications technology, equipment management and staging at each museum site, as well as substantial funding contributions. The Quest Group, Ltd., a group of private individuals highly supportive of deep-sea exploration, is underwriting part of the project costs. Turner Broadcasting coordinated the live and preproduced portions of programming at a reduced fee. The National Geographic Society, producers of a film on the project, coordinated the involvement of the 12 museums sites around the country. National Science Foundation went to the National Science Teachers Association, which wrote the science curriculum for the project, with help from the National Council for the Social Studies. The total project budget was about $7 million, approximately three-quarters of which was for equipment, curriculum development and start-up costs that would not require support in future years.

The Jason Project is inspired by the advanced technology for sea floor exploration found in the ARGO and JASON systems developed by the Woods Hole Oceanographic Institution. Together, the ARGO/JASON technologies represent a significant improvement in the speed with which oceanographers can explore the deepest parts of the ocean. Also, advancements in fiber optic technology allow for high-quality television images to be transmitted from ARGO/JASON to the surface ship via a 4,000 meter cable.

The curriculum developed for the Jason Project takes advantage of the many educational opportunities provided by such a unique and advanced scientific effort. The science of oceanography is addressed, as well as other topics surrounding such exploration that fall under the physical sciences, biology, history and geography. In addition, the Jason curriculum includes lessons on the telecommunications technology used to bring the picture to the students, and the robotics needed to build and operate the ARGO/JASON exploration vehicles. Short lessons in mythology and creative writing connect the myths of Jason and the Argonauts to the current effort.

The Jason Project seeks to combine the power and reach of the media with the experiences of live, see-it-as-it-happens scientific research. Such experiences, built into a valid pedagogical framework, have the potential to broaden and invigorate the educational experience for children.

■ THE WHERE FROM...

Bradshaw, D. and P. Brown. The Power of Distance Learning. Far West Laboratory Policy Briefs No. 8, 1989.

"Classes Turning on Television." Cerritos/Artesia Community Advocate. Cerritos, CA. May 7, 1992.

Daley, A. "Improving Education with Broadband Networks." Tech Trends, 1991.

Nupoll, R. and Salinas, K. "Filling a National Need: Distance Learning in the Elementary Classrooms." ESN Program Guide, March 1992.

Portway, P. "ED Teleconferencing and Distance Learning: An Orientation". United States Distance Learning (USDLA). San Ramon, California: October 1991.

Roberts, Linda "Linking For Learning." Washington, D.C.: Office of Technology Assessment (OTA), Congress of the United States, June 1989.

Roscher, Ted "Distance Learning Case Studies." Linking For Learning, OTA Contractor Report, June 1989.

Rundle, Doug "Distance Learning Case Studies." Linking For Learning, OTA, Contractor Report. June 1989.

Van Wickler, Freeman "Distance Learning Case Studies." Linking For Learning, OTA Contractor Report. June 1989.

Technologies For Learning At A Distance, Special Report: Office of Technology Assessment (OTA). Washington, D.C.: 1988.

Van Horn, R. "Educational Power Tools: New Instructional Delivery Systems." PHI DELTA KAPPAN, March 1991.

Sandra Welch is responsible for all of the educational programs, materials and services PBS provides to its member stations nationwide.

Welch joined PBS in February 1991 after 20 years with Kentucky Educational Television (KET), where she rose to deputy executive director and chief operating officer. She supervised all operating divisions including programming, production, development, education, promotion, business affairs, engineering, operations and enterprise.

Among her main accomplishments at KET, Welch guided the development of the "distance learning" system in which live, interactive satellite television transmissions are combined with computers, telephone and specially designed response keypads to deliver high school courses to Kentucky and 24 other states.

Welch's responsibilities include supervising the PBS K-12 Learning Services department which makes public TV programming and teaching materials available to its local stations which reach more than 29 million students in grades K-12 nationwide; the Adult Learning Service (ALS), the leader in providing college-level telecourses (more than 300,000 adults annually earn college credits though ALS telecourses); and the Adult Learning Satellite Service (ALSS). Her responsibilities also include The Business Channel, a specialized programming service that provides colleges and businesses with timely programs on training, education and business-related topics and PBS MATHLINE, a new interactive service for helping the nation's teachers and students reach the National Educational goals in math.

Welch holds a bachelor's degree in elementary education, a master's degree in library science from the University of Kentucky, and an honorary doctorate degree in business administration from Robert Morris College. Among her affiliations, she is a member of Cable in the Classroom, the U.S. Distance Learning Association and Literacy Volunteers of America.

THE EDUCATIONAL SATELLITE NEIGHBORHOOD

For over 25 years local public television stations through their membership organization, PBS, have used the latest technological delivery systems to provide educational programming resulting in equity and excellence to schools and colleges, teachers and learners nationwide. PBS, the Public Broadcasting Service, pioneered the use of satellite delivery of broadcast programs when in 1978 it began distributing video programs to its member stations via the Westar satellite. In 1989, Congress authorized funding for public television to replace its nationwide satellite distribution system and upgrade its original technical infrastructure at local stations, regional networks and PBS.

At about the same time, there were major technological advancements in the digitization of video and compressions systems. PBS and its member stations are ready to be among the first to employ digital video compression on six of the seven new satellite transponders it purchased on AT&T's new Telstar 401 satellite. By using the new compression technology, PBS has the capacity to deliver many more simultaneous video channels than ever before possible. (Initially, PBS plans to compress up to 8 channels on each of its Ku-band transponders). In addition to the transponders, Congressional funding for public television made possible the purchase of a two-way data interactive network using Very Small Aperture Terminals (VSAT's) and servers at every local public television downlink site nationwide with the hub located at PBS.

Public television and its educational partners recognize the growing interest and user benefits in aggregating educational and distance learning programming onto one satellite. Consequently, PBS allocated a portion of its capacity and working with AT&T, encouraged other independent non-profit educational programmers to take advantage of this new and expanded capacity to help form an "educational satellite neighborhood" on Telstar 401. The advantages are many:

• The capacity an independent programmer

acquired is under its own control, providing the flexibility needed to meet its own needs and that of its customers.

• Non-profit distance learning providers benefit from substantial cost-savings for long-term, full-time digitized channel capacity.

• By combining satellite-delivered educational services onto one satellite, schools, colleges, libraries, state agencies, hospitals and others have "one-step" access to many educational offerings provided by the nation's foremost leaders in distance-learning and other educational services. Users of services need only one satellite receive system to receive a multitude of educational services.

• Telstar 401, located at 97 degrees West longitude, provides complete 50-state coverage and service to Puerto Rico and the U.S. Virgin Islands.

• And Telstar 401 is a state-of-the-art satellite with many advanced technical features including: both C-band and Ku-band transponders; numerous built-in redundancy features; and uniform broadcast quality.

Telstar 401 was launched in December 1993 and became operational in February 1994. Several major distance learning providers have already moved or are in the process of moving to Telstar 401 including:

• SERC: a non-profit partnership, between public television stations and state departments of education which provides live, interactive high school and middle school courses and staff development directly via satellite.

• National Technology University (NTU)is a private, non-profit, accredited institution that delivers, via satellite, continuing education and graduate courses for engineer, scientists and technical managers. Its courses are taught by top faculty located at 45 of the nation's leading engineering colleges and universities.

• SCOLA is a non-profit corporation that distributes to schools, colleges, universities and businesses, television news programming around the clock from about 40 counties, including Russia, China, Kenya, Vietnam, Egypt, Japan, etc. The foreign-language programs are used in classrooms for language instruction and courses in international studies, political science, business and geography.

In addition to these national distance-learning programs, several states have bought or leased capacity on Telstar 401 in order to provide statewide distance learning and educational services. So far Georgia, South Carolina, Louisiana and Florida are in the "neighborhood."

The Air Force Institute of Technology (AFIT) and the Army Logistics Management College (ALMC) will also be on Telstar 401.

PBS is continuing to expand its own distance learning and educational programs provided to its member stations with new services such as PTV: The Ready to Learn Services; PBS MATHLINE; PBS ONLINE; and the associate college degree service "Going the Distance." These are in addition to the wide range of resource programs and live and interactive videoconferences PBS has been delivering directly to colleges for the past decade.

The last ten years have been tremendous growth years for distance learning. Public television is proud of its partnership with schools and colleges, departments of education, professional associations, businesses, funders and other independent distance learning providers. As a result of congressional support which made possible public television's advanced technical infrastructure, those educational partnerships can continue to build and service increasing numbers of teachers, child care providers and learners of all ages nationwide.

Chapter 13

George Connick has been President of the University of Maine at Augusta (UMA) since 1985. Prior to assuming his position at UMA he spent 19 years at the University of Southern Maine holding a number of different positions, including Director of the Division of Basic Studies and Instructional Television and Vice President for Academic Affairs.

George is a historian by training with his bachelors degree from Stanford University (1957), a masters from San Jose State University in California (1960) and his Ph.D. in history from the University of Colorado at Boulder (1969).

He has taught at San Jose State University and Sonoma State College in California, the University of Colorado (Denver) and the University of Southern Maine. At UMA, he has taught an honors course in "The First Amendment and New Electronic Technologies" while serving as president.

George is the author of numerous articles and reports and he has given more than 300 presentations on topics ranging from accreditation to the uses of technology and telecommunications at all levels of education and training. He served in 1989-90 as Chairperson of the Advisory Panel on "Rural America at the Crossroads: Networking for the Future" for the Office of Technology Assessment of the U. S. Congress.

For the past several years, George has been responsible for the development of a plan to extend educational opportunities to all the people of Maine using a combination of off-campus centers and very sophisticated telecommunications technologies. The plan for the Education Network of Maine was approved by the University of Maine Trustees in November, 1987 and the State Legislature allocated funding of approximately $5.0 million for the operation of the Network. The capital expenditures for the telecommunications network to date are approximately $15 million.

During the winter of 1993, George, accompanied by his wife Joan, completed a three month sabbatical adventure. They drove from Maine across the northern tier of states (and several Canadian provinces) to the west coast and then back to maine via the Southwest, visiting more than 50 schools and colleges which are using technology creatively for instruction and distance learning. George is working on several publications related to these visits.

More recently, the Board of Trustees of the University of Main System approved a proposal presented by Dr. Connick to separate the Education Network of Maine from UMA and to create an electronic university within the University of Maine System. Dr. Connick is on leave for the coming year in order to inaugurate the new institution (the separation from UMA was effective on July 1, 1994) and to prepare a telecommunications and technology plan for the University of Maine System to be submitted to the Board of Trustees no later than July 1, 1995.

HIGHER EDUCATION AT A DISTANCE

In 1983, a firebell in the night was sounded with the publication of the report titled "A Nation at Risk: The Imperative for Educational Reform." Although the report focused on the problems and declining quality of the K-12 sector of education, it also signaled the enormous challenges that all levels of educa-

tion are going to face in the future. The report quoted Paul Copperman's conclusion that "Each generation of Americans has outstripped its parents in education, in literacy and in economic attainment. For the first time in the history of our country, the educational skills of one generation will not surpass, will not equal, will not even approach, those of their parents (p.11)."

The report goes on to point out that the (educational) deficiencies of students comes at a critical point in American history when the demand for "highly skilled workers in new fields is accelerating rapidly. Computers and computer-controlled equipment are penetrating every aspect of our lives — homes, factories and offices. One estimate indicates that by the turn of the century millions of jobs will involve laser technology and robotics. Technology is radically transforming a host of other occupations (p. 10) ...."

At the very time when education needs more funding to address issues of quality and needed reform, resources are contracting significantly and educators are being told to do more with less. In the past several years, the American educational system, public and private, has faced greatly increased competition for financial resources from a variety of other important public policy areas (e.g. health care, social security, housing, urban decay, etc.). At the very time that American and world society is becoming much more complex and that an educated citizenry is being recognized as the essential basis for a competitive economy and individual hope for a productive and rewarding life, the resources to support the needed changes in the educational system are contracting.

In June 1992, Yale University President Benno C. Schmidt Jr. announced that he would become the president and chief executive officer of The Edison Project. Founded in 1991, the Edison Project has as its goal building the first national, private school system in the U.S. It is a partnership of Whittle Communications L.P., Time Warner Inc., Philips Electronics N.V. and Associated Newspapers Holdings Limited. In explaining why he joined the Project, Schmidt said, "The schools of America ... are in difficulty. Like so many Americans, I have been dismayed by mounting evidence that our schools are not working well. America's schools need fundamental structural change, not tinkering around the edges."

The key words in Schmidt's statement and in so many other proposals for reforming American education at all levels, is the emphasis on the need for "fundamental structural change."

There have certainly been calls for reform in American education in the past. Most often, if change occurred, it was concerned with curriculum reform, changes in the administration of schools and revision of teacher preparation programs. Fundamental changes in infrastructure, organizational approach and local control were very difficult to achieve and, in general, have not been accomplished. Many in America feel change is moving too slowly. Indeed, educational inertia is seen increasingly as a threat to America's future.

For many, the American educational system, which was the envy of the world for over a century, is faltering and appears to be in decline. But the news from the educational front is not all bad. Pockets of substantial reform are taking place at all levels of education and, in some areas, change is beginning to accelerate.

For the past 15 years, as we have wrung our collective hands over the frustratingly slow pace of reform in education, some very profound and far-reaching technological developments have been occurring in the American and world business communities. Revolutionary technological breakthroughs have begun to transform how we live and work now and in the future; more importantly, they have begun to transform our competitiveness as a nation. Computers (affordable, powerful and networked); software (affordable, diverse and user-friendly); electronic highways (satellites and fiber optics which allow for the cost-effective transmission of massive amounts of voice, video and data); and an expanding core of creative and sophisticated developers and users of technology have changed forever the way we think and work. The Information Age is, in fact, developing rapidly.

In education, the changes resulting from the introduction of technology are beginning to have an important and accelerating impact. The same issues that led to the introduction of technology in business (i.e. — need for greater efficiency, productivity, quality and service) are present in the widespread introduction of technology in colleges and universities.

Historically, American education has been labor intensive. Depending upon the specific college situation, between 75 percent and 90 percent of a college budget would usually be in salaries. As long as budgets expanded through inflation or to accommodate the growth in the number of students, this system worked reasonably well. But, when resources contract, the labor-intensive system falters. Budget cuts (or even level budgets) lead to inevitable reductions in the area comprising the vast majority of the budget — personnel — without a comparable reduction in the number of students to be served. Staff morale suffers, faculty/student ratios increase and quality often declines as the curriculum contracts to accommodate fewer offerings from fewer teachers.

During the past 10 years, we have witnessed a rapidly expanding interest in and use of, technology in higher education. One major consequence of this technological expansion has been to increase access to learners in a wide variety of ways. In fact, a new field, most often referred to as distance education, has emerged. In its broadest form, distance education may be defined as "education which spans the time and distance separating faculty, learners and a variety of educational resources."

Clearly, there are numerous educational issues which have been highlighted by the planning, development and inauguration of distance education projects across the country. And most such ventures, regardless of scope, must address the same issues. Like most non-traditional educational undertakings, distance education projects must simultaneously work with the existing structure while attempting to chart unfamiliar territory. The power and scope of technology (fiber optics, computers, video, etc.) has allowed education for the first time in its entire history to ask different types of questions and to address educational issues that seemed insoluble only a decade ago. For example, distance education offered a number of alternative solutions to the question about how to significantly increase access to unserved or under-served populations, cost-effectively and with high quality. And, in ways that we may not anticipate, distance education and technology projects lead to major structural change, not only in the organizations that promote them but also in those that must necessarily deal with the new institutions.

Not surprisingly, there is a great deal of discussion and concern among traditional educators about the implications of distance education. This new field challenges much of the conventional wisdom that has existed since the beginning of free public education. It allows questions to be raised about the notion of educational space (i.e. that students must gather in a central location in a specific room for learning), about educational time (i.e. that students must be aggregated at a specific time so that they may interact with a faculty member), about the necessity for centralized student services (i.e. that, for example, adult students must travel to campus to secure admission, financial aid, academic advising, etc.) and many other issues.

In distance education discussions, traditional definitions of mission, organizational structure, governance, methods of delivery, student service operations, library support strategies, regional accreditation, faculty compensation and workload and other issues, do not fit the traditional and understood patterns of education. But, for the most part, distance education projects were not created for the purpose of reforming American education. It is only coincidence that the evolution of powerful technologies have intersected with the need to increase educational productivity and cost-effectiveness.

Across the country, new distance education projects are developing and existing systems are growing, at a rapid rate. Three examples of distance education projects will serve to show the purposes and uses of this new educational development. Although there are dozens of initiatives taking place in the country that might have been used as examples, the three programs chosen are illustrative of different approaches to providing educational opportunity at a distance. Mind Extension University serves as a delivery vehicle for over 20 universities in a national effort using satellite and cable television to deliver degree programs directly to people's homes; California State University, Chico, as a single institution, uses a variety of technologies, both terrestrial and satellite, to serve populations in northern California and also nationally; and the Educational Network of Maine is a statewide initiative using fiber and ITFS to deliver courses and degrees (and a variety of non-credit programming, meetings, workshops, etc.) from more than a dozen institutions to high schools, off-campus centers, government offices, college and university campuses and other locations across the state.

■ MIND EXTENSION UNIVERSITY

Mind Extension University (ME/U): The Education Network is a national broker and unique delivery service primarily of educational programming and degrees from other institutions. Founded in 1987, the network combines the technology of satellite and cable television to deliver a variety of educational opportunities to students at their workplace or directly into their homes. Through contracts with cable companies across the country, ME/U is available to more than 20 million cable subscribers.

ME/U is affiliated with colleges and universities, the Library of Congress (a joint venture with ME/U named the Global Library Project) and TI-IN Network, Inc., a significant developer and broadcaster of live, interactive secondary instruction for students, faculty and staff using satellite technology.

ME/U offers approximately 150 credit courses per year at the undergraduate level and 45 at the masters level. In the non-credit area, ME/U offers approximately 150 faculty/staff development programs (60 for professionals other than teachers), 30 supplementary K-12 programs and 30 programs for college and university.

Degree programs are offered at the undergraduate and graduate level. The University of Maryland University College, in cooperation with the National Universities Degree Consortium (NUDC includes Colorado State University, Kansas State University, University of New Orleans, University of Oklahoma, Oklahoma State University, University of South Carolina, Utah State University and Washington State University) offers a bachelor's completion degree in management. The student is required to complete 30 semester hours of University College courses and the remainder may be taken from any of the NUDC members at a distance or from any accredited institution. The University of Maryland University College offers the management degree program curriculum and the other eight NUDC universities provide related upper-level courses, outside of the primary concentration area, that are integral to the completion of the bachelors degree program.

George Washington University offers a Master of Arts in Education and Human Development consisting of 36 credits of coursework. The coursework is all available over the ME/U Network. The Master of Business Administration (MBA) is available from Colorado State University. This 33 credit hour program is accredited by the American Assembly of Collegiate Schools of Business.

Recently available from the Network is the majority of courses required in the Master of Library Science degree offered by the University of Arizona. This degree does require that 12 credits be taken in residence at the University of Arizona.

Other institutions affiliated with ME/U and offering courses are California State University, Long Beach; Emporia State University; Governors State University; New Jersey Institute of Technology; Pennsylvania State University; State University of New York (SUNY)/Empire State College; University of California, Santa Barbara; University of California Extension, Berkeley; Utah State University; Western Michigan University; Regis University; and the University of Colorado, Colorado Springs.

Traditional support services in the areas of admissions, financial aid, counseling and academic advising are provided by the institutions offering courses and degree programs. Basic information about programs, national marketing, processing a generic admission application, textbook orders, videotape rentals and other student support services, including a national toll-free number are provided by ME/U.

Currently, approximately 2,500 students are enrolled for credit and 30,000 for non-credit programming.

The ME/U approach is a powerful new model for delivering educational programming at a distance. Rather than develop their own programming initially, ME/U has offered a platform for a variety of distance education providers, whether individually or in partnership, to deliver a wide array of programming. Prestigious higher education institutions have agreed to collaborate in a variety of new ways to provide high quality educational opportunity for people who, for whatever reasons, are not able to enroll on-site. Unlike single institutions which encounter high upfront costs for technology and the challenge of putting together a range of course offerings in order to program consistently, ME/U, working with a number of institutions, is able to offer an expanding list of courses and full degree programs. The ME/U model addresses many of the most vexing problems encountered by institutions wishing to enter the distance education arena.

■ CALIFORNIA STATE UNIVERSITY, CHICO

California State University, Chico (CSU) is one of the pioneers in the field of distance education. Beginning in 1975, CSU, Chico inaugurated its ITFS/Microwave system. In 1984, it added a C band uplink and a Ku band uplink was installed in 1986. Using all of these technologies, Chico delivers courses and programs not only to its northern California service region but also to locations across North America.

California State University, Chico was established in 1887 as the second Teacher's Normal School in California. One hundred years later, Chico is a comprehensive university offering more than 80 programs in 52 fields leading to both bachelor's and master's degrees. The University is located in a relatively small metropolitan area of about 64,000 people in the northern Sacramento Valley, about 100 miles north of the state capital. The University's service area covers 33,000 square miles (an area the size of the state of Maine) and serves approximately 700,000 people. Providing educational access and opportunity to the citizens of this northern California region in a cost efficient fashion is one of the missions of CSU, Chico.

The first of Chico's three major distance education initiatives began in 1975 when the University, using system-wide funds and federal funding from the Department of Commerce (NTIA), constructed an ITFS/Microwave link between CSU, Chico and the University of California, Davis. The 92-mile link was constructed to enable Chico computer science courses to be sent to Davis for their developing Ph.D. program in computer science. The initial two sites have expanded to the existing configuration of 16 sites throughout northern California. The high school in Yreka, 173 miles north of the campus, is the most distant point on the system. There are approximately 600 enrollments in 20 courses per semester. ITFS for Chico students means Instructional Television for Students. The programs available by ITFS include: Business Administration Minor, Career and Life Planning Minor, Family Relations Minor, GAIN certificate, Paralegal certificate, Political Science Minor, Psychology Minor, Sociology Minor, B.A. in Liberal Studies, B.A. in Sociology and B.A. in Social Science. The courses are taught "live" by regular CSU faculty who simultaneously teach a scheduled class on the Chico campus and are broadcast from a state-of-the-art classroom to students at a distance. The system is one-way video and two-way audio. Although licensed for four channels, only one channel is currently used. The channel is used from 8 a.m. to 10 p.m., Monday through Friday and from 9 a.m. to 3 p.m. on Saturday. Televised classes are broadcast 2,160 hours each year through this northern California network.

Administrative support for distant students is provided in several ways by the Center for Regional and Continuing Education (RCE), which has responsibility for distance education. RCE hires and trains Registration Monitors during the first two weeks of class at all off-campus learning centers. Students are provided registration packets, student information sheets, library instructions, financial aid information, program information, course syllabi, book order forms and other support. Book orders are mailed directly to students at their homes via UPS. RCE insures exam security and also hires proctors for each exam. Marketing is provided through brochures, the extension bulletin and various northern California newspapers.

Technical support for ITFS is provided by the Instructional Media Center. The staff who support the faculty and transmit the courses as well as the technical support for the installation and maintenance of equipment is provided by this unit.

The second major initiative began in 1984 when the University installed a 10 meter C band Scientific Atlanta transmit earth station on campus. In the fall of that year, CSU, Chico offered its first courses leading to the master's degree in computer science, live via satellite, to Hewlett-Packard employees at locations in the five western states of California, Colorado, Idaho, Oregon and Washington. In 1986, Hewlett-Packard donated their Roseville uplink to Chico, which became one of the few institutions in the United States to own and control both C and Ku band uplink facilities. Since 1984, the Computer Science program has been extended to 18 additional companies and into a total of 15 states (Arkansas, California, Colorado, Connecticut, Idaho, Illinois, Iowa, Nevada, New Jersey, Oregon, Pennsylvania, Tennessee, Texas, Virginia and Washington). Five computer science courses are offered live, via satellite, each semester and the program operates on a five-year cycle in order to allow corporate employees to plan their long-term program of study.

Following requests from several of the participating companies, Chico also offers its bachelor's degree in computer science live via satellite. Academic courses taken from other accredited institutions are combined with 14 courses (41 credits) offered from CSU, Chico to allow participating company employees to complete a bachelor's degree from Chico. The third initiative in distance education is Chico's extensive involvement in receiving and broadcasting teleconferences. Since 1984, CSU, Chico has received more than 100 teleconferences and produced and broadcast from the campus more than 71 teleconferences.

California State University, Chico is a national leader in the use of technology to reach distant learners in California and the nation. It has been creative and comprehensive in meeting the needs of individuals and companies when distance prevented the learner from reaching the campus.

■ THE EDUCATIONAL NETWORK OF MAINE

Maine provides an example of a statewide network, developed and operated by the University of Maine System, for the benefit of a large number of educational providers and students. The Educational Network of Maine (formerly named the Community College of Maine, the name was changed by the Board of Trustees in July 1992; hereafter, in this chapter, it is referred to as the network) was proposed to the Board of Trustees of the University of Maine System in April 1986 and approved for implementation in May 1988. The plan was developed by the University of Maine at Augusta (UMA) in cooperation with the University of Maine System campuses, the Technical College System and the State Department of Education. Each of these organizations participate in offering courses and other programming.

The Education Network of Maine began broadcasting 40 courses over two closed-circuit ITV channels to 47 locations statewide in September, 1989. There were over 2,000 students enrolled for that initial semester. Today, the Network interconnects more than 100 locations in Maine to over 65 credit courses and five full associate degrees (Business, General Studies, Social Services, Liberal Arts and Library and Information Technology) and two masters degrees (Industrial Technology and Library and Information Sciences). The Network serves over 4,000 credit students and approximately 20,000 other individuals in meetings, seminars and non-credit courses each year. The network is an integral part of the University of Maine at Augusta, which administers the network on a statewide basis on behalf of the university system. UMA develops and coordinates the course schedule, prepares the statewide course guide mailed to 450,000 homes each semester, produces the television and radio marketing ads, distributes course materials statewide, serves as the statewide bookstore, maintains the network record systems and provides numerous other services. An 800 number is available statewide, 24 hours each day (an answering machine is reached from 10:00 p.m. to 7:00 a.m.) to assist students and interested individuals in reaching the network with questions and requests.

URSUS, the on-line library catalog of the seven university campuses, Bowdoin, Bates and Colby colleges, the Maine State Library, the Maine Law and Legislative Library and the Portland and Bangor Public Libraries, is one example of the academic services available to students statewide. URSUS enables a student at an off-campus center to access a computer terminal and conduct author, title and subject searches of the collections of the major repositories in Maine. URSUS also contains a number of databases of information from other libraries in the U.S., including CARL, the Colorado Associated Research Libraries collection of periodicals. In addition, The Network, through the UMA library, provides periodical searches through INFO-TRAC, on compact disk, which indexes 400 periodicals. INFO-TRAC can search more than five years of issues and the most recent two months of The New York Times. The database is updated monthly. The original articles are available on microfilm and students may request printed copies, which are mailed or faxed to them.

The three channel, statewide interactive television system (ITV) of the network connects over 90 locations in Maine with "live" courses transmitted from a variety of University and Technical College campuses. In Fall 1992, more than 60 credit courses will be offered "live" over ITV and additional courses, with instructors on site, will be offered at the centers. Four associate degree programs (general studies, liberal arts, business and social services) are available statewide to students currently. Courses are broadcast almost continuously on three channels from 7:00 a.m. until 10:00 p.m., Monday through Saturday. The fourth channel, which interconnects the university campuses, is reserved for meetings and non-credit activities during the day and after 4:00 p.m. is used for more specialized classes (e.g. graduate engineering, graduate teacher education). The ITV system broadcasts approximately 250 hours of "live" interactive courses, training, workshops and meetings per week throughout the year.

To connect the seven university campuses, which are located from one end of the state to the other, the ITV system uses fiber optic cable, leased from New England Telephone, as the spine of the system. Off the spine, the system uses point-to-point microwave and instructional television fixed service (ITFS). There are eleven broadcast classrooms available in various parts of the state. Students communicate with the broadcast classroom by auto-dialing telephone. Four students can be handled by the telephone bridge simultaneously.

Fully interactive (audio and video) meetings may also be held between the broadcast locations. The ITV system may either be used statewide or it may be regionalized so that several programs or courses may be offered simultaneously but to different regions. Master Control for the entire system is located at the University of Maine at Augusta. All campuses and off-campus centers are connected by the University System computer network and all 90 locations are connected by an 800 number and fax machines.

The network completed three full years of operation in the Spring 1992, with the sixth consecutive semester of increasing enrollments. The average class size of ITV classes is 117 students and the average age of students taking courses over ITV is 32. Seventy-five percent of the students who enroll in ITV courses are women. Evaluations by students of the ITV system and instructors have been conducted every semester and the results are uniformly positive. In terms of student performance, the network research on course grades documents that the ITV students do as well or better than the students in traditional classes.

With the assistance of an Annenberg/CPB grant, UMA has established a Center for Distance Education. The center focuses primarily on faculty and staff development activities related to the restructuring of courses which integrate multiple technologies to reach distant students.

The Maine experience indicates the speed with which a sophisticated technological system can be planned and implemented on a statewide basis. The success of Maine is a useful model for others in the planning and design stage since Maine addressed virtually every challenge that any new project of this scope will encounter — intercampus rivalries, faculty and staff concern about the loss of jobs to technology, securing the necessary up-front funding to invest in the new technologies, obtaining the approval of the Board of Trustees of the University System for a number of important new approaches for providing educational opportunity and numerous other issues.

■ SIGNIFICANT ISSUES RAISED BY THE USE OF DISTANCE LEARNING

Distance education and the use of related technologies is a topic of great discussion, pro and con, within higher education. There are a number of reasons that this is so.

First, the discussion about the merits of technology and telecommunications to provide greater educational access and curriculum diversity for underserved populations has increased as the national recession has had a significant impact on the reduction of college and university budgets. Colleges are being asked to do more with less. Across the country there have been cuts in faculty ranks at the same time that enrollments have remained steady or grown. Classes are getting larger and there are fewer course choices. Serious questions are being raised about faculty who have little or no teaching load and about the necessity for courses and degree programs with few students enrolled. There are increasing demands for greater productivity—a real strength of distance education delivery systems.

Second, the regional, statewide and

national/international scope of an ITV system with multiple receive locations makes it the most important educational tool to date for reaching large populations quickly and effectively.

Third, distance education and technology are visible and, coupled with their scope, allows for and encourages, public discussion in ways in which traditional, local (or campus-based) education rarely does.

Fourth, the scope and visibility of an ITV system have changed the context of much of the discussion about access and quality. The traditional parameters of that discussion were known and limited before the introduction of technology. Today, the almost unlimited capacity provided by technology (i.e. through interactive television, computers and audio conferencing, etc.) allows us to consider multiple ways to address traditional barriers to education (e.g. geography, distance, time, student numbers, student preparation, course and degree options, student services, etc.).

Fifth, distance education introduces new pedagogy on a large scale primarily because the technical capacity is available to do so for the first time. There are questions raised about the quality of education provided using technology-based approaches when compared to the traditional classroom format with the instructor on-site.

Sixth, it raises questions about our traditional concepts of educational time and place. For the most part, formal education has taken place at a given location at a specific time. This approach has relied on aggregating resources at a central location where the faculty member and students are brought together. On the other hand, technology overcomes the traditional barriers of time and space and eliminates the need to aggregate participants to achieve an economical critical mass of students. The choice of where and when to learn, therefore, shifts to the student rather than remaining under the exclusive control of the institution. With ITV, students are now able to participate in a course at any number of locations. Likewise, the issue of time has changed. People using ITV, whether for credit courses or non-credit continuing education, are finding that the use of videotapes, at their convenience, is a powerful option which allows them to secure the information they need in a fashion that fits better into the lives of busy adults.

Seventh, distance education addresses issues of productivity in instruction and administration and suggests that there are new ways to achieve quality and efficiency that were not explored in the past because the technical capability, at reasonable cost, did not exist. For example, an ITV system allows an institution to deliver many additional course sections using the same number of faculty. Moreover, it is not necessary for each section to enroll a minimum number of students to justify its offering. A critical mass of students is reached by aggregating the students at multiple sites taking the same course.

Eighth, the scope of distance education has exacerbated historically difficult academic issues, such as transfer of courses, transfer of GPA, tuition and fee charges, residency requirements, uniform entry-level testing and assessment, common developmental studies curriculum, etc. Ninth, the limited channel capacity of an ITV system raises inter-campus governance questions about who ought to have the academic authority for the scheduling and actual offering of courses and degrees on a network. Prior to the development of an ITV system, campuses have traditionally had academic responsibility for the geographic region surrounding their campus. Normally, other campuses offered courses within another campus's jurisdiction only by invitation or with their approval. The new electronic technologies (i.e. ITV and computer linkages) make those former geographic boundaries obsolete and irrelevant.

Tenth, an overarching issue relates to the perception that there are winners and losers as significant change occurs. For example, the residents of rural areas who have never had access in their own community to "live" college level courses, broad-based K-12 curriculum, continuing education, workshops, training, public hearings and interactive meetings are suddenly presented with a cornucopia of opportunities. They are clear winners as change unfolds. On the other hand, students on a campus often see the benefits of ITV and other technology as primarily benefiting those at a distance from the campus and, therefore, they often view themselves as losers in two ways: the campus students may feel that they lose their exclusive relationship with the faculty member in the classroom and, in addition, they may resent the questions and participation of the increasing number of students at remote sites across the state. The ultimate trick in public policy terms is to determine how the greater good for the most citizens is achieved.

Eleventh is the issue of accreditation. Approval by one of the six regional accrediting bodies or one of the national professional associations has been an essential component of higher education throughout this century. In recent years, institutions with multiple campuses (some in different accrediting regions) and distance education (whether using off-campus instruction or distance education technologies) have posed new questions for accrediting bodies. The fact that a single institution, using satellite technology, can deliver degree programs nationally raises a variety of questions for regional accrediting bodies, not the least of which is what organization has accrediting jurisdiction over a higher education institution delivering degrees to multiple states or nationally. Many of these questions remain unresolved.

Twelfth, it is uncomfortable for century-old academic institutions to experience fundamental change. Shifting demographics (i.e. fewer traditional students and more adult learners), declining economic support for education, competing public policy agendas and probing questions about the relevance and worth of existing curriculum, degrees and pedagogy have led to difficult times on campuses across the country. The introduction of new technologies to meet multiple public and private needs have exacerbated the discussions about these topics. A new paradigm has developed which, to varying degrees, challenges the comfortable and familiar and presages a new option for doing the public's business in education.

Finally, the emergence of powerful and pervasive electronic technologies suggests that new discussions must address the appropriate educational role and mission of campuses. The era of geographic educational monopolies is ending and by the beginning of the next century, people who were historically place-bound and educationally isolated will be served by an expanding number of institutions using information technologies and modern telecommunications (e.g. broadcast television, microwave, direct broadcast satellite, cable television, computers, etc.) to address the most difficult problems of educational access. This changing technological capacity to serve learners wherever they live or work raises new questions about the need for clearly differentiated roles between and among campuses.

In "Powershift: Knowledge, Wealth and Violence at the Edge of the 21st Century (1990)," Alvin Toffler's last book in his trilogy on change in modern society, he writes that "out of this massive restructuring of power relationships, like the shifting and grinding of tectonic plates in advance of an earthquake, will come one of the rarest events in human history: a revolution in the very nature of power. A 'powershift' does not merely transfer power. It transforms it" (p. 4). Distance education is initiating significant structural change and it is having a growing impact on the entire spectrum of education. Issues that were difficult to resolve in the past — productivity, access and quality — may now be addressed in comprehensive ways that leverage benefits for institutions, learners and other organizations. Distance education and its related technologies — the modern tools of fundamental change — offer a tangible, cost-effective and high quality approach to addressing the educational issues central to the Information Age in the last decade of this century.

Distance education and technology not only have the potential for transferring power and achieving structural change but also for transforming, indeed reforming, the basic academic structure of the world of education.

PBS ADULT LEARNING SERVICE

by Sandra H. Welch, Executive Vice President, PBS

The PBS Adult Learning Service (ALS), pioneered the widespread use of for-credit telecourses by colleges and universities. Since PBS began operating the service in 1981, nearly three million students have earned academic credit for telecourses broadcast by 96 percent of the nation's public television stations. ALS works in partnership with more than half of the nation's 3,500 colleges and universities. Current enrollment for PBS's 60 plus college-credit telecourses – ranging in subject areas from art history to algebra, from business to psychology to writing – stand at more than 300,000 annually, a five-fold increase since ALS's first year.

The typical student enrolled in a PBS telecourse is between 25 and 40 years old, has family responsibilities and holds either a full- or part-time job. According to the Carnegie Foundation for the Advancement of Teaching, this group of working adults, at-home mothers and other adult learners is growing faster than any other segment of American education.

Students enroll in a telecourse through the college of their choice, pay the school the proper tuition and fulfill the course requirements as specified by the local faculty member assigned to each course. Students usually watch the telecourses when broadcast by their local public television stations. This year alone, ALS will deliver 1,000 hours of telecourse programs to stations and colleges.

In 1988, ALS established the Adult Learning Satellite Service (ALSS) to provide colleges, universities, business, hospitals and other organizations with a broad range of educational programming that they can acquire directly via satellite. Organizations with satellite dishes can tape the programs off-satellite when they ar fed by ALSS – a cost effective and convenient way to acquire programs through ALSS. During the 1994-95 academic year, ALSS will offer over 2,000 hours of programming.

The business programming strand of ALSS, The Business Channel (TBC), offers business programs specifically selected for their timelines and cutting edge content.

Through ALS, PBS and its member stations help colleges, universities and public television stations increase learning opportunities for distance learners; enrich classroom instruction; update faculty; training administrators, management and staff; and provide educational services for local communities.

ALS Programs Include:

• Going the Distance is a new educational initiative of the PBS Adult Learning Service and public television licenses in response to the growing population of adults who want to earn a college degree through distance learning. For over 13 years, public television licensees in partnership with local colleges have offered telecourses to over 2.8 million adults. Going the Distance (GTD) takes that effort to its logical next step – development by these local college-station partnerships of an actual college degree at a distance using telecourses and other offerings. Beginning in mid 1994, GTD will launch 20 pilot licensee markets with over fifty participating colleges.

• Telecourses, which can be offered as comprehensive college-credit courses for distance learners. In some cases, the television programs from telecourses can also be used as audiovisual resources for classes and libraries.

• Audiovisual resource programs, including some of PBS's finest prime-time programs, which can be used as supplementary learning resources for classes, libraries and training activities.

• Live interactive videoconferences that give organizations the opportunity to host local conferences featuring national experts on live programs that cover cutting-edge issues in education, business, health and other fields.

Chapter 14

CORPORATE TRAINING

by Patrick S. Portway and Carla Lane. Ed.D.

On October 26, 1992, at the annual TeleCon conference, we presented training via distance learning as the critical component in our nation's economic recovery. In a dramatic departure from its traditional format, the TeleCon conference in San Jose, California, opened with a significant message about video teleconferencing's role in the retraining of America's workforce. TeleCon proposed business television as the key to the education and retraining of 50 million American workers.

Dr. Anthony Carnevale, the chief economist of the American Society for Training and Development (ASTD), presented the economic justification for this massive training effort to equip America's workforce to be productive and competitive in the new world economy. Dr. Carnevale is the author of "America and the New Economy,"S a study by ASTD and the U.S. Department of Labor. This report is described in detail later in this chapter. The primary finding of this study was that the characteristics of this new world marketplace are dramatically different. Increased worldwide competition requires a workforce trained in different skills than our current workforce. Dr. Carnevale's conclusion is that we must undertake a massive retraining effort in this country if we are to remain competitive. According to the studie's finding, the criteria for economic success is no longer simply measured in terms of productivity. While productivity will still be an important quality, variety, customization, convenience and timeliness have grown in importance.

The workforce required to carry out repetitive tasks on a traditional production level is not the same as the kind of workers we need today. Customization, for example, may require adjustments in equipment, technical and math skills the average American high school graduate lacks. The increased concern for quality and worker involvement for continual improvement of products and services requires new supervisory and teamwork skills that were not a part of the experience or training of most of today's workers.

The traditional way America has tried to recover from recession is to reduce interest rates to encourage investment in new plants and equipment. Now there is a growing awareness that far greater results can be obtained by investing in improvements to the workforce. America has the lowest percentage of spending on training compared to all the nations we are required to compete with in the European Economic Community or Japan. This need for greater investment in training has been recognized by all three presidential candidates including Ross Perot.

The TeleCon presentation took this recognized and documented requirement and proposed a solution based upon video teleconferencing and satellite communications.

Electronically mediated instruction, or distance learning technology, is the only feasible way to train this vast number of people by the year 2000. There are just not enough overhead projectors, trainers and meeting rooms to do the job with traditional training techniques.

Two specific success stories were presented at TeleCon to prove the cost effectiveness of distance learning by satellite.

Hewlett-Packard's use of business television and response terminals to deliver corporate training was presented by Willem Roelandts, vice president and general manager of Hewlett-Packard's Networked Systems Group. In his presentations, he showed how Hewlett-Packard recovered its entire investment in its training network in its first training session. We have included more information on H-P's program in this chapter.

What sets Hewlett-Packard apart from other business TV users is how their use of distance learning technology has become such an integral part of their corporate strategy. Essentially, Hewlett-Packard could not introduce products as rapidly as competition requires without satellite training.

The second dramatic example was presented by Dr. Lionel Baldwin, the president of the National Technological University, (NTU).

NTU provides graduate and continuing education for engineers directly into the workplace from over 40 universities throughout the United States. NTU uses the latest state-of-the-art in digital compressed satellite communications to deliver multiple channels of video over a single satellite transponder. Over 2000 courses are delivered to students who might otherwise be unable to continue their engineering education.

The significance of this presentation by three of

the nation's leading spokesmen for high tech delivery of corporate training led TeleCon to implement something long-suggested by professionals in the teleconferencing field. TeleCon uplinked this significant opening program nationally via satellite including the NTU network. For the first time, it was possible for executives and public policymakers across the nation to join the 2000 people in the San Jose Convention Center to hear this opening session of TeleCon XII.

The TeleCon program began to emphasize education and training as our application of videoconferencing five years ago and, Via Satellite magazine, has been one of the publications recognizing the significance of this application as a new market for satellite communications.

Corporate training is a $33 billion dollar a year market in the United States today! If proposed tax incentives and tax credits for increased corporate training are implemented, the American Society for Training and Development estimates that corporate training expenditures by 1995 would grow to $100 billion a year.

The United States Distance Learning Association, which holds its annual meeting every year in conjunction with TeleCon XII, has proposed several public policy initiatives to encourage the use of electronically mediated instruction for training. USDLA has endorsed the ASTD proposals for tax credits for corporate training and is actively supporting Dr. Carnevale in his efforts to expand America's retraining effort.

Patrick Portway closed the national broadcast from TeleCon with the conclusion that "Satellite based training may be the key to America's economic success in the future."

■ RETRAINING 50 MILLION AMERICANS: THE ELECTRONICALLY MEDIATED SOLUTION

Corporate America is spending millions of dollars on remedial training of its workers, much of it at the college level. Eighty-five percent of America's work force in the year 2000 is already in the workplace now. Fifty million of those workers will need to be retrained, and 37 million will need to be trained at the entry level.

At the same time, the U.S. is moving to a global, information-based and communications based economy. The U.S. needs knowledge workers; highly educated professionals who use computers and communications systems to dramatically increase productivity. It's no longer enough to master simple office skills, operate a machine, lift objects or follow orders.

According to the National Center on Education and the Economy (NCEE, 1990), at least 70 percent of the jobs in America will not require a college edu-

cation by the year 2000. These jobs are the backbone of our economy, and the productivity of workers in these jobs will make or break our economic future. No nation has produced a highly qualified technical workforce without providing its workers with a strong general education. "But our children rank at the bottom on most international tests — behind children in Europe and East Asia, even behind children in some newly industrialized countries (NCEE, 1990)." Unlike virtually all of our leading competitors, we have no national system capable of setting high academic standards for the non-college bound or of assessing the achievement against any standards. "America may have the worst school-to-work transition system of any advanced industrial country (NCEE, 1990)."

In the future, more jobs will require at least a college degree and the ability to creatively solve problems in software-defined working environments. The American post-secondary education and training system was never designed to meet the needs of our front-line workers. Without standards set by employers, students can not be sure that there is a market for the courses they pursue. "Education is rarely connected to training, and both are rarely connected to an effective job service function (NCEE, 1990)." The NCEE reports extensive occupational preparation programs in central Europe, combining general education with work site training, which provide foreign employers with high skilled, work ready youth and offer young people a smooth transition from school to working life. Employers, knowing that students who graduate from the system have the skills they seek, are glad to hire them. Students, seeing a direct relationship between school and work, are motivated to learn.

The U.S. graduates 700,000 functionally illiterate students nationwide every year. Another 700,000 drop out. We spend the same amount on each student as Japan does, yet Japan graduates 98 percent of its high school students and the United States has a dropout rate of 20 percent for the 16- to 19-year-old population, and which soars to more than 50 percent in the inner cities. Unlike the central European systems, the Japanese emphasize general education. Although vocational schools are available to Japanese students, the majority complete high school general education programs. Many Japanese companies hire for life, and as a result Japanese employers tend to place greater emphasis on a student's general learning ability and performance in school. Specific job related skills are provided by the company throughout the individual's working life. Substantial training may last for years. Virtually all Japanese students are handed over from a school family to a work family in a seamless transition requiring little external assistance.

As employers discover basic skills deficiencies in their workers, they need assistance to establish training programs to meet specialized needs. Little practical information has been available on how to set up workplace basic programs. The American Society for Training and Development (ASTD) and the U.S. Department of Labor (DOL) undertook a major project to research this problem which resulted in the "ASTD Best Practices Series: Training for a Changing Work Force."

Through technology, participative management, sophisticated statistical quality controls, customer service, and just-in-time production, the workplace is changing, and so are the skills that employees must have to be able to change it. Many current workers do not have the basic skills which are essential to acquire more sophisticated job-related skills. The shrinking 16-24 year-old age group will force employers to hire less qualified entry-level workers.

This volatile mix of demographic, economic, and technical forces is propelling the U.S. toward a human capital deficit which threatens the competitiveness of economic institutions, and acts as a barrier to individual opportunities.

Electronically mediated instruction can be effectively applied to the massive need to retrain 50 million American workers. New technologies do not become a part of business until a key application is identified. Training is a key application of electronically mediated instruction. Surveys of existing business television networks document training as the primary application of existing networks accounting for 80 percent of their use.

Video compression, audio conferencing and computer conferencing provide low cost training sessions and in many situations will allow delivery of training and education to the desktop. While education and training have traditionally not utilized high-tech solutions, this is changing. Innovative leaders like IBM, Tandem Computer, Hewlett-Packard, and others have proven that technology is an effective means to train employees and customers quickly, cost effectively and with greater access than any other alternative. Technology and distance learning techniques have been perfected to the point where they are ideally positioned to play a critical role in responding to this vital national crisis.

■ TECHNOLOGY USE IN TRAINING

In a 1990 study ASTD forecast major growth in distance learning technologies for training. ASTD conducted a survey of 200 executives from Fortune 500 companies, asking them about technologies they currently use for training. The survey clearly showed that technologies for training were among the fastest growing areas in the training market. While many of the companies used computer and interactive video

for training, less than one third of the 153 firms that responded indicated that they use videoconferencing for training.

In 1990, 54 percent of the companies surveyed anticipated using video teleconferencing for training in the future. Chemical/allied products and retail industries reported the least use of teleconferencing, but planned the greatest increase in use. Twelve percent of the panelists in the chemical/allied products industry said they used teleconferencing for training at that time. Fifty-nine percent of the non-users said they expected to begin using this technology over the next one to three years. Nineteen percent of the retail companies reported some current use of teleconferencing and projected an increase to 44 percent.

The 1990 study showed that computer-based training was the most widely used technology for training, but that the use of all technologies was going up (Lane, 1990). The survey found that 81 percent of the respondents used some computer-based training and 93 percent planned to use it during the next one to three years. A majority of the respondents rated computer-based training as the most cost effective technology. The greatest use for computer-based training appeared to be in manufacturing, where 86 percent of the companies reported some use. While retail companies reported the least use at 69 percent, almost all retail companies in the survey said they planned to use computer-based training in the next one to three years. Approximately 50 percent of the companies reported some or substantial current use of interactive video, with public companies showing more use than private companies (57 percent versus 44 percent).

In mid-1991, ASTD released new figures and confirmed that the training of the American workforce is going to have to become a priority in the next ten years if the U.S. is to keep up with the growing world economy.

In the 1991 study, ASTD found that 42 percent of the workforce, or 49.5 million workers needed additional training during the next decade. The number was significantly higher than any released to date, and it included those who will need additional training and will not get it under present practices to keep up with the demands of their jobs.

Of the 49.5 million, there are four categories of workers who will need training.

Technical and Skills Training: Technical workers include scientists, doctors, engineers, technicians and technologists (in both manufacturing and health care), and craft workers (concentrated in the construction trades).

Skilled workers are those who will lose their present jobs with the introduction of technology and need retraining including machine operators, assembly workers, transportation workers, mechanics, repairers and precision production workers (e.g., tool and die makers).

Sixteen million of the 49.5 are workers who will need skills and technical training.

Executive management and supervisory training employees include the decision makers in the workforce such as executives, managers, and supervisors.

Executive level employees and those who will need management or supervisory training account for 5.5 million of the 49.5.

Customer service training will be needed for those employees who have contact with customers. Often entry-level positions have customer contact including sales occupations, fast food cashiers, retail store workers, receptionists and bank tellers.

Employees needing customer service training add up to 11 million. Basic skills training will be needed for those employees who are typically low paid and low skilled. Skills needed are narrowly defined to be basic reading, writing, and math skills. Job classifications include cooks, maids, taxi-cab drivers, helpers, laborers, handlers, garbage collectors, service station attendants, data entry clerks, and file clerks. Seventeen million employees will need basic skills training such as basic reading, writing and mathematics skills.

These totals do not include the approximately 37 million workers who annually need entry level or qualifying job training.

ASTD predicts a workforce crisis in the United States because more than four out of ten workers who are on the job are not being trained to do the work that today's economy demands.

Training solutions are costly. The estimated cost to close the training gap is $15 billion per year in addition to the $30 billion presently spent yearly on corporate training. It is estimated that $475 billion is spent yearly for capital improvements to plant and equipment, and by comparison the $15 billion for human resource improvement while significant, is relatively modest. ASTD says that it is well documented that the gains in productivity from workplace learning exceed the gains from capital investment by more than two-to- one.

While it is clear that a sizable investment is needed, it is also clear that technology will be a part of that investment, including all forms of electronically mediated instruction.

Video, either one-way with interactive audio or two-way, face-to-face, allows trainers to reach learners with a clear and focused message. New technologies such as digital compression and low bandwidth broadcasts have made video options feasible both in cost and convenience.

Audio interaction also plays a very important role in the process of learning over long distances. Learners call in their questions or answers to a trainer via a phone line. Computer conferencing can be used for additional questions, sending assignments, and involving others in the discussion.

In 1992, ASTD raised the figures for the cost of training to $33 billion a year. ASTD reported that most of the training is carried out on a face-to-face basis with little more technology than an overhead projector. Because of the changing workforce requirements and the failure of our education system to provide the necessary skills required in the new economy, a massive influx of funding will be required to upgrade the workforce. ASTD has proposed tax credits and tax incentives to increase corporate training. If any of these proposals should be approved by law, expenditures for corporate training could reach $100 billion dollars a year.

All of the surveys by ASTD have found significant increases in the use of teleconferencing, computer aided instruction, interactive laser disks and other technology in corporate training. However, for a number of reasons it still represents a small percentage of the total potential use for training.

New technologies which are making significant inroads in training include multimedia and audiographic systems. Trainers can easily combine graphic and computer data element with their video transmission providing valuable visual for learners in distance locations. In distant classrooms, learners can provide feedback, ask questions, and take tests and quizzes via audience response terminals. Each student is equipped with a numbered terminal and is able to answer questions with coded keys or signal the trainer. Trainers receive a signal on their monitor and know which learner in a specific location has a question or needs help.

Other accessories to video systems such as touch screens and graphics tablets combine with the basic elements to provide better and more understandable presentations. Computer conferencing allows the learner to have access to the trainer 24 hours a day—even after the class is completed.

Training at a distance is as effective as training face-to-face according to a number of studies. IBM and NEC jointly developed an Interactive Satellite Education Network (ISEN) in the 1980s. Some of the innovations introduced in this corporate training network (compressed video and response terminals) are only now being widely applied in education. IBM has realized millions of dollars in savings over face-to-face training without any loss in training effectiveness. In a study conducted by IBM over their ISEN video network, multiple classes in several courses were conducted both over the network and in person. In the first course, learners over ISEN had an average grade of 1.5 percent higher than those who took the course in person. For the second course, those who took it in person had a one percent higher average score than

ISEN students. But in the final course tested, learners who had taken the course over ISEN, had a seven percent better course average grade than those who had taken it face-to-face. IBM officials are convinced that training employees over video is at least as effective if not more so than training them in person.

Hewlett-Packard's cost savings have been significant, but their move into distance learning was part of a new corporate strategy rather than a cost cutting effort. In order to accelerate the development of new products, the interval between releases of new products had to be cut dramatically. Field engineering was required to learn more in less time. Learners could no longer come to class, the class had to go to them. Distance learning became an integral part of Hewlett-Packard's new competitive strategy. Other companies are also convinced that distance learning works including Federal Express, Domino's Pizza,

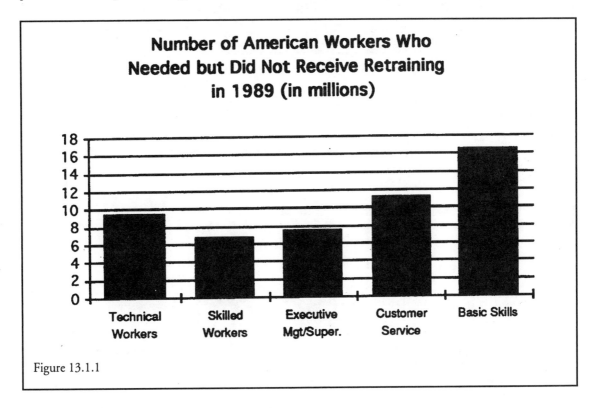

Figure 13.1.1

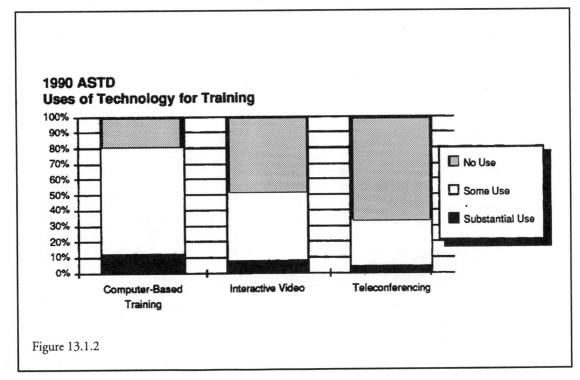

Figure 13.1.2

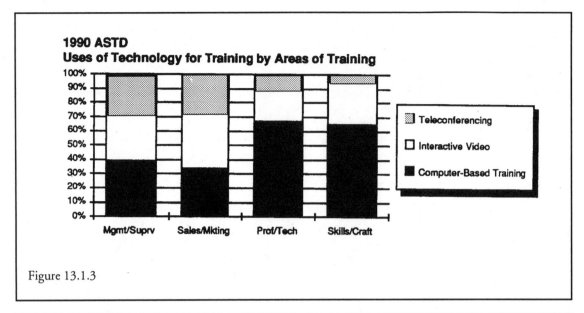

Figure 13.1.3

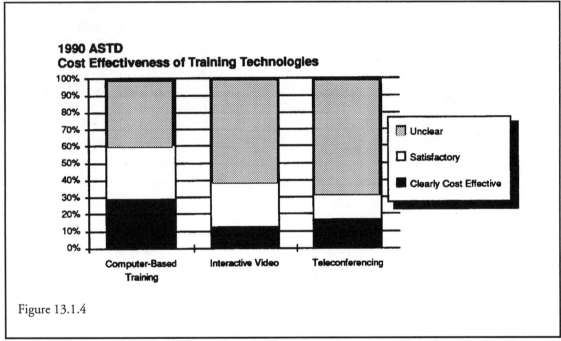

Figure 13.1.4

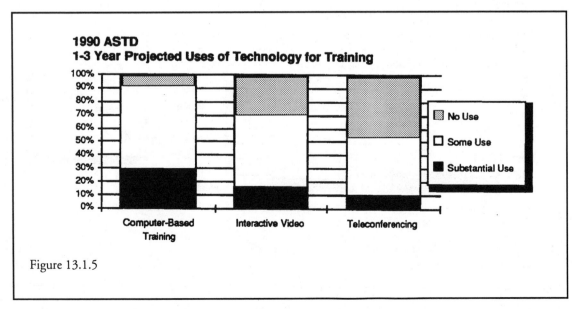

Figure 13.1.5

Xerox and General Motors Corporation.

General Motors started the GM Satellite Network in 1984, but network programming has remained ad hoc to fit the need for a wide variety of applications. In 1990 General Motors and the UAW joined forces to establish the UAW-GM Television Network. The network offers 12-15 hours of programming monthly to 300,000 hourly employees in 150 U.S. plants.

With the technology available, American business should be able to meet the rigorous training demands that lie ahead in the next years. However, technology alone will not solve the problem. The answer lies in knowing the content that needs to be taught, when it needs to be taught, and to whom it should be taught. Once these critical questions are answered, the next question involves knowing what technology is available and how to use it best so that it is cost effective. We are still not reaping the benefits of technology as a nation and we are not using technology as effectively as it might be used to train employees.

■ ASTD/DEPARTMENT OF LABOR STUDY

"American companies are lagging behind, struggling to compete with a workforce that is inadequately prepared to do the job demanded by our rapidly changing business and technological environment," according to Anthony Carnevale, Ph.D., ASTD chief economist and principle ASTD/DOL study author. He contends that America's competitors in other countries commit significant resources to building a well trained workforce. "Productivity used to be the only competitive standard, but the new economy introduces a whole new set of standards to add to productivity: quality, customization, variety, timeliness and convenience. America's future ability to compete is dramatically dependent on our ability to prepare our workforce to meet these new standards."

The ASTD/DOL study identified 16 basic skills (Carnevale, 1990) that the workplace of the future would need in the employee of the future. The skills that employers need include reading, writing and computation (mathematics), knowing how to learn, problem solving, personal management, and interpersonal skills as well as the abilities to conceptualize, organize, and verbalize thoughts; resolve conflict; work in teams and share leadership. The research confirmed that there is an undeniable link between work force basics and the competitive life cycle of any new strategy, technology, product or service deficiencies in basic skills undermine the cycle and cause delays, defects, and customer rejections. The research revealed that the most effective methodology to provide training in workplace basics is the applied approach, which links learning outcomes directly to job performance.

Carnevale (1991) conceptualized in his book "America and the New Economy," how America would remain competitive in the New Global Economy. Carnevale makes the case that U.S. companies can leverage an investment in training into a competitive advantage. Human resources emerged as the leading-edge competitive tool because employees had not kept up with the technology required for the global economy. We have reached a stage in which we need a workforce that knows more to be able to do more. Brain power is rapidly becoming a critical resource.

"The standards of competition have changed and work itself is changing as a result, but the way most organizations train people for work has not changed," Carnevale contends. "It is as if they are trying to move from old-fashioned assembly lines to just-in-time production without the benefit of just-in-time learning."

Most reports state that a full one-third of Americans cannot read or write English with the competence of an eighth grader. Computer illiteracy figures are staggering. Mathematics and computational skills are poor. One-half of all 18-year olds have not mastered basic language, mathematics or analytic skills. "We need workers with skills that will allow us to be competitive in the next century," according to Sam Ginn, Pacific Telesis chief executive. Research by ASTD confirms that "it is the lower half of the workforce that is not well prepared and this is where America is losing the competitive race."

The basic tool that could change this situation is an individual's knowledge of learning how to learn. Carnevale's research clearly showed that Americans did not know how to learn, even though it is the foundation skill for all other skills. According to a 10-year plan for IBM's employee training, "above all, education will be self-directed shifting control of the learning to the student with distance education technology playing a key role in supporting this vision."

A secondary benefit of learning how to learn is that it empowers the learner to become a self-directed learner. The self-directed learner is capable of identifying a deficiency, finding resource materials that will provide the information, and conduct his or her own study based on methods specifically directed at his or her learning style. As the length of learning time is shortened and the act of learning becomes easier, learning becomes less frustrating for the learner. As the frustration decreases, learning becomes less of a burden and can become enjoyable. The self-directed learner can also alleviate the logjam of learners waiting for trainers. Self-directed learners are capable of directing their learning with minimal guidance from an instructor and also benefit greatly from group study that is not facilitated.

The corollary to employees learning how to learn is to train the trainer to facilitate and using learning style methods. Many training programs use people who are content experts as trainers; yet many content experts have no formal training in adult education methods. While it is not suggested that content experts be dropped from any training program, it would be a clear advantage for the students in the class to have an instructor who is a trained adult educator as well as a content expert. It would also be a clear advantage for the training function to have all faculty members trained in adult education methods which value facilitation, learner motivation, and self-directed learning.

As learners move from being dependent learners to independent or self-directed learners, another useful method is collaborative learning. Sometimes called the leaderless group, collaborative group learning is a natural grouping for organizations that are collapsing hierarchy. Group learning lends itself to groups composed of learners who are geographically dispersed but have a need to learn the same material. A training facilitator can guide the group in its work as well. Audio, video, and computer conferencing technologies are well suited for the collaborative learning group. Leaderless group skills also transfer well to the workplace which is based on collaborative teams.

From the employer's perspective, the skill of knowing how to learn is cost-effective because it can mitigate the cost of retraining efforts. When workers use efficient learning strategies, they absorb and apply training more quickly, saving their employers money and time. When properly prepared, employees can use learning-to-learn techniques to distinguish between essential and nonessential information, discern patterns in information, and pinpoint the actions necessary to improve job performance.

Employers, particularly those dealing with rapid technological change, see the learning-to-learn skill as an urgent necessity for their workers. Productivity, innovation, and competitiveness all depend on developing the learning capability of the work force. Machinery and processes are transferable between companies and countries, but it is the application of human resources to technology and systems that provides the competitive edge.

The leaner, smaller work force will heighten the importance of training and adaptability. If present trends continue, the gap between need and capability will expand and will prevent the U.S. from increasing its competitive edge in some industries and retaining it in others. The company that is successful is a very different breed says Harvard University professor, Rosabeth Moss Kanter. It is designed so that "the work force is multi-skilled and multi-trained and can be redeployed quickly as issues change."

■ THE WORKFORCE CHALLENGE

That U.S. business faces a work force crisis has been accepted. The question is what can turn around the problem in the shortest time? Work-based training would meet the challenge of increasing the skills and knowledge of employees and bring with it increased productivity and a respectable position in the global economy.

ASTD points out that for companies that face the following facts, the competitive edge will grow sharper:

1. The average 1.4 percent of payroll that U.S. companies invest in training reaches only 10 percent of the work force. Japanese and European-owned companies based in the U.S. spend three to five times more on employee training than American companies.

2. The K-12 education system will not be able to help most of the people who will be working in the year 2000. Ninety percent of the people who will be working then are already working now.

3. Within the next ten years, 74 percent of Americans working today will need retraining. In manufacturing alone, as many as 15 million jobs will require different skills than those required today.

4. The surplus labor force of a few years ago has dwindled to almost nothing. By the year 2000, there will be too few trained and knowledgeable workers to satisfy the nation's economic needs.

5. The process of bringing innovations to the marketplace takes too long in the U.S. The Japanese can get a new car to market in 40 months; it takes Americans five years. The United Kingdom is able to get pharmaceuticals to market in half the time it takes American companies. American business has the technology it needs to be more productive, but it cannot afford to wait for the school system to produce a new generation of workers capable of capitalizing on new technology. American business can, and must, afford its human capital investment ... its investment in training.

Work-based training would meet the challenge of increasing the skills and knowledge of employees and bring with it increased productivity and a respectable position in the global economy. Importantly, training directly helps employees to develop the vital skills that are needed ... if the training is based on the foundation skills of learning how to learn. Training directly helps employees develop the needed vital skills ... if the training is based on the foundation skill of learning how to learn.

ASTD reports that "the investment in learning on the job has contributed more than half of all increases in the nation's productive capacity over the last 40 years. That is almost three times more than the investment in machine capital has produced. The most globally competitive companies are already making the employee training investment. Through their investment in human capital and strategic development, these companies have been able to build a work force that can make more effective use of technology, develop collaborative and efficient managers and employees, and are more readily able to solve problems through creative solutions that capture the imagination of the marketplace. In 1984, IBM restructured its education function, giving it a central role in corporate strategies for growth. Jack Bowsher, then the director of education for IBM, led a two-year study in preparation for the restructuring. He says, "The important question is, does your organization have a training program that's adequate for preparing a competitive workforce? Do you have a workforce with some training or competitive workforce? There is a difference."

The investment in human capital creates an additional educational by-product. Training that is based on the foundation skill of learning how to learn will increase the work force's ability to quickly learn as the job changes. It is the basic tool for lifelong learning. David T. Kearns, deputy secretary, U.S. Department of Education, states: "People have to continue learning new techniques and new ways all the time. Having a well-educated and empowered work force is going to change how companies are run." In recent years, Xerox spent $125 million just on training to achieve its strategic goal of quality. Kearns says that training should integrate the strategy, direction, and vision of the company, and the skills and behavior that people need in order to get the job done.

Training makes a real difference for companies committed to building quality, improving customer service and capturing greater market share, according to ASTD. "To succeed in today's competitive world economy, the U.S. must achieve leadership in technological innovation and leadership," says CEO James E. Burke, Johnson & Johnson. "Key to achievement of these goals is a renewed commitment to a vastly improved education and training system." CEOs in all U.S. companies can do something within their own companies to create positive change for their employees and their businesses.

Just as the bureaucracy of the American educational system has been criticized for producing a poor educational product, the same criticism has been leveled at the corporate commitment to training. Carnevale concludes that most employers' current commitments to training and development are insufficient. While the $33 billion that is spent annually on training (about 1.4 percent of the national payroll) is enormous, the impact has not been. The funding per employee is not enough to conduct the massive retraining to maintain America's competitive position in the global economy past the year 2000.

Employers who are committed to training do substantially more, with training investments averaging anywhere from three to six percent of payroll. To upgrade their skills, only 10 percent of employees receive employer-provided formal training and 14 percent receive informal training from their employers. Research also shows that people who are trained formally in the workplace have a 30 percent higher productivity rate after one year than people who are not formally trained. Formal training is four to five time more productive than informal training and three times more productive when done in-house. Those who receive formal training on-the-job also enjoy an earnings advantage of 30 percent or more over those who do not.

Between 1929 and 1989, learning-on-the-job accounted for more than half the productivity increases in the U.S.; learning was twice as important as technology in boosting productivity and twice as important as formal education.

The U.S. productivity rate is still the world standard, but is increasing much faster among our competitors. If current trends continue, Japan will surpass the U.S. in 2003. The U.S. is already losing the productivity race to Japan in chemicals, steel and other primary metals, electrical machinery, and transportation equipment.

Training is not evenly distributed. The majority of training is limited to white collar, technical elites. Of the nation's 3.8 million companies, only one half of one percent do 90 percent of the training, yet 80 percent believe their employees have adequate skills.

The frequency with which mediated instruction is utilized to train employees is low given the enormous need to retrain millions of employees. Too frequently, the training function is conducted by a content expert who is not an adult education facilitator. Trainers are no more experienced with the selection, use and evaluation of electronically mediated instructor than their educational counterparts. This is particularly true for those who work in training because of their content expertise. Too frequently the class is conducted in the traditional manner and without audio-visual equipment. Too frequently, the concepts of learning style, continuing learning or learning how to learn are not addressed. Too frequently, even the term "distance learning" falls on the ears of trainers who do not understand the term's definition or the efficiency it can bring to the training function. The U.S. work force is not ready to meet the new competitive standards - quality variety, customization, convenience and timeliness. It is not adequately prepared to take advantage of new technology, especially information technology. It has not learned how to achieve results through the work of teams and does not have sufficient skills to leapfrog ahead of its competitors to deliver on the new competitive standards. It has not learned how to shift its focus to the customer, how to engage in partnerships with its competitors, or how to think in terms of global markets. In short, it has a great deal to learn in a hurry.

■ ELECTRONICALLY MEDIATED INSTRUCTION

One section of the ASTD study demonstrates the important role that distance education will play in solving the problem as it describes the training and education provided through distance education technologies for technical, engineering, scientific and health workers. The enormity of the basic skills problems strongly suggests that a basic skills training network could be used by many corporations to provide the skills. The fifty million Americans who need to be retrained now can be reached through distance learning. One content expert who is also a master facilitator can reach thousands of employees. Distance learning is cost effective for the organization because the cost is spread over many employees rather than just a few. It is efficient because the learner does not have to travel great distances to receive the benefit of learning from an expert. Study guides are written to guide learners through the class with exercises and activities that emphasize content application. Interaction is built into the class through two-way audio, phone lines, or student response keypads. It is an effective way for employees to learn because programs are written by instructional designers who build an instructional package to reach all learning styles - whether the student learns by seeing, hearing, reading, interacting, or doing.

High-technology delivery methods — such as satellite networks and interactive video — make it faster and more efficient for employers and providers to work together to train employees. Satellite networks, are being used to aid employers in sharing programs and information with each other and allow universities to send programs easily and directly to employer sites.

As a result of this technology, the National Technological University (NTU) was founded, with courses accessed by employees solely through satellite networks. NTU is a fully accredited, advanced degree awarding institution that delivers its programs by means of satellite delivery systems transmitted from universities; it offers programs leading to ten different master of science degrees. NTU's board of directors is composed of industry representatives who provide consultation on curriculum design and development. The courses are in-depth and focus on the cutting edge of technology. NTU has converted to a digitized format that is transmitted over satellite.

Texas Instruments, a manufacturer of electronics products, including peripheral computer parts, with revenues exceeding $4 billion, and Du Pont, a diversified chemical and energy company with profits exceeding $27 billion, are two companies that offer satellite courses to their employees and to employees of other organizations. Both offer day-long seminars on development in electronic technology to other companies via satellite. Sharing courses through satellite networks makes advancements available to other organizations but also means that the costs of training can be shared as well.

There are a number of high-technology course delivery options that are already in place throughout the U.S. Networks at Federal Express, Xerox, General Motors, Domino's Pizza, and Hewlett-Packard serve as models which can be adapted by others.

■ AMERICA AND THE NEW ECONOMY

America's economy was built on mass production systems. The competitive standard was productivity - producing high quantities of goods and services at low prices. We were able to do this faster than anyone else. In the New Global Economy, the U.S. can no longer compete on the basis of mass production alone. Consumers now demand quality, variety, customization, convenience and timely products and services.

Globalization

It is increasingly clear that our ability to stabilize domestic markets is no longer enough (Carnevale, 1991); the New Economy is global and the tendency is for global economic events to affect and impinge on the domestic economy. Other nations face many of the same obstacles, but we move into the New Economy with the additional burden of our past success. Old and once successful habits die hard. The U.S. set the standards in the old economy, but we labor on toward the New Economy dragging the dead weight of our past industrial successes behind us. We cannot ignore the growth of the international competitive market. U.S. corporations now shop the world for labor, resources and business partners. The presence of foreign companies and money in the U.S. will continue to increase. U.S. companies must rise to the needs of this expanded market–or someone else will.

Technological Advancemnets

If U.S. education and training is viewed as an intellectual house of cards that has been identified as needing change through gradual incremental processes, it can be seen as collapsing from the weight of real-world competition rather than an assault on the established order. Carnevale states that "the interplay between theory and practice is one factor that sets a deliberate pace for technical change. The state of the technical art is almost always ahead of the technical practice because there is an inevitable hiatus between the acceptance of new ideas and their embodiment in new technology."

Carnevale notes that the role of change agent is also hampered because once invented, "new technologies are not immediately adopted"... as "fear, superstition, vested interest and instability give the past and present a powerful hold on the future. Legislative conservatism is currently preventing the easy adoption of learning technologies for education, but business also has an administrative equivalent which prevents its use of learning technologies. "The inability to swallow the sunken cost in a current technology and its accompanying infrastructure is a persistent cause of the competitive edge lost to those who are willing to push technical frontiers in mature industries," Carnevale warns.

In the most fundamental sense, what distance learning systems try to do is to connect the trainer with the learners when physical face-to-face interaction is not possible. Just as highways move vehicles, telecommunications systems carry instruction, moving information instead of people. The technology at distant locations, including computers, video cassette recorders (VCRs) fax machines, television monitors cameras and the telephone, are critically important. Together, these technologies affect how interaction takes place, what information resources are used, and how effective a distance learning system is likely to be.

Many technologies are being used to provide education over a distance which include satellite, fiber optics, microwave, the public telephone system, and coaxial cable. Any of these technologies can be interconnected to form "hybrid" systems. No one technology is best for all situations and applications. Different technologies have different capabilities and limitations, and effective implementation will depend on matching technological capabilities to educational and training needs.

The technologies for accessing, storing and manipulating information have more impact on the distance education experience than the technologies for transmitting signals. Personal computers, display technologies, optical memory systems, facsimile (fax) machines and graphics scanners expand the use of information and resources at distant location. Future developments in transmission, processing, and storage technologies promise even greater capabilities and benefits for training at the same or lower costs. Advances in digital compression technology will greatly expand the number of channels that can be sent over any transmission medium, doubling or tripling channel capacity. Telecommunications systems may also enable new and different uses beyond the traditional training setting. Public education systems can also serve the needs of adult learners, continuing education at home or in the workplace and the community. Educational institutions want to build partnerships with business and shared telecommunications systems may be a solution, particularly for small businesses.

The base of telecommunications infrastructure available for distance education is wide and expanding, giving schools an opportunity to utilize existing local resources and forge innovative partnerships. Local, regional, and state distance education efforts can be linked with telecommunications networks operated by colleges and universities, local business, public broadcasting stations and state governments. Increasingly, the private sector, including the telephone and cable companies are becoming active in helping schools expand their teaching and learning opportunities, while helping themselves.

New Entrants into the Labor Force by the Year 2000

Today, the labor pool for entry level employees is shrinking. History has encouraged us to believe that educated and trained workers are abundant, but demographics will play a critical role in dictating the priorities of corporate America. The overall size of the work force will decline especially at the entry-level, as the baby boomers move into older age. The rate of labor force growth will be slower than during the past 12 years. A smaller work force must be more skilled and use more technology simply to maintain output.

There is an increase in work force diversity as we enter the New Economy. Every group from rank-and-file workers to senior management will be more diverse in race, culture, and gender than ever before. To make this multicultural environment productive, managers need to manage differently. Managers and employees need to learn new skills and how to deal with conflict between groups.

The New Competitive Standards

Variety: Consumers demand variety. In the last 10 years, the number of items carried on supermarket shelves rose from 12,000 to 24,000. There are 271 kinds of cereal and 16 kinds of noodle soup. Americans choose among 572 different models of cars, trucks and vans. Consumer banking has expanded from six basic services to over 100.

Customization: American organizations must have the ability to customize products and services for the consumer. Both the Japanese and the Italians are heading toward a system in which a consumer can specify fabric, style and size to receive customized apparel in a matter of weeks. Financial services are also being customized with the help of information technology.

Convenience: More and more consumers are able to afford convenience. Built-in convenience is found in products and services such as automatic teller machines and remote controls. Convenience in the form of successful customer relations can also be a powerful selling tool. Unsuccessful relations, on the other hand, can prove fatal. It is estimated that a dissatisfied customer will relate his or her unhappiness to ten other people.

Timeliness: The United States must respond faster to market needs and dramatically shorten product cycle times. The early bird will get the market share in the New Economy. America still leads in the development of technology, but is being caught by other countries and is often beaten in the commercialization of the technology. Americans take about five years to move from auto design to market. The Japanese take only take three and one-half years. The United States takes four to fives years to design and build a new blast furnace. Japan takes about three years and Korea takes two years. In the apparel industry, it takes most American companies 66 weeks to get from fiber to finished garment. Many European and Asian companies reach the customer in 23 weeks; at least one Japanese manufacturer hopes to reduce the time to a few weeks.

■ ORGANIZATIONAL REQUIREMENTS OF THE NEW ECONOMY

The new competitive standards affect organizational structures, requiring a move away from top-down systems and toward more flexible networks and work teams. Technical changes result in new work processes and procedures. These require constant updating of employer-specific technical knowledge. In a world of rapid change, obsolescence is an interminable danger. As technology replaces more of the hands-on work, more employees will be dedicated to service functions where they will spend more time face-to-face with co-workers and clients.

Organizational formats in the New Economy require more general skills. Interpersonal skills, communications skills and effective leadership skills are required by more and more non-supervisory employees. Managers in the New Economy relinquish control of work processes to work teams and will need to provide integration through leadership and monitoring.

The requirements for basic skills will increase and deepen because of the growing complexity and scale of jobs performed. Emphasis will shift from specialization to a broader range of competencies. In the past, the work force was considered a cost to be controlled. In the New Economy, it is an asset to be developed. Our international competitors have already recognized the value of the work force and the impact of training on economic growth. Sixty percent of German youths get three years of formal apprenticeship training in the workplace. France, Ireland, Sweden, Korea, Japan and others all have national workplace training incentives. Japanese auto workers get 3.5 times as much training as U.S. auto workers.

A whole set of skills and a wider range of knowledge is required in the New Economy; training becomes a critical tool to enable U.S. organizations to survive and compete. Moving an organization toward the new competitive standards requires a work force with solid grounding in hard competencies and job knowledge. "Soft" skills, the ability to interact and influence others, take on great importance. The ability to take responsibility for the organization and its goals will be required. Education and experience become more important in getting and keeping jobs. Workers are valued more for skills and less for organizational time and loyalty.

Training makes a demonstrable difference in the ability to meet the new competitive standards. Its impact can be measured at several levels - the economy, the organization and the individual. The U.S. companies that have made significant investments in their work force have enjoyed increased economic success. ASTD projects that 50 million U.S. employees (42 percent of the workforce) need training right now and are not getting it; 16 million will need skills and technical training; executive level employees and others needing management/supervisory training; 11 million need customer service training; and 17 million need basic skills training (reading, writing and math skills). Annually, an additional 37 million will need entry level or qualifying job training.

■ THE WORKFORCE WE HAVE VS. THE WORKFORCE WE NEED:

The New Skills

The ASTD study showed that through technology, the workplace is changing, and so are the skills that employees must have to be able to change with it. The research confirmed that there is an strong link between work force basics and the competitive life cycle of any new strategy, technology, product or service. Deficiencies in basic skills undermine the cycle and cause delays, defects and customer rejections. The research revealed that the most effective methodology to provide training in workplace basics is the applied approach, which links learning outcomes directly to job performance.

Increasingly, employers have been discovering that their work forces needs skills that seem to be in short supply, skills over and above the basic academic triumvirate of reading, writing, and computation. The skills that employers want include problem solving, personal management and interpersonal skills as well as the abilities to conceptualize, organize and verbalize thoughts; to resolve conflict; and to work in teams - all of these skills are critical but often lacking.

Basic workplace skills are of interest because rapid technological change, participative management, just-in-time production, and other workplace innovations have created a demand for more flexibility, adaptability, and a higher base level of skills from all workers, including those at the non- supervisory level. While it is recognized that a percentage of Americas workers have always done well in the workplace despite skills deficiencies, it is apparent that future success will be illusory for many workers if they continue to be ill-equipped in a broad spectrum of basic workplace skills.

Some of the key findings reveal that in addition to reading, writing and computation, the thirteen other skills that employers have identified as basic to success range from learning to learn to shared leadership. There is an undeniable link between work force basics and the competitive life cycle of any new strategy, technology, product, or service. Deficiencies undermine the cycle cause delays and defects, and customer rejections.

The most effective method to provide basics training is the applied approach, which links learning outcomes directly to job performance. Using this approach, employers can fill in employee skill gaps and build individual competence in workplace basics. There are 16 basic skills which employees must have if U.S. companies are to meet the new global competitive standards in the New Economy.

The Sixteen Basic Skills

1. Foundation Skill: Learning how to learn
2. Reading Competence
3. Writing Competence
4. Computation (mathematics) Competence
5. Communication - Listening (interpersonal skill)
6. Communication - Oral (verbalize thoughts) (interpersonal skill)
7. Adaptability: Creative Thinking (and conceptualize)
8. Adaptability: Problem Solving (and organization)
9. Personal Management: Self Esteem
10. Personal Management: Goal Setting/Motivation
11. Personal Management: Personal/Career Development
12. Group Effectiveness: Interpersonal Skills
13. Group Effectiveness: Negotiation (resolve conflict)
14. Group Effectiveness: Teamwork
15. Influence: Organizational Effectiveness
16. Influence: Leadership (and shared leadership)

Learning to Learn

Learning is an integral part of everyday life at work. The skill of knowing how to learn is a must for every worker and is the key to acquiring new skills and sharpening the ability to think through problems and to surmount challenges. It opens the door to all other learning and facilitates the acquisition of other skills. It is safe to say that at least 95 percent of Americans do not know what their learning style is or how to address it. What we do know, we learned from watching the "A" students in grade school and high school. The concept of "study smarter—not harder" applies here. Few of us know how to "study smarter" based on our individual learning style. A secondary benefit of learning how to learn is that it empowers the learner to become a self-directed learner capable of identifying a deficiency, finding resource materials, and doing the work based on methods appropriate for his or her learning style.

From the employer's perspective, the skill of knowing how to learn is cost-effective because it can mitigate the cost of retraining efforts. When workers use efficient learning strategies, they absorb and apply training more quickly, saving their employers money and time. When properly prepared, employees can use learning-to-learn techniques to distinguish between essential and nonessential information, discern patterns in information, and pinpoint the actions necessary to improve job performance. Many employers - particularly those dealing with rapid technological change see the learning-to-learn skill as an urgent necessity. Productivity, innovation, and competitiveness all depend on developing the workers' learning capability. Machinery and processes are transferable between companies and countries, but it is the application of human knowledge to technology and systems that provides the competitive edge.

Many workers do not have the basic skills which are essential to acquire more sophisticated job-related skills. The shrinking 16-24 year-old age group will force employers to hire less qualified entry-level workers. This volatile mix of demographic, economic, and technical forces is propelling the U.S. toward a human capital deficit which threatens the competitiveness of economic institutions, and acts as a barrier to individual opportunities.

Because science, medicine or psychology has not yet provided us with the information about how we learn, we are still only guessing at it based on what seems to work. Yet, because there is a keen interest in the process, research is ongoing and there is a growing body of information that is giving us insight into our abilities to memorize, retain, assimilate, and create.

Largely, because we have so little scientific fact, what we do know about how we learn has not been taught as part of the core curriculum in K-12, higher education, or training. During the last several years, it has been taught at the graduate level in some schools of education.

In a learning how to learn class, a learner's learning style is identified through instruments such as the Canfield Learning Style Survey. The learner is given the survey and is guided in the interpretation of the results. Basic and secondary learning styles are identified, and the learning methods that will best support the learning style are pinpointed.

Learners are encouraged to ask their trainers to present information to them using these methods. For example, many people learn most efficiently from visual materials, yet most instruction is delivered verbally. The mismatch leads to confusion for the learner, and a longer learning process. Because the learner has not learned the material, the training dollar was not well used. Learners can self grade the instrument, but it is most helpful in discussing learning styles if the scores for the entire class are shown together so that the learners and the instructor discuss the wide range of styles which can be represented in a group of only six people.

■ Canfield Learning Style Instrument Interpretation

Conditions

Peer: Working in teams; good relations with others; having friends; feeling positive about working and building something together. Clearly a high priority in the 90s Organization: Work logically and clearly organized; meaningful assignments and sequence of activities.

Goal Setting: Setting one's own objectives; using feedback to modify goals and procedures; making one's own decisions on objectives. This is an important element of being self-directed and proactive. They need to know how they fit in with the larger company goal.

Competition: Desiring comparison with others; needing to know how one is doing in relations to others. America fosters this - but competing does not automatically foster excellence. Competition is an extrinsic reward ... it is better replaced with an intrinsic reward system.

Instructor: Knowing the instructor personally; having mutual understanding; liking one another. Give plenty of eye contact and positive non-verbals.

Detail: Specific information on assignments, requirements, rules, etc. People who want minimal amounts of detail are "right" brain conceptual thinkers. They need to understand the concept first ... and then will sit through the detailed explanation - remembering only the details that are important to their conceptual understanding. People who want details in a sequential order are "left" brain linear thinkers. Take them through the process in an orderly, chronological process.

Independence: Working alone and independently; determining one's own study plan; doing things for oneself.

Authority: Desiring discipline and maintenance of order; having informed and knowledgeable instructors and superiors.

Content

Numeric: Working with numbers and logic; computing; solving mathematical problems; etc. Provide with charts, spreadsheets.

Qualitative: Working with words or language; writing, editing, talking. Provide a report to them prior to a meeting or the need to make a decision. Lengthy question and answer period will give them time to formulate the idea in their own words

Inanimate: Working with things; building, repairing, designing, operating. Provide a physical model or way to work with the idea in question

People: Working with people; interviewing, counseling, selling, helping.

Mode

Listening: Hearing information; lectures, tapes, speeches, etc.

Reading: Examining the written word; reading texts, pamphlets, etc.

Iconic: Viewing illustrations, movies, slides, pictures, graphs, etc.

Direct Experience: Handling or performing; shop, laboratory, field trips, practice exercises, hands-on, etc.

Expectancy Score: The predicted level of performance.

Sample Class Learning Style Scores
(All Figures are Percentiles)

Student	#1	#2	#3	#4	#5	#6

Conditions:

Peer	65	75	88	82	47	47
Organization	20	60	30	99	99	82
Goals	45	52	35	12	15	03
Competition	75	07	37	37	05	10
Instructor	25	83	38	94	10	65
Detail	17	53	45	65	99	90
Independence	88	55	57	45	01	40
Authority	65	10	65	18	08	10

Content:

Numeric	72	72	52	53	22	90
Qualitative	30	40	90	60	30	30
Inanimate	30	12	55	20	90	90
People	52	62	95	14	71	05

Mode:

Listening	25	80	73	65	12	95
Reading	55	63	40	60	90	47
Iconic	25	25	07	80	01	08

Direct:

Experience	88	25	87	10	98	10
Expectancy	97	95	75	60	59	89

Trainers who teach to learning styles report that it takes a little longer to prepare for the class the first time it is taught, but the results are worth the effort. Basically, the trainer prepares information for the visual graphic learner (graphs, charts, drawing, video, computer), visual linguistic learner (printed word), experiential "haptic" learner (hands on demonstrations), auditory learner (spoken word, tapes,) and the collaborative (verbal) learner (question and answer period, discussions, interaction). Because most classes are made up of a combination of learners, the use of all of the methods will reach the majority of the learners.

For example, in the class scores shown on the chart, the scores for content and mode vary so widely, the trainer could not be sure to reach all the learners just by providing reading assignments or just by lecturing. Only by providing content in all four modes and in all four types of content, could the trainer be sure to reach a preference of every learner in the class. If only a lecture were provided to cover the content, only one of the six learners would be actively engaged in his or her learning preference (at the 90th percentile). If a reading assignment were included, another learner would be engaged. If a discussion were added, another learner would be engaged.

Trainers working at a distance are often not sure how much feedback students require. Because time is always at a premium, it would be helpful to know which students need more feedback. The Canfield Learning Style Inventory also provides this information. In the same class shown on the chart, students 2, 4 and 6 require more feedback. Student #4 requires the most feedback and would benefit greatly from instruction in becoming a self-directed learner. Student #1 is already an independent learner.

Many good trainers have been using all of these methods for years, because they intuitively understood that people learn differently and not everyone benefits from the lecture only method. During the last several years, science has been reporting back through its studies that all of these methods "spark" a different spot within the brain which will show on brain scans. This leads to the theory that each learning method may induce the brain to commit the memory of the information to different section of the brain. The more places the information is stored, the better chance that the information can be retrieved correctly.

The same theory has been at work in video production for many years and is the basis for the rule of thumb not to use just a "talking head." We frequently admit that talking heads are boring on video and clearly go to extremes to provide graphics, demonstrations, interviews, visiting experts, well designed video study guides and plenty of secondary footage (B-roll) in our training videos. Video has taken advantage of the concept of learning styles for decades, again - perhaps intuitively. This attention to producing "good" video may be the reason that over 300 (and growing) studies have shown that students learn as well - if not better–from video than they do in the traditional classroom. Dr. Howard Gardner of Harvard has identified seven distinct learning styles. A more complete discussion of learning styles appears in Chapter 10.

The much discussed concept of interaction as a major component of videoconferencing is also a way of addressing the collaborative learning style in which the learner learns most efficiently when allowed to process the information verbally.

■ BASIC SKILLS COMPETENCE: READING, WRITING AND COMPUTATION

The inability of large numbers of new workers to meet reading, writing, or computational (simple mathematics) standards is an economic and competitive issue. This forces employers to spend more on these critical competence skills. The majority of workers are literate and numerate but frequently, cannot use these skills effectively because they are rusty when called upon to use mathematical principles they have not used for 20 years, because they must use the skills in a context different from the one in which they originally learned them, or because they do not understand how to expand or apply the skill.

Reading has historically been considered the fundamental vocational skill for a person to get, keep, get ahead, or to change jobs. One educational assessment by Kirsch and Jungeblut in 1986, indicates that there is a large nationwide population of intermediate literates who only have fourth to eighth grade literacy equivalency (but are high school graduates) and who have not obtained a functional or employable literacy level. This group will make up as much as 65 percent of the entry-level work force over the next 15 years. In the ASTD-DOL study, employers generally found the fewest deficiencies in reading. Only in secretarial/clerical and technical positions did a significant portion of the respondents find deficiencies. This is a concern for all employers - either now or in the future.

Writing is consistently ranked among the highest priorities for job applicants and employees. One study states that more than 50 percent of the business respondents identified writing skill deficiencies in secretarial, skilled, managerial, supervisory, and bookkeeping personnel.

Because of technology, simple mathematical computation is important as employers focus on an employee's ability to compute at higher levels of sophistication. The introduction of sophisticated management and quality control approaches demand higher mathematical skills. Ironically, as occupational skill-level requirements climb, higher educational dropout rates and worsening worker deficiencies in computational skills are appearing (Brock, 1987; Kirsch and Jungeblut, 1986; Semerad, 1987). Employers complain particularly about miscalculations of decimals and fractions, resulting in expensive production errors. Employees must calculate correctly to conduct inventories, complete accurate reports of production levels, measure machine parts or specifications so that medium-to-high levels of mathematics skills are required across job categories. The business effect of math skill deficiencies is bottom line losses.

■ COMMUNICATION SKILLS: ORAL COMMUNICATION AND LISTENING

Formal education in communication has been directed at reading and writing skills that are used least in the workplace. Most have only one or two years in speech related courses ands no formal training in listening. Workers who can express their ideas orally and who understands verbal instructions make fewer mistakes, adjust more easily to change, and more readily absorb new ideas than those who do not. Thus career development is enhanced by training in

oral communication and listening because these skills contribute to an employee's success in all of the following areas: interviewing, making presentations at or conducting meetings; negotiating and resolving conflict; selling; leading; being assertive; teaching or coaching others; working in a team; giving supervisors feedback about conversations with customers; and retraining. Employees spend most of the day communicating, and the time they spend will increase as robots, computers, an other machines take over mundane, repetitive jobs.

Skill in oral communication is a key element of good customer service. More than 76 million workers are in the service sector and companies that provide excellent service tend to stay far ahead of their competitors. To provide good service, all employees (not just designated sales and marketing employees) must learn how to talk and listen to customers, handle complaints and solve their problems. Only about 33 percent of organizations with 50 or more employees provide listening skill training. Workers spend 55 percent of their time listening, yet listening skills are appalling:

We use only 25 percent of our listening capacity.

We use only 10 percent of our memory potential.

We forget half of what we have heard within eight hours. Eventually, we forget 95 percent of what we have heard unless cued by something later on.

We distort what little we do remember (Nichols and Stevens, 1957; Barker, 1971).

For an eight hour workday that breaks down as follows:

We spend about four hours in listening activity.

We hear for about two hours.

We actually listen for an hour.

We understand thirty minutes of that hour.

We believe only 15 minutes' worth of what we listen to.We remember just under eight minutes worth (Elsea, 1986).

As workers go up the corporate ladder, the listening time increases so that top managers spend as much as 65 percent of their day listening (Keefe, 1971). Because most people have had no training in this critical skill, poor listening habits cost hundreds of millions of dollars each year in productivity lost through misunderstandings and mistakes. At the rate of one $15 mistake per U.S. employee per year, the annual cost of poor listening would be more than a billion dollars.

Adaptablility: Creative Thinking and Problem Solving

Problem-solving skills include the ability to recognize and define problems, invent and implement solutions, and track and evaluate results. Creative thinking requires the ability to understand problem-solving techniques but also to transcend logical and sequential thinking and make the leap to innovation.

New approaches to problem-solving, organizational design, and product development all spring from the individual capacity for creative thinking. At work, creative thinking is generally expressed through the process of creative problem solving. Increasingly, companies are identifying creative problem solving as critical to their success and are instituting structured approaches to problem identification, analysis, and resolution.   Unresolved problems create dysfunctional relationships in the workplace. Ultimately, they become impediments to flexibility and to dealing with strategic change in an open-ended and creative way. Creative solutions help the organization to move forward toward strategic goals. Organizational strategy is an example of creative thinking.

Personal Management: Self-Esteem, Motivation/Goal Setting, Employability/Career Development

Another key to effectiveness is good personal management. Self-esteem, motivation/goal setting, and employability/career development skills are critical because they impact individual morale which in turn plays a significant role in an institutions ability to achieve bottom line results.

Employers have felt the pressure to make provisions to address perceived deficiencies in these skill areas because they realize that a work force without such skills is less productive. Conversely, solid personal management skills are often manifested by efficient integration of new technology or processes, creative thinking, high productivity, and a pursuit of skill enhancement. Unfortunately, problems related to these skill areas have increased primarily because entry-level applicants are arriving with deficiencies in personal management skills. On the job, the lack of personal management skills affects hiring and training costs, productivity, quality control, creativity, and ability to develop skills to meet changing needs. This presents a series of roadblocks that slow or halt an organizations progress. An organization with such difficulties cannot plan accurately for its future to integrate new technology, establish new work structures, or implement new work processes.

Group Effectiveness: Interopersonal Skills, Negotiation, and Teamwork

The move toward participative decision making and problem solving inevitably increases the potential for disagreement, particularly when the primary work unit is a peer team with no supervisor. This puts a premium on developing employees group effectiveness skills.

Interpersonal skills training can help employees recognize and improve their ability to determine appropriate self-behavior, cope with undesirable behavior in others, absorb stress, deal with ambiguity, structure social interaction, share responsibility, and interact more easily with others. Teamwork skills are critical for improving individual task accomplishment because practical innovations and solutions are reached sooner through cooperative behavior.

Negotiation skills are critical for the effective functioning of teams as well as for individual acceptance in an organization. Change strategies are usually dependent upon the ability of employees to pull together and refocus on the new common goal. Carnevale wrote in a previous book that there are two ways to increase productivity. "The first is by increasing the intensity with which we utilize (human) resources (working harder), and the second is by increasing the efficiency with which we mix and use available resources (working smarter)."

Influence: Organizational Effectiveness and Leadership

To be effective, employees need a sense of how the organization works and how the actions of each individual affect organizational and strategic objectives. Skill in determining the forces and factors that interfere with the organizations ability to accomplish its tasks can help the worker become a master problem solver, an innovator, and a team builder.

Organizational effectiveness skills are the building blocks for leadership. A proactive approach toward increasing organizational effectiveness skills through training reflects the commitment to shared leadership concepts operating in the organization. Implementing shared leadership values has a positive impact on productivity. When leadership functions are dispersed, those who perform in leadership roles willingly take on the responsibility for creating and communicating the vision of the organization and what its work groups should accomplish. By their proximity, they are also better able to create and communicate the quality of the work environment necessary to realize that vision.

One approach is the superteam which is defined as a high performing team which produces outstanding achievements. Leaders of superteams spend as much time anticipating the future as they do managing the present by thinking forward to, and talking to others about their goal, for it is this that provides the team with its purpose and direction (Hastings, Bixby, and Chaudhry-Lawton, 1986). Deploying visionary leaders improves institutional response time to changing and increasingly complex external environment factors that affect the organization's ability to operate effectively.

At its most elementary level, leadership means that one person influences another. An organization that supports the concepts of shared leadership encourages employees at all levels to assume this role where it is appropriate. The function of leadership include stating basic values, announcing goals, organizing resources, reducing tensions between individuals, creating coalitions, coalescing workers, and encouraging better performance. There is a direct correlation between the implementation of shared leadership practice and product improvement, higher morale, and innovative problem solving, which leads to a more hospitable environment for instituting change.

Top management cannot make the system work without employees taking on shared leadership roles. A great many people must be in a state of psychological readiness to take leaderlike action to improve the functioning at their levels. Historically, the roots of business failure can often be traced to inadequate training in and attention to the importance of leadership as a basic workplace skill. Too frequently, companies designate leaders without providing proper evaluation and training to ensure that they are qualified to assume leadership roles.

■ THE ORGANIZATION AND STRATEGIC ROLE OF TRAINING

The ASTD/DOL study (Carnevale, et al, 1990) reported on how training is structured, managed, financed, and coordinated with organizational strategy. It details employer-based training, who gets training, how training funds are spent, and how approaches to training differ. Discussions, facts, and figures explain the economic implications training has for industry and the economy. It explores how training can be used to achieve the organizations strategic goals and examines why and how companies use outside resources to provide training. Practical examples and case studies illustrate the characteristics of learning situations and capture the essence of the training industry's contributions to institutions. It sets forth a comprehensive training and development policy that expands and integrates the roles of government, industry, and educational institutions.

Employers realize that to be competitive, they must accelerate learning and integrate it rapidly. Therefore, designing and implementing training which supports the employer's institutional culture and strategic goals take precedence over broad-based courses unconnected to the employer's central agenda. The employer's ultimate goal in providing workplace learning opportunities is to improve the companys competitive advantage. Employers are driven to identify and use learning approaches that rarely stray from the reality of the workplace and are linked to the

employee and the employer's bottom line. More cost-effective than broad-based training, the applied approach provides training that responds to the employer's specific needs and triggers rapid integration of learning with actual job requirements, resulting in higher employee productivity.

Research and experience in adult learning show that linking learning to a worker's job pays off because employees are more likely to retain job-related information because they realize that they will immediately and repeatedly use the new knowledge. Job-related learning is flexible and frequently easier for employees than were their earlier learning experiences. Workplace learning is supported by a powerful motivator; when learning experiences are based on actual job needs, employees frequently work to increase their proficiency in the expectation that they will trigger immediate rewards in terms of achievement, status, and earnings. Thus, employers and employees are jointly motivated to make the workplace learning experience a success. Training delivery trends indicate three dominant scenarios:

1. Training design and development is controlled centrally but delivered decentrally to the plant level,

2. Executive, management and organizational development training are designed, developed, and delivered centrally but technical and skills training are entirely decentralized, and

3. Operational units have the authority to make training decisions and purchases (decentralized) but with the requirement that a central training department participate in the training selection process or approve the selection.

According to the survey, the decision to make-or-buy training is composed of the following criteria.

1. Expertise: how specialized the training need is.

2. Timeliness: does the staff have time to develop and deliver the program within the time frame.

3. Population Size: economies of scale - a large group or training that will be delivered. Frequently leads to in-house development. One-time in-house training for small groups   cannot be justified.

4. Sensitive or Proprietary (to gain a competitive advantage) Content: likely to be in-house regardless of other factors.

5. Cost: secondary to other criteria.

6. Employer Conditions: companies may not have a large training department, expertise, resources or time to meet specialized needs.

7. Other factors: outside providers can bring new ideas to rejuvenate dull training thus motivating employees to attend and learn.

The research confirmed that the lecture format is still dominant for sales training but that videotaped presentations are popular because of flexibility. It enables training immediately after hiring, introduces new products without bringing all salespersons to one location, and demonstrates products consistently. Custom video is expensive but cost-effective compared with using untrained salespersons in the field.

Organizational Strategies and Training Roles

1. Where is the corporation now?

2. If no changes are made, where will the corporation be in one year, two years, five years, and ten years? Are the answers acceptable?

3. If not acceptable, what specific action should the corporation undertake? What are the risks and payoffs? The gap analysis is essential to strategic planning and while the opportunity may exist to participate, the skill to provide that input may be absent.

Strategic Concepts

The understanding of strategic concepts has become increasingly sophisticated at the top levels of organizations, but that sophistication has not filtered down to managers and workers who implement it. Following are questions to guide trainers in gathering information and begin participating in the strategic process They are organized into two broad groups:

1. Common considerations which are relevant to all strategies, and

2. Considerations essential to specific strategies.

Common Considerations

1. Understand the environment in which the organization operates. Is the industry evolving or stable? What are the growth trends? Who are the main domestic and foreign competitors and what is the organization's competitive advantage over them? How can the organization capitalize on competitors' strategic vulnerabilities? Is the organization capable of widening the competitive gap in its favor?

2. Why has the organization been successful in the past? What strategies has it successfully employed? What was learned during that can be applied under the new strategy? What forces have driven the organization to select a new strategy?

3. What technology does the organization plan to use? If new, when will it come on line?

4. Are industry innovations anticipated that could change the market? Will these be radical breakthroughs or modifications? What effect would this have on the organization's product and its competitive position?

5. What new management philosophies or procedures will be instituted? When?

6. What regulatory issues could influence strategic considerations?

7. What functional strategies will be employed by the operating units to effect the strategy? Why? How?

Human Resource Issues

1. Organization's work place profile. What are the current strengths and weaknesses of the work force? Is the work force technical? Skilled? Flexible? Adaptable? What is the educational background? What do they need to stay current?

2. What changes must occur in the job(s), organizational culture, and skill levels of the work force?

3. Is the organization's decision to pursue an umbrella strategy likely to result in layoffs or other turnover? How much is anticipated?

4. How will union contract agreements be affected? What is the strategic role of the union?

5. What human resource development policies should be reviewed/modified for the organization's strategic emphasis?

6. What are the training implications of the strategy? Can training help the organization reach strategic goals?

7. What training programs are needed? Basic? Technical? Product management? Motivation? In-house training or outside experts?

8. How has training been regarded by the workers? By management? How credible are the programs? Trainers? How will these views affect future training efforts?

9. What delivery mechanisms are most cost-effective and practical for each training program?

10. Do employees have tuition reimbursement? Take advantage of it? How can it be used to enhance worker skills?

11. What training evaluation is used? Provides information on return on investment (ROI)?

12. Is there a procedure to ascertain if training is appropriate for the new strategy or to identify new training needs?

13. Do human resource management functions other than training need to be reviewed? Modified?

When the Strategic Emphasis is on Innovation

1. What are the technological, marketing and distribution implications of product development?

2. What resources are set aside for R&D? How does R&D staff interact with line managers? Line workers?

3. Does the organization create an atmosphere that encourages employees to think innovatively? Is this a safe environment? Is risk taking encouraged? How do supervisors provide feedback? How do they

help employees balance ideas? Are employees rewarded for their innovations?

4. How does management first react to the new and unfamiliar? How do they differentiate between practical ideas and off-the-wall ideas? Is training needed?

5. If an entrepreneurial unit approach is used, how is the remaining work force encouraged to view that team? Is the team accessible? How does it relate to R&D?

6. Does the training and development department function as an observer, reactor, or catalyst to innovation? What is the feedback loop?

When the Strategic Emphasis is on Product Development

1. What are the technological, marketing and distribution implications of product development?

2. How will technical training be provided? Will OEMs assist?

3. What creates an atmosphere that nurtures product improvements and spin-offs?

4. Will new products fill a niche or break new ground? If new ground, how will sales force get customers to realize they need the new product?

5. What training is needed? In-house or outside experts?

6. Will subject matter experts need to be identified and trained to assist in the training effort? What are the train-the-trainer considerations (time, cost)?

When the Strategic Emphasis is on Market Development

1. Will the organization need to add to its existing work force? Are workers available in the area? What is the supply and demand forecast?

2. Will employees take on new responsibilities?

3. Can training handle the integration of many new employees?

4. Will content experts need to be identified and trained to assist in the training effort? Identify train-the-trainer considerations.

5. If physical market expansion such as the construction or purchase of new facilities is planned, are training and development staff involved to offer insights? Will they be on site as new facilities open?

6. The sales force is pivotal; what qualifications must new hires have? How will the sales force be trained/retrained?

7. If foreign expansion is planned, what are the training and development implications? What and who are the resources? Can in-house trainers handle?

When the Strategic Emphasis is on Turnaround

1. How has the contextual picture changed since the organization began retrenchment? What does the organization's industry look like? Stable or evolving? Main competitors?

2. How has the organization redefined itself and its goals?

3. How will resources be reallocated to support turnaround?

4. Does the organization plan to use new technologies or processes?

5. Leadership: changed? Implications?

6. What shifts in culture, behavioral norms, values, philosophies, or procedures are required?

7. Are there any new regulatory considerations?

8. What changes must occur in jobs and skill levels?

9. Can workers meet the challenges? What training or help is needed?

10. Will furloughed workers be recalled?

11. What kinds of new hires are needed? Would a different kind of worker be suitable?

12. Suggestions for employee orientation of new hires/re-hires?

Trainers must gather quite a bit of information about their organizations before they can credibly advance the notion of integrating training with other strategic considerations. This knowledge base is what top executives mean when they say "You've got to know the business your company is in before you can be a strategic player." The next step is to begin the process of influencing decision-makers by being a trainer and in-house lobbyist.

One approach to lobbying is to build support from the bottom up by establishing visible and measurable links between training and the organization's business goals. This can be done by quantifying return on investment, contribution to productivity, or role in mitigating the costs of integrating new technology and processes. Gather information course by course, write a report and send it to the top decision makers (including accounting).

Seek support in the formal and informal leadership structures. All departments hold opportunity for influence. Create a paper trail by asking other departments about their objectives and what they see as the potential training implications of moving toward those goals. Leadership is more likely to respond to proposals that training be connected to the strategic process if support comes from a variety of sources. In the absence of CEO support, training considerations can only be part of the strategic process when they have a strong institutional underpinning. Sustained integration of training hinges on institutionalized support. In short, a training infrastructure must be built.

Because of the importance of compliance with local, state or federal laws, work to see that training has a direct line with the technical departments that conduct activities regulated by law. Work together to plan for training needs concurrently with on-line and emerging regulatory requirements.

Training the Technical Work Force

"Training the Technical Work Force" (Carnevale, et al. 1990) examines the nature and role of technical training within the competitive and technologically shifting workplace and sets forth specific guidelines for conducting effective technical training programs. The technical work force - includes professionals such as scientists, doctors, and engineers; technicians such as hygienists and draftsmen; and skilled trade or blue-collar workers. Technical workers represent about 18 percent of the American work force and receive about 30 percent of the $210 billion annually spent on training. Technical workers are especially important to competitiveness because they produce the lion's share of internationally traded products and services. They invent and produce technologies that result in the upgrading all workers. The continuous integration of new technologies with more highly skilled labor is widely recognized as the true source of American competitiveness.

Most employers agree that this nation's future prosperity depends on the energy, flexibility, and creativity of a well-trained work force that is knowledgeable, innovative, efficient, and dedicated to quality. However, they are concerned that securing the quality and quantity of technical training required to build the work force is the greatest challenge of all. Ironically, as the workplace becomes more technologically complex, the rising pool of available workers is lacking in many of the simplest and most basic skills, including reading, problem solving, computation, and knowing how to learn.

Technical work requires more education and training than any other work and must be preceded by a grounding in basic skills that prepare workers to understand and acquire the more sophisticated constructs of technical work. This places a premium on the individual who has the basics and is therefore equipped to handle higher-level technical training. Given the demographic picture, building tomorrow's technical worker may mean doing so from the basics up - a costly and time consuming endeavor. All workers tend to receive more training to qualify for their jobs than they do to upgrade their skills. Between 67 and 79 percent of technical professionals and technicians receive more qualifying training than upgrading training. Approximately 50 to 75 percent of technical professionals and technicians participate in upgrading training, but fewer than 50 percent of skilled trade workers participate in skill upgrading.

New, high-technology delivery methods — such as satellite networks and interactive video — make it faster and more efficient for employers and outside providers to work together to train technical employees. The most significant of the high-tech delivery methods, satellite networks, are being used to aid employers in sharing programs and information with each other and allow universities to send programs easily and directly to employer sites.

As a result of this technology, an entire university has been founded, with courses accessed by employees solely through satellite networks. The National Technological University (NTU) is a fully accredited, advanced degree awarding institution that delivers its programs by means of satellite delivery systems transmitted from universities. NTU offers programs leading over ten different master of science degrees. NTU's board of directors is composed of industry representatives who provide consultation on curriculum design and development.

A number of corporations have also joined together to transmit continuing education programs for engineers by satellite. These programs are usually developed internally or with the aid of universities and professional societies. They offer quality, indepth courses that are on the cutting edge of technology. Course content focuses on current engineering job responsibilities. Technical professionals are among the most highly educated and best trained of the nation's employees. They tend to receive substantial amounts of formal education and employer-provided formal and informal training to qualify for their jobs and upgrade their skills once they are on the job.

Texas Instruments, a manufacturer of electronics products, including peripheral computer parts, with revenues exceeding $4 billion, and Dupont, a diversified chemical and energy company with profits exceeding $27 billion, are two companies that offer satellite courses to their employees and to employees of other organizations. Both offer day-long seminars on development in electronic technology to other companies via satellite. Sharing courses through satellite networks makes technological advancements available to other organizations and encourages the spreading of costs of training and development among organizations.

Health care professionals receive the most upgrading, relying upon schools more than their employers for both qualifying and upgrading training. This suggests that skill needs are not employer-specific and that a stronger bond exists between health care professionals and their professional specialty than between the professional and a specific employer. The Hospital Satellite Network and AREN provide programming in these areas.

Traditionally, companies have played a relatively passive role in training technical professionals, but in recent years competitive pressures have forced companies to integrate the development and design of innovations with production and marketing. This attempt to build more integrated structures is leading employers to play a more active role in the development and delivery of training and human resource development for their technical professionals.

While specific curricula for technical professionals vary widely according to professional discipline and specific application, employer-provided training falls generally into three broad categories:

1.   New technologies and processes within the specific field of study,

2.   New applications of existing technologies, and

3.   New product demonstrations.

The average amount of time spent training technical professionals each year varies by industry with ten days a year reported by industries with very rapidly changing technologies and zero to five days reported by industries with stable technologies. In the survey, the highest average reported was 26 days by a leading manufacturer of semiconductors and microprocessors.

Technicians include employees whose primary expertise lies in a particular technical specialty area and lack the breadth of knowledge in the theoretical aspects of their specialties that is required of technical professionals. They usually receive training that applies directly to their jobs, with little emphasis on theory. Most technical training is sequential and can be divided into three categories:

1.   Principles of new technologies (equipment and processing techniques).

2.   New applications for existing technologies.

3.   Courses required for licensing and certification or refresher courses required for license renewal or re-certification.

There is a growing consensus among employer organizations that technical training is critical for successful operation within an economic environment characterized by rapidly advancing technology and complexity. Although technical training is considered to be most effective when it is aligned with organizational operations and an integrated part of the corporate strategic planning process, there is no consensus as to the optimal structure and organization for technical training. Disagreements revolve around how to structure training, under which function/discipline it should be organized, what degree of participation it should have in overall planning and what relationship it should have with line operations.

Generally, there are three ways in which employers can structure technical training: centrally, decentrally, or a combination of the two. There is a clear trend toward centralizing control over technical

courses. Trainers usually have adult learning or technical backgrounds or are composed of teams with both backgrounds.

■ ASTD TRAINING MODEL

The ASTD model (Carnevale, 1990) for a workplace basics training program links three operational components to create a new training model in which the whole performs more efficiently and effectively than its parts. The components include:

1.   A plan: an in-house marketing plan to convince management and union leadership of the need to be active in linking workplace basics training programs to strategic planning and organizational goals.

2.   A design: a modified instructional system design for developing and implementing training programs.

3.   A learning method: a job-specific, performance-oriented learning methodology for training delivery.

Combining these three components results in an applied approach to training in business organizations. The approach takes into account the constraint of resources and the rapid changes in technology and marries them to a state-of-the-art thinking about how to ensure that training is appropriate and relevant, providing the best return on investment.

Carnevale states that research and experience in adult training tell us that an applied approach works best because it:

1.   Motivates learners by linking learning to improved job performance, which in turn may lead to improvement in learners' careers and earnings.

2.   Encourages learner retention by requiring immediate and repeated use of newly gained knowledge.

3.   Improves job performance by creating learning experiences based on actual job needs.

This is a pragmatic, work-based program development and implementation system that can effect positive changes in employees. This practical and systematic training method is the result of two factors: emloyers' needs for training programs that will consistently improve employee job performance, and emergence of a systematic method for the design, development, and delivery of training.

The recommended approach first proceeds in a step-by-step fashion to measure the gap between job requirements and employee skills. Then human resource professionals create training programs that, when fully operational, are able to translate each job separate duties and tasks into practical learning experiences that successfully reduce or eliminate employee skill deficiencies.   The following are brief descriptions of the applied approach activities:

1.   Investigate in broad, comprehensive terms which jobs and workers need training because of

changes in the nature of the work or as a result of emerging workplace problems.

2. Advocate (a two-step process) support of a training program to management and senior union officials as an integral part of the strategic planning and goal-setting process.

3. Analyze jobs and tasks to determine where the need is and thereby determine what training should focus on.

4. Design the program's instructional content, related performance objectives, and criterion-referenced tests. Determine the content's structure and sequence, decide on documentation, and plan for program evaluation.

5. Develop objectives that represent the actual learning activities workers need to master and develop documentation and evaluation instruments that measure training's impact on   improving an employee's job effectiveness.

6. Evaluate the program: first, through program monitoring that provides continuous feedback on how well learners are meeting training objectives on a day-to-day basis; and about three months later, through a program evaluation procedure activated after learners are back on their jobs and have had opportunities to put their newly gained skills into operation.

The following chart graphically displays the applied approach to creating a basic workplace skills training program.

■ ASTD BLUEPRINT FOR SUCCESS

Step 1. Identify Job Changes or Problems Related to Basic Workplace Skills
- Assess the extent of the need for training because of job changes or problems
- Form a company-wide representative advisory committee
- Perform a job analysis for selected jobs
- Document employee performance deficiencies on the selected jobs
- Identify population to be targeted for training
- Build cooperation with unions

Step 2. Build Support for Training Through Alliances with Management and Unions
- Make the case for skills training in work place basics
- Build support for skills training in work place basics

Step 3. Present the Strategy and Action Plan for Approval

- Present the strategy and action plan for training
  - Select a training program architect: in-house staff versus external providers

Step 4. Perform a Task Analysis
- Perform a task analysis
- Determine whether to select a quick route through task analysis and determine which process is most appropriate

Step 5. Design the Curriculum
- Design performance-based, functional context instructional program
- Design evaluation system
- Design documentation and record-keeping system
- Obtain final budget approval to implement

Step 6. Develop the Program
- Prepare the instructional format
- Select instructional techniques
- Select facilities site and designate equipment requirements
- Develop evaluation and monitoring instruments

Step 7. Implement the Program
- Select and train the instructional staff
- Develop a learning contract: yes or no?
- Run pilot test (optional)

Step 8. Evaluate and Monitor the Program
- Carry out initial evaluation
- Begin on-going program monitoring
- Advise and consult with management on program status

■ ADULT LEARNING CONCEPTS

Research into adult learning has identified a number of findings that when utilized, will increase the learner's learning (Knowles, 1987).

Relevance: Adults need to know why they should learn the material. They learn best when they understand how the new knowledge will be immediately useful in their work or personal lives.

Self-Directed: Adults need to be self-directed in their learning. Just as they want to be in charge of their lives and responsible for the decisions they make, they need to have the same power in their continuing learning. They should be encouraged to participate in choosing and planning their own learning activities. Self-Directed learning contracts have been useful to help the learner become self-directed and to help the trainer release his or her traditional authority and turn it over to the learner, thus empowering the learner. Adults will make a voluntary commitment to learn when they experience a real need to know or to be able to do something. They do not respond to an authority figure saying it will be good for them.

Respect: Adults need to have their experience respected and considered as a resource during the learning process. Methods should build upon the experience that the learner already has. Trainers should acknowledge by their speech and action that the learner has other valuable information and that what is about to be Learned will enhance it. There should be emphasis on hands-on techniques that draw on the learner's accumulated skills and knowledge (such as problem solving, case studies, or discussion) or techniques that provide learners with experiences from which they can learn (such as simulation or field experiences).

Problem-Centered: Adults have a task-centered or problem-centered approach to learning. Learning should be organized around real tasks. Rather than teaching writing, the course should be centered around how to write effective business letters and reports.

Motivation: Adults are motivated to learn. They will respond to extrinsic motivates such as higher salaries and promotions, but will respond even more to intrinsic motivation such as the need for recognition, responsibility, achievement and self-esteem.

Process: Adults need to have the process of learning considered carefully. Training should be learner centered. The focus should be on the learner acquiring the learning rather than the instructor transmitting the information. Learning centered learning focuses on transformational learning rather than mimetic (mimicking the instructor). Transformational learning enables the learner to make a change that will have an impact on his/her job or life.

Feedback: Many adults need regular feedback about their learning particular as they embark upon becoming self-directed learning. They want clear learning objectives and want to know regularly the extent to which their objectives have been achieved. As they become self-directed, they will benefit if they have learned how to set the objectives themselves, provide their own feedback through methods which they create which may or may not involve the trainer providing feedback. Learner centered training fosters the independence of the learner while removing dependence on the trainer.

When new material is introduced, the learner may be dependent. This is because learning that involves new work situations, languages, and some sciences may not be based on content with which the learner is familiar. When the learner has very little prior information, he or she will be more dependent upon the trainer. During this period the learner may need more content structure and feedback until a learning framework is put into place. During this period, the trainer should provide direction in what to study, more external reinforcement and encouragement that the customary amount of learning is taking place and that the student is not "stupid." During this period the trainer may take on the aspect of an expert or authority in the content. The trainer may lecture, demonstrate, assign, grade and test the learner's work.

This is a temporary period of dependency for the learner and the trainer should make every effort to move the learner into the next stage of learning as quickly as possible.

The next stage of learning is marked by collaboration between the learner, other learners, and the trainer. The basis for this stage is that the learner has some knowledge or ideas. By sharing the ideas and working with others, the information can be tested and validated by the learners. The methods that are most useful in this period are interaction, practice, probing of self or others, observation, participation, peer challenge, peer esteem and experimentation. The trainer takes on a peer role and sets an informal learning environment in which the above methods can be utilized. In this stage the trainer can interact, question, provide feedback, coordinate, evaluate, manage and grade the work that is being done.

As the learner's knowledge base increases, the learner will move increasingly toward independence. This stage may also be entered immediately by learners who have previous knowledge or are confident about their ability to learn. The learner will want to continue to search for information on his or her own and may need the ability to experiment with the information, require time to process the informationand may require nonjudgmental support from the trainer. Internally, the learner is aware of his or her needs and the process that is being undertaken. It is at this stage that the trainer's role becomes that of a facilitator and provides feedback when it is requested, provides resources and otherwise may act as a consultant by listening, evaluating, delegating or negotiating with the learner.

As an adult educator, one must be able to slip easily between the roles of trainer and facilitator. The temptation always is to remain in the role of trainer and authority figure. This traditional role is easy to maintain because the authority rests with the trainer. It is a hard role to relinquish because it feels good to be in control of the situation and the learners. The hardest role to fulfill and be successful is that of facilitator. Because the trainer/facilitator has been through the training time and again, it is always easier to "tell" the learner what to do rather than to help the learner become a self-directed learner who is capable of telling him or herself what to do. As a facilitator, the trainer has to relinquish authority and control over the situation and the learners in order to empower the learners and to help them become self-directed.

■ ASTD RECOMMENDATIONS: HOW TRAINING CAN BE IMPROVED

Although the investment in training is huge at an estimated $33 billion annually, or 1.4 percent of the national payroll, analysis by the authors led them to conclude that most employers current commitments to training and development are insufficient. To prepare for their jobs, only 11 percent of American employees get employer-provided formal training, and only 14 percent receive informal training from their employers. To upgrade their skills, only 10 percent receive employer-provided formal training and 14 percent receive informal training from their employers. ASTD makes the following public recommendations for employers, educators, and government to improve training.

■ ASTD POLICY RECOMMENDATIONS FOR EMPLOYERS

1. Set overall national targets for employer spending to the levels of human resource development characteristic of successful enterprises. The overall national targets should be increased slowly in two phases:

a. Set an interim target of spending two percent of national payroll (an increase of $14 billion over current expenditures for training and development). This increases total commitments to $44 billion and coverage from the current ten percent of employees to almost 15 percent.

b. Set an ultimate goal of spending four percent of payroll nationwide, (increase of $58 billion over current commitments) or an annual $88 billion increase from the current ten percent to almost 30 percent of employees.

2. Integrate HRD into the employer institution; the CEO must make training a priority and make the training and development executive a full member of the senior management team. Make line managers responsible for training and development of subordinates. Make training available to all employees, not just white-collar workers or technical professionals.

3. Use an applied approach to developing workplace curricula. Embed learning in work processes.

4. Link workplace learning to the performance of individuals, work teams, strategic change processes, and reward systems by communicating work require-

ments to educators and hiring based on academic performance. Reward employees for learning and contributing new knowledge that results in cost efficiencies, quality improvements, and new applications and innovations.

5. Create two learning systems:

a. A training and development structure that teaches employees the required new skills.

b. A training and development system that allows employers to learn from employees to capture cost efficiencies, quality improvements, new applications, and innovations that employees discover during the production, testing, and use of products and services.

6. Develop employer strategies and government policies to link employers closely to their networks of suppliers and external education, training, and R&D institutions. Set performance standards linked to learning systems for supplier institutions; require suppliers to provide quality training to customers to ensure effective product/service use; and work with the government to provide resources to conduct R&D on best learning practices that link employers to suppliers and external education and training institutions.

7. Communicate new knowledge and changing skill requirements to educators.

8. Embed schooling in the career development process by giving more weight to educational attainment and achievement in hiring decisions.

9. Work with educators to develop and provide learn/earn curricula that combine academic and applied learning experiences.

■ ASTD POLICY RECOMMENDATIONS FOR EDUCATORS

1. Teach future employees how to make decision, solve problems, learn, think a job through from start to finish and work with people to get the job done.

2. Link academic subjects to real-world applications.

3. Strengthen the link between learning in school and on the job.

4. Schools, parents and employers should work together to provide students with opportunities to earn and learn through work experiences structured to complement academic programs.

5. The 45 percent of high school students who are tracked into the watered-down general curriculum and the 19 percent who are in vocational courses should have a new curriculum that mixes solid academic basics and applied learning.

6. Strengthen the high school vocational system but not in narrow or dead-end job categories, but by preparation leading to further higher education.

7. High school vocational education should include a mix of campus learning and carefully struc-

tured applied learning at work to accommodate different learning styles and to allow students to learn and earn.

8. Bring employers into the education structure by involving them in curriculum development and by providing student records that assess academic performance and behavioral attributes. Bring students into the employer structure by focusing learning and performance evaluation and by de-emphasizing pure reasoning in favor of learning experiences that imitate real-world situations and involve physically manipulating objects and tools.

■ ASTD POLICY RECOMMENDATIONS FOR GOVERNMENT

1. Assign employers a substantial role in planning and oversight of Job Training Partnership Act (JTPA) programs; to emphasize human capital development through work and learning rather than income maintenance; make job performance a key operational component; provide one-stop shopping for clients by coordinating services at the state and local levels.

2. Move JTPA away from its "one-size-fits-all" eligibility, treatment, and accountability system. Separate clients, treatments, and evaluative standards into four groups:

a. Target major resources for people who are poor, unemployed and demonstrate significant human capital deficits. Emphasize human development. Base accountability on measured skill changes. Fully fund all services.

b. Transitional services for poor and unemployed people with marginal human capital deficits should include job-search assistance and subsidies, to move them into the workplace. Base accountability on transitions into the workplace. Fully fund all services.

c. Employed workers who need upgrading to keep their jobs should be given retraining jointly funded by public authorities and employers. Programmatic accountability should focus on increased employability.

d. Employers allotted public funds to improve their competitive performance should provide matching funds. Funding should be available for management development supervisory training, and technical training. Disallow funding for executive development or sales training. Base accountability on matching-funds from employers.

3. Provide human capital development at resource levels that can improve employability of the disadvantaged; establish eligibility requirements that distinguish between people with developmental deficiencies and those requiring less extensive services; establish programs that offer a sequence of treatments from basic human capital development to transitional services, such as job search assistance and hiring and

training incentives for employers: base developmental programs accountability standards on skill acquisition and employability; and base transitional service accountability on job placements and tenure.

4. Utilize resources efficiently and provide comprehensive services, by delivering in coherent packages tailored to the needs of individual clients through common intake and eligibility criteria to provide one-stop shopping; base accountability on client progress rather than service delivery.

5. Incorporate three principles in crafting training programs for the dislocated: set a higher hitch in the safety net for dislocated employees and help them avoid a free fall from middle-class status into poverty; prior notification, counseling, job search assistance, and out-placement should be encouraged while employees are still on the job. Dislocated employees should receive counseling and job search assistance first and then training when a job prospect is evident or in hand. If possible they should receive training on the job.

6. Improve access to training through:

a. A mix of loans and grants for skill improvements paid for by taxes.

b. Use investment incentives to increase the standing of workplace learning.

c. Encourage experimentation and partnerships between employers and government to promote better job-related information and more effective transitions from school into the workplace.

d. Encourage experimentation and partnerships between employers and government to promote curricula development that mixes academic and applied learning delivered in class and work; research and development on curriculum and training delivery in particular occupations; collection, evaluation, and dissemination of best practices in training for specific occupations; and development of performance standards for individual occupations. The institutions receiving these grants should be trade and professional associations, unions, schools, and other institutions that represent members of occupations, provide training in occupations, or represent industries with a concentration of employees from particular occupations.

e. Encourage efficient use of learning institutions through dissemination of model practices and provide incentives for employers to off-load generic training.

7. Establish infrastructure to conduct R&D; to inventory, analyze, evaluate and model best practices in job-related learning; and to disseminate results to employers.

8. The demand-side approach to improving opportunities for job-related training should be accompanied by a supply-side strategy to increase the capability of suppliers to provide high-quality training to employers and employees.

9. Encourage experimentation with training programs intended to upgrade employees in the inter-

est of their own career development and to improve the competitiveness of state and local employers.

■ DESKTOP TRAINING AND MULTIMEDIA

There's a revolution coming for corporate training in the form of multimedia for training. A Department of Defense study on interactive video instruction stated that interactive video instructions improves achievement by an average of 38 percent over more conventional instruction, while reducing time to competency by 31 percent. The study included all instructional setting and applications. The study also found that in almost every case, interactive video was less costly than conventional instruction (Fletcher, 1990).

Low-cost digital video is a reality along with authoring tools that will enable trainers to assemble easy to use desktop training modules for just-in-time training. Desktop videoconferencing products are available which will allow easy access to trainers. The revolution in desktop training is enabled through wideband infrastructure throughout the building and between company sites. Tremendous compression abilities are needed.

Desktop video provides the ability to easily capture audio and video clips, text, still images, graphics, charts and other materials and assemble them into a multimedia piece that can be accessed from the desktop and played when necessary.

Besides the ability to provide training when it is needed at the employee's workstation, multimedia has the very great advantage of providing the training in a mode that is likely to meet the employee's preferred learning style.

■ CASE STUDY

The IBM Approach to Training Through Distance Learning: A Global Education Network by the Year 2000

For IBM, worldwide pressure to contain costs and provide education when and where it is needed has fueled the move away from traditional classroom delivery to use of technology such as satellite and self-study. IBM spends over a billion dollars on training annually, an average of 10 days per employee. Satellite education accounts for about 10 percent of this total, self-study for about 30 percent and traditional classroom education for the rest.

The company's two educational satellite systems evolved as changing conditions within the company and a highly competitive business environment created pressures to use the most effective teaching strategies. IBM wanted to educate customers and train employees quickly, effectively and at reduced cost.

In 1983, the Interactive Satellite Education Network (ISEN) grew out of the IBM marketing organization as an extension of a local closed-circuit television capability created to handle large classes and broadcast to customers, and branch-office employees. ISEN's success and the need to offer education to IBM plant, lab, and headquarters employees across the country led to the operational version of the Corporate Education Network (CENET) in 1987. ISEN and CENET use different satellite networks, but have a common strategy and management system. Over the last three years, the system and the number of students has doubled in size; the savings in travel and living expenses amounts to over $15 million and represents a return on investment of over 30 percent. The network also generates several million dollars of customer-education revenue for IBM annually.

The network broadcasts 12 concurrent courses to 238 classrooms at 49 locations. Each day, about 100 students view one of nine courses on the networks, with up to three courses offered at each location. Audiences include a full range of employees. The networks use an interactive digital multimedia technology that converts two full-motion analog video signals into compressed digital form, encrypts them, and transmits them to the satellite. At a receive site, the signal is decoded and converted back to an analog TV display. The system includes two one-way, full-motion color video pictures; high-quality two-way audio; and digital keypad response units over a secure network. The compression allows a reduction in satellite bandwidth by a factor of 10 compared to similar analog satellite systems. The system was designed to make it easy for students to ask questions and for instructors to effectively use multimedia in their courses. Due to the network's success, IBM units worldwide have installed education networks. Japan, Australia, the United Kingdom, France, and Germany have networks in place.

All of IBM's successful satellite courses incorporate critical factors in the development process which include effective visuals, interaction that holds the students' attention and reinforces key learning points, all based on instructor and administrator training. They are constantly learning from their satellite efforts and each step takes them closer to their vision of the future for distance learning.

The next decade will allow for a global education network utilizing interactive digital multimedia. Workstation-to-workstation communication may replace studio-to-classroom communication. Education courses will be stored for easy student access, allowing learning to take place where and when it is needed. Education will be more modular in order to focus on a single learning point, accommodating small chunks of education in less than one-hour sessions. As automated knowledge capture and instructional design emerge, the plan is to tie them to student learning history and needs profiles.

IBM has defined its long-range plans for corporate education in "A Vision of IBM Human Resource Performance in the Year 2000." The rationale for the changes reflects the competition of global economy and the multicultural work force. Above all, education will be self-directed shifting control of the learning to the student with distance education technology playing a key role in supporting this vision. The changes that are predicted at IBM will drive the future of IBM internal education:

- Diverse student base–different ages, cultures.
- Diverse skill base–varying levels of expertise/experience.
- Changing skill requirements–new technology and changing customer needs.
- Information overload–pace of technological change.

The critical characteristics of an education system designed to address these drivers are:

- Distributed–learning takes place when and where needed.
- Modular–focus on a single skill.
- Multisensory–stimulates multiple senses in a variety of ways.
- Nonlinear–fixed sequence of modules.
- Transferable–easy movement across language/cultures.
- Responsive–short development cycle.

The prediction of the report is that technology will play a key role in meeting the education needs and that IBM's education system in the year 2000 will include computer managed libraries of digital education modules delivered through the employee workstation. "Because learning activities must be so closely linked to performing, there must be little distinction between systems used for formal education and those used in the execution of work duties. In fact, learning will typically take place as part of the job process by means of embedded advisors, online consultations, or database searched in addition to learning that results from formal education modules." The major characteristics of IBM's Performance/Learning Support Systems are:

- Supports individual and team ability to function internationally.
- Encourages employees to be innovative and creative in their work.
- Makes learning a central and integral aspect of every person's job.

- Supports changing market and organizational priorities.

IBM's goal is to have students involved in classes that meet electronically by way of video/computer conferences. Assignments will be given and completed using electronic mail, and information resources will be the databases available through the network. Guest instructors (internal or external) will be "patched" into the network for lectures or interaction. Automatic translation tools will facilitate the interaction of students and instructors speaking different languages. At any hour of the day or night, a significant proportion of the IBM work force will be using the Global Education System. The information-handling skills acquired by using the Global Education System will be an "important aspect of learning to 'learn' - a key skill in the year 2000."

■ CASE STUDY

Hewlett-Packard's Distance Learning System Delivers Training at One-Half the Cost of Traditional Classes

"Traditional training methods imply a logjam of manufacturing and logistical delay," according to Tom Wilkins, Distance Learning Systems manager for Hewlett-Packard. "Between completion of course materials and delivery, a critical gap develops, sometimes as long as six months. The longer it lasts, the more likely it is that they will be outdated before they can be delivered to the students." The interactive classes developed for Hewlett-Packard's Information Technology are designed to close that gap and bring the interactive intimacy of classroom training to the field worldwide.

Hewlett-Packard's Information Technology Education Network (ITE-Net) is a pace-setter. In an industry-wide race to master this powerful medium, Wilkins has developed a network that can leverage limited expertise and can greatly reduce the time to retrain a large, geographically diverse population.

At the touch of a button, students from over one hundred classrooms worldwide are in direct contact with their teleclass instructor. The system provides instantaneous two-way voice communications between instructors and students. Instructors also receive immediate feedback from compiled student responses to numeric or multiple choice questions through a student response keypad system.

Spacious field classrooms can accommodate 1200 students in comfortable learning environments. Most major metropolitan areas have a conveniently located Hewlett-Packard office with the capability of receiving interactive instruction. A secure communication channel ensures privacy for marketing, support management, engineering, and customer training. The experienced ITE-Net staff can advise instructors on all aspects of teleclass delivery - from initial course development through to final production. Instructors are free to teach, with all aspects of the television production handled by the ITE-Net crew who view the class through a one-way mirror. Professional graphics are produced by talented technical illustrators and graphic artists.

The ITE-Net Uplink facility, located in Cupertino, CA, is a state-of-the-art system. For the instructor it features a fully integrated system which can be accessed by touching the screen of a video display. This consistent interface with the system enables instructors to easily use and control a vast array of educational tools. which include stored graphics, remote cameras, videotape roll-ins and drawing tablets. The instructor is positioned at a command center console. Through the console, the instructor can link to any of the remote classrooms and can involve students at remote locations, either with voice or data feedback systems. The console empowers the instructor to answer a question asked a continent away, to immediately display the slide that illustrates the issue, to zoom in on a precise portion of that slide and annotate it, and to prove the point with an online example. Seven television cameras are used to display the activities occurring in the classroom as they happen.

Students have equal access to the system and it gives them the sense of personal involvement that is critical to successful learning. The student response system places students at each reception site in immediate person-to-person communication with the instructor and fellow students through linked microphones embedded in their desks. They can electronically raise their hands, signaling a desire to be heard, ask questions, and enter into discussions at will. Another feature, a data collection ability, enables the instructor to ask questions of groups of students and display the tallied results on screen.

Wilkins feels that the ITE-Net's powerful facility is the "best available in the industry today" and that it places Hewlett-Packard "well ahead of the competitors who are struggling to keep up with the training demands of a dynamic industry." Wilkins says that the ITE-Net can "deliver educational programming for about one-half of the cost per student contact hour compared to centralized training. ITE-Net lets the instructor step out of the past and into the future of technical education, and to gain strides on the competition. Educational technology "presents the only clear answer to the emerging demands of the nineties, to constant, transformational change in products, in customer requirements, and in the skills and knowledge needed by Hewlett-Packard employees to support tomorrow's customers."

After the production of a course, video tapes are prepared for shipment. ITE-Net's lending library was created to make tapes available within two weeks of broadcast. A master set of course materials accompanies each borrowed videotape.

During its first two weeks on the air in 1987, ITE-Net broadcast training to 20 U.S. sites and a simultaneous taped class in Bristol, England for 300 commercial support engineers. Wilkins observes that traditional training classes would have taken 18 months to achieve the same results.

Wilkins believes that in the not-too-distant future, Hewlett-Packard training will be delivered to the employee's workstation. Wilkins designed ITE-Net and his projections of just-in-time learning delivered directly to the user is based on the needs he had for immediate information in his former career as an Hewlett-Packard support engineer. He envisions that a future workstation system would enable employees to access information needed to complete a current task. Through an educational center, the employee would request educational modules to be shown on the workstation monitor. Text or content materials would be printed as needed and tests would be included in the module which could be tracked over time to build up the history of effectiveness of the module as well as track the learning history of the student.

The ITE-Net has changed the way that Hewlett-Packard provides training to its worldwide group of employees. The promise for the future is that it will continue to evolve to meet the demands for just-in-time training delivered to the workstation enabling its employees to meet the demands of an increasingly competitive global marketplace.

■ CASE STUDY

NTU digital compressed video an Intetgral Step Toward On-Demand Delivery of Education at the Workstation

The National Technological University (NTU) made a giant technological leap by replacing analog transmission with digital compressed video delivered via satellite. The conversion is the first of its kind in the U.S. to integrate modern instructional technologies over a large, nationwide system. NTU designed an enhanced and proprietary ITV delivery system using the SpectrumSaver Satellite Based Broadcast Television System developed by Compressed Labs, Inc. (CLI). It is an integral step to on-demand education at the workstation. NTU offers master of science degrees in ten engineering fields. It has 41 participating universities and 385 sites in 130 organizations.

While the primary interest in the conversion has

centered on the digital technology, Dr. Lionel Baldwin, NTU president observes that it is an especially powerful example of human networks employing a complex package of technology networks to address a pressing need; the continued advancement of knowledge within the community of technical professionals and managers." Enormous forces have been "buffeting even the most established and respected institutions," Baldwin states. "Because of these conditions, new ideas and new ways of doing things are not only acceptable, they are being sought out by public and private leaders." Because NTU is a relatively new idea and a new way of doing things for many organizations, it "continues to grow 'in spite of' and also 'because of' today's tough environment."

The NTU digital satellite network is an ideal testbed for advanced instructional technology. Its human networks and technology networks form an infrastructure of outstanding instructors supported by technical staffs and state-of-the-art ITV facilities and, in many instances, workstation and computer networks. The NTU digital satellite network links these resources to a distributed learning environment.

Several NTU sponsors share a vision of an interactive, on-demand delivery of enhanced educational services nationwide by the year 2000, and will support the movement of NTU toward this goal with expert advice, operational personnel resources and special equipment. They see it as education to compete in the new global economy. "As the technological network goes all digital, as our human networks increase and improve, as our member universities become regional NTU distributors, and as NTU becomes a strategic partner with its corporate counterparts, NTU will become an increasingly significant force for strengthening U.S. engineering, management and technology in the global economic competition," according to Thomas L. Martin, Jr., former chairman of the NTU Board of Directors.

Martin, said that collectively, the trustees, staff and member organizations have begun to have a new vision for NTU as a "transnational university, a significant force promoting U.S. technological competitiveness in the global economic system through strategic partnering with the transnational corporations which are NTU's principal clients. To this end, international initiatives accelerated in Italy, Canada, Mexico, Australia, Japan, Korea, Turkey and Austria. Baldwin says that their goal is to "provide a two-way linkage so that U.S. technical professionals can have ready access to best practices abroad."

Marvin Patterson, director of Engineering, Hewlett-Packard and an NTU trustee, says that the goal of strategic partnering is to achieve "just in time" delivery of information by knowing what information is needed, when it is needed and getting it there on time in the most convenient and accessible form.

For this to occur, Martin said, NTU member companies must "include key NTU officers, program directors and faculty in their strategic planning processes so that the University can focus its resources along congruent and intersecting paths." Corporate officers must seriously regard NTU "as another critical asset at their disposal in achieving corporate objectives and must deploy that important asset constructively in their strategic plans. Parochial shackles limiting transfer of graduate course credits between NTU universities will have to go. NTU universities that are regional ITV operators should become regional NTU distributors so that all regions of the country can have full access to the entire range of program offerings available through NTU, both for degree credit and for professional development." Martin feels that the conversion may "pull the regional ITV systems into compatible digital systems. The resulting national digital delivery system with video, audio, data, facsimile and computer compatibility would provide an unparalleled national resource for the career-long education of engineers and managers" and a testing ground for new educational technology applications.

Gerald D. Prothro, assistant general manager of IBM U.S. Education, points out that: "As a customer, I want the best education, broadcast signal quality, educational delivery and education content, all at the lowest possible cost. NTU is working on these objectives Prothro said, "and the network digital upgrade promises to improve their capabilities ever further. NTU is today, and will be tomorrow, a vital adjunct to our own educational offerings." Prothro added that "since NTU and its member associated schools are the only university system which meets or is planning to meet these requirements, IBM strongly supports NTU and their efforts."

"Because of the very large financial investment and some potential technical uncertainties, this is an enormous, 'bet the company' kind of decision," Martin said, but they are convinced the potential benefits outweigh the risks.

NTU will become an increasingly significant force for strengthening U.S. engineering, management and technology in the global economic competition." The CLI system provides breakthrough economics in satellite based broadcast television and promises to revolutionize opportunities for NTU and other distance learning organizations which want to offer a broader curriculum more efficiently, flexibly and cost effectively.

NTU converted to digital because its future growth was being impacted by the technical and economic constraints caused by a lack of satellite transponder time. NTU has been a pace-setter for institutions delivering instructional television via satellite and will continue this role as a compressed video "testbed." A pioneer of the half-transponder format in distance learning, NTU has been broadcasting four channels of programming simultaneously over two satellite transponders which was not enough and led to course exclusions and tape-delayed classes, explained Tom McCall, director of the NTU Satellite Network. "Initially, we will support four channels," McCall says, "but by next summer we plan to move all night time and weekend broadcasts into weekday-daytime hours — all on one satellite transponder."

"The elimination of NTU's second transponder will cut its transmission costs in half, saving more than $1 million annually, "according to McCall. By tripling the network's capacity, NTU can broadcast all credit courses live, and provide faculty-student interaction over telephone lines during the program. "This frees the evening hours for a greater variety of seminars on demand, helping us do a better job of meeting the needs of our large numbers of continuing education students."

The NTU System

Baldwin said that the goal "is to convert NTU transmission to a state-of the-art digital system. We will use this breakthrough to improve the quality and timeliness of NTU service." He expressed confidence in the hardware NTU has selected and said he believes that although the technology is rapidly evolving, the NTU system is robust and will serve as a standard until long after a more powerful satellite comes on line for NTU in 1995. The technology will assure the continued technical vitality of the NTU network and create opportunities to expand and improve service. The total cost of the project is expected to exceed $5 million. NTU was awarded a $1.5 million grant by the Defense Advanced Research Projects Agency (DARPA) as matching funds for the installation of a state-of-the-art, high-bandwidth, digital compressed video and data system at NTU government sites and member universities.

NTU receive sites acquired a four-channel integrated receiver/decoder (IRD) from NTU. The unit was dubbed the NTU subscriber unit — NSU. In most cases, all other components in their downlink system will remain the same. Each NTU uplink will install an encoder to complete the network conversion.

The SpectrumSaver Encoder digitizes and compresses the video so it can be transmitted in just over 2 MHz of transponder bandwidth, a fraction of that required for analog video. Reducing the bandwidth also reduces the transmission costs. This will allow NTU to fit as many as 15 channels on a single satellite transponder (systems can be configured with up to 18 digitized channels as compared to a maximum of one to two analog channels). The NSU gets its input signal from the satellite antenna which contains the digital data stream that the IRDs convert to analog television with associated audio. The system control chan-

nel, an NTU proprietary feature, enables the transmission of facsimiles, VCR commands, and computer data. NTU has designed into the system a capability to set and control receive frequencies of all NSUs from NTU headquarters in Fort Collins, CO. All of these value-added features are delivered via satellite.

The 60 pound NSU is portable and uses 110 volt, 60 cycle power. It requires no special environmental temperature or humidity controls. It costs just under $9,000, but the cash purchase price to an NTU site is $8,500. The receiver cost per channel has dropped by a factor of 15 to about $2,000 per channel.

The enabling technology was created in early 1990, when U.S. innovators developed three new VLSI chips that make it possible to transmit video in an enhanced compressed digital format which provides a much clearer video image that is immune to ghosting, drop-outs, color smearing, and "snow" associated with standard analog broadcast techniques and older digital compression technologies. The new VLSI chip design employs greatly improved discrete cosine transform processes and, with better motion compression algorithms, it is now possible to produce compressed digital video that is equivalent to existing consumer analog video. The new compressed video format is a very robust, noise free, high-quality, full motion medium. Intensive tests have been conducted and McCall says the quality of the video images received at the downlink sites have been "consistently superior" to the corresponding analog signals, even under varying atmospheric conditions.

Each receiver is uniquely addressable via the satellite, with 32 discrete, changeable program keys that allow unlimited, highly secure cross-networking. A conditional access system with digital encryption, controls which programs are received at a network site (any combination of video, audio and data). A number of independent networks can be simultaneously supported. Programming can be uplinked from many different locations simultaneously through a common transponder, then downlinked to any SpectrumSaver receive site. By placing multiple channels on a single transponder, programmers will find it easier to implement cross network applications. Other system features include simultaneous facsimile transmission and a 19.5 Kbs data transmission line. System development and tests began in January, 1991.

NTU trustee Marvin L. Patterson, director of corporate engineering, Hewlett-Packard sees NTU as the underpinning for the just-in-time education work station and paving the way for education on demand. "A way of wrapping the reasons behind my endorsement for this transition can be best expressed in the form of a vision. Imagine, say, five years from now, an engineer is working on a project. Suddenly, one afternoon, he discovers that he needs knowledge about a new design technique and also the standards related to that new design technique — information he's never worked with before in his career, but it's absolutely in the critical path to his project. He puts in a request to NTU for all of the courses that relate to this area of information. A library search occurs on his behalf, and the information is downloaded, either via satellite or some other communications mechanism, but in digital compressed form, and directed right to the engineer's workstation. The engineer happens to be away at the time, but the workstation automatically stores the courseware on local optical storage. On his way to the airport, the engineer swings by his office, picks up the CD that has been created with this courseware on it, and then on the airplane en route to his destination he reviews these courses and gets the information he needs using his laptop computer. This might sound a little bit far-fetched, but the shift to digital compressed video that NTU is currently undertaking is an underpinning that could enable all of the things that I've described."

The goal of NTU's strategic partnering is to achieve "just in time" delivery of information and knowledge by NTU to its industrial partners. Knowing 'what' information and knowledge are needed, 'when' they are needed and getting them there on time in the most convenient and accessible form is the objective of NTU strategic partnering.

■ CASE STUDY

Pacific Bell

Cyril Tunis, Pacific Bell's executive director of Education and Training, has revitalized training and education by adding distance learning to the methods used by the telephone company. He combined open customer communications with cutting-edge technology. It has been a success despite budgets cuts and downsizing.

Now, there are over 100 classes and each has an average attendance of 100 students. This equates with 10-15 percent of Pacific Bell's total training hours now and it may rise to 30 percent in the future. With distance learning, students can learn at any of the company's 16 classroom sites. Everyone can have the training when it is offered on the first day; new training can be put into effect immediately across the state and employees don't have to wait for the instructor to get to their site. Essentially, the courses are the same ones that were taught by traveling instructors. Some hands-on training did not adapt to the medium, but most courses adapted readily and were redesigned for television.

According to Tunis, It's better for the instructor to stay in one place and interact with 100 people than to go around to 10 places and talk to 10 people at a time because 100 people get to interchange their experiences with one another. "We're finding that we are, in a sense, building a technical communications link. Each broadcast is more than a course because we can solve problems in the course. For example, ... we were giving a course on a new product that had some installation problems. The group talking the course was able to discuss those technical problems and broaden everybody's understanding of how to repair and avoid the trouble. It was like a 100-way telephone call."

Without the distance learning program, instructors would be required to go to the students throughout the state and talk with a groups that average ten or fewer students. Tunis reports that distance learning is much more cost effective.

The courses are transmitted via the company's fiber network. Classes are one way video and two way audio. Students can see and hear their instructor, as well as talk with him or her. Most programs are two to four hours. Full-day programs tend not to hold students' attention.

Tunis evaluates the programming according to the program's importance to key business strategies. Courses that pertain directly to business priorities are rated for return on investment (ROI). Others are rated on attendee reaction, attendee skill and knowledge and applicability. Tunis believes that this has helped build acceptance for the program.

■ REFERENCES

Barker, L. (971) "Listening Behavior." Englewood Cliffs, NJ, Prentice Hall.

Brock, W. E. (1987). "Guture Shock: The American Work Force in the Year 2000." American Association for Community, Technical and Junior Colleges Journal, 57 (4), pp. 25-26.

Carnevale, Anthony (1991). "America and the New Economy." American Society for Training and Development, and the U.S. Department of Labor, Employment and Training Administration. Washington, DC.

Carnevale, Anthony; Gainer, Leila J.; and Villet, Janice (1990). "Training in America: The Organization and Strategic Role of Training." ASTD Best Practices Series: Training for a Changing Work Force. San Francisco, Josey-Bass.

Carnevale, Anthony; Gainer, Leila J.; and Schulz, Eric. (1990) "Training the Technical Work Force." ASTD Best Practices Series: Training for a Changing Work Force. San Francisco, Josey-Bass.

Carnevale, Anthony; Gainer, Leila J.; and Meltzer, Ann S. (1990). "Workplace Basics: The Essential Skills Employers Want." ASTD Best Practices Series: Training for a Changing Work Force. San Francisco, Josey-Bass.

Carnevale, Anthony; Gainer, Leila J.; and Meltzer, Ann S. (1990). "Workplace Basics Training Manual" ASTD Best Practices Series: Training for a Changing Work Force. San Francisco, Josey-Bass.

Elsea, J. G. (1986) "First Impressins, Best Impression," Simon and Schuster/Fireside Books, New York, NY.

Fletcher, J. D. (1990). "Effectiveness and Cost of Interactive Videodisc Instruction in Defense Training and Education." Alexandria, VA, IDA Paper P-2372, Institute for Defense Analyses.

Hastings, C., Bixby, P., and Chaudhry-Lawton, R. (1986). "The Superteam Solution." Hampshire, England, Gower Press.

Keefe, W. F. (1971) "Listen, Management! Creative Listening for Better Managing."
New York, McGraw- Hill.

Kirsch, I. S. and Jungeblut, A. (1986) "Literacy: Profiles of America's Young Adults," Educational Testing Service, National Assessment of Educatinal Progress, Princeton, NJ.

Knowles, Malcolm, (1987). "Adult Learning." In R. L. Craig (ed.), Training and Development Handbook. (3rd ed.) New York: McGraw-Hill.

Knowles, Malcolm (1975). Self-directed learning: A guide for learners and teachers. New York, Cambridge

Knowles, Malcolm (1983). How the media can make it or bust it in education. Media and Adult Learning, vol. 5, no. 2 Spring. In Gueulette, David G. ed. (1986). Using technology in adult education. Glenview, IL. American Association for Adult and Continuing Education, Scott, Foresman/AAACE Adult EducatorSeries. pp. 4-5.

Lane, Carla (1990) "ASTD Sees Major Growth in Distance Learning." Ed, Vol 4 No3, pp. 15-17.

Nichols, R. G., and Stevens, L. A., (1957) "Are You Listening?" New York, McGraw-Hill.

Semerad, R. D. (1987) "Workers in the Year 2000: Why We're in Trouble." American Teacher, 71 (8), pp. 7-12.

Chapter 15

Dr. Virginia Pearson-Barnes is an internationally known speaker and consultant in communications and distance learning. As a former secondary school teacher, corporate trainer and university administrator, Ginny brings to her training and consulting a depth and breadth of experience in instructional design, teaching strategies for faculty who teach on TV and institutional strategic planning for implementing distance education programs. As an administrator and faculty member at Oklahoma State University and the University of Missouri she was responsible for the planning, development and growth of distance learning development and delivery of the high school courses delivered by satellite over the Arts and Sciences Teleconferencing Service (ASTS) to students in more than ten states.

At the University of Missouri Ginny was responsible for administering the delivery of credit courses by satellite over the National Technological Network to engineers at more than 250 corporate sites, and utilizing the University of Missouri Fiber Network for the delivery of credit courses between the four campuses.

For more than 12 years she has been an advocate for improving educational delivery and curriculum access by using improved technologies. Her research includes the study of barriers to implementation of distance education programs and a descriptive study of national distance education programs in all levels of education. Her current continuing research is in the area of strategic planning for the implementation of distance learning programs in higher education. She is now President and Founder of G. B. Communications Incorporated in Fort Smith, Arkansas. In her private practice she continues to support, encourage and assist educators in assessing their needs and improving service to clients by using distance learning technologies. Ginny has consulted and presented seminars, teleconferences and keynote addresses for government, industry and education worldwide, including organizations in the United States, Canada, Europe and south America. She is the author of numerous articles on communications and distance learning.

In 1988, Ginny was listed as one of the "Outstanding Women in Teleconferencing" by TeleCon Magazine. she has an A.B. in English/Education from Georgia State University, an M.A. in Communications from Purdue University and an Ed.D. in Human Resource Development/Adult Education from Oklahoma State University.

ORGANIZATIONAL PLANNING: OVERCOMING THE BARRIERS TO IMPLEMENTING TECHNOLOGY FOR INSTRUCTIONAL USES

■ ABSTRACT

Thirty administrators representing leadership of successful long distance training/instructional programs in organizations were interviewed and asked to identify the critical factors that should be considered in the development of a plan for the implementation of distance education programs for like organizations. Using a three round Delphi Technique this group of experts defined the twenty critical factors that they acknowledged and agreed were the critical planning issues in the implementation of long distance programming. Those factors addressed the need for human and fiscal resources, as well as the process of diffusion of barriers and produced a plan for the successful implementation of distance education programs.

INTRODUCTION

Technological advances in the telecommunications industry have fostered a rapid growth in the capability of the public and private sectors to use those capabilities for educational purposes. Advances in a variety of electronic distribution alternatives in video, audio, voice, data and facsimile communications, have allowed providers to choose technologies that can best serve their educational needs.

Institutions of higher education, community colleges and secondary schools are utilizing these varieties of telecommunications technologies to distribute instruction by long distance.

Across the United States, there are examples of institutions sharing programs of instruction offered for credit or non-credit. For example, the National Technological University in Fort Collins, Colorado, grants Masters of Science in Engineering degree programs by offering credit courses by satellite from participating universities to students located at member industry sites from coast to coast (Baldwin,1988).

The Texas Instructional Interactive Network (TI-IN) provides high school credit courses by satellite in math, sciences, foreign languages, as well as staff development for teachers. Oklahoma State University Arts and Sciences Teleconferencing Services (ASTS) delivers high school credit courses in math, science, foreign language and staff development for teachers to high school sites across the United States (Walters, 1988). The National University Teleconference Network (NUTN), a consortium of 200 universities and community colleges provides adult and continuing education by satellite (Grantham, 1988).

At Eastern Washington University, the STEP program provides high school credit courses by satellite to students (Cooper, 1988). The Missouri Educational Satellite Network (MESN) provides programs for secondary students and teachers through the auspices of the Missouri School Board Association (Gardner, 1988). In Iowa, community colleges, the public television station, the public schools and the institutions of higher education are sharing a network developed for the state's educational programs (Patten, 1987). In addition, faculty at Iowa State University are sharing information on teaching techniques through the "Teacher on TV" Program (Hoy, 1988).

Technologies of compressed video and fiber optics allow similar delivery of instruction with the additional component of two-way video interaction. The University of Missouri Video Network connects the Columbia, Rolla, St. Louis and Kansas City campuses by a fiber optic network that allows the distribution of credit courses from one campus to another (Sarchet, 1986). At California State University-Chico, credit and non-credit programs are being offered to business and industry by satellite and over a fiber optic/ITFS network, as well as sharing courses among the state campuses (Meuter, 1988).

This study focused on distance education provided via interactive, televised instruction delivered by satellite, fiber optics, ITFS, microwave or cable. The term distance education implies the long distance delivery of instruction through one or more of the afore-mentioned delivery systems.

THE PROBLEM

Despite the apparent growth in telecommunications capabilities that are available for education, and the current groups and networks that are delivering programming, the implementation of distance education programs remains an arduous process for people and institutions. Recent studies have indicated barriers to implementation of distance education programs; as well as advantages for implementing the technology in education (Barron, 1987; Evans, 1982; Rockhill, 1980; Seidman, 1986; Wilson, 1987; Lewis, 1985; and Wagner, 1984). However, no studies have been conducted to indicate the critical issues in distance education planning that need to be addressed prior, during and following implementation. Concentration on the advantages and disadvantages of implementing distance education programs was the approach, rather than structuring a strategic plan for implementation.

Boles et. al. cited Irvine (1983) in his discussion of the planning process for education and its future:

Imaginative planning and vigorous action are necessary to maintain a viable educational system. The educational system of the future will be shaped by men in purposive fashion, or it will by default, be shaped by accident, tradition and the senseless forces of environment (p. 383).

While the proliferation of telecommunications technologies has provided the means to deliver long distance instruction in a variety of ways, strategic planning for implementation of those programs is critical for initial and continued success (Souchon, 1985).

REVIEW OF THE LITERATURE:

Early issues in diffusion of innovation and barriers to implementation of technology in education were centered around cost, type of equipment, quality, difficulty to use, multiple uses, evaluations, comfortability and culture.

Olgren and Parker (1983) commented on user acceptance in a review of technology and its applications. They concluded that "user acceptance and sustained applications are two of the most important human factors...and they require as much, or even more planning than the technical design" (p. 238).

Bell and Weady (1984) cited the importance of human factors in the adoption and successful implementation of teleconferencing in an organization. According to the authors, human factors need to be considered as systems are developed.

People are not going to use a technology simply because we think it is a good idea, or because we think it will save them time and money. If we want people to accept and adopt teleconferencing we must design the technical structure and the human interface to be both initially and lastingly rewarding (p. 299).

Lawry (1986) stated that decision making for the educator can reflect similar processes to those for business and industry and to the early innovation decision process of Rogers. Yet, according to Souchon (1986) educators had difficulty in that decision process. He stated: "Much remains to be done before education defines its objectives and the world of communicators, in turn, opens its mind to the problems of education" (p. 159).

Kaye and Rumble (1981) cited the problems in decision making for educational institutions as they proceeded with distance learning systems. Armsey and Dahl (1973) concluded that problems in decision making and acceptance of the technology needed certain conditions for success. Among those were the 1) recognized existence of a need, 2) articulation of a purpose and guide, 3) identification of a structure, 4) leadership of the innovation, 5) teacher participation and support, 6) appropriate technology, 7) evaluation mechanism and 8) adequate resources for the beginning and duration of the project (pp. 101-103).

Pacey (1983) referred to the "machine mysticism" barrier to acceptance or adoption of a new technology (p. 24). According to Pacey, people tend to think that technical advance leads progress, rather than using technology to answer new patterns of problems that arise. Moreover, he stated that this point of view leads people to believe in the "myth that cultural lag occurs in every community as people try to keep up with their progressive technology" (p. 24).

Barron (1987) reviewed the literature on the study of barriers to implementation of technology in education and concluded that, while little research had been conducted, the acceptance of teleconferencing in higher education had become widespread, but televised delivery of classes, especially in graduate education, was "considered with more hesitation and suspicion by some educators" (p. 3). Barron cited Dirr's major barriers to the implementation of courses as 1) lack of money to support the effort, 2) lack of

faculty commitment and 3) lack of trained support staff (p. 4). Barron (1987) found that faculty had concerns for the students, the size of the classes, discussion and face-to-face involvement and lack of support for themselves from peers and instructors. In a second study Barron (1987) asked faculty to rank the technologies in terms of use and then secured data on the perceived barriers to that use. He, furthermore, reiterated the need for additional studies of delivery media and modes of instruction, including all aspects of distance education.

Evans (1982) studied faculty responses to the use of television as a delivery method. He addressed the decision-making process and the psychological barriers to using the technology. Rockhill (1980), cited by Barron (1987), indicated in his study of barriers to implementation of distance education programs at institutions, similar reasons for non-use as Evans. Those cited included finance, compatibility and comfort.

Additionally, Seidman (1986), Wilson (1987), Lewis (1985) and Wagner (1984) studied the advantages of implementing distance education telecommunications technologies. They found that cost efficiency, access to programming and enrichment were reasons for use cited by teachers. Similarly, Roark (1985), Lamp (1985) and Hassan (1984) studied the adoption and satisfaction of telecommunications technology by teachers and institutions.

■ IMPLEMENTATION: FACTORS IN STRATEGIC PLANNING

In a recent study conducted through Oklahoma State University's College of Education in cooperation with the University of Missouri, thirty key leaders and administrators in the field of telecommunications, distance education and television production were interviewed through the Delphi process and asked to assist in designing a strategic plan for the implementation of distance education programs in organizations. Because each delphi panel member had administered a successful, ongoing distance education program, they had experienced the process of planning for implementation. The group of participants defined the ten steps and factors that they acknowledged and agreed were the critical planning issues in the implementation of long distance programming.

Those factors followed specific adoption processes as well as a definitive strategic plan to ensure that distance education programs would be successful once they were implemented. Ranked in order of priority of planning, those participants designed the following plan and commented as follows:

Decide to Plan for Change: Awareness

Key Administrators must plan to provide an environment that is flexible for change. Understanding elements of change is a necessary process.

This philosophical stage is frequently ignored. It is an often unidentified, unspoken message, nevertheless it is the first real step in the willingness to be innovative. Usually, a "super leader" in the organization ensures this ongoing flexibility. It is this person who is enthusiastic and gathers support.

Recognizing a Real vs. Perceived Need: Interest

Identifying the recipient of the programming. Asking why the program should happen. Is it to jump on the technology bandwagon? or provide a needed service? Looking at what other competition is already doing. Is the duplication a justified, needed, improvement?

Understanding the Real Reason for Implementation: Advantage

What is the value to the organization? Is it to compete? Make money? Help clients? Political issues involved — who wants the program and why? Is it competition driven for competition's sake? Philosophy of the program must be in place. Why implement the program? Culture of the organization affects the programs. Is the program congruent with ongoing belief systems? Mission of the Organization: Evaluation. Does the programming fit the goals and objectives of the organization? Is the quality of the programming up to the standards of the institution? How will this help the organization: and if it won't, don't! What is the driving force to market the program: Will it make money? Will it be self sufficient? How large do we want it to become? What is the return on investment? Who is supporting the program outside the institution.

Planning the Program: Trial People

Who will be involved? — Do we need more positions and staffing? Space — How many buildings or rooms do we need? Production — How capable is the present TV center? Money — How much to invest? — How "grand" a program? Equipment — What is the right technology for this program? Is equipment available?

Review what the Organization does now? Observability — Will distance learning duplicate ser-

vices? Is the organization working well in training and education? Does the organization support education and training? Does the organization support change and technology? Do we have enough people and support to add change? What are the organizations strengths and weaknesses?

The GAP: Compatibility

How far do we come to get to a successful program? Will the organization be able to change and adjust? Analysis: Subtract the difference between where we are now and where we want to be.

Contingency: Pre-adoption — Pilot

What happens if it is different than expected? How flexible can we be to make adjustments? Concerns for client needs and institutional perceptions surface again. Is the institution prepared for success? for growth? What happens if it is more, better, can, does...! Implementation: Adoption — an ongoing process ... a cycle including leading people designing programming training in production. Continued training in technology and faculty support and training are considered necessary, as are financial support, auxiliary materials and continuing resources. Planning for growth, for change and believing in the program will garnish support continually. Evaluation is an ongoing process.

These issues for planning distance education programs require elaboration and personalization by organizations for each stage of the strategic planning process. The strategic plan requires that each step be taken sequentially to ensure that appropriate planning is continued. This planning assures that implementation occurs with good reason and justifiable needs. Without it, the barriers and disadvantages resulting from combining technology and education will continue to surface.

■ SUMMARY

Many educators remain hesitant, for the most part, to use the new technologies for educational purposes. Studies in both innovation and decision making theory, as well as research in the use/non-use of the technologies indicate that more research needs to be conducted to bring together the educator and the appropriate technology for the message. While there is evidence of the barriers related to cost, compatibility, comfortability, communication and support, research also indicated that the positive presence of those same factors act as "advantages" for use.

Strategic planning to implement distance education programs enables organizations to plan for these barriers and optimize the advantages of the partnership between education and technology.

■ REFERENCES

Acker, Stephen R. (1985). The teleconference trialability lab: fostering the diffusion process through organizational learning. In Lorne A. Parker and Christine H. Olgren (Comp.), Teleconferencing and Electronic Communications IV: Applications, Technology and Human Factors, 210-215. Madison: Center for Interactive Programs, University of Wisconsin-Extension.

Armsey, J. W. and Dahl, N. C. (1973). An Inquiry Into the Uses of Instructional Technology, New York: The Ford Foundation.

Baldwin, Lionel (1986). An alternative concept in engineering education. IEEE Potentials, 5 (2).

Barron, Daniel D. (1987). Faculty and student perceptions of distance education using television. Journal of Education for Library and Information Science,27, (4).

Barron, Daniel D. (1987). The use and perceived barriers to use of telecommunications technology. Journal of Education for Library and Information Science, 27, (4).

Bell, Bonnie and Weady, Cakier. (1984) Human organizational factors in implementing technology. In Proceedings of International Teleconference Symposium, 299-306.

Benson, Gregory M. and Hirschen, William (1987). Distance learning: new windows for education. T.H.E. Journal. 63-67.

Boles, Harold W. (1983). Introduction to Educational Leadership. London: University Press.

Bretz, Randall G. (1985). Satellite teleconferencing in continuing education: what lies ahead? In L. Parker and C. Olgren (Comp.) Teleconferencing and Electronics. Communication II: Applications, Technology, and Human Factors, 387-389.

Cooper, Penelope (1988). Personal Interview. STEP. Seattle, WA.

Evans, R. I. (1982). Resistance to innovation in higher education: a social perspective. In B.S. Sheehan (ed.). Information Technology: Innovations and applications. New York: Josey-Bass.

Gardner, Hal (1988). Newslink. Stillwater: Oklahoma State University, June.

Gubser, Lyn (1986). National task force on education technology. TechTrends, 11-23.

Grantham, J.O. (1988). Personal interview. Oklahoma State University.

Hawkridge, David (1983). New Information Technology in Education. Baltimore: The Johns Hopkins University Press.

Hoy, Mary (1988). Teacher on TV. Teleconference 7 (2).

Kaye, Anthony and Rumble, Greville (1981). Distance Teaching for Higher and Adult Education. London: Open University Press.

Knapper, Christopher (1980). Evaluating Instructional Technology. London: Halsted Press.

Lamp, E. Joseph (1985). Predicting levels of satisfaction as a consequence of innovation adoption. (University of Maryland.)

Lawry, Constance (1986). Staff development teleconferences for teachers: a case study of their use and suggestions for improving their effectiveness. (Oklahoma State University.)

Lewis, R. J. (1985). Instructional applications of information technologies. Postsecondary Adult Learning Telecommunications Market Research Study. Columbia, S.C.: Center for Instructional Communications Southern Educational Communications Association.

Meuter, Ralph (1988). Personal interview. California State University-Chico.

Olgren, Christine H. and Parker, Lorne (1983). Teleconferencing Technology and Applications. Mass: Artech House, Inc.

Osborne, Denis G. (1984). Science, technology and educational change. In Reflections on the Future Development of Education, 123-134. Paris: Unesco.

Pacey, Arnold (1983). The Culture of Technology. Cambridge: The MIT Press.

Parker, Lorne (1984). Teleconferencing applications and markets. In Proceedings of the International Teleconference Symposium, 101-103.

Patten, Larry (1985). Report on Telecommunications. Iowa State Assembly. Ames, Iowa.

Pease, Pamela (1988). Personal interview. TI-IN, Austin, Texas.

Portway, Patrick (1987). Educational applications for teleconferencing. Teleconference. San Ramon, CA.

Roark, Denis (1985). Factors affecting the implementation of new educational technology on higher education. (The University of Arizona.)

Rockhill, K. (1983). Academic Excellence and Public Service: a History of University Extension in California. New Brunswick: Transaction Books.

Rogers, Everett (1988). Diffusion of Innovations (3rd ed.) New York: The Free Press.

Sarchet, Bernard (1986). Engineering courses by fiber optics. Engineering.

Schaeffer, Pierre (1984). The impact of the media on general education. In Reflections on the Future Development of Education, 165-174. Paris: Unesco.

Seidman, Steven (1986). A survey of school teachers' utilization of media. Educational Technology, 19-23.

Souchone, Michel (1984). Education and the media: prospects for cooperation. In Reflections on the Future Development of Education. Paris: Unesco.

Wagner, L. (1984). Cost efficiency of distance technologies. In C. Levinson (ed.) Report from the Regional Forum on Distance Learning. Austin, Texas: Southwest Educational Development Laboratory, 29-33.

Walters, Leigh B. (1988). Personal interview. ASTS: Oklahoma State University.

Wedemeyer, Dan J. (1986). The new age of telecommunications: setting the context for education. Educational Technology, 7-13.

Willis, Norman E. (1985). Educational technology: support for improvement of learning. Council for Educational Technology.

Wilson, Savan (1987). The sky's the limit: a technology primer. Journal of Education for Library and Information Science, 27 (4), 239-244.

Chapter 16

TRENDS

by Carla Lane, Ed.D and Patrick Portway

A flood of new communications devices and services are on the way; legislation and rulings has opened up the way for telephone and cable companies to compete. The introduction of legislation for the National Information Infrastructure (NII) and the Communications Act of 1994 had been introduced but not passed as this edition went to press. Most of the squabbles involve splitting the limited number of available radio frequencies among new and existing communication services. Advances include videos on demand, delivered electronically to the home over phone and cable lines; interactive newspapers and TV shows, which allow users to custom-tailor the information and entertainment they receive.

■ THE MERGER OF COMPUTERS AND VIDEO

The most significant trend is still the merger of the computer and video into multimedia desktop terminals. The technologies that are converging are computing, television, printing and telecommunications. Bringing them together results in the whole having greater impact than each individual part and is one of the industry' most significant developments. The convergence of digital technologies and their use will impact the future of teleconferencing, distance learning, business, and entertainment. By joining television and computers, the best aspects of each technology are combined. The result is a powerful communications and information system that joins TV's ability to introduce and highlight a subject with the computer's ability to provide in-depth information tailored to immediate needs. The computer changes existing media by helping one find, store, search, and re-use many kinds of information. The movement is still toward digital high definition television.

■ AGE OF COMMUNICATIONS

Based on our review of the industry, we believe that yet another shift has occurred. It was largely accepted that the Information Age began in 1985 and has probably ended in 1994. We have now moved into the Age of Communication. This age is strongly defined by telecommunications used for gathering and disseminating information. The public's discovery of the Internet was prompted by discussions of the National Information Infrastructure and flamed by mergers. The Communications Age is characterized by the general acceptance of the public that want to be able to use telecommunications interactively as a personal tool. In the Information Age, the public was content to receive one-way communication. This represents another paradigm shift. Now employees are empowered with access to two-way communications, access to information and by companies that have downsized and are enlightened about employee empowerment. In education, the paradigm shift has been enabled through instructional methods that empower the student such as facilitation, two-way communication networks, and access to information through electronic networks. Because of funding schools are slightly behind business, but Goals 2000 funding will decrease the gap. The Communications Act of 1994 should include a universal access clause for education that was left out of the 1934 Act.

■ ASYNCHRONOUS TRANSFER MODE - ATM

Current packet switched LAN technology is not friendly to video transmissions. The ATM protocol would handle video data and voice in the LAN/WAN environment as well as internationally. ATM is widely supported as the direction of the future.

■ CHIP DEVELOPMENT

Chip development will provide faster and lower cost videoconferencing on the computer screen. Intel plans to offer two-way videoconferencing on computers free — except for the cost of a small camera ($100) by the year 2000. The capability will become a standard feature of all computers.

■ INTERACTIVE TELEVISION — ENTERTAINMENT

This is a system that connects with the cable television system. Computer programmers set up programming to work with game shows, etc. People with the system can play along with the game. One version of the system works on a small radio transmitter for which the FCC has allocated spectrum. Other versions operate over the cable or telephone lines.

■ VIDEO ON DEMAND

Digital video stored on servers can play out movies or other programming whenever the consumer demands and is accessed over cable or phone lines. This service in competition with videotape rental stores is expected to be a $30 billion business in the late 90s.

■ INTERNET/INTERACTIVE TELEVISION — EDUCATION

CU-See Me is a software program that provides two-way interactive video and audio over the Internet. School children are already using it to share data. Viewpoint, BBN and others have introduced software that works over the Internet.

■ CELLULAR COMPUTER NETWORKS

Major computer marketers, regional Bell operating companies and other technology vendors have targeted wireless, mobile data communications as the next hot growth area. The market potential for cellular-based data services is huge; the field is expected to attract 2.6 million customers nationwide by 1997. The wireless data market will hit $175 million in 1995, up from $18 million in 1992. Five factors will drive the growth; increased use of laptop computers; availability of small notebook and palm-top machines; the perfection of personal digital assistants for the mass market; reduced costs of transmission.

■ COMPUTER MOVIES

Short movies that play in a screen on your computer have become another way for users to create their own media. Using digitized video footage, morphing programs, animation, or video stills, movies can be easily made. They can be the new family scrapbook or the way to deliver product information

without mailing a video tape. The digital fusion in the movie industry is at a peak. When John Candy died in the middle of shooting a movie, his film image was digitized and inserted into the remaining scenes.

■ COPYRIGHT AND MULTIMEDIA

Intellectual property rights in the multimedia environment are going to be a major problem. To create the perception of choice in multimedia requires much more material than linear media. Some industry professionals feel that acquiring intellectual property is so costly and problematic that multimedia developers should produce everything themselves. This is an infant industry with an enormous hunger for content but no easy way to pay the bills. Multimedia developers hope that licensing will give them the ability to obtain existing intellectual property and spare themselves the cost, time and effort of creating the content. They hope to acquire only the rights they need at minimum cost, but are not sure what rights they need to acquire — and content owners are not sure what rights they are willing to license.

Technologies are changing, market practices are still evolving, the size of the market is unknown and the relationship of multimedia markets to traditional markets is undefined or ambiguous. Today multimedia developers primarily use stand alone storage-based publishing devices. As wideband transmission becomes easily available, publishing via networks will be commonplace. This is the emerging model in higher education. It discourages new users and the experimentation and exploration that is needed to stimulate and build demand. Nobody likes to hear a meter ticking. There is resistance to metered information as it is hard to budget; fixed costs are preferable.

■ DESKTOP VIDEO

Desktop video will transform the video post production industry and computers are already having a huge impact on producers of video programming. The recession forced people to take a look at alternatives for producing video. The traditional video production studio with several rooms and $2 million in equipment may be history soon. Breakthroughs in digital storage technology and better video compression techniques will continually advance desktop video. Dial-up digital networks are becoming the rule. The need for total solutions has had an effect on the technology being developed. Document and information sharing solutions like document cameras, user cameras, annotation tablets and computer interfaces are demanded by customers. The customers are becoming increasingly more sophisticated in their use of video and are demanding that vendors develop complete solutions that encompass more than just audio and video. It must be high quality and make working across distances easier. Is the desktop destined to become the major video battleground of the late 1990s? The race will be won by those who can provide the same high level of functionality and quality on workstations and videophones as exists on group conferencing systems. The systems must add value to the installed base and not create incompatible islands of technology. The winners will focus on building value throughout a family of products while teaming with partners that can focus appropriate video solutions into the right market channels. If there's a lesson to be learned from the past seven years in the industry, it is that networks will follow the desktop applications. Increased use of multimedia information and desktop videoconferencing will drive the demand. The other major development that will spur desktop video is the widespread support that is building for ISDN under the aegis of the national ISDN-1 (NI-1) program sponsored by Bell Communications Research, the common research and engineering arm of the seven Regional Bell Holding Companies. NI-1 is a standard that has been endorsed by the major producers of ISDN-related network equipment, such as switches, terminals and telephones. This will end the incompatibility problems that have been ISDN's bane. Add NI-1 to the arrival of low-cost videoconferencing products, and you have the makings of a paradigm shift in communications.

■ DIRECT BROADCAST VIDEO BY SATELLITE

Direct broadcast satellite companies n ow offer to broadcast entertainment directly to the home. Low subscription and equipment fees will define the success of this new delivery method. They have the jump on the cable companies and telcos which are just coming out of test phases for video on demand.

■ DISTANCE EDUCATION

The use of telecommunications technologies for distance education will continue to increase as educators deal with increasing numbers of students. It will become even more apparent that the ability to share resources through technology is a viable alternative to building more buildings. The need to retrain 50 million American workers and military personnel who have been mustered out will be a driving factor in the continued adoption of distance education. Distance education will be used to bring credit and continuing education programming into the school, workplace, and home. The impact of the new technologies will be felt in all areas of education and will take distance education to a different kind of level with the new desktop video conferencing systems lead students into more involvement with one another. It will help students develop a better sense of the world. As the new technologies stabilize, the expense will drop and make access to others as well as learning resources very cost effective. Computer and audioconferencing for all educational organizations will become an important part of distance education. Instructional designers will need to learn how to weave the use of the technologies into their methods.

■ ELECTRONIC MAIL - (E-MAIL)

If you've viewed e-mail as a convenient way to communicate, get ready for an explosion of e-mail into a strategic business resource the way voice mail did a few years ago. Mail-enabled applications, which promise to revolutionize a range of business procedures, will help organizations drastically reduce paperwork and errors and increase efficiency, on top of the savings in reduced phone and mail charges. E-mail has already evolved into a store-and-forward technology that handles large data files and documents, and can be extended beyond the enterprise to suppliers, customers, and trading partners worldwide. Just as you wouldn't consider installing a phone system without connecting everyone, you should apply the same thinking to e-mail. Commercial providers of access to the Internet will continue to flourish. As more people become connected to the Internet, faxing will fade away as we increase our use of computer faxing or attach formatted documents to e-mail messages over commercial services. More private services will arrive to handle private electronic meetings through proprietary software. As large organizations such as associations or trade groups find that they need ways to connect their members. The medical profession may find these services appropriate to coordinate patient treatment. Small companies which do not want to invest in servers to connect widely dispersed employees may also find the solution appropriate. These are just a few of the ways that private computer conferencing networks will provide services. FCC In July, 1992, the FCC took sweeping steps toward changing the ways Americans use their televisions and telephones. The regulation changes also took an important step to bring competition to the cable industry. Video dial tone will provide significant competition to cable television companies which operate as monopolies in most communities. The rulings allow local phone companies to pursue some combination of two options: The phone company could act as a carrier for cable companies and information service providers, bringing their signals into the home on its high-capacity fiber optic cable. Or it could offer movies and television programs on its own, competing directly with cable companies. The FCC is moving closer to giving broadcast television stations two channels; one for regular transmissions; one for high-definition broadcasts. It has also sold bandwidth for pocket telephones and hand-held, wireless computers.

FIBER AND BROADBAND TO THE CURB

Fiber optic cable is being installed by many phone companies because it has the ability to carry voice, data and video. The installation has been lagging, but with the FCC regulation change the telcos now have a major incentive to speed up the process and begin delivering programming. Fiber from the curb into the home will be the last obstacle for this transition. Fiber will pass by the home. It will be up to the homeowner to decide whether to pay for the final few feet of installation.

HIGH DEFINITION TELEVISION (HDTV)

The U.S. HDTV standard will be digital. Zenith's digital approach has been chosen by the FCC as the model for development of a U.S. standard.

INTERACTIVE NETWORKS

New interactive devices will allow viewers to play along with their favorite game show, sporting event, or murder mystery. Interactive Network of Mountain View, CA, has been testing a product in Northern California. The heart of the system is a $200 portable control unit.

While TV shows are being broadcast, Interactive Network employees sit at computers and program the information that is sent to the control unit. For example, during "Jeopardy," when an answer is shown on the TV screen, the Interactive Network technicians send four possible questions over an FM radio signal that is picked up by the handset. The questions are shown on a small liquid crystal display screen and one answer is chosen by pressing a button on the control unit. When the correct answer is given on the show, the Interactive Network technicians immediately send the information over the radio waves to the handset. If you are correct the handset adds points to your score. When the games are finished, the control unit adds up the points. By connecting the telephone cord that comes with the unit to the phone outlet and calling the

score to Interactive Network, a player can compete with other interactive players for prizes.

The company has been loaning out the units at Giants and Athletics games in an effort to get people familiar with the technology. To play, you first predict the outcome of the batter's trip to the plate. If you guess an out, you then predict how that will happen. If you guess a fly out but the batter grounds, you'll receive points for being half-right. Throughout the game the control unit displays the latest scores from other baseball games, much like the scoreboard at the ballpark. You can also get information on a particular player's batting statistics — number of hits, strikeouts, etc. — and team statistics at any time during the game.

Here's how it works. Interactive Network producers watch the telecast and enter game calls and statistical information. From the central computer of the network, game control data is shipped to FM stations and Interactive Network game data is simulcast along with the television broadcast. The control unit uses a telescoping antenna to receive the FM radio signal that carries the information (it may be necessary to use an FM booster). At the conclusion of an event, subscribers connect the handset to their phone cord for a 20-second call which is transmitted over a telephone digital switching network. All participants' scores are collected, results and standings tabulated, and then broadcast back to each subscriber in four minutes. The control unit has a long-lasting rechargeable battery.

KNOWBOTS (SHORT FOR KNOWLEDGE ROBOTS)

As more information becomes available knowbots will search for new information according to guidelines set by the user. The knowbot will automatically and regularly search. This software program will be the equivalent of having a reference librarian at your fingertips which can quickly sort through the vast amounts of electronic data. Knowbots will be able to assemble the equivalent of a personal daily newspapers, magazines, or information on any topic. Searches will be done on public databases as well as private or premium pay services to which the user subscribes. At its extreme, the knowbot or intelligent agent, becomes an intelligent partner in mediating human communication.

LANs AND WANs

The LAN industry will continue to shape and mature in the late 90s as users continue to grapple with interoperability, service and support. One LAN expert says that the LAN industry is experiencing massive market upheaval. The new corporate networks will feature heterogeneous multivendor, multiprotocol environments. They will be mainly PC LAN based as users continue to downsize or rightsize their computer operations from host terminals connections to LANs. Client/server applications will become a reality as users demand more comprehensive, robust applications. Instead of mainframes, networks will be the center of focus. Users will have a choice of an expanding array of less costly products (because of stiffer competition), and continue to grapple with interoperability of new and old products. At the same time, they've lost the security of knowing where they will be in five years.

■ LOW EARTH ORBITING SATELLITE — LEO

Satellites orbiting a few hundred miles up will form a moving set of communications cells passing overhead. There is negligible delay and a small surface can be used as an antenna (i.e. the lid of a laptop computer). Some systems, called limited LEOs, will cover only populated areas. Others, like Motorola's, Iridium, will cover the entire earth with phone and data services. This is called a full LEO.

■ MULTIMEDIA

The promise of multimedia is to move more information more easily by doing it electronically and to provide more resources to everyone. It is no longer a technology in search of an application. The enabling technologies are not all in place, but it is becoming clear that the true multimedia platform is more likely to be something different. It will house a microprocessor, but we probably won't think of it so much as a computer as we will think of it as a telecommunications instrument. To date, it has been suggested that multimedia will become a market only when the communications providers have a national fiber optic infrastructure capable of handling the massive bandwidth that each of us will need. On the other hand, we may not have to wait years for multimedia to become a telecommunications reality. Multimedia will become a preferred communications vehicle for entertainment, advertising, and education. Animation, video, and sound will proliferate throughout new interactive applications.

■ INTERNET AND THE NATIONAL INFORMATION INFRASTRUCTURE (NII)

The Internet is an existing worldwide system for linking smaller computer networks together including governmental institutions, military branches, educational institutions, and commercial companies. There is no surcharge to send or receive messages through Internet. Only ASCII messages up to 50,000 characters can be sent through this system. Thousands of users are being added to the Internet on a weekly basis as the commercial services such as CompuServe, America Online, e-World, and Delphi open connections to Internet. Using the Internet can be daunting for new users. However new White and Yellow Pages for the Internet will make finding groups and information easier. Still, many users prefer seeing the Internet through an easy to use graphic interface. More direct graphic interfaces are being developed such as Mosaic and GINA which telnet and ftp at the click of a button. Audio interfaces are also coming.

MBONE, the Multicast Backbone of the Internet, can carry live audio and video to sites around the world. The idea is to construct a semi-permanent IP multicast testbed to carry the IETF transmissions and support continued experimentation between meetings. The MBONE is a virtual network. It is layered on top of portions of the physical Internet to support routing of IP multicast packets since that function has not yet been integrated into many production routers. The network is composed of islands that can directly support IP multicast, such as multicast LANs like Ethernet, linked by virtual point-to-point links called "tunnels". The tunnel endpoints are typically workstation-class machines having operating system support for IP multicast and running the "mrouted" multicast routing daemon. The NII will be a direct descendent of the MBONE.

■ PDAs - PERSONAL DIGITAL ASSISTANT

The first round of personal digital assistants (PDAs) was not successful because the technology wasn't ready; but the idea is still appealing. Apple Computer's Newton was the most notable of the releases that weigh about a pound and use a special unattached pen for data entry. While Newton seemed to be a handy gadget, it couldn't get the hang of recognizing its owner's handwriting and failed primarily because of that flaw. New and improved software was released, but the market was too disappointed to notice.

■ PHOTONICS

Photonics may be the final answer to fiber optic. If you're installing fiber, make sure your planned installation can include photonics. Photonics are enabled by gallium arsenide integrated circuits for optical interconnections within and between computer and communication equipment.

■ FIBER OPTICS

Present fiber optic computer and communication links are limited by discrete component electronics. In development is a projected 32-channel parallel monolithic IC connector which could vastly increase performance and drive down costs to open up long-sought new fiber optic markets which can replace current copper-wired connections. The cost of fiber optic cable is not drastically more expensive than copper or coaxial cable. The cost of fiber is driven up by the cost of the connectors.

■ STANDARDS

There will be a continued enhancement of industry standards as technology progress, applications evolve and customer's grow more sophisticated. Standards are due soon for high resolution graphics, as well as encryption and multipoint video. Storing the video signal is a problem as it takes a 300 megabyte hard disk to store just 10 seconds of digital video. Compression is the answer to storage problems. Future digital-video products will offer compression ratios of 50: 1 to 500:1. JPEG (Joint Photographic Expert Group) is an industry standard for still-image compression that is moving into full-motion video. MPEG (Moving Pictures Experts Group) has a three-part compression standard for professional and consumer application — digital video, digital audio and systems compression. MPEG compression compresses similar frames of video, track elements which change between frames and discards the redundancies. This allows full-motion video to be sent at CD-ROM data rates — around 160K per second.

■ MPC

A multimedia personal computer standard describes a PC that can run Microsoft's Windows efficiently because the system software beneath multimedia would be "Windows with Multimedia Extension. (The specification calls for added audio and CD-ROM hardware).

■ TEAMWARE OR GROUPWARE

The most successful organizations use employee teams to solve problems with minimal management intervention. The groups are organized on an as-needed basis and will draw on the expertise of many employees. The critical factor in the success of the team's competitive solution may be the speed with which they are able to communicate and make themselves understood. Electronic mail and desktop videoconferencing promises is part of the team solution, as it gives a company a competitive edge. Teamware is works on decentralized LANs, and ties together loosely organized groups of people to allow them to work together more effectively. With it the group can easily share information, track their work and collaborate on team projects by sharing documents, audio and video. Everyone has to use it, so teamware must be intuitive and easy to use. Teamware is an educational solution for distance learning through computer conferencing but it hasn't reached the adoption stage yet. Few companies are using the team-approach effectively. Those who seize this opportunity early will have the competitive edge.

■ TELECOMMUTING

The move toward telecommuting will continue to increase as large metropolitan areas such as Los Angeles realize the benefits of telecommuting for the environment and businesses realize the benefits of productivity from telecommuters. People can telecommute in three ways: by working from their homes, by driving to a satellite officer operated by their employer or by driving to a neighborhood telecommuting work center that provides facilities for a variety of employers. Telecommuting offices, centers where employees can go that are close to their homes will continue to increase. Employees with the ability to telecommute will spend more time conducting business this way. The use of cellular and portable equipment such as telephones and computers will continue to drive this trend.

In California, telecommuting has made major strides in the state's government and education centers in the past several years. It has been spurred largely by air quality regulations requiring employers to cut down on traffic congestion and lengthy commutes or face fines of as high as $25,000 per day. The energy saved by six or seven telecommuters in one year is equivalent to the average U.S. annual household energy consumption. If only five percent of commuters in Los Angeles County telecommuted only one day each week, they would save 9.5 million gallons of gasoline in a year.

Telecommuting in the San Francisco Bay Area could reduce emissions of carbon dioxide, the major cause of global warming, by 100 million pounds per year. It would take about 10 million new trees to absorb that much carbon dioxide from the atmosphere. The neighborhood Telecommuting WorkCenter of Riverside County, CA, reports that the average commute for its telecommuters is 10 minutes, compared with two or three hours. Each company provides equipment for its employees and pays for their long distance, modem, Fax and photocopy charges. Riverside provides private offices, cubicles, a conference room, Pacific Bell Centrex telephone and voice mail, secure data transmission service, free parking, a lunch room, and even an exercise room. The Riverside WorkCenter, November 1991, is a three-way partnership among the State of California, the Riverside County Transportation Commissions and private businesses. TRW, Southern California Edison, Pacific Bell and CalComp supply the 40 telecommuters. The center is the largest of its kind in the country. The California Community Colleges, the largest educational system in the world with over 150 sites, has a pilot Telework Project funded by CalTrans.

Telecommuting brings a host of quantifiable benefits to employer and employee alike. For the employer, it can reduce the cost of office space (30 percent for the State of California), increase productivity (from 3 to 60 percent or more when clear objectives are set), decrease turnover and absenteeism and boost hiring and retention. By removing geographic boundaries, employers can improve recruiting efforts and retain employees with scarce expertise or talents. They can also better utilize specialized labor pools such as people with disabilities who find it easier to navigate their own homes several days a week than to go into the office. Women on maternity leave report being able to return to work faster when they can work from their homes. It's also a boon for single parents where telecommuting can complement day care or provide a means to continue working when children are ill. Freed from commuting schedules, telecommuters can work in synch with their own body clocks, which produces higher energy levels. Fewer distractions is often the principal reason for higher productivity, but the emotional benefits of reduced stress, increased family interaction and decreased commuting time and cost are also major benefits to telecommuting.

It has been estimated that telecommuting two days per week from home saves employers $8,000 per employee per year. (This assumes an annual salary of $20,000, a productivity increase of 20 percent, reduced personnel costs of 10 percent, parking at $500 per year reduced by 40 percent, and use of central office facili-

ties of 150 square feet at $30 per square foot rent per year reduced by 40 percent.) Telecommuters who do come to the main office will find that their office space is now assigned in much the same way that a hotel assigns a room to a guest. With all their materials available by computer, the desk site doesn't matter. Managers also become better at managing and supervising because they start thinking in terms of deliverables. With telecommuting, the manager emphasizes the work product rather than the work process. Telecommuting has the best chance of working if managers and telecommuting employees identify measurable work objectives to be achieved. This also promotes strong communications and planning skills. The best choices for telecommuters are self directed employees who work well alone.

■ TRANSMISSION

In general, the latest technologies make it more efficient to shift voice communications from wires to radio signals, cellular phones and more complex television images to wires from the broadcast spectrum.

■ TELEVISION

Televisions will be much smarter and will be built with microchips able to store a billion bytes of data — 250 times the capacity of most personal computers — so that they can hold and sort entire two-hour movies delivered in a few seconds over high-capacity fiber-optic wires.

■ VIDEOCONFERENCING

The trend will continue toward the miniaturization of videoconferencing to computers. There will be a continued move towards integrating multi-media with video conferencing. The second generation of desktop videoconferencing products is making its way into the market. There will be a continued drop in video conferencing system prices due to lowered equipment costs and stabilized network rates.

Glossary

Video, data and audio have developed separately as media. This means that the inventors and engineers of the technology haven't been speaking to each other until the last several years. Even then, the language that has evolved is hard to pronounce, much less remember.

Multimedia is bringing them together and they are creating groups of letters that don't mean anything unless you speak the original technocrat that spawned them. Until that digital fusion is complete, we have to contend with different languages.

Teleconferencing is the overall word for academic use for the technology. After a brief foray into the language development, the educational technologists reverted to the technocrat. The language may very well be one of the reasons that it is hard to learn about the new technologies, let alone adopt them as a personal digital assistant.

If you don't speak technocrat, this glossary is for you.

A

Access Channels: Dedicated channels giving nondiscriminatory access to a local cable system by the public, government agencies or educational institutions.

ACCUNET Switched Digital Services: High-speed dial-up digital data services offered by AT&T for full duplex digital transmission at speeds of 56, 64, 384 and 1536 kbps. Uses include data, voice and video services.

Acoustic Coupler: A device that allows a conventional telephone handset to feed its signal into a modem, as opposed to direct couplers, which feed the modulated/demodulated signal directly into the phone line.

Acoustic Echo Canceller: All speakerphones have some form of adaptive echo canceller that produces a synthetic replica of the potential echo to subtract from the transmit audio. Most units have a center clipping echo suppresser to remove the residual echo from the transmit signal. The goal of the acoustic echo canceller is to reduce the amount of direct and reverberant loudspeaker coupling to the microphone to prevent echo. To achieve this, the algorithms used in today's devices require an audio system that is feedback stable.

Acoustic Echo Return Loss - AERL: The minimum loss experienced by a sound in traveling from the loudspeaker to the microphone in a conference room. It is expressed in dB or decibels. A 0 dB loss corresponds to a perfectly reflective room or to very close coupling between loudspeaker and microphone. In practice, AERL figures can range from 0 to -30 dB, with a poor room having the former figure.

Acoustic Echo Return Loss Enhancement - AERLE: The maximum echo cancellation provided by the acoustic canceller. Typical figures will vary from 6 to 18 dB. The larger the number the better. It is important to note whether the figure is quoted with the center clipper enabled or disabled. If quoted with center clipper disabled, it is a true measure of the cancellation provided by the echo canceller rather than the attenuation provided by the center clipper.

Acoustic Modem: A modulator-demodulator unit that converts data signals to telephone tones and back again.

Active Satellite: A satellite that transmits a signal, in contrast to a passive satellite that only reflects a signal. The signal received by the active satellite is usually amplified and translated to a different frequency before it is retransmitted.

Ad Hoc: Teleconferencing technology and sites assembled for an event; equipment may be rented or permanently installed; sites are not always part of the network.

Addressable: The ability to signal from the headend or hub site in such a way that only the desired subscriber's receiving equipment is affected. This makes it possible to send a signal to a subscriber and effect changes in the subscriber's level of service such as the ability to receive a program.

ADPCM — Adaptive differential pulse code modulation: A method of compressing audio data by recording the differences between successive digital samples rather than full value of the samples. There are many different types of ADPCM standards; this refers to the standard as defined in the CD-ROM XA and CD-I standards.

ADSL: Asynchronous Digital Subscriber Loop.

Affiliate Network: Group with their own satellite receive equipment; routinely receive same programming.

Agent: See knowbot.

Algorithm: 1. Rule of thumb for doing something with a semblance of intelligence. For example, a descrambling algorithm will yield a clear, unscrambled message from an apparently meaningless one. 2. The procedure used for performing a task.

America Online - AOL: Commercial information service with a graphical interface.

Amplifier: A device used to increase the strength of video and audio electronic signals.

Amplitude: The size or magnitude of a voltage or current waveform; the strength of a signal.

Analog: Information represented by a continuous electromagnetic wave encoded so that its power varies continuously with the power of a signal received from a sound or light source.

Analog-to-Digital - A/D Conversion: The conversion of an analog signal into a digital equivalent. An A/D converter samples or measures an input voltage and outputs a digitally encoded number corresponding to that voltage.

Analog Transmission: Transmission of a continuously variable signal as opposed to a discrete signal. Physical quantities such as temperature are described as analog while data characters are coded in discrete pulses and are referred to as digital.

Animate: To effect motion of any sort; e.g. to animate a person's presentation.

Animation: A video, film, and computer production technique utilizing cartoon-type artwork to create the illusion of movement.

ANSI - American National Standards Institute: ANSI is one of these "terminal emulation" methods. Although most popular on PC-based bulletin-board systems, it can also be found on some Internet sites. To use it properly, you will first have to turn it on, or enable it, in your communications software.

Answerback: The response of a terminal to remote control signals.Antenna (dish) The device that sends and/or receives signals (electromagnetic) from the satellite.:

Antenna Power: The product of the square of the broadcast antenna current and the antenna resistance where the current is measured.

Aperture: A cross section of the antenna exposed to the satellite signal.

Application: The use of a technology to achieve a specific objective.

Application Software: In computers, programs used to interact with and accomplish work for the user. Application software is usually written in a higher computer language such as Basic, COBOL, FORTRAN or Pascal, and may be written by the user or supplied by the manufacturer or a software company.

Applications Program: A computer program dedicated to a specific purpose or task. Applications programs which produce discernible results and can sometimes be machine-independent, are distinct from systems programs, which are designed to drive particular electronic devices and are always machine-dependent.

Archie: A system which allows searching of indexes of files available on public servers on the Internet.

Archival: A medium that is readable and/or writable for an extended period.

Armored Cable: Coaxial cable that can be direct buried without protective conduit, or used in underwater applications.

ARPANet: A predecessor of the Internet. Started in 1969 with funds from the Defense Department's Advanced Research Projects Agency.

ASCII: American Standard Code for Information Interchange; pronounced "Askee." An eight-level code for data transfer adopted by the American Standards Association to achieve compatibility between data services.

Aspect Ratio: The ratio of picture width to height (4 to 3 for North American NTSC broadcast video).

Asynchronous Communication: Takes place in different time frames and accessed at the user's convenience. Synchronous communication takes place in the same time frame such as a live teleconference.

Asynchronous Time-Division Multiplexing: An asynchronous signal transmission mode that makes use of time-division multiplexing.

Asynchronous Transmission: A technique in which the time interval between characters may be of unequal length. Transmission is controlled by start and stop elements at the end of each character. Used for low-speed terminal links.

ATM - Asynchronous Transfer Mode: ATM switching protocol can handle all types of traffic — voice, data, image, and video.

Attenuation: The difference between transmitted and received power due to loss through equipment, lines, or other transmission devices; usually expressed in decibels. The loss in power of electromagnetic signals between transmission and reception points.

Attenuator: A device for reducing the amplitude of a signal.

ATSC -Advanced Television Systems Committee.

ATV - Advanced Television: An agglomeration of techniques, based largely on digital signal processing and transmission, that permits far more program material to be carried through channels than existing analog systems can manage. In this sense, HDTV (high definition television) is a subset of ATV. ATV does not automatically signify improved picture or sound performance. Those are things that can be accomplished with ATV in systems designed for such purposes, but it can also carry ten somewhat lower- quality signals where only one could exist previously, or permit ghost cancellation for ordinary NTSC signals. In each case, the new features derive from the use of digital techniques of one form or another.

Audio Bridge: An audio bridge connects the telephones at remote sites, equalizes the noise distortion and background noise for a live audio teleconference.

Audio Frequency: A frequency lying within the audible spectrum (the band of frequencies extending from about 20 Hz to 20 kHz).

Audio Presentation: Often overlooked, but just as important as the video (perhaps more so) is the sound portion of the program. Without the audio, nothing is understood while a video failure could be tolerated if the sound portion is not affected. Use a good audio system to augment the video display. Do not use the built-in speaker of the TV monitor or room PA system. Rather, employ a high quality stand-alone system, with the speakers positioned adjacent to the TV screen. This affords the best audio experience. The audio quality coming off the satellite signal is true high fidelity, and its reproduction further enhances program presentation.

Audio Teleconferencing: Two-way electronic voice communication between two or more groups, or three or more individuals, who are in separate locations.

Aural Cable: Services providing FM-only original programming to cable systems on a lease basis.

Azimuth: Angle between an antenna beam and the meridian plane, measured along a horizontal plane. How far east or west in the southerly sky the satellite is located in relation to the local meridian, or north-south plane. It is measured in degrees, clockwise from true north.

Audiographic: Teleconference system which uses narrow band telecommunications channels (telephone lines or subcarriers); transmits audio and graphics. Graphics can be transmitted by facsimile transceivers (transmitter-receiver), computers (text or graphic display), or electronic drawing systems (such as electronic blackboard) which allow a participant to draw or write on an electronic screen which is transmitted to a remote site where participants can see it.

Audio Response: A form of output that uses verbal replies to inquiries. The computer is programmed to seek answers to inquiries made on a time-shared on-line system and then to utilize a special audio response unit which elicits the appropriate prerecorded response to the inquiry.

Audio Response Unit: Device that provides a spoken response to digital inquiries from a telephone or other device. The response is composed from a prerecorded vocabulary of words and can be transmitted over telecommunication lines to the location from kwhich the inquiry originated.

Audio-Subcarrier: Frequency which transmits audio for an accompanying video signal or independent audio (such as a radio program). Audio is sent along with the video signal, but on a different frequency.

Authoring System: Computer software that allows one to develop the framework for an interactive multimedia presentation. Authoring software enables the use of multiple data types as well as the controls needed to play-back information on the computer from devices such as CD-ROMs, computer hard disks and videodiscs.

Automatic Number Identification - ANI: The automatic identification of a calling station, usually for automatic message accounting. Also used in pay-per-view automated telephone order entry to identify a customer for billing and program authorization purposes.

B

B-Mac: A method of transmitting and scrambling television signals where MAC (multiplexed analog component) signals are time-multiplexed with a digital burst containing digitized sound, video synchronizing, authorization, and information.

Backbone: A high-speed network that connects several powerful computers. In the U.S., the backbone of the Internet is often the NSFNet, a government funded link between a handful of supercomputer sites across the country.ofconsidered

Backbone Microwave System: A series of directional microwave paths carrying common information to be relayed between remote points. The backbone microwave system is engineered to allow the insertion of signals, the dropping off of signals and the switching of signals along its length at designated relay points. In order to maintain the signals in the highest possible quality, the equipment used in the backbone microwave system is normally of a higher technical performance level than other microwave electronics in the network. Antennas are always directional.

Backhaul: A term used for the transmission of a signal (normally video) from the ends of transmission systems such as microwave to a central point. For a satellite videoconference, a backhaul refers to a signal brought in from a secondary site to the origination site, mixed with the primary signal, and sent out over the program out satellites.

Bandwidth: Determines the rate at which information can be transmitted across that a medium. The rates are measured in bits (bps), kilobits (kbps), megabits (Mbps), or gigabits per second (Gbps). Typical transmission services are 64 kbps, 1.544 mbps (T1), and 45 Mbps (T3). The space between the top and bottom limit of air-wave frequencies that are transmitted over a communications channel. The maximum frequency (range), measured in Hertz, between the two limiting frequencies of a transmission channel; the range of frequencies that can be carried by a transmission medium without undue distortion. Narrowband uses lower frequency signals such as telephone frequencies of about 3,000 Hertz and radio subcarrier signals of about 15,000 Hertz. Broadband uses a wide range of frequencies (broadcast and cable TV, microwave and satellite; carries a great deal of information in a short time; more expensive to use. C band is in the 4 to 6 giga-Hertz (gHz) Ku Band is 12 and 14 gHz .14.0 and 14.5 gHz are used to uplink; 11.7 and 12.2 gHz are used to downlink. A receiver with dual band capability can receive C and Ku band signals.

Base Band: The unmodulated signal that is delivered from a satellite receiver.

Base Band Distribution Systems: Usually used when the viewing areas are close together, and when TV monitors are used for viewing. The base band audio/video output from the satellite receivers is fed directly into the monitor. This form of wiring uses several twisted pair wires which can be very expensive when wiring more than 50 feet because of the need for many amplifiers and splitters. The picture quality is much sharper using a base band system, than with any other system.

Basic Rate Interface - BRI: The basic subscriber loop for one or two users, which delivers two 64 kpbs B channels and one 16 kbps D channel over a standard twisted pair loop. Each circuit-switched B channel can transmit voice or data simultaneously. The D channel transmits call control messages and user packet data.

Batched Communication: The sending of a large body of data from one station to another station network, without intervening responses from the receiving unit.

Baud: A unit of digital transmission signaling speed derived from the duration of the shortest code element. Speed in bauds is the number of code elements per second. 300 Baud is low, 2400 Baud and 9600 Baud are much faster and common for transmitting data by computer.

BBS - Bulletin Board System/Service: The BBS is an area within a network where users can "post" information for public display, in much the same way one posts information on a regular bulletin board. Most networks dedicate a bulletin board to special interest areas, such as education or computer care.

Beyond the Horizon Region: That physical region beyond the optical horizon with which line-of- sight radio communications is not normally possible, but can occur if atmospheric conditions are such to cause beam bending or forward scattering of the radio signal.

Bicycle Tapes: The process whereby video tape material is distributed by sending or "bicycling" the tape after presentation to the next site for its scheduled presentation.

Bidirectional Flow: A pathway allocating two-way data or communication exchange; flow in either direction represented on the same flow line in a flowchart.

Binary: Numbering system with two possible states on or off as designated by 0 and 1.

BISDN - Broadband ISDN: In 1995-96 is expected to offer dedicated circuits, switched circuits and packet services at rates of 155 Mbps and above.

BISDN is currently in the conceptual stage, and the term refers to a family of services being defined by the standards organizations. The goal of BISDN is to take advantage of the immense amount of raw bandwidth being made available due to the proliferation of fiber cable plant, and to enable customers to send data, voice, and video at high speeds and in an integrated manner. BISDN is expected to be fully defined in 1993-95, and deployment will take place in the latter half of the decade.

SONET-based fiber will serve as the delivery vehicle for BISDN services. BISDN will employ the concept of cell relay (Asynchronous Transfer Mode - ATM), which uses a transmission scheme based on small, fixed-sized

(53-byte) cells. These cells carry address and raw information, and the carrier networks will use address information to route the cells to the appropriate destination. As discussed above, frame relay is an interface; in contrast, cell relay is broader in scope and defines the size of the packets and the process for carrying packets across a network. BISDN is expected to encompass different types of services, including datagram service, switched circuits and permanent circuits, and to run at speeds ranging from 155 to 622 mbps. Some services, like SMDS and frame relay, will be in operation before BISDN is introduced, and the BISDN specs are expected to incorporate these preexisting services.

Binary Files: Those containing information that is not represented in the file by ASCII characters. These may be graphics, formatted files, or even executable programs. In order to send these files, special up- and downloading protocols must be used. Base Two, a number system comprised of zeros and ones, which represent off and on, absence or presence of a pulse. Used to store data.

Bit: A contraction of the words "binary digit," the smallest unit of information. A code element of digital transmission. One bit per second equals one baud "binary digit" single unit of information 0 or 1. See kbps or mbps.

Bit Density: A measure of the number of bits received per unit of length or area.

Bit Error Rate: Fraction of a sequence of message bits that are in error. A bit error rate of 10-6 means that there is an average of one error per million bits.

Bit Rate: Speed at which bit positions are transmitted, normally expressed in bits per second (see Baud.)

Bit Stream: A continuous string of bit positions occurring serially in time.

BITNET: Another, academically oriented, international computer network, which uses a different set of computer instructions to move data. It is easily accessible to Internet users through e-mail, and provides a large number of conferences and databases. Its name comes from "Because It's Time." BITNET is linked to Net North, the Canadian equivalent, and EARN, the European Academic and Research Network, as well as Internet/NREN.

Blanking (picture): The portion of the composite video signal whose instantaneous amplitude makes the vertical and horizontal retrace invisible.

Blanking Level: The level of the front and back porches of the composite video signal.

Blanking Pulse: 1. A signal used to cut off the electron beam and thus remove the spot of light on the face of a television picture tube or image tube. 2. A signal used to suppress the picture signal at a given time for a required period.

Blanking Signal: A specified series of blanking pulses.

Block: A group of bits, or characters, transmitted as a unit. An encoding procedure is generally applied to the group of bits or characters for error control purposes.

Block Downconverter - BDC: Located at the antenna. The multi-conversion process of converting the entire band to an intermediate frequency (4 GHz to 1 GHz) for transmission to multiple receivers, where the next conversion takes place. The BDC receives the signals from the Low Noise Amplifier (LNA) and converts them from the extremely high 4 GHz range to a much lower range, usually around 1 GHz. This range is less critical to signal loss, and permits the use of inexpensive long-run cable to interconnect with the receiver. Perhaps the biggest advantage of the BDC is the manner in which it handles the "block" of signals. It can be thought of as a passive device, converting and passing on to the receiver all of the channels on the satellite (of the selected polarity). This allows for the installation of multiple receivers through signal splitters, and simultaneous program viewing or taping. Older installations used downconverters that operated on only one channel, tuned by the receiver. These downconverters converted the LNA signals to 70 MHz, which provided considerable flexibility in quality and length of the connecting cables. The BDC method is used by Kuband systems and is also compatible with C band receivers.

Block-Error Rate: The ratio of the number of blocks incorrectly received to the total number of blocks sent.

Bounce: What your e-mail does when it cannot get to its recipient — it bounces back to you — unless it goes off into the ether, never to be found again.

Branch Cable: A cable that diverges from a main cable to reach some secondary point.

Branching: A computer operation, such as switching, where a choice is made between two or more possible courses of action depending upon some related fact or condition.

Bridge: Device which interconnects three or more telecommunication channels, such as telephone lines. A telephone conference audio bridge links three or more telephones (usually operated assisted). Usually a meet-me audio bridge or provides ateleconference direct dial access number. Both connect remote sites and equalize noise distortion.

Bridges, Gateway, Routers: Devices that convert LANs to other LANs, computers and WANs by allowing systems running on different media (copper wire, fiber optics, etc.) and protocols (rules to communicate).

Bridging Amplifier: An amplifier connected directly into the main trunk of the CATV system. It serves as a sophisticated tap, providing isolation from the main trunk, and has multiple high level outputs that provide signal to the feeder portion of the distribution network. Synonymous with bridger and distribution amplifier.

Broadband: Communications channels that are capable of carrying a wide range of frequencies. Broadcast television, cable television, microwave and satellite are examples of broadband technologies. These technologies are capable of carrying a great deal of information in a short amount of time, but are more expensive to use than technologies like telephone which require less band width. Broadband (Wideband) distribution systems. A telecommunications medium that carries high frequency signals; includes television frequencies of 3 to 6 megahertz. Broadband distribution systems work like cable TV, in that up to twenty channels are available from a single coaxial cable. A main trunk cable will originate at the control room, and run down the hallways of the viewing area. Smaller cables can tie into the main cable at any point along its length. Any room that is near the main cable run can have access to all of the channels on the system. Normal television sets are used, and a variety of channels can be received by simply changing channels on the television set.

Broadband Network: A loal area network (LAN) residing on coaxial cable capable of transporting multiple data, voice and video channels.

Broadcasting: The dissemination of any form of radio electric communications by means of Hertzian waves intended to be received by the public. Transmission through space, utilizing preassigned radio frequencies, which are capable of being received aurally or visually by an audience.

Brokers: Organizations which maintain primary leases or ownership of communications satellite time and provide subleases to teleconference originators.

Buffer: Temporary storage facility used as an interface between system elements whose data rates are different; Memory area in computer or peripheral device used for temporary storage of information that has just been received. The information is held in the buffer until the computer or device is ready to process it. Hence, a computer or device with memory designated as a buffer area can process one set of data while more sets are arriving.

Bug: A system or programming problem. Also refers to the cause of any hardware or software malfunction. May be random or non-random.

Bundle: A package that includes several products for one price. For example, a CD-ROM drive, with controller card, cable, software, and one or more CD-ROMs.

Bus Interface: An electronic pathway between CPUs and input/output devices. A bus interface for a CD-ROM drive consists of a controller card and cable.

Business Television — BTV: Corporate use of video transmission for meetings/training via satellite.

Burst: 1. In data communication, a sequence of signals counted as one unit in accordance with some specific criterion or measure. 2. A color burst.

Burst Modem: In satellite communications, an electronic device used at each station that sends high-speed bursts of data which are interleaved with one another. These bursts must be precisely timed to avoid data collisions with multiple stations.

Burst Transmission: Data transmission at a specific data signaling rate during controlled intermittent intervals.

Bus: A circuit or group of circuits which provide an electronic pathway between two or more central processing units (CPUs) or input/output devices.

Bus Controller: The unit in charge of generating bus commands and control signals.

Byte: A group of bits treated as a unit used to represent a character in some coding systems. The values of the bits can be varied to form as many as 256 permutations. Hence, one byte of memory can represent an integer from 0 to 255 or from -127 to +128.

Byte: Primary and secondary memory (RAM and magnetic media) are measured in kilobytes (1,024, or 210 bytes) and megabytes (one million bytes).

C

Cable/Cable Television: A broadband communications technology in which multiple television channels as well as audio and data signals are transmitted either one way or bidirectionally through a distribution system to single or multiple specified locations. Uses coaxial cable to transmit programs. Direct-by-wire transmission to homes from a common antenna to which these homes are linked. Cable companies provide the service in most cases. Distinguished from television reception through a roof-top antenna that picks up the broadcast signal. The only acronym was CATV, denoting community antenna television.

Cable Television Channel Classes: Class I Source is a television broadcast signal that is being presently transmitted to the public and conveyed to the cable system for retransmission to the public, direct connection, off-the-air or obtained indirectly by microwave or by direct connection to a television broadcast station.

Class II: A signaling path provided by a cable television system to deliver to subscriber terminals television signals that are intended for reception by a television broadcast receiver without the use of an auxiliary decoding device and which sinals are not involved in a broadcast transmission path.

Class III: A signaling path provided by a cable television system to deliver to subscriber terminals signals that are intended for reception by equipment other than a television broadcast receiver or by a television receiver only when used with auxiliary decoding equipment.

Class IV: A signaling path provided by a cable television system to transfer signals of any type from a subscriber terminal to another point in the cable television system.

Cablecasting: Origination of programming, usually other than automated alphanumeric services, by a CATV system.

Cable Communications Policy Act of 1984: This act, passed by Congress in 1984, updated the original Communications Act of 1934. The primary changes dealt with cable television regulation, theft of service, equal employment opportunity (EEO) an various licensing procedure changes.

Cable Compatible: Generally refers to consumer devices, such as television sets and videocassette recorders, that are designed and constructed to allow direct connection of a CATV subscriber drop to the device. Frequently, they have a tuner capable of receiving cable channels other than 2-13 (e.g., midband, superband, and hyperband channels). Even though a device may be cable compatible, it may still require an external descrambler to receive scrambled channels such as the premium pay channels or pay-per-view channels.

Cache: In a processing unit, a high-speed buffer storage that is continually updated to contain recently accessed contents of main storage. Its purpose is to reduce access time. A holding area for data within the CD-ROM drive itself or on its interface board, that allows the system a method for matching data transfer rates and presentation speed requirements.

CAI: Computer Assisted Instruction.

Camera: In television, an electronic device utilizing an optical system and a light-sensitive pick-up tube to convert visual images into electrical impulses.

Camera Control Unit — CCU: An electronic device that provides all the operating voltages and signals for the proper set up, adjustment and operation of a television camera.

Candle Power: A measure of intensity of a light source in a specific direction.

Carrier-to-Noise Ratio: In cable television, the ratio of peak carrier power to root mean square (RMS) noise power in a 4 MHz bandwidth.

CATV- Community Antenna Television: A broadband communications system capable of delivering multiple channels of entertainment programming and non-entertainment information from a set of centralized antennas, generally by coaxial cable, to a community. Many cable television designs integrate microwave and satellite links into their overall design, and some now include fiber optics.

Carrier: Vendor of transmission services operating under terms defined by the FCC as a common carrier. Owns a transmission medium and rents, leases or sells portions for a set tariff to the public via shared circuits.

CAV - Constant Angular Velocity: A disk that rotates at a constant rate of speed. Examples are hard drives, floppy disks, magneto-optical discs and some videodiscs. A CAV videodisc permits access to video within seconds, allows for up to 54,000 still frames, or may contain up to 30 minutes of full motion video (or any combination of stills and video). (See CLV)

C Band: A category of satellite transmissions which transmit from earth at 4.0 to 6.0 GHz and receive from the satellite at between 3.7 and 4.2 GHz which are also shared with terrestrial line-of-sight microwave users. This band of transmissions has less path loss than the other standard used for satellites (Kuband) but must have a large antenna for the same receiver input power level due to its use of longer wavelength frequencies. Other problems relating to the use of C band include the shared use of these frequencies with terrestrial microwave transmission which cause interference with the weaker satellite signals in certain areas.

CBT - Computer Based Training: The use of interactive computer or video programs for instructional purposes.

CCITT: Consultative Committee on International Telephony and Telegraphy; An international standards group.

CCITT Standard: Transmission rate Px64 or multiples.

CCL - Connection Control Language: A scripting language that allows the user to control a modem.

CCTV: Closed-Circuit Television.

CD-Audio: Also called CD-DA for Compact Disc-Digital Audio. The use of CDs to record music in digital audio format. The disc holds a sequence of audio tracks. Each can be a very high-fidelity stereo recording. These discs can be played on conventional CD players, CD-I systems and at least some CD-ROM drives. Standards for this are called the Red Book.

CD audio jack: An outlet on a CD-ROM drive that provides audio playback through speakers or headphones. Only Red Book, or true CD-Audio sound can be heard from the audio jack on a CD-ROM drive.

CD-ROM Compact Disc - Read Only Memory: CD-ROM discs can store a variety of data types including text, color graphics, sound, animation and digitized video that can be accessed and read through a computer. A disc can store up to 600 megabytes of data, much more information that can be stored on a 3.5 inch compute disk, which hold up to 1.4 megabytes. This makes CD-ROM an inexpensive medium for storing large amounts of data. Because CD-ROM was not designed to store digitized, full-motion video, compression technology is important in compressing data to fit on a disc as well as decompressing data for playback.

CD-I: 1. Compact Disc-Interactive. Stores text, audio, video, images and animation. Requires a CD-I player and will not work on a regular CD-ROM player. 2. This interactive multimedia system, developed by Philips and Sony, connects to a television and stereo audio system. The standards for this are called the Green Book.

CD-R: CD-Recordable Term used to describe special players and media which enable the creation of a single CD-ROM, written from the PC as if it were a magnetic disk drive. The end product, however, is a read-only disc: it cannot be erased or written over. Therefore, this technology is also known as "write-once CD." See CD-WO.

CD-ROM: A laser-encoded optical memory storage medium, defined by the Yellow Book standard.

CD-ROM Drive: A computer peripheral that plays CD-ROMs.

CD-ROM XA: CD-ROM Extended Architecture: A compact disc standard that permits the interleaving of compressed audio and video tracks for sound and animation synchronization. Based on the Yellow Book, it also uses some elements of the Green Book (CD-I).

CD-WO - Compact Disc-Write Once: A term that describes compact discs that can be written to directly (rather than mass produced) with a laser recorder. Recent developments allow the CD-WO to be appendable. CD-WO media is physically defined by the Orange Book standard, Part II, and a proposal for the logical format has been submitted to ECMA by the Frankfort Group.

Center Clipper: Variable attenuator which is used to eliminate any residual echo left by the echo canceller. A key difference between one canceller and another is the manner in which this center clipper operates. In a high quality canceller, the center clipper will operate very rapidly and smoothly, resulting in no residual echo during double-talk and no clipping of syllables. The center clipper is in essence a level- activated switch. Signals above the threshold level are passed unaltered and signals below the threshold are blocked. When speech is present in both directions, the center clipper tends to mutilate the speech signal, adding audible amounts of harmonic and intermodulation distortion. This distortion is often referred to as "gritch" and sounds remarkably like its name when it occurs. The transmit signal can be totally chopped out if the level of the transmit signal drops below the estimated level of returning echo.

Central Office: The physical location where communications carriers terminate customer lines and locate the switching equipment that interconnects those lines.

Central Processing Unit - CPU: The unit of a computer that includes circuits controlling the interpretation and execution of instructions.

Channel: A signal path of specified bandwidth for conveying information. 1. A half-circuit; 2. A radio frequency assignment (which is dependent upon the frequency band and the geographic location). Channel capacity in a cable television system is the number of channels that can be simultaneously carried on the system. Generally defined in terms of the number of 6 MHz (television bandwidth) channels.

CFDA - Catalog of Federal Domestic Assistance: The CFDA is a government-wide compilation of federal programs, projects, services and activities that provide assistance or benefits to the American public. The primary purpose of the CFDA is to assist users in identifying programs that meet specific objectives of a potential applicant, and to obtain general information on federal assistance programs. The catalog is published once yearly, usually in June. An update occurs around December.

CFR - Code of Federal Regulations: The CFR is the "book" of federal laws and regulations. Usually referenced like this, 34 CFR 74.137. The "34" indicates that the subject of the regulation is education. The "70.137" refers to a specific paragraph.

Character Generator: An alphanumeric text generator, a typewriter like device, commonly used to display messages on a television set. Chyron is a brand name for a character generator which is often mistakenly used to cover all character generators generically. Some sophisticated models also include color, graphics, and mass memory for text storage.

Charge-Coupled Device - CCD: A solid-state device used in many television cameras to convert optical images into electronic signals. These imagers are organized into rows and columns called pixels. The charge pattern formed in the CCD pixels when light strikes them forms the electronic representation of the image.

Chip: A thin silicon wafer on which electronic components are deposited in the form of integrated circuits; the basis of digital systems.

Chip Sets: Application-specific integrated circuits (ASICs) are being developed for use in video application products such as codecs, desktop video, and home satellite entertainment. ASICs operate more like computer hardware. Programmable chips operate much like computer software. The chip sets meet the CCITT H.261 compression standard and will be the driving force in the widespread use of video communications technology because they will lower the cost and open up the technology to a much larger group of users.

Chroma Key: In color television, an electronic matting process of inserting one image over a background. Used very commonly with weathercasters who are standing in front of a blank wall painted process blue. The electronics remove the blue and insert the weather map so that on the television screen the two images merge and the weathercaster appears to be standing in front of a large map painted on the wall.

Chrominance Signal: The color signal component in color television that represents the hue and saturation levels of the colors in the picture.

Circuit: Means of two-way communication between two or more points. 1. In communication systems, an electronic, electrical, or electromagnetic path between two or more points capable of providing a number of channels. 2. Electric or electronic part. 3. Optical or electrical component that serves a specific function or functions.

Circular Polarization: A mode of transmission in which signals are downlinked in a rotating corkscrew pattern. A satellite's transmission-capacity can be doubled by using both right-hand and left-hand circular polarization.

Closed Circuit Television - CCTV: A private television system in which signals are sent usually via cable, to selected viewing points throughout the distribution system but are not broadcast to the public. The signal does not have to meet FCC commercial specifications.

CLV - Constant Linear Velocity: A disc that rotates at a varying rate of speed. Examples are CD-Audio, CD-ROM, CD-I, CD-ROM XA, and some videodiscs. CLV videodiscs may contain up to one hour of full motion video, but still frames and quick access time are forfeited. (See CAV.)

CMC: Computer mediated communication.

Command Line: On Unix host systems, this is where you tell the machine what you want it to do, by entering commands.

Communications Software: A program that tells a modem how to work.

Compatible: Describes different hardware devices that can use the same software or programs without modification, or with appropriate software.

CompuServe - CompuServe Information Service - CIS: One of the oldest and largest commercial services.

C/N - C/NR - Carrier to Noise Ratio: Refers to the ratio of the satellite carrier (or signal) to noise level in a given channel. Usually measured in dB at the LNA output.

Coaxial Cable - Coax: A type of metal cable used for broadband data and cable systems. It has excellent broadband frequency characteristics, noise immunity and physical durability. Consisting of a center conductor in the form of a tube which carries broadband signals by guiding high-frequency electromagnetic radiation, insulating dielectric, conductive shield, and optional protective covering.

Co-Channel Interference: Interference on a channel caused by another signal operating on the same channel.

Codec: A COder-DECoder converts analog signals, (voice or video), into digital form (1 or 0) for transmission over a digital medium and, upon reception at a second codec, re-converts the signals to the original analog form. Two codecs are needed - one at each end of the channel.

Color Bars and Tone: A color standard test pattern used by the television industry to adjust equipment to standard levels. The tone is generated at a certain preset frequency so that audio levels can be set.

Color Burst: In NTSC terminology, refers to a burst of approximately nine cycles of 3.58 MHz subcarrier on the back porch of the composite video signal. This serves as a color synchronizing signal to establish a frequency and phase reference for the chrominance signal.

Color Signal: Any signal at any point in a color television system for wholly or partially controlling the chromaticity values of a color television picture.

Color Subcarrier: In NTSC color, the 3.58 MHz subcarrier whose modulation sidebands are interleaved with the video luminance signal to convey color information.

Color Transmission: A method of transmitting color television signals which can reproduce the different values of hue, saturation, and luminance which together make up a color picture.

Combining Network: A passive network which permits the combining of several signals into one output with a high degree of isolation between individual inputs; commonly used in CATV headends to combine the outputs of all processors and modulators into a single coaxial cable input. Synonymous with combiner.

Common Carrier: Usually a telecommunications company that owns a transmission medium and rents, leases or sells portions for a set tariff to the general public via shared circuits through published and nondiscriminatory rates. In the U.S., common carriers are regulated by the FCC or various state public utility commissions.

Communications Satellite Corporation - COMSAT: A common carrier service that provides commercial communications services.

Communications Satellite: Relay system in orbit above earth for telecommunications signals (voice, video, data); require earth stations to transmit and receive signals at the ground locations. Commonly called a "bird."

Compression: The application of any of several techniques that reduce the amount of information required to represent that information in data transmission. This method reduces the required bandwidth and/or memory.

Compressed Video: Processes video images; transmits changes fro one frame to the next which reduces the bandwidth to send them over a telecommunications channel; reduces cost. Also called bandwidth compression, data compression or bit rate reduction. The most publicized compression techniques are proposed by two expert groups, that of JPEG (Joint Photographic Expert Group) and MPEG (Moving Picture Expert Group), who are defining methods for image compression in still frame and real-time video. The algorithm used by these two groups is called discrete cosine transform (DCT). DCT transforms a block of pixels into a matrix of coefficients and estimates redundancy in the matrix. The advantage of JPEG and MPEG is that the algorithms are symmetrical; that is, the same amount of processing is required for the encode and decode functions. These are ideal for two-way applications such as videoconferencing.

CompuServe Information Services - CIS: A computer network service.

Computer: A functional unit that can perform substantial computations, including numerous arithmetic operations or logic operations, often without intervention by a human operator.

Computer-Aided Design - CAD: A computer system whereby engineers create a design and see the proposed product in front of them on a graphics screen or in the form of a computer printout.

Conferencing: A term used to indicate when several network users communicate on a particular subject. Conferences can be "live" or conducted via a BBS (see above).

Connect Time: Time period during which a user is utilizing a computer on-line - or directly connected with the computer.

Connect Time Charges: Most networks charge users for the time they spend on-line. These are referred to as connect time charges. The amount charged depends on the network's fee schedule. Users must also pay a separate fee if the call to connect is toll.

Control Room: A room separate from a studio in which the director, the technical director (TD), the audio engineer, and other technical and program assistants control program production.

CONUS: Contiguous United States.

Consortium: Voluntary group affiliated for a purpose. Consortia is plural.

Continuous Presence Video: Simultaneous and continuous pictures of participants.

Convergence - Digital Fusion: The merging of video, audio, and data communications through digitization of the media. The equipment to receive the signals is projected to be a telecomputer.

Courseware: Software used in teaching. Often used to describe computer programs designed for the classroom.

Crash: An abrupt, unplanned computer system shutdown caused by a hardware or software malfunction.

Crawl: A visual technique; electronically generated words or graphics that move horizontally or crawl across the screen, usually at the bottom.

Crawl Space: Space for textual messages usually at the bottom of the television screen.

Credits: The names of people on whom the production can be blamed. Electronically generated words that usually move horizontally up the screen like a scroll, or inserted a page at a time.

CREN: Corporation for Research and Educational Networking.

Crosstalk: 1. Undesired transfer of signals from one circuit to another circuit. 2. The phenomenon whereby a signal transmitted on one circuit or channel of a communications system is detectable or creates an undesirable effect in another circuit or channel.

Cue: Signal to start, pace, or stop any type of production activity or talent action.

Cursor: A symbol on the display of an editing or display terminal that can be moved up, down, or sideways and indicates where the next character is to be located or where "home" or beginning is located.

Cut: A command that stops all action in actual production; or a visual technique for changing abruptly from one picture to an entirely different one; for example, quick cuts in which many different visuals appear rapidly one after another on the screen.

Cyberphobic: A person who is fearful of working in cyperspace.

Cyberspace: Coined by science fiction writer William Gibson in the 1970s, it describes the virtual place of computer memory, networks, and multimedia.

D

D1 and D2: Digital tape component and composite formats (respectively) used for professional video recording. D1 is costlier than D2. Both can go through many generations of dubbing without visible loss of picture quality.

Daemon: An otherwise harmless Unix program that normally works out of sight of the user. On the Internet, you'll most likely encounter it only when your e-mail is not delivered to your recipient - you'll get back your original message plus an ugly message from a "mailer daemon" saying the message was undeliverable.

DARPA: Defense Advanced Research Projects Agency of the Pentagon. Replaced ARPA.

Data: Any and all information, facts, numbers, letters, symbols, etc. which can be acted on or produced by the computer.

Database: Organized collection of files and information stored on a computer disk available for update and retrieval.

Data Communications: 1. The movement of encoded information by means of electrical or electronic transmission systems. 2. The transmission of data from one point to another over communications channels.

Data Compression: A technique that saves storage space by eliminating gaps, empty fields, redundancies, or unnecessary data to shorten the length of records or blocks.

DBS - Direct Broadcast Satellite: Service uses high power satellites to broadcast multiple channels of TV programming to inexpensive home small-dish antennas.

DCT - Discrete Cosine Transform: Compression algorithm.

Debug: To detect, trace, and eliminate mistakes in computer programs or in other software.

Dedicated Lines: Leased telecommunications circuits that are devoted to a specific application; a circuit designated for exclusive use by two users; i.e., for interactive portion of a teleconference.

Dedicated System: Videoconferencing equipment, transmission circuits, and teleconferencing facilities that are permanent and used on a regularly scheduled basis as opposed to rented for a one-time or ad hoc event.

Default: A standard setting or action taken by hardware or software if the user has not specified otherwise.

Definition: Also called resolution. The fidelity with which detail is reproduced by a television system ranging from a fuzzy to a sharp appearance.

Degausser: 1. Demagnetizer. 2. A device for bulk erasing magnetic tape.

Delay: Time it takes for a signal to go from sending station through the satellite to receiving station.

Demodulate: To retrieve an information carrying signal from a modulated carrier. A demodulator is a device that removes the modulation from a carrier signal.

Dial up - Dialup: To call another computer via modem. A connection or line reached by modem, as in "a dialup line."

Dial-Up Teleconferencing: Using public phone line to connect with a teleconference, either with or without operator assistance.

Dielectric: A non-conductive insulator material between the center conductor and shield of coaxial cable. The dielectric constant determines the propagation velocity.

Digital: Discrete bits of information in numerical steps. A form of information that is represented by signals encoded as a series of discrete numbers, intervals or steps, as contrasted to continuous or analog circuits. Digital signals can be sent through wire or over the air. The method allows simultaneous transmission of voice and data. All digital technology is emerging as the primary transmission mode for voice, video, data and facsimile; Information represented by signals encoded as a series of discrete numbers, intervals or steps. Can be sent through wire or over the air. Allows simultaneous transmission of voice, video and data.

Digital Computer: A computer that operates on discrete data by performing arithmetic and logic processes on these data.

Digital Media: Refers to any type of information in digital form including computer-generated text, graphics and animations, as well as photographs, animation, sound, and video.

Digital Transmission: The transmission of information in the form of "1s" and "0s." Information customarily sent in this form is related to computer data traffic which is already in digital form. Other communications include audio and video.

Digital Video Effects - DVE: Video effects accomplished through digital devices that manipulate the video; e.g., page turns, revolves, boxes that zoom into and out of the picture, images that turn into pixels, etc.

Digitizer: A device that converts an analog signal (either images or sound) into a digital signal that can be manipulated on the computer. Video capture boards convert video images from vdeo sources such as the VCR or video camera, while sound digitizers take any sounds, the spoken word as well as music off of a cassette or CD player, and turn them into digital data. That data can be edited using sound editing and multimedia software.

Diode: An electronic device used to permit current flow in one direction and to inhibit current flow in the other.

Directional Microphone: A microphone that detects and transmits sound from only a certain direction. Useful in preventing unwanted sound from being transmitted.

Direct Read After Write - DRAW: A laser based technology for recording data on a videodisk.

Disc: Preferred usage (spelling) of the term for reference to optical storage media, such as CD-Audio, CD-I, CD-ROM, videodisc, or WORM.

Disk: Preferred usage (spelling) of the term for reference to magnetic media, such as floppy and hard disks.

Disk - Disc: A record-like magnetic-coated piece of material that can store digital information; may be a hard disk or pliable floppy disk.

Disk Drive: A computer data storage device in which data is stored on the magnetic coating (similar to that on magnetic tape) of a rotating disk.

Dish: Parabolic antenna. Primary element of a satellite earth station; sends and/or receives satellite signals. Usually bowl-shaped; concentrates signals to a single focal point. The antenna cross section exposed to the signal is the aperture.

Display: The visual presentation on the indicating device of an instrument.

Dissolve: Gradual transition from one television picture to the next by fading out one picture and simultaneously fading in another.

Distortion: An undesired change in wave form of a signal in the course of its passage through a transmission system.

Distributed Data Processing: Data processing in which some or all of the processing, storage, and control functions, in addition to input-output functions, are situated in different places and connected by transmission facilities.

Distributed Function: The use of programmable terminals, controllers, and other devices to perform operations that were previously done by the processing unit, such as managing data links, controlling devices, and formatting data.

Distribution: A way to limit where your Usenet postings go. Handy for such things as "for sale" messages or discussions of regional politics.

Distribution Systems: Any program that can be received by the satellite antenna, can be distributed into several viewing areas. The distribution system is one or more wires that run from the earth station control room, into several classrooms or conference rooms. Broadband wiring systems use a single coaxial cable, while baseband systems use several twisted-pair wires. Both systems must be custom designed for each location, using high output amplifiers and exact cable lengths.

Domain: The last part of an Internet address, such as "news.com." The zones include:

    edu-education
    mil-military site
    com-commercial organizations
    gov-government body or department
    net-networking organization
    int-international organization (mostly NATO)
    org-anything that doesn't fit elsewhere, such as a professional society

Domestic Satellite: A satellite that provides communication services primarily to one nation.

DOS: Disk Operating System.

Dot: To impress the Net veterans encountered, parties say "dot" instead of "period." For example: "My address is john at site dot domain dot com."

Dot file: A file on a Unix public-access system that alters the way the user or the messages interact with that system. For example, a user's .login file contains various parameters for such things as the text editor used when a message is sent. When an ls command is done, these files do not appear in the directory listing; do ls -a to list them.

Double speed drive: Refers to a CD-ROM drive that will read certain kinds of data faster than the standard requires (155KB/sec). Many drives now have 300KB/sec transfer rates (also known as twice the standard, or 2X); at least one claims 600KB/sec (4X).

Double-Talk: The situation where parties at both ends of a conference are speaking simultaneously. A quality echo canceller will provide a continuous speech path in both directions during double-talk.

Down: When a public-access site runs into technical trouble, and you can no longer gain access to it, it's down.

Downconverter: A device used to lower the frequency of any signal.

Downlink: Transmission of radio frequency signals from a satellite to an earth station. A satellite receiving station.

Download: 1. Transfer data from a main computer or memory to a remote computer or terminal. 2. There are several different methods, or protocols, for downloading files, most of which periodically check the file as it is being copied to ensure no information is inadvertently destroyed or damaged during the process. Some, such as XMODEM, only let you download one file at a time. Others, such as batch-YMODEM and ZMODEM, let you type in the names of several files at once, which are then automatically downloaded.

Drive bay: The opening in a computer chassis designed to hold a floppy drive, hard drive, CD-ROM drive, tape drive or other device. May be half-height or full-height, exposed or internal.

Drop-Outs: Black or white lines or spots appearing in a television picture originating from the playback of a video tape recording.

DS1: Digital signal level 1; a digital transmission format in which 24 voice channels are multiplexed into one T1 channel.

DS3: Digital signal level 3; a telephony term describing the 45 mbps signal carried on a T3 facility. It is most often associated with broadcast video transmission. Although the broadcast purest will rightfully point out that as a digital signal it is not a true broadcast quality RS-250B standard signal, it is the nearest approximation to a broadcast signal in a digital environment.

DSP: Digital signal processing.

DTMF - Dual Tone Multiple Frequencies: Standard telephone signaling technique which can be used through any transmission medium of voice grade or better. he technique is often used for remote switching control functions.

Dual Band Capability: Many receivers are capable of both C and Kuband operation.

Dub - Dupe - Duplicate: The duplication of an electronic recording. Dubs can be made from tape to tape in video, or from record to tape in audio. In video, one generation of quality is usually lost between each duplicate except when using high grade broadcast equipment and one inch wide tape. Usually a video tape that appears fuzzy and the colors have lost clarity, it is a 3rd, 4th, or 5th generation tape - in other words, a copy, of a copy, of a copy, of a copy, of a copy.

Duplex: In a communications channel the ability to transmit in both directions.

DVI - Digital Video Interactive: DVI is a programmable (variable bit and frame rate) compression and decompression technology developed by Intel offering two distinct levels and qualities of compression and decompression for motion video. Both PLV and RTV use variable compression rates. Production Level Video (PLV), a proprietary asymmetrical compression technique that is well suited for encoding full motion, color video requires compression to be performed by Intel at its facilities or licensed encoding facilities set up by Intel. PLV emulates MPEG. It has a very high image quality. Real Time Video (RTV) provides comparable image quality to framerate (motion) JPEG and uses a symmetrical variable rate compression. To provide expanded still image editing features, future versions of Intel's DVI will be JPEG compliant.

E

E-Layer: A heavily ionized signal reflecting region location 50-70 miles above the surface of the earth, within the ionosphere.

E-mail - Electronic mail: The term for private messages sent as files from one computer to another, either over a local area network (LAN), or via modem over the phone lines. E-mail is like having your own private mail box on a network. Used as both a noun and verb. Mail can be sent between Internet and commercial services as follows:

AMERICA ONLINE: "user@aol.com" Use all lower case and remove spaces.

APPLELINK: "user@applelink.apple.com".

AT&T MAIL: "user@attmail.com"

BITNET: "user@host.BITNET" (Note that the bitnet host name is not necessarily the same as the Internet host name.) If this fails, try directing your mail through a gateway such as "cunyvm.cuny.edu", pucc.princeton.edu", or "wuvmd.wustl.edu". The address would be as follows: "user%domain.BITNET@pucc.princeton.edu" (or cunyvm or wuvmd). This should help those with SMTP servers that are not quite up to date.

BIX: "user@bix.com"

COMPUSERVE: "userid@compuserve.com". Use the numeric CompuServe identification number, but use a period instead of a comma to separate the number sets. For example, for CompuServe user 17770,101 - mail to "17770.101@compuserve.com".

CONNECT: "user@dcjcon.das.net"

DELPHI: "user@delphi.com"

e-WORLD "username@eworld.com

EASYLINK "username@easy"

FIDONET: "firstname.lastname@p#.f#.n#.z#.fidonet.org". To send mail to a FidoNet user, you not only need the name, but the exact FidoNet address s/he uses. FidoNet addresses are broken down into zones, net, nodes, and (optionally) points. For example, the address of one Fido BBS is "1:102/834". The zone is 1, the net is 102, the node is 834. A user's address could include a point as well: "1:102/834.1" - the final 1 is the point. So, to send mail to "John Smith" at Fido address "1:102/834", e-mail to "John.Smith@f834.n102.z1.fidonet.org". To send mail to that user at Fido address "1:102/834.1", e-mail to "John.Smith@p1.f834.n102.z1.fidonet.org".

GENIE: "user@genie.geis.com" where "user" is their mail address. If a user tells you their mail address is "xyz12345" or something similar, it isn't. It usually looks like "A.BEEBER42" where A is their first initial, BEE-BER is their last name, and 42 is a number distinguishing them from all other A.BEEBER's.

INSTITUTE FOR GLOBAL COMMUNICATIONS (IGC, or "PEACENET"): "user@igc.org"

INTERNET: send mail to "user@domain", where user is the recipient's login name, and domain is the full name and location of the computer where s/he receives e-mail. Examples are "savetz@rahul.net" and "an017@cleveland.freenet.edu".

MCI MAIL: send your mail to "user@mcimail.com". "User" can be a numeric identification (which is always seven digits long, or three zeroes followed by seven digits,) their account name (which is one word) or first and last names separated with an underline. (E.g. "1234567@mcimail.com", "123-4567@mcimail.com", or "John_Vincent_Jones@mcimail.com".)

NVN (National VideoText Network): user@nvn.com

PC LINK: "user@aol.com". Incoming mail is limited to 27K. (There is no pclink.com domain. All mail to the America Online, Inc. owned systems goes to aol.com.)

PRODIGY: Use "abcd12a@prodigy.com" where "abcd12a" is the member's PRODIGY service ID. If you experience mail delivery problems that may require action by the administrators of this system, write to "admin@prodigy.com". In order for Prodigy members to receive Internet mail, they need to download Prodigy's offline mail reading software, called "Mail Manager".

QuantumLink: "Q-Link", a Commodore 64/128-based service offered by America Online, Inc., is not on the Internet for technical reasons.

WELL: "user@well.sf.ca.us" or "user@well.com"

WWIVnet: You must have the user number or user name of the person you want to send mail to. as well as his/her WWIVnet node number. WWIVnet addressing looks similar to: "1@9010" or: "DAN Q@9010". First replace the "@" with a dash: "1-9010". If you use the user name as the recipient, replace the spaces with an underscore: "DAN_Q-9010". Take that address and use it as the account name in one of the following examples: "wwiv!1-9010@tweekco.uucp" or "1-9010%wwiv@tweekco.uucp". Of course, replace "1-9010" with your recipi-ent's address.

Service Providers:

America Online 800-827-6364 voice

ANS Advanced Network and Services, Inc. 800-456-8267 voice

AT&T: 800-248-3632 voice

BARRNet Bay Area Regional Research Network: 415-725-1790 voice

Bix: 800-695-4882 modem, 800-695-4775 voice

BBN Technology Services, Inc 617-873-8730

CERFNet CA Education and Research Federation Network 800-876-2373 voice

CICNet Committee on Institutional Cooperation Network, Inc.: 800-876-2373 voice

Clark Internet Services, Inc. 800-735-2258: ask Op to dial 410-730-9764Connect 408-973-0110 voice

CompuServe 800-848-8990 voice

Delphi 800-695-4005 voice

Dialog 800-334-2564 voice

Dow Jones News/Retrieval 800-522-3567 voice

e-World 800-775-4556 voice

Genie 800-638-9636 voice

Global Enterprise Services, Inc. 609-897-7300 voice

Institute for Global Communications 415-923-0220 voice

NEXIS/LEXIS 800-227-9597 voice Gov't Customers:513-865-7223MCI Mail 800-444-6245 voice

MIDnet Midwestern States Network, Inc. 402-472-7600 voice

Northwest Academic Computing Consortium 206-562-3000 voice

PC-Link 800-827-8532 voice

PSI Performance Systems Int'l., Inc. 800-827-7482 voice

Prodigy New account information 800-766-3449 voice   Membership services 800-759-8000 voice

Sprint 800-817-7755 voice

SURAnet Southeastern Universities Research Assoc. Network 800-787-2638 voice

UUNET Technologies, Inc. 800-488-6383 voice

Well: 415-332-4335 voice. E-mail: support@well.sf.ca.us

Earth Station: The location antenna used to send or receive signals to satellites normally located in the geostationary orbit. A parabolic antenna and associated electronics for receiving or transmitting satellite signals.

Echo: The reflections of signal energy that cause it to return to the transmitter or to the receiver.

Echo Canceller: Eliminates audio transmission echo. A telephone line echo canceller produces a synthetic replica of the echo it expects to see returning and subtracts it from the transmitted speech. The replica it creates is based on the transmission characteristics of the telephone cable between the echo canceller and the telephone set.

Echo Reduction: A newer method of echo control, developed in 1988, uses attenuation in a new way to subjectively reduce the returned echo without the mutilation (choppiness, level drops, distortion) found in suppressers or center clippers. It rapidly and momentarily applies a variable amount of attenuation in between transmitted speech peaks (where the echo would be audible). It compares the transmit and receive signals to determine the likelihood of objectionable echo in the transmit signal, then calculates and inserts the appropriate amount of attenuation for that instant to control the echo. During outgoing speech peaks, the echo is masked by the strong local speech and rendered inaudible to the listener so that no attenuation is required.

Editor: A computer program used to edit (prepare for processing) text or data.

Educational Access Channel: A cable television channel specifically designated for use by local education authorities.

Edutainment: Multimedia designed for teaching. It's based on the theory that learning doesn't have to be boring.

EFM: Eight to fourteen modulation.

Electrically Alterable Read Only Memory - EAROM: A type of memory that is nonvolatile, like ROM, but can be altered, or have data written into it, like RAM.

Electromagnetic Interfrence: Any electromagnetic energy, natural or man-made, which may adversely affect performance of the system.

Electromagnetic Spectrum: The frequency range of electromagnetic radiation that includes radio waves, light and X-rays. At the low frequency end are sub-audible frequencies (e.g., 10 Hz) and at the other end, extremely high frequencies (e.g., X-rays, cosmic rays).

Electronic Blackboard: A device that looks like an ordinary blackboard but that has a special conductive surface for producing free-hand information that can be sent over a telecommunications channel, usually a telephone line.

Electronic Editing: The process by which audio and/or video material is added to a previously recorded tape in such a manner that continuous audio and/or video signals result.

Electro-Mechanical Pen: A device that has an electronic pen with a mechanical arm for producing free-hand information that can be sent over a telecommunication channel, usually a telephone line.

Electronic Editing: In videotapes, a process by which picture and sound elements (live or pre- recorded) are joined together without physically cutting the tape. In sophisticated editing suites, this is done by computer.

Electronic Mail - E-Mail: A system of electronic communication whereby an individual sends a message to another individual or group of people; includes computer mail and facsimile (FAX).

Elevation: The location of the satellite in the sky from your viewing site. How high above the horizon the satellite is, which is called elevation or altitude which is measured in degrees.

EMACS: A standard Unix text editor preferred by Unix types that beginners tend to hate.

EMI: Electronic mediated instruction.

Emoticon: See "smiley".

Encoder: A device that electronically alters a signal (encrypts) so that it can be clearly seen only by recipients that have a decoder which reverses the encryption process.

Encryption: An encoder electronically alters a signal so that it can be clearly seen only by recipients who have a decoder to reverse encryption. Selective addressability/scrambling designates receivers to descramble a signal. Each decoder has a unique "address."

End of Tape Sensing: A form of sensing (optically or mechanically) that automatically stops the tape transport at the end of tape or upon breakage of tape.

End User: The ultimate last user of a telecommunications system whether or not it is a student within a school, business or a subscriber on a cable television system.

ENG: Electronic news gathering.

Enter: To place on the line a message to be transmitted from a terminal to the computer.

EPROM: Erasable-Programmable Read-Only Memory.

EPS - Electronic Performance Support System: A computer supported just-in-time information system that might hold instruction manuals and other information on how to perform the tasks at hand.

EROM - Erasable Read-Only Memory: In a computer, the read-only memory (ROM) that can be erased and reprogrammed. Synonymous with erasable-programmable read-only memory (EPROM)

ESEA: Elementary and Secondary Education Act of 1965. This acronym is used mainly when referring to programs by their legislative authorization. The most common example is the "Chapter 1" series of programs.

Ethernet: Baseband protocol and technology developed by Xerox and widely supported by manufacturers; a packet technology that operates at 10 mbps over coaxial cable and allows terminals, concentrators, work stations and hosts to communicate with each other.

ETV: Educational television.

Eudora: E-mail program.

Execute: To perform the operations required by an instruction, command or program.

F

F2F - Face to Face: When you actually meet those people you been corresponding with/flaming.

Facilitator: In adult education (androgogy), the person responsible for a class who acts as a guide and resource to the students. The person responsible for the local component of a video teleconference site is normally called a facilitator.

Facsimile - FAX: A devce which uses a form of electronic transmission allowing movement of hard-copy documents from widely separated geographic areas via a telecommunications channel, usually a telephone line. Usually called a FAX machine now but previously was called a telecopier.

FAQ - Frequently Asked Questions: A compilation of answers to these. Many Usenet newsgroups have these files, which are posted once a month or so for beginners.

FDDI Fiber Distributed Data Interface: 1. Transports data up to speeds of 100 Mbps. 2. FDDI is a high-speed (100Mb) token ring LAN.

FDMA - Frequency Division Multiple Access: Refers to the use of multiple carriers within the same satellite transponder where each uplink 0has an assigned frequency slot and bandwidth.

Federal Communications Commission (FCC): An independent government agency established by the Communications Act of 1934 to regulate the broadcasting industry. The Commission later assumed authority over cable. The FCC is administered by seven commissioners and reports to Congress. The FCC assigns broadcasting frequencies, licenses stations, and oversees interstate communications.

FEC - Forward Error Correction: Adds unique codes to the digital signal at the source so errors can be detected and corrected at the receiver.

Feedback: In video; wild streaks and flashes on the monitor screen caused by re-entry of a video signal into the switcher and subsequent over-amplification. In audio, piercing squeal from the loudspeaker caused by the accidental re-entry of the loudspeaker sound into the microphone and is over-amplified. Feedback can also occur when using a conference telephone while the TV volume is too loud.

Feeder Cables: The coaxial cables that take signals from the trunk l0ine to the subscriber area and to which subscriber taps are attached. Synonymous with feeder line.

Fetch: Macintosh program for retrieving files via FTP.

Fiber Optics: Communications medium based on a laser transmission that uses a glass or plastic fiber which carries light to transmit video, audio, or data signals. Each fiber can carry from 90 to 150 megabits of digital information per second or 1,000 voice channels. Transmission can be simplex (one-way) or duplex (two-way) voice, data, and video service.

Field: One-half of a video frame two fields equal one frame or a full video screen. One field will contain all of the odd or even scanning lines of the picture.

Field Blanking Interval: The period provided at the end of the field picture signals primarily to allow time for the vertical sweep circuits in receivers to return the electron beam completely to the top of the raster before the picture information of the next field begins.

File: An organized collection (in or out of sequence) of records related by a common format, data source or application.

File Server: A component of a local area network, or LAN, which stores information for use by clients, or workstations.

Filename Extension: A three-letter (usually) code at the end of a filename that give some indication as to the type of file in non-Macintosh environments that lack icons or other methods of identifying files. Common extensions include .txt for text files, .hqx for BinHexed files, .sea for a self-extracting file, and .sit for Stuffit files.

Film at 11: One reaction to an overwrought argument: "Imminent death of the Net predicted. Film at 11."

Film Chain: Also called film island, or telecine. Consists of one or two film projectors, a slide projector, a multiplexer, and a television camera. Converts film and slides to television signals.

Finger: An Internet program that lets you get some bit of information about another user or computer, provided they have first created a .plan file.

FIPS: Federal Information Processing Standard.

Fixed Satellite Service - FSS: The earth stations are not mobile. This service generally provides telephone and TV distribution.

Fixed System: A permanent satellite receive and transmit system. The fixed system is put in place for regular use and broadcasts are made to the same sites repeatedly. The fixed systems are used for employee training, product introduction, meetings and other needs. Most fixed networks are owned by corporations such as Merrill Lynch, Hewlett-Packard, Sears, J. C. Penney, and General Motors and are not available to outsiders for video-conference use. However, they can receive teleconferences which are of interest to them.

Flame: On-line yelling and/or ranting directed at somebody else. Often results in flame wars, which occasionally turn into holy wars.

Floppy Disk - FD: Out-of-use term for diskette.

FM Microwave Radio: Ultra-high frequency often used to provide the return link in fully interactive systems (simplex). It can also be used in duplex to provide two-way full-motion video and audio interactivity.

FM Broadcast Band: The band of frequencies extending from 88 to 108 MHz.

FM-TV: Frequency modulated TV.

Follow-up: A Usenet posting that is a response to an earlier message.

Font: A complete set of characters for one style of one typeface (and traditionally, in metal type, in one size), including upper and lowercase letters, numerals, punctuation marks, and special characters. Often used to mean the software that renders a particular typeface. Sometimes used interchangeably with typeface.

Typeface The full range of letters and other characters of a given type design. Usually includes all the weights and styles, but is sometimes used to mean just one weight and style.

Type Family A collection of related typefaces, designed to work together attractively. Also used to mean the collection of weights and styles of a single typeface.

Foo/foobar: A sort of on-line algebraic place holder, for example: "If you want to know when another site is run by a for-profit company, look for an address in the form of foo@foobar.com."

Format: An established system standard in which data is stored.

Footcandle: The unit of illumination equal to 1 lumen per square foot.

Footprint: Earth area covered by a satellite beam.

Format - Videotape: Designated by the width of tape and method of recording e.g., 2-inch Quad, 3/4-inch U-Matic, 1/2-inch VHS, 1/2-inch Beta, 1/2-inch BetaCam.

Format - Programming: Type of program (drama, documentary, newscast, interview, etc.).

Fortune Cookie: An inane/witty/profound comment that can be found around the Internet.

Four-Wire Circuit: A circuit that has two pairs of conductors (four wires), one pair for the send channel and one pair for the receive channel; allows two parties to talk and be heard simultaneously.

Fractal Compression: Compression technique which uses real-time adaption of the numbers of bits allocated to different colors based upon the present scene.

Fragmentation: Storing parts of a file in disparate available space on a disk, rather than contiguously.

Frame: Full screen or frame of video is made up of two fields. Thirty frames is one second of video.

Frame Relay: A high speed interface between switches and T1 or T3 multiplexers. Frame relay is a connection-oriented interface that initially will be incorporated into private T1 and T3 multiplexers. While some carriers have committed to offer public frame relay service, others consider frame relay to be an "interim technology" and are focusing on cell relay (see BISDN below). T1 and T3 multiplexers equipped with frame relay will provide a packet-oriented, HDLC-framed interface to routers and X.25 packet switches. The packets will be routed to the proper destination by the multiplexers. Minimal protocol processing enables frame relay multiplexers to achieve high throughput. Initially, permanent virtual circuits will be supported; later, it is likely that switched virtual circuits services may also be provided by frame relay. The major advantage of frame-relay-equipped multiplexers is that only a single connection is required from the customer premises equipment (routers or X.25 packet switches) to the multiplexer. Also, with frame relay support in multiplexers, users contend for bandwidth provided via the multiplexer, and thus line cost efficiencies can be improved.

Frame Store: A video storage and display technique where a single frame of video is digitized and stored in memory for retrieval and subsequent display or processing. An electronic device used to store still pictures; a highly sophisticated slide projector used to insert pre-produced still materials into a live production for visual enhancement. The graphic material can be words, graphs, quotes, or photographs.

Frame to Frame Differencing: Compression technique which encodes only the information that represents the difference between successive frames.

Franchise: Authorization issued by a municipal, county, or state government entity which allows the construction and operation of a cable television system within the bounds of its governmental authority. The franchise area is the geographical area specified by a franchise where a cable operator is permitted to provide CATV service.

Freenet: An organization whose goal it is to provide free Internet access in a specific area, often by working with local schools and libraries. The first and preeminent example is the Cleveland Freenet. Freenet also refers to the specific Freenet software and the information services that use it.

Freeware: Software that doesn't cost anything. It can be distributed freely. However, the author still holds the copyright which means that the software can't be modified.

Freeze Frame: Repeating or holding one frame so that it appears that the action has stopped.

Frequency: The number of times a complete electromagnetic wave cycle occurs in a fixed unit of time, usually one second. The rate at which a current alternates, measured in Hertz on a telecommunications medium.

Frequency Modulation: The range of frequencies within which an audio device will function.

FTP - File-transfer Protocol: A system for transferring files across the Internet. Anonymous FTP is a conventional way of allowing you to sign on to a computer on the Internet and copy specified public files from it. Some sites offer anonymous FTP to distribute software and various kinds of information. You use it like any FTP, but the username is "anonymous". Many systems will allow any password and request that the password you choose is your userid. If this fails, the generic password is usually "guest".

Full Duplex Audio Channel: An audio channel which allows conversation to take place interactively and simultaneously between the various parties, ithout electronically cutting off one or more participants if someone else is speaking. With a Half Duplex Audio Channel, only one party can speak at a time without cutting off the other end.

Full-motion Video: Not compressed. A standard video signal of 30 frames per second, 525 horizontal lines per frame, capable of complete action.

Fully Interactive Audio/Video: Two or more video conferencing sites can interact with one another via audio and video signals. Two sites may be fully interactive without necessarily being full-motion sites.

G

Gain: An increase in signal power in transmission from one point to another; usually expressed in decibels.

Gateway: A machine that exists on two networks, such ass the Internet and BITNET, and that can transfer mail between them.

Gateway: A network element (node) that performs conversions between different coding and transmission formats. The gateway does this by having many types of commonly used transmission equipment to provide a means for interconnection.

GB - Gigabyte: A unit of data storage size which represents 2^{30} (over 1 billion) characters of information.

Gb - Gigabit: 2^{30} bits of information (usually used to express a data transfer rate; as in, 1 gigabit/second = 1Gbps).

Generational Loss: Reduction in picture quality resulting from copying video signals for editing and distribution.

Genlock: Ability of a device that handles video signals to synchronize itself to an external signal, as for overlaying graphics onto the incoming signal.

Geostationary Orbit - Geosynchronous - Clarke Belt: An orbital path approximately 22,300 miles above the earth. This unique satellite orbit has the characteristic that objects located in it rotate at the same relative speed as the surface of the earth. Objects placed in this orbit such as communications satellites can be considered fixed with respect to antennas located on the surface of the earth which are oriented towards them. Satellites in this orbit are always positioned above the same spot on the earth and from the earth, they appear fixed in space. Microwave transmission from these earth located antennas can be sent to the relatively fixed satellites in this orbit which serve as microwave repeaters back to the surface of the earth. British physicist and science fiction writer, Sir Arthur C. Clarke, invented satellite communication in his 1954 paper Wireless World, which explained this east-west orbit, 22,300 miles above the equator; three satellites based in this orbit could provide world-wide communications.

Get a life: What to say to somebody who has spent too much time in front of a computer.

GHz: GigaHertz. See Hz - Hertz.

Glass Master: A highly polished glass disc, coated with photoresist and etched by a laser beam, that is used at the start of the compact disc manufacturing process.

GIF - Graphic Interchange Format: A format developed in the mid-1980s by CompuServe for use in photo-quality graphics images. Now commonly used everywhere on-line. The filename extension generally given to GIF files is .gif.

Glitch: 1. A narrow horizontal bar moving vertically through a television picture. 2. A short duration pulse moving through the video signal at approximately reference black level on a wave-form monitor. 3. A random error in a computer program. 4. Any random, usually short, unexplained malfunction.

GNU - Gnu's Not Unix: A project of the Free Software Foundation to write a free version of the Unix operating system.

Gopher and Gopher Server: An Internet information retrieval system. Software following a simple protocol for tunneling through a TCP/IP Internet, and running errands, especially the retrieval of "documents." This information system is technically known as a Gopher Server and is part of an international network of Gopher Servers. The Gopher concept was created and initially implemented at the University of Minnesota. The software they created has migrated around the Internet and is now serving the public at large. Since its initial conception, many other organizations have contributed software to this effort.

Graphics: Visual data. This includes photographs, line drawings, computer-generated artwork, and graphs. Graphics can be entered into the computer using scanners, drawing programs, cameras, and graphics tablets.

Green Book: The specification for the CD-I standard. See CD-I.

Groupware: Groupware is an interactive collaboration of workers or students via networked applications on the computer. It provides audio, video conferencing and data sharing among a group of users using the network at the same time. Examples of programs/equipment that foster the concept of groupware is CLI's Cameo, Northern Telecom's Visit, and IBM's Person-to-Person.

GUI - Graphical user interface: The underlying principle of client/server computing is empowerment of the end-user through the delivery of information services to the desktop. Services are delivered across the network to a graphical user interface where data is massaged, merged, and maximized. GUIs are designed to juggle multiple applications in windows, through icon-driven commands that standardize application usage and optimize the underlying flow of information across the network.

H

H.261: CCITT standard for video compression. It is used to transmit video at rates between 64 Kilobits per second and T1 speeds. It is also referred to as Px64. Px64 supports intra-coded frames (JPEG-like compression techniques) or "p-frames" (predictive frames, typical of temporal compression and decompression techniques like MPEG). Px64 is an evolving multi-dimensional video telephone conferencing standard that defines compression of audio and motion video images at resolutions of 288 lines by 360 pixels or 144 lines by 180 pixels. Complying with the CCITT's recommendation H.261 Px64 incorporates multiplexing, demultiplexing and framing of multimedia data, as well as transmission protocol and bandwidth congruence, and call setup and teardown. Px64 supports intra-coded frames (JPEG-like compression techniques) or "p-frames (predictive frames, typical temporal compression and decompression techniques like MPEG).

Hacker On the Net: Unlike among the general public, this is not a bad person; it is simply somebody who enjoys stretching hardware and software to their limits, seeing just what they can get their computers to do. What many people call hackers, net.denizens refer to as crackers.

Half-Duplex: A communications channel over which both transmission and reception are possible but only in one direction at one time; e.g., a two-wire circuit.

Handshake: Two modems trying to connect first do this to agree on how to transfer data.

Handshaking: Exchange of predetermined signals when a connection is established between two data-set devices.

Hang: When a modem fails to hang up.

Hard Copy: 1. Any physical document. 2. Computer printout on permanent media such as paper.

Hard-Wired: The direct local wiring of a terminal to a computer system.

Hardware: Collectively, electronic circuits, components and associated fitting and attachments. The physical parts, components and machinery associated with computation.

HDSL - High-Bit-Rate Digital Subscriber Line: A method of providing high-speed data services over unconditioned copper wires at a top speed of 1.544 mbit/s. VHDSL (for very high-bit-rate digital subscriber line) is double that at 3 mbit/s. The key advantage of HDSL and VHDSL is that they allow telcos to provide services like frame relay, SMDS, and high-quality compressed video over existing telephone lines which is much less expensive than pulling fiber or installing additional repeaters. VHDSL is based on carrierless amplitude/phase modulation (CAP) and has applications beyond HDSL and VHDSL such as a video-on-demand service (being tested by Bell Atlantic) using CAP in which subscribers can interactively request videos, which are then transmitted over high-speed lines.

HDTV: Higher (than normal) definition TV. HDTV is generally defined as a system that offers, as a minimum, certain specific features and characteristics. These are Wide aspect ratio (now agreed as 19:9 or 1.778:1.; effectively doubled horizontal and vertical resolution (compared to existing systems); absence of encoding/decoding artifacts (requires component operation); and compact disc quality stereo sound. The technology applied to make HDTV transmittable in existing 6 MHz channels is essentially the same as the technology necessary for multichannel operation in those same channels.

Headend: Electronic control center that receives and re-transmits broadcast TV signals or original signals to receiving locations in a cable system or satellite network. System usually includes antennas, preamplifiers, frequency converters, demodulators, modulators, processors and other related equipment.

Header: The part of an e-mail message or Usenet posting that contains information about the message such as who its from, when it was sent, etc.

Helical Recording Format: A recording format in which the tape is unwrapped around a cylindrical scanning assembly with one or more recording heads.

HFS - Hierarchical File System: Used on the Macintosh platform for directory structure. The hierarchical directory structure allows a volume to be divided into smaller units known as directories and, in turn, sub-directories. The hierarchical directory structure uses a graphical metaphor of folders containing files or additional folders. Macintosh interface elements, like color icons, are embedded with file structure information.

High Band: That portion of the electromagnetic spectrum from 174 to 216 MHz, where television channels 7 through 13 are located.

High Sierra Format: The original format proposed by the High Sierra Group for organizing files and directories on CD-ROM. A revised version of this format was adopted by the International Standards Organization as ISO 9660.

High Sierra Group: An ad hoc group of CD-ROM researchers and developers who first gathered at the High Sierra Hotel in Lake Tahoe, CA, to propose a standard CD-ROM file format. This proposal was later amended and approved as the ISO 9660 standard for CD-ROM.

Hollywood Syndrome: Tendency to base ones video behavior on a model that includs a highly polished presentation rather than interaction and the use of fast-paced visuals for effect rather than substance.

Holy War: Arguments on the Internet that involve certain basic tenets of faith, about which one cannot disagree without setting one of these off. For example: IBM PCs are inherently superior to Macintoshes or Macs are inherently superior to IBMs.

Homes Passed: The number of living units (single residential homes, apartments, condominium units) passed by cable television distribution facilities in a given cable system service area.

Homogeneous Network: A network of similar host computers such as those of one model of one manufacturer.

Horizontal Blanking: The blanking signal at the end of each scanning line that permits the return of the electron beam from the right to the left side of the raster after the scanning of one line.

Horizontal Resolution: The maximum number of black and white vertical lines that can be resolved within a horizontal expanse of raster equal to one picture height. NTSC television pictures normally have 300 lines of resolution or less.

Horizontal Retrace: The return of the electron beam from the right to the left side of the raster after the scanning of one line.

Host System: A public-access site; provides Internet access to people outside the research and government community.

Hosts: Computers (not terminals) that process data, act as data sources or destinations in a communications network.

.hqx: The filename extension used for BinHex files.

HTML - HyperText Markup Language: The language used to mark up text files with links for use with World-Wide Web browsers.

Hub: 1. A signal distribution point for part of an overall system. 2. The master station through which all communications to, from and between micro terminals must flow.

Hue: The attribute of color perception that determines whether the color is red, yellow, green, blue, purple, etc.

kHz - Kilohertz: 1,000 Hertz

MHz - Megahertz: 1,000 kHz-one million Hertz

GHz - Gigahertz: 1,000 MHz-one billion Hertz.

Hyperband: The band of cable television channels above 300 MHz.

Hypermedia: Software that allows the user to interactively manipulate text, images, animation, graphics, sounds, digitized voice, and video.

HYTELNET: Stands for HyperTelnet. HYTELNET is essentially a database of Telnet sites and other Internet resources that can link to other programs when you want to connect to a site you've found. Not as useful as Gopher. A system that provides access to libraries around the world through the Internet.

Hz - Hertz: Basic measure of frequency with which an electromagnetic wave completes a full cycle from its positive to its negative pole and back again. Hertz is a unit of frequency equal to one cycle per second. Normal house current is 60 Hertz (60 cycles per second).

I

I/O: Input/output.

IAB - Internet Architecture Board: The coordinating committee for Internet design, engineering and management.

ICN: Iowa Communications Network.

Icon: A pictorial, symbolic representation of a function or task. Used in GUIs (Graphical User Interfaces) such as Windows and Apple Macintosh Finder. See GUI.

IEEE: Institute of Electrical and Electronic Engineers.

IETF: Internet Engineering Task Force.

IHE - Institution of Higher Education: A postsecondary educational institution college, university, and other such schools.

IM: Intermodulation distortion occurs when two or more signals are passed through a nonlinear device such as an amplifier.

IMAP: A new protocol for the storage and retrieval of e-mail. Much like POP - the Post Office Protocol.

Inbound: The direction of a signal relative to the hub of a local area network (LAN) or other telecommunications system. Inbound signals would be traveling from originating points other than the primary hub in the reverse direction to the hub.

Information Agent: A software program (currently only an interface to frequently updated databases) that can search numerous databases for information that interests you without your having to know what it is searching. Archie and Veronica are current examples of information agents.

Infrared: That portion of the electromagnetic spectrum just below visible light; infrared radiation has a wavelength from 800 nm to about 1mm. Fiber-optic transmission is predominantly in the near-infrared region, about 800 to 1600 nm.

Initialization: The process carried out at the commencement of a program to test that all indicators and constants are set to prescribed conditions.

Intelsat: The International Telecommunications Satellite Organization operates a network of satellites for international transmissions. The stated purpose is the design, development, construction, establishment, maintenance, and operation of the space segment of the global communications satellite system.

Interactive: Any application that allows the participants at distant locations to communicate with each other; may indicate two-way video and two-way audio; one-way video and two-way audio through a normal telephone call placed to the origination site; asynchronously (not in real time) as through computer conferencing such as an electronic mail system; or through interaction with a teaching machine such as a computer which is programmed to respond to the user with messages on the screen, voice or other sounds to indicate that an answer is right or wrong.

Interactive Television: Lets owners of ordinary TVs order movies, home shopping, mutimedia packages and other digitized products from electronic jukeboxes.

IMHO: Internet shorthand for "In my humble opinion."

Instructional Design: The methodology used to deliver information in a manner that achieves learning.

Integrated Circuit - IC: An electronic circuit made by manipulating layers of semiconductive materials.

Integrated System: A system in which all components including the various types of amplifiers and taps have been designed from a well-founded overall engineering concept, to be fully compatible with each other.

Interactive: The active participation of the user in directing the flow of the computer or video program.

Interactive video: The capability to transmit and receive two-way video transmissions between two or more sites.

Interactive Cable System: A two-way cable system that has the capability to provide a subscriber with the ability to enter commands or responses on an in-home terminal and generate responses or stimuli at a remote location. An example of an interactive system would be order entry for Pay-Per-View the order information is transmitted upstream on the cable from the subscriber's terminal to the headend, processed by a billing/authorization computer, and authorization to view a specific Pay-Per-View event is sent downstream to the subscriber's terminal.

Interactive Multimedia: A multi-level multimedia presentation that allows you to access information randomly and nonsequentially.

Interconnect: The connection of two or more cable systems. 2. The connection of a headend to its hubs.

Inter-Exchange Carrier - IXC: Carriers that can carry inter-LATA traffic. Long distance telephone companies such as AT&T, MCI, and US Sprint.

Interface: The link between two pieces of disparate equipment, such as a CPU and a peripheral device. Also, a method of translating data from computer to user. For Internet, the user interface is difficult for the uninitiated to use. Software programs have been written which change the look of the screen by provide pull-down menus, buttons, hierarchical files folders or hypertext to use and move around the Internet. Software program names include Mosaic, Lynx, Internet in a Box and GINA.

Interference: A scrambling of the content of signals by the reception of desired signals.

Interlaced Video: Process of scanning video frames in two passes, with each pass painting every other line of the frame onto the screen. NTSC's 525-line frame scans in two fields of 262.5 lines each that take 1/60 second to paint; a frame takes 1/30 second to paint. Noninterlaced video scans complete video frames in one pass usually producing a higher image quality.

Interleave: A method of storing information in an alternating sequence of frames.

International Telecommunication Union - ITU: Organization composed of the telecommunications administrations of the participating nations. Focus is the maintenance and extension of international cooperation for improving telecommunications development and applications.

Internet: A worldwide system for linking smaller computer networks together - governmental institutions, military branches, educational institutions, and commercial companies. Networks connected through the Internet use a particular set of communications standards to communicate, known as TCP/IP. Internet is the name given to the overall connectivity of all its various sub-networks, including USENET, APRAnet, CSnet, BITNET, etc. There is no surcharge to send or receive messages through Internet. Only ASCII messages up to 50,000 characters can be sent through this system. With a lowercase "i", an internet is a group of connected networks.

InterSLIP: A free program provided to the Macintosh Internet community by InterCon Systems. In conjunction with MacTP, InterSLIP enables users with modem and a SLIP account to use excellent software like Fetch and TurboGopher.

IP - Internet Protocol: The main protocol used on the Internet.

IRC - Internet Relay Chat: A service where users can "talk" via typing to people around the world.

Iridium: Motorola's $3 billion worldwide direct cellular project which will enable users to have one worldwide number. This will be accomplished by a system of low earth orbiting satellites (LEOs).

ISDN - Integrated Services Digital Network: A set of standards provide a common architecture for the development and deployment of digitally integrated communications services. A set of standardized customer interfaces and signaling protocols for delivering digital circuit-switched voice/data and packet-switched data services. ISDN is designed to provide standard interfaces to custom premises equipment such as computers, telephones, and facsimile machines through basic rate interface (BRI) to PBXs, host computers, and LANs through primary rate interface (PRI); to the pubic switched network through SS7; and to local packet data terminals and the public packet-switched network through X.25 and X.75/X.75' packet services. The key to ISDN is out-of-band signaling which permits the users' equipment and the network to exchange control and signaling information over a separate channel from that which carries user information. A digital telecommunications channel that allows integrated transmission of voice, video and data. ISDN lines used to access network services are divided into bearer, or "B" channels, and a supervisory, or "D" channel, for out-of-band signaling. B Channels carry digitally encoded customer information suchas voice and data traffic, while the D channel provides the information required to set up, route and disconnect calls on B Channels. D channels can also carry other information such as caller identification. Twenty-three B channels and one D channel form a Primary Rate Interface or "23B+D". PRI B channels can be used for any combination of voice, data, and image transmission at 64 kpbs. In addition, B channels can be grouped together to create wider bandwidths for applications like video transmission.

ISO: International Standards Organization.

ISO 9660: The international standard for directory structures and file layout on CD-ROMs, a logical, structural standard compared to the physical standards for manufacturing called the "Yellow Book." This standard specifies, for single sessions, exactly how information is stored on a CD-ROM to be accessible in any CD-ROM drive running on a variety of common operating systems.

ISOC - Internet Society: ISOC is a membership organization that supports the Internet and is the governing body to which IAB reports.

ITFS Antenna System - Instructional Television Fixed Service: Local (up to 25-mile radius) one-way, over-the-air block of TV channels operating at microwave (very high) frequencies reserved for educational purposes; can be received only by TV installations equipped with a converter to change signals back to those used by a TV set. One-way audio and full motion video. The antenna may be omnidirectional or shaped to cover a specific geographic area. In rare instances the ITFS antenna system can be found to be very directional for special repeater applications or to serve a series of co-linear receive sites. The ITFS television transmission system was first authorized in 1963 by the FCC for educational television in the 2.5 to 2.686 GHz band. The ITFS band has subsequently been re-allocated for shared operation among multipoint distribution services, multichannel multipoint distribution services, operational fixed services, and ITFS users.

IVDS - Interactive Video and Data Services: Name for license which will be granted by the FCC to devices called Interactive TV Appliances (ITAs). ITAs include TVAnswer, a two-way television service for consumers for game shows, sporting events and respond instantly to news polls and interactive advertising as well as participate in distance learning. The system will also let viewers shop, bank, pay bills, organize TV programming and order a pizza.

IVR: The IVR unit answers the call, greets the caller, and guides the caller through possible responses with a series of voice prompts. The desired information is provided via prerecorded voice fragments (words) or computer-generated speech.

J

Jack: A connecting device to which a wire or wires of a circuit may be attached and which is arranged for the insertion of a plug.

JANET - Joint Academic Network: JANET is Great Britain's national network. JANET addresses work backwards from normal Internet addresses (largest domain to the smallest). Mot gateways to JANET perform the necessary translations automatically.

Janus disc: A CD-ROM that contains data tracks in two or more different formats, such as ISO 9660 and HFS (Macintosh Hierarchical File Structure).

JPEG - Joint Photographic Expert Group: JPEG is an industry standard for still-image compression that is moving into full-motion video. Storing the video signal is a problem as it takes a 300 megabyte hard disk to store just 10 seconds of digital video. Compression is the answer to storage problems. JPEG is a compression technique based upon intraframe encoding technology. It allows full restoration of symmetrically compressed images. Symmetrical compression means that the image takes an equally long time to be compressed as it does to be decompressed. An asymmetrical scheme takes longer to compress an image than to decompress it and typically compresses the image on a computer other than the one to be used for decompression. Relying on a newly adopted format to encode and decode digital images based on independent, non-temporal (intraframe) data, JPEG typically divides an image into 8 by 8 pixel blocks. These 64 square pixel matrices, called a "search range," enable the aggregate quantization of the image and color data store by the pixels within each of the blocks. Advanced JPEG algorithms that use larger search ranges, up to as much as 32 by 32 pixels, called "super blocks," enable significantly faste encoding and decoding (up to 40 to 1 compression at about 4MB per second) but demand exponentially more processing power to maintain the same degree of image quality as the smaller pixel block.

Originally, JPEG was intended to compress only still images. However, video is nothing more than a quick presentation of successive still images. Thus, a form of JPEG, known commonly as motion JPEG, is being used to compress motion images, particularly in applications like video editing where it is necessary to access individual frames of video and to scrub forward and backwards through source material.

JPEG, MPEG and Px64 specifications use a Discrete Cosine Transform (DCT), an encoding algorithm that quantifies the human eye's ability to detect color and image distortion. DCT parses color content data thereby enabling the use of a higher pixel depth sampling rate (typically 24 or 32 bits) than non-DCT compression techniques. JPEG typically controls 24 bits per RGB pixel, retaining a high quantization of luminance and color resolution.

JPEG System Highlights: Used to encode still images. It compresses about 20 to 1 ratio before visible image degradation occurs. Compression ratios exceeding 100 to 1 attainable but image degrades excessively. At very low compression ratios of 5 to 1 maximum JPEG maintains absolute resolution. It excludes audio compression. Symmetrical (compresses at same rate as it decompresses,) uses the same hardware to encode and decode. It compresses redundant data occurring within each frame (intraframe). Compresses comparatively slow, depending on computer speed, about 1 to 3 seconds for a 1 MByte image. Decompresses a full sized image in .5 to 1 second or reduced sized image in real time. It has good quality at maximum compression.

Jukebox: CD-ROM drive with a disc changing mechanism, capable of playing multiple discs.

K

Kaleida: Joint venture between Apple Computer and IBM to usher in a new type of interactive communications that combines video, sound, text, graphics, and animation. The new venture will develop, license, and make available specifications and technologies to promote the exchange of multimedia information between a variety of computing and consumer electronics devices. It seeks to create a multimedia standard that will cross existing platforms, along with distribution networks for the data.

Kbps, kb/s - Kilobits per Second: A unit of measure of data of 1,000 bits per second or 1,000 Baud.

Kermit: A file transfer protocol named after Kermit the Frog. Kermit is generally slower than XMODEM, YMODEM, and ZMODEM.

Kerning: Reducing the horizontal space between characters of type. Originally, casting a letter so that part of it (e.g., the top of the f) extends beyond the body of the letter, into the space occupied by the next letter. Sometimes used to mean adding or removing space between letters.

Keyboard: An alphanumeric, input/output, peripheral device used to communicate with a computer.

Killfile: A file that lets you filter Usenet postings to some extent, by excluding messages on certain topics or from certain people.

Kilobaud: The measure of data transmission speed a thousand bits per second.

Kilobyte: A unit of measurement equal to 1024 bytes.

Kine Recording: The technique of converting a video image to motion picture film.

Knowbot: Short for knowledge robot. Embedded machine intelligence capable of automatically and regularly searching for new information on parameters set by the user. At its extreme, the knowbot becomes an intelligent partner in mediating human communication. Sometimes called an agent.

Knowledge Navigator: An information agent popularized by an Apple video about working with computers in the future.

Ku Band: A category of satellite transmissions higher in frequency than those used as "c band" which are being transmitted from satellites placed in the geostationary orbit. The group of microwave frequencies from 12 to 18 GHz and the band of satellite downlink frequencies from 11.7 to 12.2 GHz. The higher frequencies (12 GHz versus 4 GHz) have created the possibility of smaller receive antennas and the realization of direct broadcast satellite (DBS) signals to the end user without the necessity of going through a cable television sstem or other shared use receive site due to the factors of size and cost.

L

LAN - Local Area Network: Private transmission network interconnecting offices within a building or group of buildings and usually designed to convey traffic; e.g., voice, data, facsimile, video. Usually associated now with a computer network made up of computers, printers, and mass storage units. MAN Metropolitan area network. WAN - Wide Area Network.

Large-Scale Integration - LSI: The process of engraving many thousands of electrical circuits on a small chip of silicon.

Laser - Light Amplification by Stimulated Emission of Radiation: 1. A device for generating coherent electromagnetic signals (e.g., light). Low powered lasers are frequently used to transmit light signals into optical fibers. 2. Laser light contains waves that have the same phase, as opposed to conventional light, whose individual wave phases are unrelated to the phases of the others.

LATA: Local access and transport area of a telephone company.

Lavaliere: A small microphone that can be clipped onto clothing or suspended from neck cords and worn in front of the chest.

LCD - Liquid Crystal Display: A method of creating alphanumeric displays by reflecting light on a special crystalline substance. Frequently used in electronic games and watches, and in portable electronic instruments.

LEA - Local Educational Agency: (a) a public board of education or other public authority legally constituted within a state for either administrative control of, or direction of, or to perform service functions for, public elementary or secondary schools in a city, county, township, school district, or other political subdivisions of a State; or such combination of school districts or counties a State recognizes as an administrative agency for its public elementary or secondary schools; (b) any other public institution or agency that has administrative control and direction of a public elementary or secondary school; (c) as used in vocational education programs the term also includes any other public institution or agency that has administrative control and direction of a vocational education program.

Leading (pronounced "ledding"): Vertical space between lines of type, measured in points. In metal type, leading is the additional space (from inserting strips of lead between lines of metal type). In phototype and digital type, where there is no metal body determining the height of the type, leading has come to mean the total space from one line to the next, usually measured from baseline to baseline.

Leased Lines: A term used to describe the leased or rented use of dedicated lines from point to point. Lines could include fiber optic cables, telephone cables, microwave or other transmission systems.

LEC: Local exchange carrier of a telephone company.

LED - Light Emitting Diode: A semiconductor which emits light when a proper voltage is applied to its terminals.

Light Pen: A pen-like device that contains a photosensitive cell and small aperture lens that produces or detects electronic signals; can be used to write free-hand directly on a TV screen or to enter, edit and position computer text or graphics.

Linear: Video technology designed to be played from beginning to end without stops.

Links: Communication pathways between nodes.

Lip Sync: Synchronization of the sound portion with the visual portion of a television program.

LISTSERV: A powerful program for automating mailing lists.

LNA - Low Noise Amplifier: Located at the antenna. Refers to electronic equipment, used in conjunction with satellite reception, intended to amplify extremely weak satellite signals without introduction of noise. They are rated in different noise temperatures, expressed in degrees Kelvin. The lower the noise temperature figure, the higher the carrier-to-noise ratio, and the better the picture.

Local Exchange Carrier (LEC): Carriers that can carry only intra-LATA traffic. Local telephone companies such as US West, Contel, Centel etc.

Local Loop: The local loop gets the signal from the receive site to the viewing room. Microwave, fiber optics, cable and sometimes broadcast are used to distribute the signal. Also referred to as the "Last Mile".

Location - Remote: Production shooting site other than a studio.

Log On/Log In: Connect to a host system or public-access site.

Log Off: Disconnect from a host system.

Low Band: That portion of the electromagnetic spectrum from 54 to 88 MHz, where television channels 2-6 are located.

Low Earth Orbiting Satellite - LEO: Low earth orbit satelites which require 77 small, smart satellites to provide linkage around the world. The satellites move overhead in their low orbit. Motorola's Iridium (from the element Iridium which has 77 electrons) uses the concept to provide, digital, satellite-based personal communications via small, hand-held transportable receivers. With the system, voice, fax, or data calls can be made or received anywhere. The user will have one universal telephone number for the phone. Local gateways will store customer billing information, keep track of user locations, and interconnect with terrestrial carriers worldwide. The dual-mode phone will access customers' regular cellular service first, switching to Iridium only when there is no terrestrial signal, to assure least-cost routing. The system is planned to be launched in 1994 with service by the end of 1996. Inmarsat's Project 21 will provide similar services.

Low Power Television: Broadcast medium that is similar to commercial TV but limited in broadcast coverage area by its low power signal. Can air one class per time frame which can be received at multiple sites.

Lumen: Unit of light flux.

Luminance: 1. Luminous flux emitted, rejected, or transmitted per unit of solid angle per projected area of the source. 2. The photometric equivalent of brightness. 3. The brightness part of a television picture.

Luminance Signal: That portion of the television signal which conveys the luminance or brightness information.

Lurk - Lurkers - Lurking: People who read messages in a Usenet newsgroup or other public system without ever responding or contributing to the topic.

Lux: Unit of n equal to 1 lumen per square meter or approximately 0.1 candle power.

M

MacBinary: A file format that combines the three parts of a Macintosh file; the data fork, the resource fork, and the Finder information block. No other computers understand the normal Macintosh file format, but they can transmit the MacBinary format without losing data. When you download a binary Macintosh file from another computer using the MacBinary format, your communications program automatically reassemble the file into a normal Macintosh file.

Machine Language: Binary code that can be directly executed by the processor, as opposed to assembly or high-level language.

MacTCP: A Control Panel from Apple that implements TCP on the Macintosh. MacTCP is required to use programs such as Fetch and TurboGopher.

Magnetic Media: Any medium on which data is stored as variations in magnetic polarity. Usually floppy disks, hard disks, and tape.

Magnetic Tape: A mylar tape, coated with magnetic particles, on which audio, video or data can be stored.

Mailing List: Essentially a conference in which messages are delivered right to your mailbox, instead of to a Usenet newsgroup. You get on these by sending a message to a specific e- mail address, which is often that of a computer that automates the process.

Magneto-optical: An information storage medium that is magnetically-sensitive only at high temperatures. A laser heats a small spot, which allows a magnet to change its polarity. The medium is stable at normal temperatures. Magneto-optical discs can be erased and re-recorded.

Master: The original video tape, audio tape or film of a finished product. Usually stored in a vault or area protected from the environment. Dubs are made from the master. Once the master is worn out, it can not be replicated.

Master Antenna Television System - MATV: An antenna and distribution system which serves multiple dwelling complexes such as motels, hotels, and apartments. It is, in effect, a miniature cable system.

Mastering Facility: A manufacturing plant where compact disc ÒmastersÓ are created for the mass production or replication of the actual compact discs. Metal Master: A metal disc created by plating an etched glass master disc with nickel. Used in a mastering facility to create metal stampers for the mass production of compact discs.

MATV - Master Antenna Television: Centrally-located receiving system that distributes off-air signals and to multiple places in the cable transmission system.

Master Control: Nerve center for telecasts. Controls the program input, switching, and retrieval for on-the-air telecasts. Also oversees technical quality of programs.

Matte: The keying of two scenes; the electronic laying in of a background image behind a foreground scene, such as a picture of a town meeting behind the newscaster reporting on the meeting.

Matte Key: Keyed (electronically cut-in) title whose letters are filled with shades of gray or a specific color.

MBONE - Multicast Backbone: Internet. An outgrowth of the first two IETF "audiocast" experiments in which live audio and video were multicast from the IETF meeting site to destinations around the world. The idea is to construct a semi-permanent IP multicast testbed to carry the IETF transmissions and support continued experimentation between meetings. The MBONE is a virtual network. It is layered on top of portions of the physical Internet to support routing of IP multicast packets since that function has not yet been integrated into many production routers. The network is composed of islands that can directly support IP multicast, such as multicast LANs like Ethernet, linked by virtual point-to-point links called "tunnels". The tunnel endpoints are typically workstation-class machines having operating system support for IP multicast and running the "mrouted" multicast routing daemon.

mbps, MB/s - Megabits per Second: A unit of measure of data of 1,000,000 bits per second or 1,000,000 Baud.

MCC - Microelectronics and Computer Technology Corporation: An industry consortium that developed the MacWAIS software.

MCU - Multipoint Control Unit: MCU's have the ability to support multipoint videoconferences on codecs of the same brand and (in most cases) model.

Mean Time Between Failure - MTBF: A statistical quantitative value for the time between episodes of equipment or component failure.

Medium (Media): Any material substance(s) that can be used for the propagation of signals. Examples are copper, air,water, and fiber optics.

Meet-Me Bridge - Meet-Me Teleconferencing: A type of telephone bridge that can be accessed directly by calling a certain access number; provides dial-in teleconferencing. The term "meet-me bridging" refers to the use of this type of bridge.

Mega: 1. Ten to the sixth power, 1,000,000 in decimal notation. 2. When referring to storage capacity, two to the twentieth power, 1,048,576 in decimal notation.

Megabyte: A unit of measurement equal to 1024 x 1024 bytes, or 1024 kilobytes; 8 million bits.

MegaHertz - MHz: One million cycles per second.

Memory: Computer's information storage capability. RAM - random access memory; ROM - read only memory.

Menu: A list of symbols and functions that can be selected on a computer system.

Microcomputer: A relatively precise term for computers whose central processing units (CPUs) are microprocessor chips. By contrast, mainframes and most minicomputers have CPUs containing large circuitry. Microcomputers include personal computers, small business computers, desktop computers, and home computers.

Microfiche: A system of storing and retrieving information microforms, consisting of film in the form of separate sheets, that contain original text, pictures, data, or anything which has been reduced to micro- images for a greater storage efficiency and arranged in a grid pattern for location of those original images by means of Cartesian coordinates.

Microfilm: A system of storing and retrieving information microforms, consisting of film as a data medium, usually in the form of a roll or strip, that contains micro-images of the original information. The images are generally in a sequential arrangement rather than in rows or columns as on microfiche.

Microprocessor: A central processing unit implemented on a chip.

Microsecond: One millionth of a second.

Microsoft Windows: A GUI (graphic user interface) operating environment developed by Microsoft for use on PCs running under the MS-DOS operating system.

Microwave: That portion of the electromagnetic spectrum from approximately 1,000 Megahertz to 100,000 Megahertz. The microwave energy is capable of being focused in concentrated beams in specific directions due to its short wavelength characteristics and sent over long distances. Point-to-point transmission system that transmit signals through the air using transmitters and antennas attached to tall towers. Provides program audio and video plus the capacity for additional voice and data material. It is also capable of being transmitted over wide areas from a central point or shaped into specific coverage areas with special antennas (ITFS). Extreme examples of long distance focused microwave transmissions are the signals sent from a satellite uplink earth station to a satellite 22,300 miles above the earth and from that satellite back to earth.

MHz - Megahertz: Refers to a frequency equal to one million Hertz, or cycles per second.

Midband: The band of cable television channels A through I, lying between 120 and 174 MHz.

MIDI - Musical Instrument Digital Interface: 1. An industry-standard connection for computer control of musical instruments and devices. 2. Musical Instrument Digital Interface. Industry standard for exchange of musical information between computers and musical instruments or music synthesizers.

Midsplit System: A cable-based communications system that enables signals to travel in two direction, forward and reverse simultaneously with upstream (reverse) transmission from 5 MHz to about 100 MHz and downstream (forward) transmission greater than about 150 MHz. Exact crossover frequencies vary from manufacturer to manufacturer.

MIME - Multipurpose Internet Mail Extensions: A new Internet standard for transferring non-textual data, such as audio messages or pictures, via e-mail.

Minicomputer: An intermediate range computer, between full-size mainframes and 16-bit microcomputers. Historically, minicomputers have served dedicated uses, such as in scientific and laboratory work.

MIS: Management Information System.

Mixed Mode Disc: A CD-ROM that contains both CD-ROM (Yellow Book) and CD-Audio(Red Book) tracks.

MNP: Microcom Networking Protocol.

Model: 1. A representation in mathematical terms of a process, device, or concept. 2. An academic model; a program with a certain set of procedures or elements which can be duplicated by others in their institutions.

Modem - MOdulator/DEModulator: Device that connects terminals and hosts through analog links by converting data signals to analog signals and back again. Transmission rate of 300 Baud is slow; 2400 is faster.

Modem-Encryption Devices: By placing encryption units at modem interfaces, some systems have all data on the link encrypted and decrypted in a manner that is transparent to the sending and receiving stations.

Modular: Constructed with standardized units or dimensions for flexibility and variety in use; allows for easy replacement, substitution, expansion or reconfiguration of modules or sub-assemblies.

Modulator: A device which converts the video signal and audio signal onto a viewable TV channel. It takes the video and audio signals that are separated by the receiver and combines them into a signal that can be received by an ordinary TV set. This signal is called an "RF" signal, meaning radio frequency, and is usually set for either channel 3 or 4. The advantage of using a modulator is that it permits the use of standard TV receivers for displays, but the signal quality is not as good as using a direct video and audio feed to a monitor TV display. As with receivers, modulators should also be redundant.

Monitor: A television monitor is capable of projecting from an attached device such as a video tape recorder or camera; or from a cable such as that connected by cable companies or the cable from the satellite receiving unit. It is not equipped with receiver electronics which enable it to receive local broadcast channels. Studio monitors are usually high resolution so that the best possible picture is seen.

Mosaic: A free graphical front end to the Internet that supports browsing of multimedia data that includes plain and formatted text, picture, video and sound. The data is based on the hypertext document format where text or pictures can act as links to other places in the same or different documents. The document a user points to on the receiving desktop could be on the same machine or on another computer elsewhere on the Internet. Mosaic clients are currently available for Windows, Macintosh, the X Window System and many flavors of UNIX. Mosaic was developed by the National Center for Supercomputing (NCSA).

MOTSS - Members of the Same Sex: Originally an acronym used in the 1980 federal census.

MPC - Multimedia Personal Computer: A standard which describes a PC that can run Microsoft's Windows efficiently because the system software beneath multimedia would be "Windows with Multimedia Extension." The specification calls for added audio and CD-ROM hardware. MPC is a registered trademark of the MPC Marketing Council.

MPEG - Moving Pictures Experts Group: Multimedia compression standard for professional and consumer applications - digital video, digital audio and systems compression. MPEG compression compresses similar frames of video, tracks elements which change between frames and discards the redundant information. This allows full-motion video to be sent at CD-ROM data rates - around 160kbps. MPEG, which is now being called MPEG1 in some circles achieves increased compression through the use of a combination of interframe and intraframe (sometimes called MPEG 1-Frame) image and audio compression algorithms, including predictive and interpolated technologies. These techniques analyze the degree of motion present in a search range and predict the "temporal redundancy" (anticipated repetitive motion) occurring between adjacent frames. As in JPEG, pixel search blocks form the basis of the representative sampling unit.

MPEG differs from motion JPEG (sometimes called "frame-rate" or motion JPEG). For frame rate JPEG to achieve the same data rates as MPEG, it must resort to bandwidth reduction strategies, such as chroma subsampling, reducing data rates either through an averaging of color information, or through decimation, which discards alternate lines or pixels of image information. The compression ratios obtainable with MPEG, which can be as high as 200 to 1 with decimation and chroma sampling, are directly dependent on the amount of data redundancy present in a given image. MPEG is designed to deliver data in the 1 to 2mbps range making it suitable for replaying full motion video stored on CD-ROMs.

MPEG2, which is still under development, will define a compression and decompression technology suitable for delivering data at a 5 MB to 10 MB-per-second rate. It is envisioned as a data delivery system capable of delivering high-quality, high data content capacity images to computers and television.

MPEG3 is projected to deliver data at a stunning (by today's standards) 60 mbps making it suitable for complex saturated color data signals, including HDTV. MPEG3 is several years away from finalization.

MPEG System Highlights: Used to encode motion images. Will deliver decompressed data in the 1.2 to 1.5 mbps range enabling CD-ROM to play back full motion color motion images at 30 frames/second. Compresses about 50 to 1 ratio before image degradation occurs. Compression ratios are as high as 200 to 1 attainable but with observable degradation, including audio compression. Asymmetrical (compresses slower than it decompresses), uses different hardware or techniques to encode and decode. Compresses redundant data appearing in sequential frames (temporal, interframe). Compression rate is fast. Decompresses in real time. Fair quality at maximum compression.

MUD/MUSH/MOO/MUCK/DUM/MUSE: These are multi-user, text based, virtual reality games. A MUD (Multi-User Dungeon) is a computer program which users can log into and explore. Each user takes control of a computerized persona/avatar/incarnation/character. You can walk around, chat with other characters, explore dangerous monster-infested areas, solve puzzles, and even create your very own rooms, descriptions and items. There are an astounding number of variations on the MUD theme.

Multiple System Operator - MSO: An organization that operates more than one cable television system.

Multi-Link Audio: Any application allowing viewers to be connected by phone (usually by audio bridge) to the spot where the broadcast originates.

Multimedia: The combination of multiple digitized data types; text, sound, computer-generated graphics and animations, photographs and video. The merger of digital technologies based on the use of computers. The technologies that are converging are computing, television, printing and telecommunications.

Multimedia Extensions: Adds audio and video recording and playback capabilities to Microsoft Windows. Part of the MPC standard.

Multisession: A drive that has the ability to read a CD-ROM on which data was recorded in at least two different recording sessions, or a disc that contains data recorded at different times.

Multiple Access: The ability of more than one user to use a transponder. Transponders have three basic resources frequency, time and space. The frequency domain is used in FDMA. Time domain multiple access is used in TDMA by time-sharing the transponder. Space domain multiple access makes use of either the polarization discrimination, orthogonal digital codes or through spread-spectrum techniques.

Multiple Audio Subcarrier Tuning: Essential to take advantage of radio and data services riding piggy-back on video signals. Also, some programming may use nonstandard (6.8 and 6.2 MHz) frequencies.

Multiplexer - MUX: Device that uses one of several techniques to combine multiple analog or digital signals onto a single path.

Multiplexing Transmission of two or more information streams over a single physical medium at the same time such that each data source has its own channel. Allows a number of simultaneous transmissions over a single circuit. Common methods are frequency division multiplexing (FDM) where the frequency bands are split to constitute a distinct channel and time division multiplexing (TDM) where the common channel is allotted to several different information channels, one at a time.

Multiprocessor: A computer employing two or more processing units under integrated control.

Multitasking: Pertaining to the concurrent execution of two or more tasks by a computer.

N

N + 1: Created by the FCC, this formula forms the basis by which the FCC regulates expansion of channel capacity for non-broadcast use. The FCC requires that if the government, education, public access, and leased channels are in use at least 80 percent of the Monday-through-Friday period for at least 80 percent of the time during any three-hour period for six consecutive weeks, then within six months the system's channel capacity must be expanded by the operator.

Nanosecond - nsec: One billionth of a second.

Narrowband: A telecommunications medium that carries lower frequency signals; includes telephone frequencies of about 3,000 Hertz and radio subcarrier signals of about 15,000 Hertz.

Narrowcast: Transmission of programs to a specifically defined audience normally using the newer technology delivery systems. Sometimes referred to as a target audience, a limited audience, or a "narrow" audience, hence the name "narrowcast."

National Cable Teleision Association - NCTA: Washington, D.C. based trade association for the cable television industry; members are cable television system operators; associate members include cable hardware and program suppliers and distributors, law and brokerage firms, and financial institutions. NCTA represents the cable television industry before state and federal policy makers and legislators.

Net: The Net-sanctioned way to refer to the Internet for the initiated.

Net.god: One who has been on-line since the beginning, who knows all and who has done it all.

Net.personality: Somebody sufficiently opinionated/flaky/with plenty of time on his/her hands to regularly post in dozens of different Usenet newsgroups, whose presence is known to thousands of people.

Net.police: Derogatory term for those who would impose their standards on other users of the Internet. Often used in vigorous flame wars (in which it occasionally mutates to net.nazis).

Netiquette: A set of common-sense guidelines for not annoying others.

Network: 1. Two or more information sources or destinations (points or nodes) linked via communications media to exchange information. 2. It can be as simple as a cable strung between two computers a few feet apart or as complex as hundreds of thousands of computers around the world linked through fiber optic cables, phone lines and satellites.

Network Architecture: A set of design principles, including the organization of functions and the description of data formats and procedures, used as the basis for design and implementation of a user- application network.

Network Interface Card: Also known as NIC. Add-in circuit board that allows a PC to be connected to a local area network (LAN).

Network License: A license from a software vendor that allows an application to be shared by many users over a network.

Newbie: Someone who is new to the Internet. Sometimes used derogatorily by Net.veterans who have forgotten that, they, too, were once newbies who did not innately know the answer to everything. "Clueless newbie" is always derogatory.

Newsgroup: A Usenet conference.

Newsreader: A program to read news and providing capabilities for following or deleting threads.

NFS - Network File System: A network service that lets a program running on one computer use data stored on a different computer on the same Internet as if it were on its own disk.

NIC - Network Information Center: An organization which provides network users with information about services provided by the network. As close as an Internet- style network gets to a hub; it's usually where you'll find information about that particular network.

NII - National Information Infrastructure, Data Highway, Information Highway: An interoperable linking of all networks for business, government, education and consumer uses. Much of the highway already exists in phone lines, coaxial cable, satellites, and cellular networks, and already functions as the Internet. The difference between the Internet and the future NII is primarily based on more bandwidth, faster operating systems, intelligence in data routing, security for all services which are conveyed through audio, data, and video modes.

NNTP - Net News Transport Protocol: A transmission protocol for the transfer of Usenet news.

NOC - Network Operations Center: An organization that is responsible for maintaining a network.

Node: An addressable unit in a network, which can be a computer, work station or some type of communications control unit.

Noise - Audio: Unwanted sounds (static) that interfere with the intended sounds; or unwanted sound signals.

Noise Temperature: The amount of thermal noise present in a system, expressed in degrees Kelvin. The lower the noise temperature, the better.

Noise - Video: Unwanted electronic interference that shows up as snow.

Non-Composite Video Signal: A signal which contains only the picture signal and the blanking pulses.

NOS: Network operating system.

NREN - National Research and Education Network: Created by an act of Congress, this new network - still in interim stages - is replacing Internet as the national e-mail system connecting research, governmental, and high education networks and data bases. There is concern among some educators that K-12 will not have easy or immediate accessto NREN as it's implemented.

NSA line eater: The more aware/paranoid Net users believe that the National Security Agency has a super-powerful computer assigned to reading everything posted on the Net. They will "jokingly" refer to this line eater in their postings. Goes back to the early days of the Net when the bottom lines of messages would sometimes disappear for no apparent reason.

NSF - National Science Foundation: Funds the NSFNet, a high-speed network that once formed the backbone of the Internet in the US.

NTSC: National Television System Committee; defined the 52 5-line color video signal frequency spectrum which extends from 30 Hz to 4.2 MHz. NTSC video consists of 525 interlaced lines, with a horizontal scanning rate of 15,734 Hz, and a vertical (field) rate of 59.94 Hz. A color subcarrier at 3.579545 MHz contains color hue (phase) and saturation (amplitude) information. 30-frame-per-second color TV standard in use in U.S., Canada, Mexico, Japan and a few other countries.

O

OEM: Original equipment manufacturer.

Off-Line: Mode of operation in which terminals, or other equipment, can operate while disconnected from a central processor. Contrast with "on-line" where there is a direct connection to a host computer.

Off-Premises System: Refers to a teleconferencing room or equipment located outside of a user organization's facility; e.g., a video teleconferencing room operated by a vendor and available to the public for a fee.

On-Demand Bandwidth: Dialable digital bandwidth access using the public switched telephone network instead of dedicated facilities.

On-line: When a computer is connected to an on-line service, bulletin-board system or public-access site, it is on-line.

Operating System: A computer program that runs the computer and handles data traffic between the disks and memory.

One-Way Video, Two-Way Audio: People at originating location can be seen and heard by participants at other locations. The people at the originating location can hear, but cannot see participants at other locations. With two-way video, each group can see and hear other groups. Usually limited to point- to-point.

Open Systems Interconnect - OSI: Generally open systems and networks are based on standards and the OSI model, providing applications and data portability, providing interoperability between systems, having common user interfaces, providing transparency below the application level, and provided by multiple sources, with multiple sources having input on development.

Optical Character Recognition - OCR: The machine identification of printed characters through use of light-sensitive devices; often used as a method of entering data.

Optical Fiber: An extremely thin, flexible thread of pure glass able to carry one thousand times the information possible with tradiional copper wire. See Fiber Optics.

Orange Book: Colloquial name of the standard that describes CD-Recordable equipment, media and formats. An extension of the "Yellow Book" standard which includes specifications for incremental writes or multiple sessions. It specifies standards for CD-R and magneto-optical cartridge systems as well as KodakÕs Photo CD. Most CD-ROM drives today can only read Òsingle sessionÓ discs or the first session of a multisession disc.

Origination Site: The location from which video and/or audio is transmitted and uplinked in a teleconference. Other sites participating are receive sites.

OSI: Open Systems Interconnection.

OS/2: IBM operating system.

Outbound: Direction of a signal relative to the hub of a local area network (LAN) or other telecommunications system. Outbound signals would be traveling away from the primary hub in the forward direction to the extremities of the system.

P

PABX - Private Automatic Branch Exchange: A private automatic telephone exchange, usually located at the user's site, that routes and interfaces the local business telephones and data circuits to and from the public telephone network.

Packet: The unit of data sent across a packet switching network. The term is used loosely. While some Internet literature uses it to refer specifically to data sent across a physical network, other literature views the Internet as a packet switching network and describes IP datagrams as packets.

Packet Switching: A communications data transmission method that breaks down messages into smaller units of standard sized pieces called packets, which are individually addressed and routed through a network; the network link is occupied only during packet transmission. Packet switching increases efficiency in transport.

PAL: Phase Alternation by Line, the 625-line, 25-frame-per-second TV standard used in Western Europe, India, China, Australia, New Zealand, Argentina, and parts of Africa. Brazil uses PAL-M, a 525-line variant.

PamAmSat: International satellite operator.

Parabolic Dish: A satellite antenna, usually bowl-shaped, that concentrates signals to a single focal point. See reflector.

Parallel Input/Output: Inputting data to or outputting data from, storage in whole information elements, e.g., a word rather than a bit at a time. Typically, each bit of a word has its own wire for data transmission, so that all of the bits of a word can be transmitted simultaneously.

Parity Check: A check of the accuracy of data being transmitted. To accomplish this, an extra parity bit is added to a group of bits so that the number of ones in the group is, according to the specification, even or odd. Then, at the receiving end, the bits in the word are added, the parity bit needed for that total is determined, and the total is then compared with the parity bit transmitted.

Path Table: One of two tables contained in the volume descriptor of a CD-ROM, which comprise the file management system for the disc. The path table contains the names of all directories on the disc, and is the fastest way to access a directory that is not close to the root directory.

Pay-per-society: The idea that the pay-per-view video concept will work for all areas of an information society. Many on-line services already charge by the minute or have monthly rates.

Pay-Per-View - PPV: Usage-based fee structure used sometimes in cable television programming in which the user is charged a price for individual programs requested.

Peripheral: Device such as a communications terminal that is external to the system processor.

PBX - Private Branch Exchange: A private telephone exchange that serves a particular organization and has connections to the public telephone network; refers to a multi-line telephone exchange terminal with various features for voice and data communications.

PC: Personal computer, microcomputer.

PCM - Pulse Code Modulation: A methd of converting analog sound into digital representation by use of successive samples.

PDA - Personal Digital Assistant: Small, hand-held devices that combine computer power with graphics, sound, video and communication capabilities. They will take several forms including electronic note takers and portable display telephone. They hold various programs, address files and databases depending on the user's needs. Many feature a modem, fax, radio mail and computer.

PDIAL List: List of public providers that offer full Internet access.

Phase: A fraction, expressed in degrees, of one complete cycle of a wave form or orbit.

Photonics: Gallium arsenide integrated circuits for optical interconnections within and between computer and communication equipment called GaA. Present fiber optic computer and communication links are limited by discrete component electronics. In development is a projected 32-channel parallel monolithic IC connector which could vastly increase erformance and drive down costs to open up long- sought new fiber optic markets which can replace current copper-wired connections.

Picture Element: One of many monochrome or color "dots" that comprise a television picture (also called pixel or pel).

Picture Signal: That portion of the composite video signal which lies above the blanking level and contains the picture brightness information.

Picture Tube: The television cathode-ray tube used to reproduce and display an image created by variations of intensity of the electron beam which scans the coated surface on the tube interior.

Ping: A program that can trace the route a message takes from your site to another site.

Pixel: The smallest controllable element that can be illuminated on a display screen. Closely related to resolution.

.plan file: A file that lists anything one wants others on the Net to know about the person. It is placed in the home directory at the public-access site. Then, anybody who fingers (see) you, will see this file.

Platform: Refers to different computer types or operating environments; e.g., Macintosh, DOS/Windows, CD-I and Sega are different platforms.

Point of Presence - POP: The point where the inter-exchange carrier's responsibilities for the line begin and the local exchange carrier's responsibility ends. Location of a communications carrier's switching or terminal equipment.

Point-to-Multipoint: A teleconference broadcast from one location to several receiving locations (also known as downlink sites.)

Point-to-Point: Teleconference between two locations. Point-to-Multipoint - one location to many sites.

Polarization: A characteristic of the electric field on an electromagnetic wave in space. The directional aspects of a signal. Signals can have circular or planar polarization. Four types of polarization are used with satellites; horizontal, vertical, right-hand circular and left-hand circular. Electromagnetic waves have the ability to vibrate in different radial directions. Typically, satellite signal polarization is either horizontal or vertical. The signal coming from the satellite to the dish will either be vibrating along a horizontal or vertical plane. The receiving equipment must be adjusted to receive the correct polarization.

Polycarbonate: Material from which compact discs are made.

POP - Post Office Protocol: A protocol for the storage and retrieval of e-mail. Eudora uses POP.

Port: In software, the act of converting code so that a program runs on more than one type of computer. In networking, a number that identifies a specific "channel" used by network services. For instance, Gopher generally uses port 70 but is occasionally set to use other ports on various machines.

Portable Transmitter: A transmitter so constructed that it may be moved about conveniently from place to place but not ordinarily used while in motion, although some portable communications equipment does provide the capability to used while in motion.

Post: To compose a message for a Usenet newsgroup and then send it out for others to see.

Post-Production: For a program which is not a live broadcast, all the footage would be shot with the talent on a set constructed for the purpose or in remote locations. After the shooting, the post-production begins. The tape is electronically edited on video editing equipment. Music and graphics might be added. When the editing is finished, the program is complete. It then might be sold as a training video, situation comedy, drama, etc. Most commercial productions are produced this way, even though it might appear to be a live broadcast.

Postmaster: The person to contact at a particular site to ask for information about the site or complain about one of his/her user's behavior.

PostScript: A page-description language, developed by Adobe Systems, that converts any computer image - whether text or graphics - to a form that compatible output devices can interpret and print. PostScript typefaces can be printed on any PostScript compatible printer.

POTS: Plain Old Telephone Service.

PPP - Point-to-Point Protocol: PPP provides a method for transmitting datagrams over serial point-to-point links.

pps: Packets per second.

Pre-Production: The first phase of a videoconference. Pre-production includes planning, research, script writing, developing taped segments to be dropped into a live production, hiring and rehearsing talent, and anything else done up to minute the broadcast begins.

Pre-Produced Segments: Video segments done prior to the day of the broadcast/production. These are videotaped and edited segments which will be shown during the broadcast to take the audience into the field for interviews, demonstrations, or on site visits to places that somehow embody the content. For example, in a program about environmental pollution, a pre-produced segment might show a polluted stream with beer cans and dead fish. It might show the source of the pollution such as chemical or sewage processing plants.

Premastering: The process of logically formatting an authored application and database. A working application or database converted to a standardized format such as ISO 9660 ready for writing to a final compact disc.

Press-to-Talk Microphone: Microphone that is activated by pressing a bar or button.

Prestel: The British Post Offices public viewdata service.

PRI - Primary rate interface: PRI is a CCITT-defined ISDN trunking technology that delivers 64 kbps clear channels and standardized out-of-band signaling. PRI can serve customer premise equipment (CPE) such as a PBX, LAN gateway, or host computer or can serve as a trunk interface between central offices.

Prime Focus: Type of feed in a parabolic dish antenna which is positioned above the dish as the antennas focal point.

Printed Circuit (PC) Board: A circuit board whose electrical connections are made through conductive material that is contained on the board itself, rather than with individual wires.

Printer Font: The software that contains the image of a typeface in outline form; used by a laser printer or imagesetter to produce the image on paper or film. Also called outline font.

Screen Font: The software that contains the bitmapped images of a particular typeface, at various sizes; used to produce an approximation of the typeface on the screen. Also, a particular size of bitmapped image of a typeface. Also called a bitmap font.

Type Manager: A software program (e.g. Adobe Type Manager, Bitstream FaceLift) that generates images of a typeface for the screen or a printer, based on the typeface's printer font.

Program Day/Date/Local Time: Broadcast times are usually listed in Eastern Standard Time or Eastern Daylight Time depending upon the time of the year. Remember to convert this to the local time.

Programming Language: An artificial language, established for expressing computer programs, which uses a set of characters and rules whose meanings are assigned prior to use.

Projection Television: A combination of lenses and/or mirrors that project an enlarged television picture on a screen.

PROM - Programmable read-only memory: A type of read-only memory that can be programmed by the computer user. This programming usually requires special equipment.

Prompt: 1. Any symbol or message presented to an operator by an operating system, indicating a condition of readiness, location, or that particular information is needed before a program can proceed. 2. When the host system asks you to do something and waits for you to respond. For example, if you see "login:" it means type your user name.

PRO-Que Channel: PRO is a term used primarily in television transmissions to designate a separate audio voice grade signal sent with a television channel which is used for program instructions and queuing for the broadcast engineers. The voice quality 3 kHz channel may contain audio or data as may most be appropriate for the specific application. It is not received by a standard television receiver without special equipment.

Protocol: A set of rules and procedures for establishing and controlling conversations on a line. The set of messages has specific formats for exchanging communications and assuring end-to-end data integrity of links, circuits, messages, sessions and application processes. Usually associated with communications over computer. It is the language that computers use when talking to each other. The method used to transfer a file between a host system and your computer. A formal description of message formats and the rules two computers must follow to exchange those messages. There are several types, such as Kermit, YMODEM and ZMODEM. Protocols can describe low-level details of machine-to-machine interfaces (e.g., the order in which bits and bytes are sent across a wire) or high-level exchanges between allocation programs (e.g., the way in which two programs transfer a file across the Internet).

Proprietary: A device or program designed and owned by a particular manufacturer or vendor, as opposed to a standard. CD-ROM drives are manufactured to read discs that comply with the Yellow Book standard, but their controller cards may be either supplied by the manufacturer (proprietary) or based on the Small Computer Systems Interface (standard).

PTT - Post, Telephone & Telegraph Administration: Refers to operating agencies directly or indirectly controlled by government in charge of telecommunications services in most countries of the world.

Public Access Channel: A cable television channel specifically designated as a noncommercial public access channel available on a first-come, non-discriminatory basis.

Public Access Provider: An organization that provides Internet access for individuals or other organizations, often for a fee.

Public Switched Network: Any switching system that provides a circuit switched to many customers.

Pulse Code Modulation: A time division modulation technique in which analog signals are sampled and quantized at periodic intervals. The values observed are typically represented by a coded arrangement of 8 bits of which one may be for parity.

Q

QuickTime: Apple's multimedia extension to its System 7 operating software for the Macintosh. It is a time-based management system for combining text, graphics, sound, still images, animations and video. The software incorporates its own compression technology so that digitized movies can be stored and played off of a computer hard disk.

R

Rain Attenuation - Rain Losses: The attenuation (loss) of a signal due to rainfall. If you are receiving a teleconference on a Ku band dish, local rainstorms can drastically weaken the signal strength of the program. The result will be sparkles which interfere with the ability to see the program. During a heavy downpour or thunderstorm, signal reception may be lost temporarily. The noise temperature perceived by the receiving antenna may increase due to rain being present in the link.

RAM: Random-access memory. A volatile memory used by a computer's central processing unit as a chalkboard for writing and reading information. RAM is measured in multiples of 4096 bytes (4K bytes), and serves as a rough measurement of a computer's capacity.

Raster: The scanned (illuminated) area of a television picture tube.

RBOC: Regional Bell operating company.

README files: Files found on FTP sites that explain what is in a given FTP directory or which provide other useful information (such as how to use FTP).

Real Soon Now: A vague term used on the Net to describe when something will actually happen.

Receivers: Convert satellite signals into channels viewed (one at a time) on a TV monitor; designed to tune-in the format, bandwidth, and audio sub-carrier. Programs broadcast in code (encryption) are decoded at receive sites.

Basic Receivers: Lowest cost; limited (or manual) channel tuning capability; may use fixed antennas.

Multi-Format Receivers: Most versatile; adjusts for all broadcast formats; receive any satellite video program in six or more bandwidth selections, and two agile audio subcarrier switches; usually a motorized systems.

Receiver - TV: Has receiver electronics which enable it to receive local broadcast signals. A monitor may not be equipped with receive equipment.

Receiver - Satellite: Electronic unit capable of receiving video and audio signals from satellites, usually from only one satellite at a time.

Receive Site: The site receiving the transmission from the origination site. A video teleconference might have 100 or more receive sites.

Red Book: The specification for Compact Disc-Digital Audio.

Retrieval Engine: A program which finds and presents data. Same as search engine.

Redundant: A backup satellite receive system which would go into operation if the primary system failed. Although the reliability of all electronics has greatly improved, it is desirable to have backup equipment in the receive chain; i.e., a duplicate of each item except the dish. Dual LNA's or BDCs can be mounted on dual feedhorns and could easily be switched in the event of primary system failure. Likewise, two receivers could be operated simultaneously using signal splitters, and either could be switched to the viewing room.

Reflector: Antenna's main curved "dish," which collects and focuss signals onto the secondary reflector or the feed.

Repeater: A term used to describe the process of reprocessing and send a weak signal on to a more distant service area. The weak signal condition develops as the initially strong signal passes through the miles of air, moisture, rain and snow which gradually attenuates or reduces its power level.

Resolution: A measure of picture resolving capabilities of a television system determined primarily by bandwidth, scan rates and aspect ratio. Relates to fineness of details perceived.

Retrace: The return of a scanning beam to a desired position.

Retrofitting: The installation of additional - equipment or the rebuilding of sections of a system after it has been installed.

RF - Radio Frequency: Radio frequencies are generally considered as any electromagnetic signal from normal radio to microwave transmission.

RFC - Request for Comments: The Internet's Request for Comments documents series. The RFCs are working notes of the Internet research and development community. A document in this series may be on essentially any topic related to computer communication, and may be anything from a meeting report to the specification of a standard.

RFD - Request for Discussion: The part of the newsgroup creation process where you propose a group and discussion starts.

RFI: Radio frequency interference.

RFP: Request for proposal.

RGB: Method of transmitting video signals that feeds red, green, and blue channels over separate wires; provides highest-quality video signal and is the format for most computer equipment.

Rlogin: Lets you log into other computers on the Internet as though you were connected to them directly. "rsh" is a junior version of rlogin.

ROM: Read only memory. A type of permanent, non-erasable memory that plugs directly into the wiring of a computer, and contains computer programs. Some computers are supplied with some built-in ROM, whereas others have external slots for inserting ROM cartridges.

ROTFL - Rolling on the Floor Laughing: How to respond to a particularly funny comment on the Net.

ROT13: A simple way to encode bad jokes or movie reviews that give away the ending, etc. Essentially, each letter in a message is replaced by the letter 13 spaces away from it in the alphabet. There are on-line decoders to read these; nn and rn have them built in.

Routing: Selecting the minimum delay path (and/or minimum cost path) in a network for a message or packet to reach its destination.

RS-232-C: Standard interface between a piece of equipment and a telephone circuit.

RS-250B: The technical standards established for the determination of a true broadcast quality signal. All technical parameters for each type of measurable signal degradation are at a level approximately ten times that first detectable as visible in a television picture by the average viewed.

RTM - Read the manual: Often used in flames against people who ask computer-related questions that could be easily answered with a few minutes with a manual. Often RFTM.

S

S/N - S/NR - Signal to Noise Ratio: Final relationship between the video or audio signal level to the noise level. Ratio of the signal power to the noise power in a specified band width, expressed in dBW.

Sampling Rate/Frequency: The number of samples taken per second of an analog signal, expressed in Hertz. A 44.1KHz sampling rate, used for CD-Audio sound, represents 44,100 samples per second.

SAP - Supplementary Audio Program: SAP is used to designate that part of the audio signal transmitted with the standard multi-channel sound television broadcast. The multi-channel audio signal contains stereo (left and right), SAP and PRO. The SAP signal is most often used in television broadcast applications for a second language.

Satellite: An electronics retransmission device serving as repeater normally placed in orbit around the earth in the geostationary orbit for the purpose of receiving and retransmitting electromagnetic signals. It normally receives signals from a single source and retransmits them over a wide geographic area.Satellite C/ku band Domestic communications satellites operate on two frequency ranges designated C and Ku band. Each requires specific electronic equipment. C band is less expensive; operates at 4 kHz. Ku-band operates at 12 kHz. Some teleconferences are broadcast on both bands.

Satellite Earth Terminal: That portion of a satellite link which receives, processes and transmits communications between Earth and a satellite.

Satellite Footprint: In geostationary orbit, communications satellites have direct line-of-sight to almost half the earth - a large "footprint" which is a major advantage. A signal sent via satellite can be transmitted simultaneously to every U.S. city. Multiple downlinks can be aimed at one satellite and receive the same program; called point to multipoint.

Satellite Master Antenna Televisio System - SMATV: A system wherein one central antenna is used to receive signals (broadcast or satellite) and deliver them to a concentrated grouping of television sets (such as might be found in apartments, hotels, hospitals, etc.).

Satellite Receiver: A microwave receiver capable of receiving satellite transmitted signals, downconverting, and demodulating those signals, and providing a baseband output (e.g., video and audio). Modern receivers are frequency agile and usually capable of multiple band reception (e.g., C band and Ku band.)

Satellite Relay: An active or passive satellite repeater that relays signals between two earth stations.

Satellite System: The use of orbiting satellites to relay transmissions from one satellite dish to another or multiple dishes.

Fixed Position System: Low cost systems limited to reception from one satellite and one band. Motorized System Receives programs on different satellites by adjusting the dish position. Automated Systems Microprocessor controlled for instant movement to satellites (positions stored in memory).

SCA-FM: Subsidiary Communications Authorization; an electronic technique that places the radio signal on the FM spectrum; these signals can only be picked up with special tuners that distinguish the SCA from the FM signals.

Scalability: The ability to vary the information content of a program by changing the amount of data that is stored, transmitted or displayed. In a video image, this translates into creating larger or smaller windows of video on screens.

Scan-Converter: A device that converts video frequency signals to audio frequencies and vice versa; used in freeze-frame video to transmit video signals over telephone lines.

Scanner: A device for digitizing text, drawings or photographs - anything in paper form. It works like a photocopy machine, but instead of paper, the scanner converts the printed information into digital images. Scanners are used with OCR (optical character recognition) technology, which takes the scanned pages of text and graphs and converts them into the individual letters and words that make up the text and the dots that make up the image, so that the text and images can be edited using a computer.

Scanning: The process of breaking down an image into a series of elements or groups of elements representing light values and transmitting this information in time sequence.

SCPC - Single Channel Per Carrier: Signal transmission technique often used in satellite transmission which concentrates one channel of information on a single transmitted carrier for relay through the satellite. The channel may be digital, analog or multiplexed analog in nature provided that its information may be sent on a single narrow band carrier. The Single Channel Per Carrier transmission technique allows multi-channel operation in the satellite with access from any location on the earth.

SCPT: Single carrier per transponder.

Scramble: To interfere with an electronic signal or to rearrange its various component parts. In pay television, for example, the signal might be scrambled, and a decoder, also called a descrambler, might be necessary for the signal to be unscrambled so that only authorized subscribers would receive the clear signal.

Scrambler: A device that transposes or inverts signals or otherwise encodes a message at the transmitter to make it unintelligible to a receiver not equipped with an appropriate descrambling device. Synonymous with encoder.

Screen Capture: A part of communications software that opens a file on the computer and saves to it whatever scrolls past on the screen while connected to a host system.

SCSI: Small Computer Systems Interface. Pronounced "scuzzy." A standard interface used to connect peripheral devices, such as a CD-ROM drive, to a computer.

Screen Density: The maximum number of accessible screen elements in a video display.

Scrolling: A property of most alphanumeric video display terminals. If the screen of such a video terminal is filled, it will move the enire display image upward, either at a smooth pace or one line at a time, so that room is continuously made at the bottom of the screen for new information.

.sea: Self-extracting Archive: A compressed file or files encapsulated in a decompression program; needs no other program to expand the archive.

SEA - State Educational Agency: The state board of education or other agency officer primarily responsible for the supervision of public elementary and secondary schools in a state. In the absence of this officer or agency, it is an officer or agency designated by the governor or state law.

Search Engine: A program which finds and presents data. Same as retrieval engine.

SECAM: Systeme Electronique pour Couleur Avec Memoire, the 625-line, 25-frame-per-second color television -system used in France, Eastern Europe, USSR and parts of Africa.

Sector: A physical data block of a CD-ROM.

Seek: In CD-ROM drives, the act of locating requested data on a disc.

Seek Time: Usually expressed in terms of "average seek time," it provides a comparative number indicating the time required to get from one position to another, in reading a CD-ROM. Some older CD-ROM drives had seek times in excess of 1,000 milliseconds (ms), or one full second. The newest drives have seek times approaching 200ms.

Selective Addressability - Selectively Addressable Scrambling: The capacity to designate selected receivers to descramble a particular signal. Each decoder has a unique "address." First developed as the pay-per-view option for cable TV, then adopted by satellite networks.

Server: A computer that can distribute information or files automatically in response to specifically worded e-mail requests.

Servo: In CD-ROM drives, an electro-mechanical device that uses feedback to achieve precise starts and stops for movements of the optical head and focusing of the laser beam.

Semiconductor: A material whose resistivity lies between that of conductors and insulators, e.g., germanium and silicon. Solid state devices such as transistors, diodes, photocells, and integrated circuits are manufactured from semiconductor materials.

Semiconductor Memory: Computer memory using solid state devices instead of mechanical, magnetic, or optical devices.

Serial Input/Output: Data transmission in which the bits are sent one by one over a single wire.

Shareware: Software that is freely available on the Internet. If you like and use the software, you should send in the fee requested by the author, whose name and address will be found in a file distributed with the software.

Shared Visual Space: Allows participants to interact with a common graphics display area; e.g., any person can make a change which is seen by all.

.sig file: Sometimes, .signature file. A file that, when placed in your home directory on your public-access site, will automatically be appended to every Usenet posting you write.

.sig quote: A profound/witty/quizzical/whatever quote that you include in your .sig file.

Sign-On Procedure: The process of connecting with a remote computer, including the provision of identification details and security access.

Signal-to-Noise: The amount of useful information to be found in a given ratio

Single Session: A drive that can read discs on which data was recorded only once, or a CD-ROM on which data was recorded in one pass, either through CD-Recordable technology, or the standard mastering process.

Silicon Chip: A wafer of silicon providing a semiconductor base for a number of electrical circuits.

Simplex: A circuit capable of transmission in one direction only. Contrast with half duplex and full duplex.

SIMTEL20: The White Sands Missile Range used to maintain a giant collection of free and low-cost software of all kinds, which was "mirrored" to numerous other ftp sites on the Net. In the fall of 1993, the Air Force decided it had better things to do than maintain a free software library and shut it down. But you'll still see references to the collection, known as SIMTEL20, around the Net.

Site: The origination site is the location from which video and/or audio is transmitted and uplinked in a teleconference. Receive transmission from the origination site.

Skew: The angular deviation of recorded binary characters from a line perpendicular to the reference edge of a data medium.

Skewing: Horizontal displacement of video information in bands of approximately 16 lines per field producing a sawtooth effect which is most apparent on vertical picture detail of a television picture originating from the playback of a video tape recording.

SLIP - Serial Line Internet Protocol: SLIP is currently a de facto standard, commonly used for point-to-point serial connections running TCP/IP. It is not an Internet standard but is defined in RFC 1055.

Slow Scan: Uses transmitters that scan selected frames and transmit the visual information over telephone lines to receive sites where it is reconstituted as a still picture. May refer to still frame video that accepts an image from a camera or other video source one line at a time.

SMATV: Satellite master antenna television. A distribution system that feeds satellite signals to a hotel, motel, apartment complex, etc.

SMDS - Switched Multimegabit Data Service: A public network service that will enable customers to send packets between LANs at either T-1 or T-3 rates. Switched Multimegabit Data Service is offered by public network providers and is a connectionless (i.e., datagram) service. It will enable customers to exchange packets between sites at T1, T3 at potentially higher rates. This LAN-like service will be offered by local exchange carriers and will initially be available only within selected metropolitan areas. A typical SMDA customer will have a wide-area communications device - i.e., a router - connected to a campus LAN or backbone, which interface through a subscriber line to the local telco central office. The communication between the customer premises device (e.g., the router) and the telco will adhere to a protocol called Subscriber Interface Protocol (SIP). This protocol has three levels, only two of which are standardized Levels 1 and 2 are fashioned from the IEEE 802.6 MAC standard for metropolitan area networking, which is called the Distributed Queued Dual Bus (DQDB) protocol. The third layer was promulgated by Bellcore in one of its Technical Advisories. The SMDS service will support both T1 and T3 access from the user's router to the local exchange carrier's central office. SMDS is considered by the regional Bell operating companies (RBOCs) to be their first broadband service, and it will eventually be incorporate as a service offering for Broadband ISDN (BISDN) family of services in the late 1990s.

Smiley: A way to describe emotion on-line. Look at this with your head tilted to the left :-). There are scores of these smileys, from grumpy to quizzical.

:-) smile :) also a smile

:-D laughing :-} grin

:-] smirk :-(frown

;-) wink 8-) wide-eyed

:-X close mouthed :-o oh, no!

SMPTE Time Code: Society of Motion Picture and Television Engineers' system of giving each frame of video a number to allow indexing and precise tape control. EBU time code is the European Broadcast Union version of SMPTE time code.

SMTP - Simple Mail Transfer Protocol: The Internet standard protocol for transferring electronic mail messages from one computer to another. SMTP specifies how two mail systems interact and the format of control messages they exchange to transfer mail.

SNA: System network architecture.

Snail Mail: Mail that comes through a slot in your front door or a box mounted outside your house.

SNMP: Simple Network Management Protocol The Internet's standard for remote monitoring and management of hosts, routers and other nodes and devices on a network (RFC 1157).

Snow/Ice on the Satellite Dish: A significant build up of snow (4-5 inches) on the dish can interfere with signal reception. Snow can be removed with a soft broom or soft cloth. Since accurate curvature of the dish is vital to a good signal, avoid banging or hitting the dish. Ladders should not be leaned against the dish as it may warp or change the azimuth and/or elevation setting. A small amount of ice should not cause problems.

Software: A set of programs, procedures or related documentation associated with a system; materials for use with audio visual equipment; programs in contrast to equipment.

Solar Outage: If an antenna is pointed at or near the sun, the sun's high radiated noise level may be many times stronger than the desired signal.

Solid State: A class of electronic components utilizing the electronic or magnetic properties of semiconductors.

SONET - Synchronous Optical Network: Will offer dedicated point-to-point lines via fiber, with bandwidths ranging from 51.84 mbps to over 2gbps. SONET defines optical interfaces for high speed digital transmission - ranging from 51.84 mbps to more than 2 gbps in multiples of 51.84 mbps. The purpose of the SONET standard is to guarantee that fiber, and fiber terminating equipment (e.g. digital loop carrier systems) from different central office vendors, can all interface with each other. While many trials are currently under way to test the SONET central office standards, all new fiber deployment is expected to be compliant with this standard.

After the SONET CO standards are proven, carriers will begin providing SONET-compatible equipment to customers; the roll out of SONET circuits to customers will begin in 1993. With SONET, customers will be able to order "pipes" running at speeds higher than T3. SONET will be the transmission platform for other high-speed (above T3 speeds) services, such as SMDS and BISDN. SONET will be a major breakthrough for carriers, because standardization will significantly lower their equipment and operational costs, which, in turn, should result in lower cost for private networking.

Sound board: A device required by a DOS-based computer to access digital sound, exists in the form of an add-in board inserted in the computer, and accesses (and/or creates) .WAV, .SND, MIDI and other digital sound formats.

Special Event Teleconferencing: Teleconference that uses facilities that are temporarily linked for a specific event; implies a temporary satellite network for one-way video and two-way audio.

Specialized Common Carrier: 1. A company authorized by a government agency to provide a limited range of telecommunications services. Examples of specialized common carriers are the value- added networks. 2. Those common carriers not covered in the original federal communications legislation.

Spectrum: Range of electromagnetic radio frequencies used in transmission of voice, data, and TV.

Spin Up: Come up to speed. When a CD-ROM is inserted in a drive, it must reach a certain rate of rotational speed in order to be read.

SS7 - Signaling System 7: Increases both the efficiency of the telcos' interoffice trunking facilities and their opportunities for revenue generation by enabling network-wide services. With SS7 trunk signaling, premium services such as ISDN and Custom Local Area Signaling Service can be easily and efficiently extended across the network.

SSMA - Spread Spectrum Multiple Access: Frequency modulation technique.

Standard Broadcast Band: The band of frequencies extending from 535 to 1605 kHz, usually called AM.

Star Network: A network configuration in which there is only one path between a central or controlling node and each end-point node.

Station: Assigned satellite location.

Stereophonic: Giving, relating to, or constituting a three-dimensional effect of auditory perspective, by means of two or more separate signal paths.

Still-Image Video: System by which still images are transmitted over standard telephone lines, usually allowing for real-time interaction between locations.

STL - Studio Transmitter Link: Description of a type of microwave link which connects a television studio to the television station transmitter location. The designation is used by the Federal Communications Commission to differentiate a specific band of frequencies allocated for this specific application.

Studio: A specially designed room with associated control and monitoring facilities used by a broadcaster for the origination of radio or television programs.

Subcarrier: Signal which is transmitted along with the main video signal carrier. Subcarriers can transmit data, color picture information or audio.

Subscription Television - STV: The broadcast version of pay television. Not a cable service, it is distributed as an over-the-air broadcast signal. Its signals are scrambled and can be decoded only by a special device attached to the television set for a fee. STV contains no commercials.

Superband: The band of cable television channels J through W lying between 216 and 300 MHz.

Supercomputers: The fastest and most powerful computing systems that are available at any given time.

Surfing the Internet: Skimming across topics on the Internet - moving in and out of systems looking for information that is not specified. More like browsing than a true search. Also net surfing.

Switch: Mechanical or solid-state device that opens or closes circuits, changes operating parameters or selects paths for circuits on a space or time division basis.

Switched Circuit: A circuit that may be temporarily established at the request of one or more stations.

Switched Network: Any network in which switching is present and is used to direct messages from the sender to the ultimate recipient. Usually switching is accomplished by disconnecting and reconnecting lines in different configurations in order to set up a continuous pathway between the sender and the recipient.

Switched System: A communications system (such as a telephone system) in which arbitrary pairs or sets of terminals can be connected together by means of switched communications lines.

Symmetrical Compression: A compression system that requires equal processing capability for compression and decompression of an image. Used in applications where both compression and decompression will be utilized frequently. Examples include still-image databasing, still-image transmission (color fax), video production, video mail, videophones and videoconferencing. Asymmetrical Compression requires more processing capability to compress an image than to decompress an image. It is typically used for the mass distribution of programs on media such as CD-ROM.

Synchronous Communication: Communication which takes place in the same time frame. Examples are live teleconferences which must be viewed when they are broadcast. If the teleconference is taped and viewed later, it becomes asynchronous communication - communication which takes place at the convenience of the end user through the technology of video tape recording.

Synchronous Transmission: Data characters and bits are transmitted at a fixed rate with the transmitter and receiver synchronized. This eliminates the need for start-stop elements, thus providing greater efficiency.

Syntax Error: A mistake in the formulation of an instruction to a computer.

Sysadmin: The system administrator; the person who runs a host system or public-access site.

Sysop - A System Operator: Somebody who runs a bulletin board system or network; responsible for keeping the network or BBS working properly.

T

T1 (DS-1. Channel: High-speed digital data channel/carrier with a bit rate of 1.544 mbps which requires a bandwidth of approximately 2.1616 MHz to transmit in a television type cable environment. (1.4 x 1.544 = 2.1616); a general term for a digital carrier (DS-1. available for high-volume voice or data traffic; often used for compressed video teleconferencing networks. Each T1 circuit can accommodate 24 voice channels. A video codec operating at the T1 rate uses the equivalent of 24 voice channels. A codec operating at 56 or 64 Kbps is operating in the range of one voice channel. A standard video signal digitized at 90 Mbps has approximately 1400 voice channels. The compressed video signal quality and the cost decreases as the transmission speed decreases.

T3 (DS-3.: A carrier of 45 mbps bandwidth; one T3 channel can carry 28 T1 channels. Used for point-to-point digital video transmissions or for major PBX-PBX interconnection. Dedicated service delivered via fiber. The price for a T3 circuit can be comparable to seven to 12 T1 circuits. In addition to being offered by the traditional local and interexhange carriers, a number of alternative access carriers offer T3 circuits in major metropolitan areas.

Talk-back Circuit: An audio return link from a receive location to the originating video/audio point. The equipment used is generally either a leased telephone line, dedicated radio link, or special microwave equipment made for this service.

TANSTAAFL: Internet shorthand for "There Ain't No Such Thing as a Free Lunch."

.tar: The filename extension used by files made into an archive by the Unix tar program.

TBC - Time Base Corrector: An electronic accessory to a videotape recorder that helps make mixed format playback or transfers electronically stable. It helps maintain picture quality even in dubbing operations within a single tape format.

TCP: Transmission Control Protocol.

TCP/IP - Transmission Control Protocol/Internet Protocol: The combination of TCP and IP. The particular system for transferring information over a computer network that is at the heart of the Internet.: IP is the network layer protocol for the Internet. It is a packet switching, datagram protocol defined in RFC 791.

TDMA - Time Division Multiple Access: Form of multiple access where a single carrier is time shared by many users. Signals from earth stations reaching the satellite consecutively are processed in time segments without overlapping.

Telco: Generic name for telephone companies.

Telecommunications: Communicating over a distance. Use of wire, radio, optical or other electromagnetic channels to transmit or receive signals for voice, video and data communications.

Telecommuter - telecommuting: Ability to work from home, local office, or from the road because of equipment. Equipment allows the telecommuting employee to work from anywhere. The equipment includes a telephone, fax, modem and as the NII is deployed, video.

Telecomputer: Equipment used to receive digitized information in audio, video, and data modes.

Teleconference: Electronic communications between two or more groups, or three or more individuals, who are in separate locations via audio, audiographics, video or computer. Audio teleconference - two-way communication between two or more groups, or three or more individuals, in separate locations. Video teleconference - one (or more) uplink and downlink sites. May be fully interactive voice and video, two-way voice and one-way video; full-motion, compressed, or freeze-frame video.

Telemetry: The science of sensing and measuring information at some remote location and transmitting the data to a convenient location to be read and recorded.

Telenet: A public packet-switching network operated by US Sprint. Also known as "SprintNet".

Telnet: The Internet standard protocol for remote terminal connection service. Allows a user at one site to interact with a remote time sharing system at another site as if the user's terminal was connected directly to the remote computer (see "rlogin"). On the Macintosh, NCSA Telnet is the standard.

Telephone Conference Bridge: Device that links three or more telephone channels for a teleconference; usually refers to a bridge that provides only dial-up teleconferencing where an operator calls each participant. Contrast to meet-me bridge.

Telephony: the use or operation of an apparatus for transmission of sounds between widely removed points with or without connecting wires.

Teleport: A generic term referring to a facility capable of transmitting and receiving satellite signals for other users.

Teletext: Broadcast service using several otherwise unused scanning lines (vertical blanking intervals) between frames of TV pictures to transmit information from a central data base to receiving television sets. Users of a teletext service grab pages from the transmission cycle using a keypad similar to that used in videotex systems.

Television: The electronic transmission of pictures and sounds.

Telewriter: General term for an electronic device that produces free-hand inforation that can be sent over a telecommunications channel, usually a telephone line.

Terminal: 1. Generally, connection point of equipment, power or signal. 2. Any terminating piece of equipment such as a computer terminal.

Text: In terms of files, a file that contains only characters from the ASCII character set. In terms of FTP, a mode that assumes that files will be transferred containing only ASCII characters.

Terrestrial Carrier/Land Line: Telecommunications transmission system using land-based facilities (microwave towers, telephone lines, fiber optic cable).

Thread: A group of messages in a Usenet discussion group that all share the same subject and topic, so one can easily read the entire thread or delete it, depending on the specific newsreader.

TI - Terrestrial Interference: TI is normally generated as a result of relatively strong terrestrial microwave signals overpowering the weak satellite transmissions which are the primary signals of interest at a satellite earth station.

Time Code: Code electronically placed on a videotape that appears on the screen or on a counter to locate specific footage and edit tape. Logs are made of the footage on tape before editing so it can be located and viewed quickly during the editing.

Time Sharing: Pertaining to the interleaved use of time on a computer system that enables two or more users to execute computer programs concurrently.

Token Ring: A type of LAN. Examples are IEEE 802.5, ProNET-10/80 and FDDI. The term "token ring" is often used to denote 802.5

Touch Screen: A video-and/or computer monitor which responds to the user's finger touch in order to control the program.

Transceiver: Terminal that transmits and receives.

Transfer Rate: The amount of data that can be communicated from the CD-ROM drive to the CPU. Standard CD-ROM data transfer rate is 155KB/sec, (often rounded to 150KB/sec).

Transmission Channel: The medium by which a signal is sent and received between separate locations.

Transponder - Channel - Downlink Frequency: A satellite microwave repeater (receiver and transmitter) receives the signal from an uplink, amplifies it, down converts the frequency of a received band of signals, and retransmits the signal back to earth. Satellites have 12, 24, or more transponders each with the capacity for one color TV signal and two audio channels. Typically transponder with 24 transponders have twelve polarized for vertical and twelve for horizontal transmissions in order to optimize the bandwidth of the satellite and the respective transponders.

Treatment: A narrative description of a media program. In videoteleconferencing, usually describes routine of action and precedes a rundown and/or script.

Tuner: A device, circuit, or portion of a circuit that is used to select one signal from a number of signals in a given frequency range.

TVRO - Television Receive Only: Earth stations which receive (but not transmit) satellite transmissions. Normally comprised of a parabolic antenna, low noise converted (LNC) or low noise amplifier (LNA) and a satellite receiver. The antenna gathers the weak signals transmitted from the communications satellite located in the geostationary orbit which are then amplified and downconverted to a more usable portion of the spectrum by the low noise converted. From the LNC the signals may travel up to several hundred feet to a satellite receiver; the output of which is typically video and audio or modulated channel three or four. 23B+D The capability of ISDN primary rate interface (PRI) to enable data terminals served by a DMS-100 ISDN node to have fully digital circuit- and packet-switched ISDN internetworking with data terminals served by PBXs.

Twisted Pair: A pair of wires used in transmission circuits and twisted about one another to minimize coupling with other circuits. UTP - unshielded twisted pair.

Two-Wire Circuit: A typical telephone circuit on the public switched network; a circuit formed by two conductors insulated from each other to provide a send and receive channel in the same frequency.

Tymnet: U.S. based packet-switching network.

U

UHF-VHF: UHF stands for ultra high frequency television transmission channels above channel 13 (Channels 14-3.. VHF stands for very high frequency; television transmission channels 2 through 13.

Unions: The unions most closely associated with video production. Usually, if one group of people on a production are union, all will be. They may belong to the following unions:

Actors Equity represents actors in certain areas of the country.

AFM: American Federation of Musicians. This union represents professional musicians in all areas of performance - recording and personal appearance.

AFTRA: American Federation of Television and Radio Artists. A union for artists who perform on broadcast media, including tape.BMI and ASCAP: Broadcast Music Incorporated and American Society Composers and Publishers which serve the same function of licensing and collecting creative royalties on works of music played in live public performance and recordings.

IATSE: International Association of Theatrical and Stage Employees.: Normally these people are found on live theater stages working with sets, props, lighting and other theatrical gear. However, they may also be the union representing the same group of workers as IBEW.

IBEW: International Brotherhood of Electrical Workers. IBEW has a special chapter for engineers, camera operators, audio engineers, video engineers, lighting designers and technicians, video editors and any other technicians or electricians who work in television production. IBEW members are employed by television stations, some cable companies, some corporations and some production companies.

SAG: Screen Actors Guild which represents talent. Originally, this union represented actors working in the motion picture industry, but in recent years they have represented other talent areas.

UNIX: An operating system developed by Bell Laboratories that supports multiuser and multitasking operations.

UNMA - Unified Network Management Architecture: (AT&T).

Uplink: An earth station that transmits a radio frequency signal to a communications satellite. The transmitting facility, or uplink, consists of a large dish-shaped antenna and high-power amplifiers. The uplink is like the transmitter of a radio or television station, except that it concentrates it signals in one direction by means of a parabolic dish antenna that delivers a strong pinpoint signal to a specific satellite in space.

Upload: Copy a file from your computer to a host system. Upload is the term used for sending information over a network. Download refers to receiving information off a network. To save on connect time charges users often download information on to a data disk, and then work with it offline.

Usenet: An anarchic network of sorts, composed of thousands of discussion groups on every imaginable topic.

Usenet Newsgroup: Discussion group on one topic.

User name: On most host systems, the first time you connect you are asked to supply a one-word user name. This can be any combination of letters and numbers.

UUCP - Unix-to-Unix Copy Program: A method for transferring Usenet postings and e-mail that requires far fewer Net resources than TCP/IP, but which can result in considerably slower transfer times.

V

Value Added Network - VAN: A data network operated in the U.S. by a firm which obtains basic transmission facilities from the common carriers, and adds value such as error detection and sharing and resells the service to users. Telenet and TymNet are examples of VANs.:

Vector Quantization: Compression coding technique that uses block processing to exploit redundancies within a frame. For example, if the blue sky background within a frame is one constant color, one pixel of that color is all that needs to be stored. Quick duplication of the pixel by vectors (usually 8 x 8) occurs when decompressed and displayed on a monitor.

Veronica: An information agent that searches a database of Gopher servers to find items that are of interest to the user.

VF: Voice frequency.

Vi: An extremely powerful Unix editor with the personality of a junkyard dog. Much-beloved by many Unix aficionados.

Videodisc: Information stored on an optical disc is retrieved via laser technology (versus a stylus or needle). The most commonly known optical disc is the audio compact disc. 12" or 8" in diameter optical disc; requires laserdisc player; may contain up to 54,000 still frames or 30 minutes of full motion video on each side (or some variation of each); stores information in analog format. Their use and popularity has been largely eclipsed by VCRs using magnetic videotape cassettes. Levels of Interactive Videodisc Systems:

Level I: A videodisc player with the following capabilities: still/free frame, picture stop, chapter stop, frame address and two audio channels. Level I videodiscs have limited memory and limited processing power.

Level II: A videodisc player with the capabilities of Level I, plus programmable memory and improved access time.

Level III: Level I or II players combined with an external computer and/or other peripheral processing device. Level III IVD systems may have two monitors - one for the video and one for the computer - or may display the video and computer screens on a single monitor.

Level IV: Combines computer and videodisc technologies into one piece of equipment. Too expensive for most commercial uses, it is used almost exclusively by the military.

Video: A term pertaining to th bandwidth and spectrum of the signal which results from television scanning and which is used to reproduce a picture.

Video Camera: A camera which converts images to electrical signals for recording on magnetic tape or live transmission.Videodisc A record-like device storing a large amount of audio and visual information that can be linked to a computer; one side can store the pictures and sounds for 54,000 separate television screens.

Video Display: Presentation of the TV signal can be as simple as using a 19" TV receiver or as elaborate as large screen projection costing $200,000. The ideal lies somewhere in between. Analyze the room, physical layout, anticipated audience size, and AV support staff. In general, the larger the screen, the better. Projected images produce greater psychological impact, and help to dispel a viewer's feeling that he or she is watching TV. A general rule of thumb that has been suggested concerning minimum screen size is to figure no more than one viewer per diagonal inch. A 19" set would accommodate 19 viewers. This may not always be the case and it does not provide the larger-than-life experience that may be more effective in communicating the message.

Videotape: A plastic, iron oxide-coated tape of various widths from 1/4" to 2" for recording and playback of video and audio signals and additional technical code information.

Video Teleconference: A meeting involving at least one uplink and a number of downlinks at different locations. Electronic voice and video communication between two or more locations. It can be fully interactive voice and video or two-way voice and one-way video. It includes full-motion, compressed, and freeze-frame video.

Videotex: The generic term used to refer to a two-way interactive system(s) for the delivery of computer-generated data into the home, usually using the television set as the display device. Some of the more often used specific terms are "viewdata" for telephone-based systems (narrowband interactive systems); "wideband broadcast" or "cabletext" for systems utilizing a full video channel for information transmission; and "wideband two-way teletext" for systems which could be implemented over two-way cable television systems. In addition, hybrids and other transmission technologies, such as satellite, could be used for delivery of videotex services on a national scale.

Viewdata: Generic term used primarily in the U.S. and Great Britain to describe two-way information retrieval systems based on mainframe computers accessed by dumb or intelligent terminals whose chief characteristic is ease of use. Originally designed to use the telephone network, viewdata in the U.S. is being implemented over other distribution media such as coaxial. Viewdata's salient characteristic is the formatting, storing, and accessing of screens (sometimes called frames or pages) of alphanumeric displays for retrieval by users according to a menu or through use of keyboard search. A two-way form of videotex.

Virtual Private Network: Use of the public switched telephone system to provide a capability similar to that of a private network.

Virtual Reality - VR: Loosely defined as putting users into a computer-generated environment, rather than merely reacting to images on a display screen. Full immersion VR can include a helmet that senses head movement and changes the view seen through small TV screens mounted in front of each eye along with gloves that allow users to touch objects in the virtual world.

Virtual Space: Refers to a type of videoconference in which each participant is assigned a separate camera and is seen on a separate monitor, large screen or assigned spatial area.

VLSI: Very large scale integration.

Voice Actuated: Equipment activated in response to a voice. A voice-switched microphone is activated by a voice. In voice-switched video cameras are activated by voice to send a picture of the speaker.

Voice Mail: Products record, store and forward voice messages from one electronic mailbox to another by using sigle commands from any touch tone phone.

Voice-Over: Words spoken by an off-camera narrator - over the video.

Voice-Switched Microphone: Microphone that is activated by a sound of sufficient amplitude; generally allows only one person to speak at a time.

Voice-Switched Video: Type of video conference in which the cameras are activated by voice signals to send a picture of a particular person in the group. Not all participants can be seen at any one time in contrast to continuous presence video.

Volatile Memory: A storage medium in which information is destroyed when power is removed from the system.

VSAT - Very Small Aperture Terminal: Small earth stations with a satellite dish usually 4 to 6 feet (1.2 to 1.8 meters) in diameter used to receive high speed data transmission; can also transmit slow- speed data. A VSAT uplink for compressed video a C-Band frequencies is approximately 4.5 meters in diameter for most satellite applications.

VT100: Another terminal-emulation system. Originally, a dedicated terminal built by DEC to interface to mainframes. Supported by many communications program, it is the most common one in use on the Internet. VT102 is a newer version.

VTR - VCR: Video tape recorder or video cassette recorder. Equipment capable of recording video. All video equipment is not equipped to record as it requires recording heads to accomplish.

W

WAIS - Wide Area Information Service: A distributed text search system based on a standard (Z39.50) that describes a way for one computer to ask another to do searches for it. It looks at the content of files (not just the titles).

WAN: Wide Area Network (see LAN)

WATS Line: Wide Area Telecommunications Service A type of telephone service in which subscribers pay a base rate rather than a charge per call. An in-Wats line allows anyone in a designated area to phone an 800 number and pay nothing for the call. An out-Wats line allows users to place outgoing long-distance calls.

Wavelength Multiplexing: Transmitting individual signals simultaneously by using a different wavelength for each signal. Synonymous with frequency division multiplexing.

Whois: An Internet program which allows users to query a database of people and other Internet entities, such as domains, networks and hosts, kept at the NIC. The information for people shows a person's company name, address, phone number and e-mail address.

Wildcards: Special characters such as * and ? that can stand in for other characters during text searches in some programs. The * wildcard generally means "match any other characters in this spot," and the ? generally means "match any other character in this spot."

Wipe: Optical effect in which the picture appears to have been wiped from the screen; i.e. from left to right, top to bottom.

Wireless Cable: Uses microwave frequencies to transmit programming to a small antenna (about the size of an open newspaper) at subscriber homes.

World Wide Web - WWW: Presents information in a friendly hypertext format. WWW displays pages of information, with links to other pages. Mosaic is the program that really makes Web materials come alive.

Different systems display the links differently, by highlighting the link items or by putting a code (such as a number in brackets) after the item. Others put the link in boldface or in color.

WorldWindow: Offered by the Washington University Libraries in St. Louis, MO, a gateway to dozens of login services. Telnet to library.wustl.edu (no login needed).

WORM: Write Once Read Many. A type of permanent optical storage which allows the user to record information on a blank disc. Information may be added until the disc is full, but not erased or changed.

Word Processor: A computer-based typing and text-editing system.

Workstation: Computers that are generally targeted at technical users, interface over a network easily, often run UNIX, come standard with more compute power than PCs and are capable of fast graphics. Distinctions between high-end personal computers and workstations are blurring. For high-end animation work such as 3-D logs, morphing or animated characters, workstations provide the compute power and graphics performance the animator needs. Workstations are used for computer-generated imagery (often called CGI) because they take less time than PCs to render images.

Wrap: The end of a program or production sequence, as in, "That's a wrap."

Wraparound: Local activities prior to, in the midst of, or following a teleconference to focus the content toward outcomes and ideas which can directly assist the participants.

X

X.25: Set of packet-switching standards published by the CCITT. An international standard for control of data communications between two or more computers or terminals using packet-switching technology.

XMODEM: A common file transfer protocol.

Y

Yanoff: Scott Yanoff publishes a regularly updated on-line resource guide. To find out how to get a copy finger yanoff@csd4.csd.uwm.edu

Yellow Book: The physical specification for CD-ROMs. See CD-ROM.

YMODEM: Another common file transfer protocol.

Z

.zip: The filename extension used by files compressed into the ZIP format common on PCs.

ZMODEM: The fastest and most popular file transfer protocol.